Sweeteners and Sugar Alternatives in Food Technology

Sweeteners and Sugar Alternatives in Food Technology

Second Edition

Edited by

Dr Kay O'Donnell
Weybridge, UK

Dr Malcolm W. Kearsley
Reading, UK

WILEY-BLACKWELL
A John Wiley & Sons, Ltd., Publication

This edition first published 2012 © 2012 by John Wiley & Sons, Ltd.
First edition © Blackwell Publishing Ltd.

Wiley-Blackwell is an imprint of John Wiley & Sons, formed by the merger of Wiley's global Scientific, Technical and Medical business with Blackwell Publishing.

Registered office: John Wiley & Sons, Ltd, The Atrium, Southern Gate, Chichester, West Sussex, PO19 8SQ, UK

Editorial offices: 9600 Garsington Road, Oxford, OX4 2DQ, UK
The Atrium, Southern Gate, Chichester, West Sussex, PO19 8SQ, UK
2121 State Avenue, Ames, Iowa 50014-8300, USA

For details of our global editorial offices, for customer services and for information about how to apply for permission to reuse the copyright material in this book please see our website at www.wiley.com/wiley-blackwell.

The right of the authors to be identified as the authors of this work has been asserted in accordance with the UK Copyright, Designs and Patents Act 1988.

All rights reserved. No part of this publication may be reproduced, stored in a retrieval system, or transmitted, in any form or by any means, electronic, mechanical, photocopying, recording or otherwise, except as permitted by the UK Copyright, Designs and Patents Act 1988, without the prior permission of the publisher.

Designations used by companies to distinguish their products are often claimed as trademarks. All brand names and product names used in this book are trade names, service marks, trademarks or registered trademarks of their respective owners. The publisher is not associated with any product or vendor mentioned in this book. This publication is designed to provide accurate and authoritative information in regard to the subject matter covered. It is sold on the understanding that the publisher is not engaged in rendering professional services. If professional advice or other expert assistance is required, the services of a competent professional should be sought.

Library of Congress Cataloging-in-Publication Data

Sweeteners and sugar alternatives in food technology / edited by Kay O'Donnell, Malcolm Kearsley. – 2nd ed.
  p. cm.
  Includes bibliographical references and index.
  ISBN 978-0-470-65968-7 (hardcover : alk. paper)   1. Sweeteners.   2. Sugar substitutes.
3. Sugars in human nutrition.   I. O'Donnell, Kay.   II. Kearsley, M. W.
  TP421.S938 2012
  664′.5–dc23   2012010720

A catalogue record for this book is available from the British Library.

Wiley also publishes its books in a variety of electronic formats. Some content that appears in print may not be available in electronic books.

Cover images: Background: © Stockphoto.com/Lighthaunter; Cereal: © iStockphoto.com/LauriPatterson;
Stevia leaf: © iStockphoto.com/dirkr; Sweets: © iStockphoto.com/Juanmonino;
Stevia Powder: © iStockphoto.com/olm26250
Cover design by Steve Thompson

Set in 10/12 pt Times by Aptara® Inc., New Delhi, India
Printed and bound in Malaysia by Vivar Printing Sdn Bhd

1   2012

Contents

Preface	xvii
Contributors	xix

PART ONE: NUTRITION AND HEALTH CONSIDERATIONS — 1

1 Glycaemic Responses and Toleration — 3
Geoffrey Livesey

- 1.1 Introduction — 3
- 1.2 Glycaemic response in ancient times — 4
- 1.3 Glycaemic response approaching the millennium — 5
- 1.4 The glycaemic response now and in future nutrition — 6
- 1.5 Glycaemic response and adverse outcomes: both physiological and in response to advice — 7
- 1.6 Measurement and expression of the glycaemic response — 7
- 1.7 The acute glycaemic response to sugars and alternatives — 13
- 1.8 Long-term glycaemic control with sweeteners and bulking agents — 15
- 1.9 Are low glycaemic carbohydrates of benefit in healthy persons? — 18
- 1.10 Gastrointestinal tolerance in relation to the glycaemic response — 18
- 1.11 Conclusion — 19
- References — 20

2 Dental Health — 27
Anne Maguire

- 2.1 Introduction — 27
- 2.2 Dental caries — 27
  - 2.2.1 The problem — 27
  - 2.2.2 Aetiology — 28
  - 2.2.3 Control and prevention — 29
  - 2.2.4 Determining cariogenicity — 30
- 2.3 Reduced-calorie bulk sweeteners — 32
  - 2.3.1 Erythritol — 32
  - 2.3.2 Isomalt — 32
  - 2.3.3 Lactitol — 34
  - 2.3.4 Maltitol — 34
  - 2.3.5 Sorbitol — 36

		2.3.6	Mannitol	37
		2.3.7	D-tagatose	38
		2.3.8	Xylitol	38
		2.3.9	Key points from the dental evidence for reduced-calorie sweeteners and their use	42
	2.4	High-potency (high-intensity) sweeteners		43
		2.4.1	Acesulfame K	43
		2.4.2	Aspartame and Neotame	44
		2.4.3	Cyclamate and saccharin	44
		2.4.4	Sucralose	45
		2.4.5	Other sweeteners	46
		2.4.6	Key points from the dental evidence for high-potency (high-intensity) sweeteners and their use	47
	2.5	Bulking agents		47
		2.5.1	Polydextrose	47
		2.5.2	Fructose polymers	47
		2.5.3	Key points from the dental evidence for bulking agents	48
	2.6	Summary		49
	References			49

3 Digestive Health — 63
Henna Röytiö, Kirsti Tiihonen and Arthur C. Ouwehand

3.1	Introduction; prebiotics, sweeteners and gut health	63
3.2	Intestinal microbiota	63
3.3	Gut health	64
3.4	Prebiotics versus fibre	64
3.5	Endogenous prebiotics	64
	3.5.1 Milk oligosaccharides	64
	3.5.2 Secreted substrates in the gut	65
3.6	Prebiotics	65
3.7	Current prebiotics	65
3.8	Health benefits	67
3.9	Synbiotics	69
3.10	Safety considerations	70
3.11	Conclusion	71
Acknowledgements		72
References		72

4 Calorie Control and Weight Management — 77
Michele Sadler and Julian D. Stowell

4.1	Introduction	77
4.2	Caloric contribution of sugars in the diet	77
4.3	Calorie control and its importance in weight management	77
4.4	Calorie reduction in foods	78
4.5	Appetite and satiety research	80

	4.6	Sweeteners and satiety, energy intakes and body weight	81
		4.6.1 Satiety and energy intake	82
		4.6.2 Body weight management	83
	4.7	Relevance of energy density and glycaemic response	84
		4.7.1 Energy density	84
		4.7.2 Glycaemic response	84
	4.8	Legislation relevant to reduced calorie foods	85
	4.9	Conclusions	87
		Acknowledgement	87
		References	88

PART TWO: HIGH-POTENCY SWEETENERS   **91**

5 Acesulfame K   **93**
Christian Klug and Gert-Wolfhard von Rymon Lipinski

	5.1	Introduction and history	93
	5.2	Organoleptic properties	93
		5.2.1 Acesulfame K as the single sweetener	94
		5.2.2 Blends of acesulfame K with other sweetening agents	95
		5.2.3 Compatibility with flavours	98
	5.3	Physical and chemical properties	98
		5.3.1 Appearance	98
		5.3.2 Solubility	98
		5.3.3 Stability	99
	5.4	Physiological properties	100
	5.5	Applications	100
		5.5.1 Beverages	100
		5.5.2 Dairy products and edible ices	103
		5.5.3 Bakery products and cereals	104
		5.5.4 Sweets and chewing gum	104
		5.5.5 Jams, marmalades, preserves and canned fruit	107
		5.5.6 Delicatessen products	107
		5.5.7 Table-top sweeteners	108
		5.5.8 Pharmaceuticals	109
		5.5.9 Cosmetics	109
		5.5.10 Tobacco products	110
		5.5.11 Technical applications	110
	5.6	Safety and analytical methods	110
		5.6.1 Pharmacology	110
		5.6.2 Toxicology	110
		5.6.3 Safety assessments and acceptable daily intake	111
		5.6.4 Analytical methods	111
	5.7	Regulatory status	112
		5.7.1 Approvals	112
		5.7.2 Purity criteria	112
		References	112

6 Aspartame, Neotame and Advantame — 117
Kay O'Donnell

- 6.1 Aspartame — 117
  - 6.1.1 Introduction — 117
  - 6.1.2 Synthesis — 117
  - 6.1.3 Sensory properties — 119
  - 6.1.4 Physicochemical properties — 120
  - 6.1.5 Physiological properties — 123
  - 6.1.6 Applications — 125
  - 6.1.7 Analysis — 126
  - 6.1.8 Safety — 126
  - 6.1.9 Regulatory status — 127
- 6.2 Neotame — 127
  - 6.2.1 Neotame structure and synthesis — 128
  - 6.2.2 Sensory properties — 128
  - 6.2.3 Physiochemical properties — 130
  - 6.2.4 Physiological properties — 131
  - 6.2.5 Applications — 131
  - 6.2.6 Safety — 132
  - 6.2.7 Regulatory — 132
- 6.3 Advantame — 132
  - 6.3.1 Synthesis — 133
  - 6.3.2 Sensory properties — 133
  - 6.3.3 Stability — 133
  - 6.3.4 Solubility — 133
  - 6.3.5 Safety — 134
  - 6.3.6 Regulatory — 134
- References — 134

7 Saccharin and Cyclamate — 137
Grant E. DuBois

- 7.1 Introduction — 137
- 7.2 Current understanding of sweetness — 137
- 7.3 Saccharin — 139
  - 7.3.1 History, manufacture and chemical composition — 139
  - 7.3.2 Organoleptic properties — 140
  - 7.3.3 Physical and chemical properties — 144
  - 7.3.4 Physiological properties — 146
  - 7.3.5 Applications — 147
  - 7.3.6 Safety — 149
  - 7.3.7 Regulatory status — 151
- 7.4 Cyclamate — 151
  - 7.4.1 History, manufacture and chemical composition — 151
  - 7.4.2 Organoleptic properties — 154
  - 7.4.3 Physical and chemical properties — 156
  - 7.4.4 Physiological properties — 158

		7.4.5	Applications	158
		7.4.6	Safety	159
		7.4.7	Regulatory status	160
	References			163

8 Sucralose 167
Samuel V. Molinary and Mary E. Quinlan

	8.1	Introduction		167
	8.2	History of development		167
	8.3	Production		168
	8.4	Organoleptic properties		168
	8.5	Physico-chemical properties		170
	8.6	Physiological properties		174
	8.7	Applications		175
		8.7.1	Beverages	175
		8.7.2	Dairy products	178
		8.7.3	Confectionery	178
		8.7.4	Baked products	178
		8.7.5	Pharmaceuticals	179
	8.8	Analytical methods		179
	8.9	Safety		179
	8.10	Regulatory status		181
	References			181

9 Natural High-Potency Sweeteners 185
Michael G. Lindley

	9.1	Introduction		185
	9.2	The sweeteners		187
		9.2.1	Thaumatin	187
		9.2.2	Steviol glycosides	191
		9.2.3	Lo han guo (mogroside)	197
		9.2.4	Brazzein	200
		9.2.5	Monatin	201
	9.3	Conclusions		203
	References			204

PART THREE: REDUCED-CALORIE BULK SWEETENERS 213

10 Erythritol 215
Peter de Cock

	10.1	Introduction		215
		10.1.1	History	215
		10.1.2	General characteristics	215
		10.1.3	Manufacturing process	217

10.2	Organoleptic properties			218
	10.2.1	Sweetness intensity		218
	10.2.2	Sweetness profile		218
	10.2.3	Cooling effect		219
	10.2.4	Synergy with other sweeteners		219
10.3	Physical and chemical properties			219
	10.3.1	Stability		219
	10.3.2	Solubility		220
	10.3.3	Melting point and other thermal characteristics		220
	10.3.4	Viscosity		220
	10.3.5	Hygroscopicity		220
	10.3.6	Boiling point elevation and freezing point depression		220
	10.3.7	Water activity at various concentrations versus sucrose		220
10.4	Physiological properties and health benefits			221
	10.4.1	Digestion of carbohydrates		221
	10.4.2	Metabolic fate of erythritol		222
	10.4.3	Caloric value		225
	10.4.4	Digestive tolerance		225
	10.4.5	Glycaemic and insulinaemic response		225
	10.4.6	Dental health		226
	10.4.7	Anti-oxidant properties		227
10.5	Applications			228
	10.5.1	Table-top sweeteners		228
	10.5.2	Beverages		230
	10.5.3	Chewing gum		231
	10.5.4	Chocolate		234
	10.5.5	Candies		235
	10.5.6	Fondant		236
	10.5.7	Lozenges		236
	10.5.8	Bakery (pastry) products		237
10.6	Safety and specifications			239
10.7	Regulatory status			239
10.8	Conclusions			240
References				240

11 Isomalt 243
Anke Sentko and Ingrid Willibald-Ettle

11.1	Introduction		243
11.2	Organoleptic properties		244
	11.2.1	Sweetening potency versus sucrose	244
	11.2.2	Sweetening profile versus sucrose	245
	11.2.3	Synergy and/or compatibility with other sweeteners	245
11.3	Physical and chemical properties		245
	11.3.1	Stability	245
	11.3.2	Solubility	247
	11.3.3	Viscosity	247
	11.3.4	Heat of solution	247

		11.3.5	Boiling point elevation	248
		11.3.6	Melting range	249
		11.3.7	Hygroscopicity – moisture uptake at various relative humidities	249
		11.3.8	Water activity at various concentrations versus sucrose	250
	11.4	Physiological properties	252
	11.5	Applications	254
		11.5.1	Hard candies	254
		11.5.2	Chocolates	259
		11.5.3	Low boilings	261
		11.5.4	Chewing gum	263
		11.5.5	Pan coating with ISOMALT GS	264
		11.5.6	Compressed tablets	266
		11.5.7	Baked goods	267
		11.5.8	Fruit spreads	268
		11.5.9	Breakfast cereals, cereal bars and muesli	268
		11.5.10	Overview – further applications	270
	11.6	Safety	270
	11.7	Regulatory status: worldwide	271
	11.8	Conclusions	271
	References	272

12 Lactitol 275
Christos Zacharis

	12.1	History	275
	12.2	Organoleptic properties	275
	12.3	Physical and chemical properties	276
		12.3.1	Stability	276
		12.3.2	Solubility	277
		12.3.3	Viscosity	278
		12.3.4	Heat of solution	279
		12.3.5	Boiling point elevation	280
		12.3.6	Hygroscopicity	280
		12.3.7	Water activity	280
	12.4	Physiological properties	281
		12.4.1	Metabolism	281
	12.5	Health benefits	282
		12.5.1	Lactitol as a prebiotic	282
		12.5.2	Lactitol to treat hepatic encephalopathy	285
		12.5.3	Lactitol and diabetes	285
		12.5.4	Tooth-protective properties	285
	12.6	Applications	287
		12.6.1	Chocolate	287
		12.6.2	Baked goods	288
		12.6.3	Chewing gum and confectionery	289
		12.6.4	Ice cream and frozen desserts	289
		12.6.5	Preserves	290
		12.6.6	Tablets	290

12.7	Regulatory status		291
12.8	Conclusions		291
References			292

13 Maltitol Powder — 295
Malcolm W. Kearsley and Ronald C. Deis

13.1	Introduction		295
13.2	Production		296
	13.2.1	Alternative methods of maltitol manufacture	296
13.3	Structure		297
13.4	Physical and chemical properties		297
	13.4.1	Chemical reactivity	298
	13.4.2	Compressibility	298
	13.4.3	Cooling effect (heat of solution)	298
	13.4.4	Humectancy and hygroscopicity	298
	13.4.5	Molecular weight	298
	13.4.6	Solubility	299
	13.4.7	Sweetness	299
13.5	Physiological properties		299
	13.5.1	Calorific value	299
	13.5.2	Dental aspects	300
	13.5.3	Diabetic suitability	300
	13.5.4	Glycaemic index	301
	13.5.5	Laxative effects	301
13.6	Applications in foods		302
	13.6.1	The main food applications of maltitol	303
13.7	Labelling claims		305
13.8	Legal status		306
13.9	Conclusions		306
References			307

14 Maltitol Syrups — 309
Michel Flambeau, Frédérique Respondek and Anne Wagner

14.1	Introduction		309
14.2	Production		310
	14.2.1	Maltitol syrups	310
	14.2.2	Polyglycitols	310
14.3	Hydrogenation		311
14.4	Structure		312
14.5	Physico-chemical characteristics		312
	14.5.1	Chemical reactivity	313
	14.5.2	Cooling effect (heat of solution)	313
	14.5.3	Humectancy	313
	14.5.4	Hygroscopicity	313
	14.5.5	Molecular weight	314
	14.5.6	Solubility	315
	14.5.7	Viscosity	316

14.6	Physiological properties		316
	14.6.1	Calorific value	316
	14.6.2	Dental aspects	317
	14.6.3	Glycaemic index	318
	14.6.4	Toleration	319
	14.6.5	Sweetness	321
	14.6.6	Conclusions	322
14.7	Applications in foods		323
	14.7.1	Hard candy	324
	14.7.2	Aerated confectionery	324
	14.7.3	Caramels	325
	14.7.4	Sugar-free panning	326
	14.7.5	Chewing gum	327
	14.7.6	Other sugar-free or reduced-sugar confectionery	327
	14.7.7	Dairy applications	327
	14.7.8	Bakery applications	328
	14.7.9	Ketchup	328
14.8	Legal status		329
14.9	Safety		329
14.10	Conclusions		329
References			330

15 Sorbitol and Mannitol — 331

Ronald C. Deis and Malcolm W. Kearsley

15.1	Introduction		331
15.2	Production		331
	15.2.1	Sorbitol powder	332
	15.2.2	Sorbitol syrups	333
	15.2.3	Mannitol	333
15.3	Hydrogenation		335
15.4	Storage		335
15.5	Structure		335
15.6	Safety		336
15.7	Physico-chemical characteristics		337
	15.7.1	Chemical reactivity	337
	15.7.2	Compressibility	337
	15.7.3	Cooling effect	338
	15.7.4	Humectancy	338
	15.7.5	Hygroscopicity	338
	15.7.6	Molecular weight	338
	15.7.7	Solubility	339
	15.7.8	Viscosity	339
15.8	Physiological properties		339
	15.8.1	Calorific value	339
	15.8.2	Dental aspects	340
	15.8.3	Diabetic suitability	340
	15.8.4	Glycaemic response	340

		15.8.5	Tolerance	341
		15.8.6	Sweetness	341
	15.9	Applications in foods		342
		15.9.1	Gum	342
		15.9.2	Hard candy	343
		15.9.3	Tabletting	343
		15.9.4	Surimi	343
		15.9.5	Cooked sausages	343
		15.9.6	Baked goods	344
		15.9.7	Panning	344
		15.9.8	Over-the-counter products	344
		15.9.9	Chocolate	344
	15.10	Non-food applications		344
		15.10.1	Sorbitol	344
		15.10.2	Mannitol	345
	15.11	Legal status		345
	15.12	Conclusions		346
	References			346
16	**Xylitol**			**347**
	Christos Zacharis			
	16.1	Description		347
	16.2	Organoleptic properties		348
		16.2.1	Sweetness	348
		16.2.2	Sweetness synergy	349
	16.3	Physical and chemical properties		350
		16.3.1	Heat of solution	350
		16.3.2	Stability	351
		16.3.3	Solubility	351
		16.3.4	Viscosity	352
		16.3.5	Boiling point elevation	352
		16.3.6	Water activity	354
		16.3.7	Hygroscopicity	354
	16.4	Physiological properties		354
		16.4.1	Metabolism	354
		16.4.2	Suitability for diabetics	355
		16.4.3	Tolerance	356
		16.4.4	Caloric value	356
		16.4.5	Health benefits	357
		16.4.6	Other health benefits associated with xylitol	364
	16.5	Applications		366
		16.5.1	Confectionery	366
		16.5.2	Chewing gum	367
		16.5.3	Hard coating applications	367
		16.5.4	Chocolate	368
		16.5.5	Dairy products and frozen desserts	368

		16.5.6	Baked goods	368
		16.5.7	Non-food applications	369
	16.6	Safety		369
	16.7	Regulatory status		370
	References			371

PART FOUR: OTHER SWEETENERS 383

17 New Developments in Sweeteners 385
Guy Servant and Gwen Rosenberg

	17.1	Sweet taste modulators		385
	17.2	Sweet modulator targets		385
	17.3	Industry need for reduced-calorie offerings		385
	17.4	Sweet taste receptors		386
		17.4.1	Sweet taste modulator mechanism of action	386
		17.4.2	Identification and evaluation of sweet taste modulators	387
		17.4.3	Optimisation of sweet taste modulators	388
	17.5	Commercially viable sweet taste modulators		390
	17.6	Regulatory approval of sweet taste modulators		390
	17.7	Commercialisation of sweet taste modulators		391
	17.8	Future sweet taste modulators and new sweeteners		392
	17.9	Modulators for other taste modalities		392
	17.10	Savoury flavour ingredients		393
	17.11	Bitter blockers		393
	17.12	Cooling flavours		393
	17.13	Salt taste modulators		394
	17.14	Conclusions		394
	References			394

18 Isomaltulose 397
Anke Sentko and Ingrid Willibald-Ettle

	18.1	Introduction		397
	18.2	Organoleptical properties		397
	18.3	Physical and chemical properties		398
		18.3.1	Physical properties	398
		18.3.2	Chemical properties	400
	18.4	Microbiological properties		401
	18.5	Physiological properties		402
		18.5.1	Dental health	403
		18.5.2	Effect on blood glucose and insulin	403
		18.5.3	Effect on fat oxidation	405
		18.5.4	Gastrointestinal tolerance	405
	18.6	Toxicological evaluations		406
	18.7	Applications		406
		18.7.1	Beverage applications	406
		18.7.2	Confectionery applications	410
		18.7.3	Other applications	411

18.8	Regulatory status	413
18.9	Conclusions	413
References		413

19 Trehalose — 417
Takanobu Higashiyama and Alan B. Richards

19.1	Introduction	417
19.2	Trehalose in nature	418
19.3	Production	419
19.4	Metabolism, safety and tolerance	420
19.5	Regulatory status	421
19.6	Properties	421
19.7	Application in food	423
	19.7.1 Technical properties	423
	19.7.2 Stabilisation of carbohydrates	423
	19.7.3 Stabilisation of proteins	424
	19.7.4 Stabilisation of flavours and aromas	425
19.8	Physiological properties	426
19.9	Conclusions	428
References		429

PART FIVE: BULKING AGENTS – MULTI-FUNCTIONAL INGREDIENTS — 433

20 Bulking Agents – Multi-Functional Ingredients — 435
Michael Auerbach and Anne-Karine Dedman

20.1	Introduction	435
20.2	Gluco-polysaccharides	437
	20.2.1 Polydextrose	437
20.3	Resistant starches and resistant maltodextrins	449
	20.3.1 Fibersol-2	450
	20.3.2 Nutriose FB	452
20.4	Fructo-oligosaccharides	454
	20.4.1 Inulin and low-molecular-weight FOS	454
References		462

Index — 471

Preface

Indulgence in sweet foods and drinks is a human weakness and both are consumed far beyond their value in relieving hunger and thirst. Sweetness is most commonly associated with sucrose, and this is the most widely consumed sweetener in the world although it has been criticised for many years by some with regard to its contribution to obesity, dental caries and other diseases. While glucose (and fructose) syrups are widely used to replace sucrose in foods, this is largely a cost-saving exercise and does not address many of the problems associated with 'sugar' consumption.

Over the last 30 years, a wide range of sugar replacers have been developed and marketed. These include the bulk sweeteners – the polyols, which replace sucrose and glucose on a weight for weight basis, and the high-potency sweeteners where a kilogram of sugar in a food product might be replaced by a few grams of a very sweet material. In the case of high-potency sweeteners, this has led to the development of a parallel industry to provide ingredients, which can be used in conjunction with the sweetener to retain the 'bulk' of the traditional product. More recently, sugars with many of the properties of bulk sweeteners and sweet taste enhancers that increase the potency of sweet compounds have been new additions to the market.

Replacement of sugars in foods has given us a new perspective on healthy foods where claims such as 'sugar-free', 'no-added-sugar' and 'reduced calorie/sugar' are being exploited by food manufacturers.

The use of ingredients to improve the nutritional status of a food product is one of the major driving forces for new product development, and sugar replacement is still seen as an area for development. This book provides a unique reference for food scientists and technologists with information on sugar replacement options to produce foods that not only taste and perform as well as sugar-based products but also offer consumer benefits including calorie reduction, dental health benefits, digestive health benefits and improvements in long-term disease risk through strategies such as dietary glycaemic control.

This second edition of *Sweeteners and Sugar Alternatives* follows the same basic layout as in the successful first edition with many of the same authors contributing to their relevant areas of expertise. New chapters on isomaltulose, trehalose and developments in sweeteners have been included to reflect changes in the use and understanding of sweeteners and sweet taste. For ease of reference, the book is set out as follows:

Part One: Nutrition and Health Considerations. This part considers the physiological effects and subsequent health benefits of sweeteners and sugar alternatives. Subjects include improved glycaemic control, dental health, digestive health and the role of these products in calorie control and weight management.

Part Two: High-Potency Sweeteners. This part describes the properties and applications of the most commonly used sweeteners. These products are unique in that they provide high sweetness without bulk and without any major impact on calories.

Part Three: Reduced-Calorie Bulk Sweeteners. This part describes both the properties and applications of polyols and includes reference to polyglycitols, hydrogenated glucose syrups containing less than 50% maltitol. Bulk sweeteners provide the physical characteristics of sugar and glucose but with reduced calories and other physiological benefits.

Part Four: Other Sweeteners. This part describes new developments in sweeteners and specifically how they elicit a sweet taste when consumed. Also included are chapters on isomaltulose and trehalose that, although classed as sugars, exhibit many of the properties of the bulk sweeteners.

Part Five: Bulking Agents – Multi-Functional Ingredients. This chapter focuses on the oligo- and polysaccharide materials that are most applicable as sugar alternatives and that have sugar-like properties in food applications and, often, prebiotic properties. They allow greater flexibility when replacing sugar in formulations and complement the use of all types of sweeteners in foods.

The summary tables at the end of each chapter and the extensive references are meant to inspire those who wish to learn more.

A sincere thanks to all the contributors to the book.

Kay O'Donnell and Malcolm W. Kearsley

Contributors

Michael Auerbach
Regulatory Advisor
Active Nutrition, DuPont Nutrition & Health
NY, USA

Peter de Cock
Cargill
Vilvoorde, Belgium

Anne-Karine Dedman
Technical Manager
Active Nutrition, DuPont Nutrition & Health
Paris, France

Ronald C. Deis
Corn Products International
Newark, DE, USA

Grant E. DuBois
Sweetness Technologies, LLC
Roswell, GA, USA

Michel Flambeau
Application and Technical
Support Director
Tereos Syral
Marckolsheim, France

Takanobu Higashiyama
Hayashibara International Inc.
Broomfield, CO, USA

Malcolm W. Kearsley
Reading, UK

Christian Klug
Head of Quality Management
and Regulatory Affairs
Nutrinova Nutrition Specialities and Food
Ingredients GmbH
Sulzbach, Germany

Michael G. Lindley
Lindley Consulting
Crowthorne, UK

Geoffrey Livesey
Independent Nutrition Logic Ltd
Wymondham, UK

Anne Maguire
Centre for Oral Health Research
School of Dental Sciences
Newcastle University, UK

Samuel V. Molinary
Consultant
Scientific & Regulatory Affairs
Beaufort, SC, USA

Kay O'Donnell
Weybridge
UK

Arthur C. Ouwehand
Active Nutrition, DuPont Nutrition & Health
Kantvik, Finland

Mary E. Quinlan
Tate & Lyle
London, UK

Frédérique Respondek
Scientific and Regulatory Affairs Manager
Tereos Syral
Marckolsheim, France

Alan B. Richards
Hayashibara International Inc.
Broomfield, CO, USA

Gwen Rosenberg
Senomyx Inc
San Diego, CA, USA

Henna Röytiö
Active Nutrition, DuPont Nutrition & Health
Kantvik, Finland

Michele Sadler
Consultant Nutritionist
MJSR Associates
Kent, UK

Anke Sentko
BENEO GmbH
Mannheim, Germany

Guy Servant
Senomyx Inc.
San Diego, CA, USA

Julian D. Stowell
Vice President, Scientific Affairs
Active Nutrition, DuPont Nutrition & Health
Reigate, UK

Kirsti Tiihonen
Active Nutrition, DuPont Nutrition & Health
Kantvik, Finland

Gert-Wolfhard von Rymon Lipinski
MK Food Management Consulting GmbH
Bad Vilbel, Germany

Anne Wagner
Vice President
Innovation and Quality
Tereos Syral
Marckolsheim, France

Ingrid Willibald-Ettle
BENEO GmbH
Mannheim, Germany

Christos Zacharis
Technical Manager Functional Sweeteners
Active Nutrition, DuPont Nutrition & Health
Surrey, UK

Part One
Nutrition and Health Considerations

1 Glycaemic Responses and Toleration

Geoffrey Livesey
Independent Nutrition Logic Ltd, Wymondham, UK

1.1 INTRODUCTION

Sugars and sweeteners have an important role in the human diet and choosing the right ones in the right amounts can influence health. Knowledge will enable good choices, and further research and understanding of the literature will confirm or deny how good our choices are, and where improvements are possible. Choice is not simply a matter of which is the healthier or healthiest, since the technological properties and economics of sugars and sweeteners impact on which of them can be used suitably in a particular food.

A wide range of potential influence on health is offered by sugars and sweeteners when selected appropriately, as will be evident in detail from other chapters. These include the following:

- A reduced risk of dental caries.[1]
- Potential for improved restoration of the early carious lesions.[2]
- A reduction in caloric value that may contribute towards a lower risk of overconsumption, obesity and improved survival.[3,4]
- Substrate for butyrate production, and potentially reduced risk of colon cancer.[5]
- The formation of osmolytes efficacious for laxation and lower risk of constipation or accumulation of toxic metabolites.[6]
- Substrate for saccharolytic and acidogenic organisms in the colon that contribute to prebiosis and 'digestive health' potentially including improved immunological function.[7,8]

Each of these can influence the choice of sugars and sweeteners. Of particular relevance is their impact on glycaemic response and potential to contribute to low glycaemic index (GI) or glycaemic load (GL) diets.

Lowering post-prandial glycaemia and insulinaemia through an appropriate choice of sugars[9] and sweeteners,[7] together with other low-glycaemic carbohydrates,[10] fibre, protein, lower energy intake and exercise,[11] can each improve glycaemic control. In turn, this appears to lower the prevalence or risk of developing metabolic diseases including metabolic syndrome, diabetes (and associated complications), heart disease, hypertension, stroke, age-related macular degeneration and certain cancers.[12–16]

In those who are susceptible, lower glycaemic carbohydrate foods may also benefit appropriate weight gain during pregnancy,[17] limit insulin requirements in gestational diabetes,[18] potentially allow favourable foetal growth patterns and fat accretion,[19] reduce neural tube defects[20] and aid recovery from surgery.[21]

Meta-regression of interventional studies of lower GI or GL diets show a time-dependent lower body weight over a 1-year period[22] and supports weight maintenance after weight loss.[23] Reduced food intake in humans[4] may be partly responsible for weight loss and maintenance. Lowering of body weight improves survival among newly diagnosed diabetes patients,[24] and may contribute to longer survival beyond old age as seen in animal studies while lowering glycaemia with isomalt.[4]

The converse of all aforementioned is that, given the right circumstances, a poor choice of type and amount of all carbohydrates, including sugars and sweeteners, could augment ill health. Attributes of sugars and sweeteners affecting health via the glycaemic response are nutritional and need to be seen in the context of the whole diet. It is appropriate, therefore, to consider the glycaemic aspect of diet and health from ancient to the present and future times – so far as these can be ascertained, explained and envisaged.

1.2 GLYCAEMIC RESPONSE IN ANCIENT TIMES

It is often argued that our genes might not cope with diets that are substantially different from those eaten by our ancestors.[25–30] Quite what these diets were or how tolerant ancient genes have become are matters of uncertainty. Successful genes were in existence for both herbivorous and carnivorous diets prior to humankind; however, no early diet appears to have been high glycaemic. Those peoples who would normally consume 'early' or rudimentary diets, such as recent hunter–gatherers, experience low levels of diabetes and respond adversely to diets we may now consider high glycaemic.[26,31] This is consistent with the notion that early genes were unadapted to high-glycaemic responses, and also consistent with a notion of adaptation having occurred in the people of today's relatively more glucose-tolerant 'western' cultures, at least among a large proportion of them. Those not having adapted, contribute to prevalent diabetes and other conditions mentioned that are currently experienced, which is far higher than in either hunter–gatherers or rudimentary horticulturalists or simple agriculturalists or pastoralists.[26] For the people of these 'basic' cultures and for 'unadapted' westerners (easterners or southerners or northerners), a high-glycaemic response remains a health hazard, for which a variety of strategies exist to help them cope.[11] Europe has a rich culture and a documented history of its foods, and so we can obtain some idea of how the glycaemic character of diets may have developed over time.

Generally, we may assume diets to partly reflect the foods that can be found or are made available to eat. If this is so, examination of the inventory of foods identified in European history may shed some light on what was eaten and what might now be eaten for optimal health. Such an inventory is provided by Toussaint-Samat[32] from which an assessment of the development in the glycaemia character of contemporary diets has been made taking account of the protein, fat, fibre and sources of carbohydrate (Figure 1.1). The picture cannot be accurate but what is clear is a progressive increase in the GL, with a markedly rapid increase in this GL following industrialisation. We cannot be sure of the prevalence of disease in Europe throughout the whole of this timescale, but we would not likely dispute that the prevalence of obesity and metabolic disease is as high now as ever.

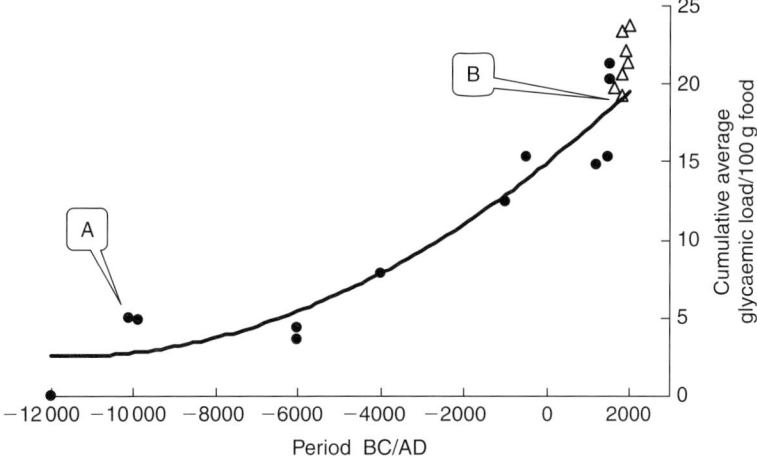

Fig. 1.1 Evolutionary adaptation to ancient diets of low glycaemic load may have left mankind genetically predisposed to non-communicable diseases provoked by today's high-glycaemic diets. Based on the history of foods in Europe,[32] with calculations by this author (A, agricultural revolution; B, industrial revolution). Open symbols show values post the industrial revolution.

Such a trend is argued to also have occurred throughout more recent times in the United States,[25] with recent emphasis on reducing the fat content of the diet, a doubling of flour consumption during the 1980s and an increase overall in sugar, corn syrup and dextrose consumption prior to the end of the millennium.[33–35] These together with a lower dietary fibre content of foods[34] imply exposure to diets eliciting a high-glycaemic response.

1.3 GLYCAEMIC RESPONSE APPROACHING THE MILLENNIUM

Much of our understanding of the interplay between health and the glycaemic response to foods has arisen from investigations into the dietary management of diabetes. Whereas very low-glycaemic carbohydrate foods such as Chana dahl were used in ancient India for a condition now recognised as diabetes,[36] nineteenth century recommendations in western cultures were for starvation diets, which were, of course, non-glycaemic. The drawback of such is obvious and in 1921, high-fat (70%) low-carbohydrate (20%) diets were recommended,[37] which by definition would be low glycaemic. A gradual reintroduction of carbohydrate into recommendations for diets for diabetic patients arose as carbohydrate metabolism came under some control using drugs, but mainly because 'dietary fat' was recognised to have a causal role in coronary heart disease, to which diabetics and glucose intolerant individuals succumb, more readily in some cases than others.[38–41] The metabolic advantages of replacing dietary fat (saturated fat) with high-fibre high-carbohydrate was lower fasting glycaemia, lower total-, HDL- and LDL-cholesterol and lower triglycerides.[42–46] Such benefits may in part be related to dietary fibre or its influence on the glycaemic response.[47,48] Certainly, the non-digestible carbohydrate in these diets would ensure some degree of lower glycaemia for a given carbohydrate intake and support beneficial effects from lower saturated fat intake.

During these times, the adverse influence of higher glycaemia or more dietary carbohydrate was either unrecognised or the risk was accepted by the medical profession in fear of (or compromise for) the adverse effects of 'dietary fat'. The adverse influence of higher glycaemia may also have been overlooked due to the apparent benefits of the non-digestible carbohydrate in the high-carbohydrate foods. Indeed, the Institute of Medicine has recommended high-fibre diets to combat coronary heart disease,[49] and this builds upon the dietary fibre hypothesis that proposed higher prevalence of diabetes, heart disease and other conditions associate with diets deficient of fibre.[50,51] An absence of fibre in high-sugar products left sugar (sucrose) vulnerable; nevertheless, this sugar remained preferable among nutritionists to high (saturated) fat, which it might displace from the diet, giving rise to the concept of the 'sugar–fat-seesaw' discussed elsewhere.[52–54]

Throughout the whole of these times, the primary purpose of recommending energy from carbohydrate was to displace the intake of energy as fat. In part, this is because carbohydrate supplies energy, but also because carbohydrate counters the insulin desensitising influence of both mobilised body fat and dietary fat.[55–57] This purpose for carbohydrate was retained in the GI concept, whereby carbohydrate of low-glycaemic response further improved glycaemic control in diabetes patients,[10] and possibly the plasma lipid profile.[58]

However, it must be considered whether carbohydrates have a long-term future as a means to displace fats from the diet. It is noteworthy that the increasing carbohydrate content of diets throughout European history, which partly explains the higher GL (Figure 1.1), has not adequately displaced 'fats' from the diet or prevented obesity. Excess of carbohydrate prevents the use of fat stores and encourages dietary fat to be stored. In general, elevating the consumption of monounsaturated and polyunsaturated (bar trans) fats is considered beneficial in respect of diabetes, coronary heart disease and a variety of conditions[59–62] and is consistent with early diets.[63] In addition, there is little or no evidence that carbohydrate ingestion can selectively limit the ingestion of saturated fats. Proponents of the Mediterranean diet (high in mono- and polyunsaturated fats) would hold that the use of carbohydrate for the purpose of limiting fat intake is unsound.

1.4 THE GLYCAEMIC RESPONSE NOW AND IN FUTURE NUTRITION

The general picture now for glycaemic control is that a high-fibre, low-glycaemic and low-saturated fat diet is optimal.[63] With obesity being a major problem and a risk factor for type-2 diabetes and heart disease, an appropriate energy balance has become of major importance.[64] Weight loss has for some time been recognised as important to the survival of newly diagnosed type-2 diabetes patients[24] and improvement in prognosis for cardiovascular disease.[65,66] These are practical examples of how caloric restriction improves survival in at-risk groups. Of course, caloric restriction implies here a diet reduced in energy via lower saturated fat and lower GL than is generally consumed.

It is clearly preferable to limit the intake of both saturated fat and high-glycaemic carbohydrate as energy sources to facilitate weight reduction, rather than simply to exchange energy sources. Prior nutritional debates of 'fat versus carbohydrate' might now be viewed as too imprecise in both the description of the food components and how the components are pitched against each other. A similar concern arises when it is argued that low-GI foods should find automatic favour over low-GL foods when in communication with the consumer.

Choosing low-GI foods does not automatically mean maintaining a lower fat intake since approximately 50% of the variance in the GI of foods can be attributed to their fat content. The nutrition debate still needs to provide greater scope for consideration of the adverse influence of 'saturated fats *plus* high-GL' together in general nutrition.

Sugars and sweeteners provoke a range of glycaemic responses related to the carbohydrate structure without the need to ask whether the glycaemic response is actually brought about by co-ingested dietary fat,[7,9,67] and so may variably promote, defer or help prevent ill health. Various research groups indicate at the time of writing that 'the concepts and methods regarding the GI [or GL] are sufficiently mature to recommend preparing the population to use GI as a way to help choose healthier foods...'[16] This is a position consistent with that over a decade ago in the WHO/FAO recommendations to primary producers and processors of foods: 'Consider how existing and new technologies can be used to help meet dietary goals regarding the quantity and nutritional properties of food carbohydrates...' and to 'provide appropriate information to the consumer on food labels.'[68]

1.5 GLYCAEMIC RESPONSE AND ADVERSE OUTCOMES: BOTH PHYSIOLOGICAL AND IN RESPONSE TO ADVICE

Advice to consume a diet of low-, in exchange for high-glycaemic foods has raised consideration about whether this would detract from other nutritional advisory messages. There are, however, no known adverse effects of choosing a diet including low-glycaemic carbohydrate foods instead of high-glycaemic ones,[69] other than for occasionally temporal gastrointestinal discomfort whenever this is accompanied by excessive low digestible carbohydrate ingestion (discussed in Section 1.10).

Occasionally claims are made that the benefits of low GI can be achieved by selecting whole grain foods, fruits and legumes, and that low-glycaemic advice would interfere with this whole food advice. However, such benefit of whole foods is hardly ever likely to be achieved optimally because the glycaemic indices of foods in these food categories cover wide ranges of GI values (Figure 1.2). Intervention choosing low instead of high-GI fruits is shown to be of benefit to diabetes patients, for example.[70]

1.6 MEASUREMENT AND EXPRESSION OF THE GLYCAEMIC RESPONSE

By 1929, the potential of carbohydrate to raise plasma glucose, some of which may spill over into urinary losses in diabetes patients, was indicated by its available carbohydrate content,[71] for which a direct assay to determine the composition of foods was later refined.[72] Fibre was suitable for diabetes patients as it provided no glucose, to either elevate plasma glucose concentrations or urinary losses. Another measure of the glycaemic potential became known as the GI.[73] Later, the quantity called GL, the product of available carbohydrate and GI, was introduced[74] and validated as a measure of the glycaemic response.[75] GL can be assayed directly and without need for knowledge of the available carbohydrate content,[75–77] about which assumptions are too often made.[4] The GI became widely known, and many GI testing centres have opened. Meanwhile, GI has received criticism as it is said to not meet

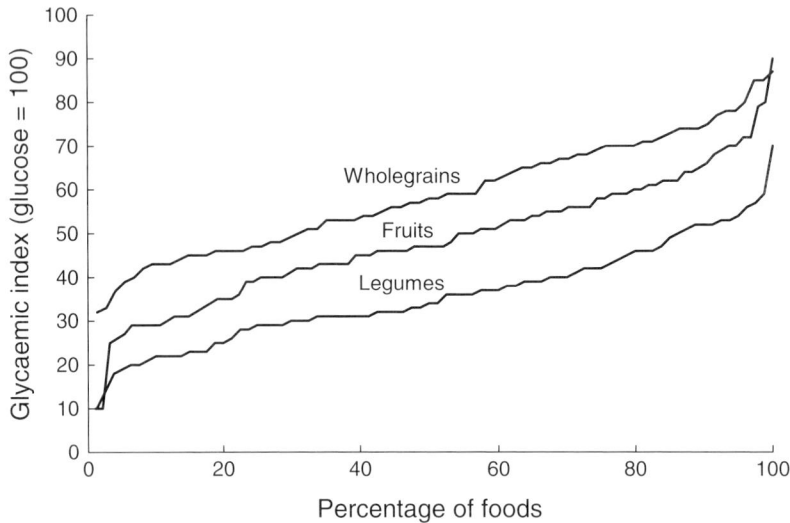

Fig. 1.2 Wholegrain, fruit and legume foods each span a wide range of glycaemic index (GI) values. Information is on 74 wholegrain food, 94 fruit foods and 80 legume foods from the 2008 International Tables of GI and glycaemic load,[79] and are presented in order from the lowest to the highest GI.

many useful criteria for inclusion in conventional food tables or in communication with the consumer,[78] though tabulation is possible and finds application nonetheless.[67,79]

The precision of the GI assay, initially examined in a study among five laboratories based on capillary blood sampling using high-carbohydrate foods,[80] has since been the topic of discussion with the aim of standardisation,[81] has subsequently been assessed among 28 different laboratories,[82] and now has Australian[83] and International[84] standardisation. The standardised protocol is only a little different now from that used in the first inter-laboratory study in particular, with regards to the precision achieved. An outstanding question is whether the methods for assessing GI and GL are adequately reproducible for communication with the consumer.

A useful point of reference when assessing a method's adequacy is one often used in regulatory enforcement for substantiation of reported or declared values in food labelling. Tests need to be able to assess whether a reported value is compliant with regulations specifying boundaries of accuracy required for labelling purposes. Such enforcement often finds it generally practical to 'accept' an 'error' of no more than 20% in a nutrient value reported on a food label in comparison with an officially analysed (or assessed) value.[85] Such an apparently large 'permitted' discrepancy ensures that differences between reported and official values do not arise simply by chance due to imprecision of the test method. However, this particular approach of using a nutrient value as the reference amount that defines the absolute size of the 20% value has limitations. One is that the 20% of nutrient value is extremely onerous when nutrient values are low, because as the value approaches zero, the percentage error approaches infinity. The second is that the 'permissible error' differs according to the nutrient amount; 20% of 1 g is 0.2 g, but 20% of 100 g is 20 g, which is 100 times higher. The third is that a constant 20% of nutrient value fails to follow the real error structure in the analytical data except if the error size is an exact proportion of the measurement size, which for biological tests is practically never. For a test such as GI, a basal

Table 1.1 Precision of glycaemic response[a] values according to the definition of the 'true' value.

Sucrose and alternatives	Glycaemic response (g GL/100 g ingredient)[a]	LSD from 'true' value (g GL/100 g ingredient)[b]	
		Definition a (true = single laboratory result)[c]	Definition b (true = combined laboratory results)[d]
Sucrose	64	11	4
Erythritol	0	5	2
Xylitol	12	10	4
Sorbitol	9	7	3
Maltitol	45	12	4
Regular maltitol syrup	52	14	5
Isomalt	9	9	3
Lactitol	6	6	2

GL, glycaemic load; LSD, least significant difference.
[a]Expressed as glycaemic load (g equivalents of glucose per 100 g carbohydrate ingredient) as would be derived using the glycaemic index protocol. The value and standard deviations of reproducibility are estimated for data on sucrose from ref. 79 and data on alternatives are from ref. 7.
[b]LSD; least significant difference between claimants analysis and the 'true value' at $P < 0.01$.
[c]The true value being that defined as true according to results from a single proficient assessment laboratory and testing of the difference in analyzed values $A - B$ from the claimant and assessment laboratory value, respectively.
[d]The true value being that defined as true according to results combined from the two laboratories involved; that is from an assessment laboratory and from a laboratory providing an original value the source of information declared on a label, and so testing the difference $A - (A + B)/2$.

or zero response does not have zero error of measurement. For a biological test, the 20% of nutrient value, if invoked, would not therefore seem to be a practical tool for assessment of compliance of a nutrient value with potential regulations – as noted previously in the first edition of this book[86] and now elsewhere.[87,88] Rather, some other reference amount would seem useful as used, for example, with vitamin C or iron for which dietary reference values (DV) can provide the reference amount, that is something other than a nutrient value can be used as reference amount.

On the basis of the data from inter-laboratory studies[80,88] and the composition of foods,[82,89] the GI methodology may appear difficult to justify from an analytical perspective alone because of its imprecision among laboratories (Figure 1.3). By contrast, this is less apparent for GL (Figure 1.3), making GL a potentially suitable measure or method of expression of the glycaemic response for communication with the consumer.

This conclusion for GL applies also to sugar and alternative sweeteners (Table 1.1) whether or not GL is expressed per 100 g available carbohydrate (e.g. the GI of sucrose) or per 100 g ingredient weight for the alternative bulk sweeteners so far examined. The difference in accuracy between GI and GL (Figure 1.3) arises simply because of the different reference amounts: 100 g carbohydrate in the case of GI (expressed g GL per 100 g carbohydrate = GI), and 100 g fresh weight of food product or ingredient in the case of GL (expressed as g GL per 100 g fresh weight). The 100 g fresh weight might equally be substituted by 100 g GL, where ~100 g GL per day (or per 2000 kcal) might be optimal for prevention of metabolic conditions such as coronary heart disease or type-2 diabetes and possibly other diseases (Livesey and Taylor, unpublished observations, 2008).

A further issue with regard to compliance is that of defining the true value with which the reported or claimed nutrient value (GI or GL) should comply. One possibility (definition 'a'

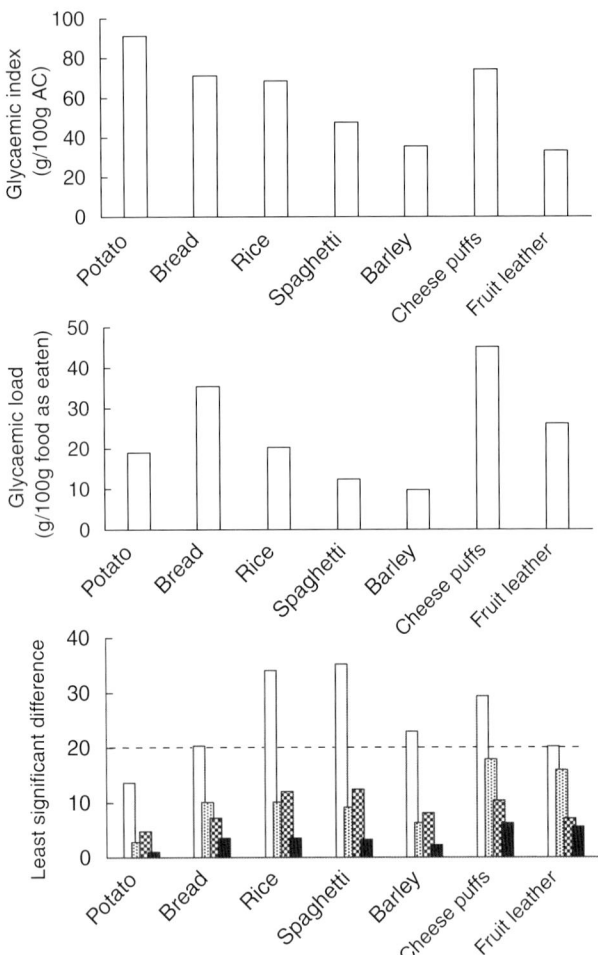

Fig. 1.3 Precision of measurement and expression of the glycaemic response. Data are based on inter-laboratory studies[80,82] and the composition of these foods.[82,89] The least significant difference (LSD) is either (a) the size of difference between a claimant laboratory value and a subsequent assessment laboratory value or (b) the size of difference between a claimed value and the true value; the last being the combined mean of the claimed and the subsequently assessed values, assuming both laboratories are proficient. LSD = $t \times SD_{reproducibility}$ (standard deviation of reproducibility) for $p < 0.01$ and t is approximated by the multiplier 2.8 as described in ISO standard 5725.[140] The validity of this approach implies that $SD_{reproducibility}$ is the apparent among laboratory standard deviation, apparent because it combines the within laboratory and the true among laboratory variabilities (and any interaction). The four bars from left to right: GI by definitions 'a' and 'b', and GL by definitions 'a' and 'b'.

in the legend to Figure 1.3) is that the analysed value provided by an assessing laboratory is considered the true value. However, this is problematic if both reporting and assessing laboratories are proficient because scientific validity would favour a combined value for the two laboratories as defining the true value. In this case, the between laboratory standard deviation would be smaller by $1/\sqrt{2}$ (definition 'b' in the legend to Figure 1.3) and the difference between the true value and the claimant laboratory value would be half that for the

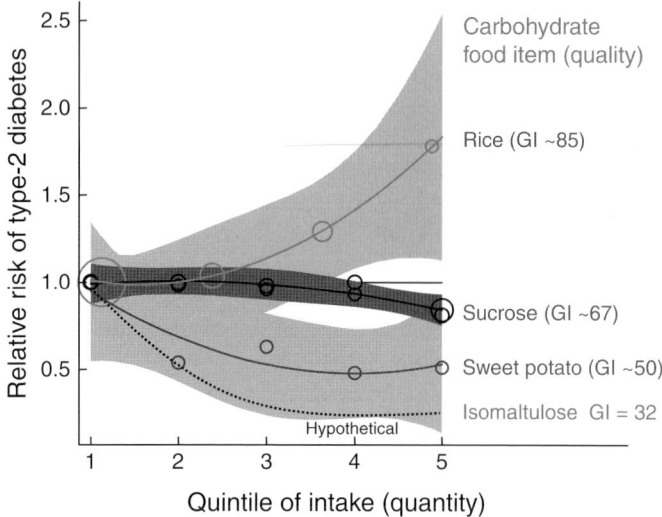

Fig. 1.4 Both the quantity and quality of carbohydrate food affect the relative risk of type-2 diabetes. Relative risk data are from large prospective observational studies for rice and sweet potato[141] and sucrose.[116,142] The hypothetical curve for isomaltulose is hand drawn based on its glycaemic index of 32 (Table 1.2). Shaded areas are 95% confidence intervals. Originally published at the Diabetes UK Annual Professionals Conference, Glasgow, March 2008.[143]

difference between the values obtained by two laboratories (all assuming both laboratories reach a similar precision). Using definition 'b', now both GI and GL might be considered sufficiently precise for communication with the consumer at a 20% level of compliance, while GL might be considered suitable at a compliance level of 10%.

In considering the foregoing, readers should be aware that there is at present no international consensus of whether GI or GL (or other approach to assessing glycaemic response) is a preferred measure for communication with the consumer, though local preferences worldwide appears to favour GI. Nor is there international consensus on the approach that should be taken for assessing compliance. Nevertheless, scope evidently exists for regulatory procedures to facilitate the communication.

In considering the choice between GI and GL as the expression for communication, consideration might also be given to prospective observational studies (Figure 1.4), which illustrated the following:

- Both the quantity and the quality of carbohydrate food or ingredient affect the risk of type-2 diabetes.
- Carbohydrate such as sucrose with a middle-of-the-road GI appears to not affect the relative risk of type-2 diabetes. Thus, also total carbohydrate, which has a middle-of-the-road GI, would not be expected to affect the diabetes risk in prospective observational studies.
- Replacement of a high-glycaemic starch staple with a low-glycaemic starch staple would lower the relative risk of type-2 diabetes.
- Although sucrose appears without effect on the relative risk of type-2 diabetes, hypothetical considerations suggest its replacement by an alternative carbohydrate such as isomaltulose of lower GI (Table 1.2) has potential to lower the relative risk.

Table 1.2 Glycaemic and insulinaemic responses to bulk sweeteners and alternatives.

Sugars or alternatives	Relative glycaemic response (RGR)	Categorisation (high, intermediate, low, very low)	Relative insulin response (RIR)	Citation
Starch hydrolysis products				
Maltodextrin	91	High	90	a, b
Disaccharides				
Maltose	105	High	–	c
Trehalose	72	High	51	d
Sucrose	68	Intermediate	45	c, d
Lactose	46	Low	–	c
Isomaltulose	32	Very low	27	a
Monosaccharides				
Glucose	100	High	100	d
Fructose	19	Very low	9	d, c
Tagatose	3	Very low	3	a, e
Hydrogenated monosaccharide				
Erythritol	~0	Very low	2	d
Xylitol	12	Very low	11	c, d
Sorbitol	9	Very low	11	d
Mannitol	~0	Very low	~0	d
Hydrogenated disaccharides				
Maltitol	45	Low	27	d
Isomalt	9	Very low	6	d
Lactitol	5	Very low	4	c, d
Hydrogenated polydispersed saccharides				
Maltitol syrup				
High maltitol	48	Low	35	d
Intermediate maltitol	53	Low	41	d
Regular maltitol	52	Low	44	d
High polymer	36	Very low	31	d
Polyglycitol	39	Very low	23	d
Hydrogenated polydextrose	~5	Very low	~5	c
Non-digestible polysaccharides				
Polydextrose	~5	Very low	~5	c
Resistant maltodextrins	~10	Very low	~10	f
Fructans	~5	Very low	~0	g

RGR, relative glycaemic response (% of that for oral glucose); RIR, relative insulin response (% of that for oral glucose).
[a]Sydney University Glycaemic Index Research Service.
[b]Macdonald and Williams[146] in article entitled 'Effects of ingesting glucose and some of its polymers on serum glucose and insulin levels in men and women'.
[c]Foster-Powell et al. and Atkinson et al.,[67,79] in articles on 'International table of glycaemic index and glycaemic load'.
[d]Livesey[7] in article entitled 'Health potential of polyols as sugar replacers, with emphasis on low glycaemic properties'.
[e]Donner et al.[97] in article entitled 'D-tagatose, a novel hexose: acute effects on carbohydrate tolerance in subjects with and without type-2 diabetes'.
[f]Ohkuma et al.[147] in article entitled 'Pyrolysis of starch and its digestibility by enzymes'.
[g]Rumessen et al.[98] in article entitled 'Fructans of Jerusalem artichokes: intestinal transport, absorption, fermentation, and influence on blood glucose, insulin, and C-peptide responses in healthy subjects'.

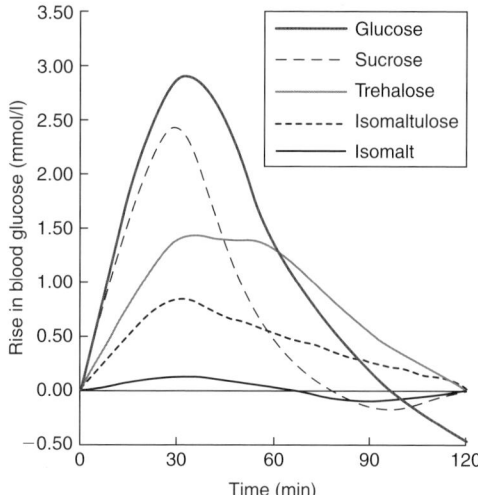

Fig. 1.5 Potential now exists to tailor the glycaemic response of sweetened foods by choosing ingredients. The response curves shown are for 50 g ingredient, relative to glucose in healthy people. Information on other low-glycaemic sweeteners and bulking agents is given in Table 1.2 together with references.

1.7 THE ACUTE GLYCAEMIC RESPONSE TO SUGARS AND ALTERNATIVES

The acute glycaemic response to glucose, sucrose, trehalose, isomaltulose and isomalt (Figure 1.5) illustrate how it is now possible to create bulk sweeteners, and so tailored foods, with almost any glycaemic response, likewise the insulin response (not shown). Other examples together with insulin responses are summarised in Table 1.2.

A numerical value for GI or GL does not itself provide information about whether the values are high or low compared with the range for foods eaten or compared with diets that associate prospectively with the incidence of disease or death. To put information about the glycaemic response of foods into perspective, the GI has been classified according to whether it is high, intermediate, low or very low[7] (www.glycaemicindex.com), as shown in Table 1.3. This classification may also help to communicate with the patient or consumer. For example, it can be suggested that a consumer or patient selects food from a lower band of glycaemic response (i.e. lower class or two lower classes where possible). Further, should a

Table 1.3 Classifications of glycaemic index (GI) and glycaemic load (GL).

Glycaemic classification[a]	GI[b] (g eq./100 g)	GL[c] (g eq./day)	GL[d] (g eq./serving)
High	>70	>120	>20
Intermediate	>55–70	>80–120	>10–20
Low	>40–55	>20–80	>4–10
Very low	0–40	0–20	0–4

[a]Based on www.glycaemicindex.com and ref. 7.
[b]Based on measurements with 25–50 g carbohydrate.
[c]Based on prospective epidemiology.[74,115–117]
[d]Based on a 10 g serving size (or exchange rates) noted in the international tables.[67,79]

high-glycaemic food be eaten at a meal for any reason (e.g. enjoyment) any other carbohydrate source eaten at the same time ought to come from a low-glycaemic band. Furthermore, diabetes patients have for years practiced carbohydrate exchange as part of dietary therapy, for which a similar glycaemic response or similar insulin requirement was (and sometimes still is) presumed to arise from any food containing 10 g of carbohydrate. Diabetes patients can now update this approach by practicing exchanges based upon the GI or GL, while also attempting to reduce saturated fat intake. The alternative may be emphasised, limit energy intake while still eating (or eating more) low-glycaemic carbohydrate and mono- or polyunsaturated fats. Some advantages of the GL over the GI have been emphasised.[4,78] Thus, GL can be used as a 'virtual nutrient' to be considered alongside all other nutrients when assessing the relation between diet and health.

The indication that GL should be limited to 120 g per day (Table 1.3) implies some 40–60% of people in western populations may be at risk of metabolic disease; this is consistent with the high prevalence of coronary heart disease, obesity and diabetes.

Intense sweeteners are consumed in such small quantities that they have no glycaemic response of their own; additionally, the structure of such sweeteners would normally not be expected to yield glucose upon metabolism. Generally too, none of the intense sweeteners have pharmacological actions to improve glycaemic control (an exception is stevioside[90]). Aspartame is a more typical example of an intense sweetener that is without acute glycaemic response,[91,92] another is sucralose.[93] Clearly however, compared with maltodextrins, maltose, glucose and sucrose, under controlled conditions marked reductions in the acute glycaemic response would be expected for intense sweeteners delivering comparable sweetness. Addition of intense sweeteners to foods or drinks that normally would not contain sugars for sweetness would, however, confer no glycaemic advantage.

The use of intense sweeteners in place of glycaemic carbohydrates wherever bulk is necessary for technological or organoleptic reasons requires the glycaemic response to bulking agents to be considered here too. The glycaemic (and insulinaemic) response to maltodextrin, bulk sweeteners and bulking agents varies considerably (Table 1.2). The causes of the lower glycaemia are numerous.[7] Compared with glucose, the lower value for sucrose is due mainly to dilution within the molecule with a fructose moiety. A similar situation occurs with maltitol, maltitol syrups and polyglycitol, where glucose moieties are 'diluted' with a sorbitol moiety. Fructose alone is low glycaemic due to both slow absorption and need for conversion to glucose in the liver prior to appearance in blood as glucose; in addition, the carbohydrate may be partly stored as glycogen rather than released into the circulation. Further still, the energy from fructose is conveyed in the circulation for oxidation in part as lactate more than is the case for glucose. A similar situation occurs for sorbitol and xylitol, though slower absorption likely gives rise to less lactate; in addition, a high proportion escapes absorption. With isomalt and lactitol, an even greater proportion escapes absorption, which gives these polyols the lowest glycaemic response of all so far mentioned. Another polyol, erythritol, is almost unique in that although most is absorbed, it is low glycaemic; this is because it is poorly metabolised in the tissues and is excreted in the urine. Mannitol behaves similarly, though is largely (75%) unabsorbed.

At the other end of the scale are maltodextrins, which can give rise to a glycaemic response as high as glucose, likewise maltose. Trehalose, an isomer of maltose, has a glycaemic response comparable to sucrose (Table 1.2) in terms of its GL (g glucose equivalents per 100 g), though it peaks less sharply and there is persistence in the raised glycaemia that would likely help protect against hypoglycaemia in susceptible individuals. Isomaltulose, derived by rearrangement of sucrose, gives a similar though lower profile, and so lower GL;

this even though all of the isomaltulose is hydrolysed and absorbed. Other low-glycaemic carbohydrates include tagatose, fructans (fructo-oligosaccharides and inulin), polydextrose and resistant maltodextrins. The reduction in glycaemia caused by sucrose replacing high-glycaemic starch is considered an advantage.[9] Greater reductions would be possible on replacing maltodextrin, maltose, glucose and sucrose with alternative sweeteners, either partially or completely depending upon the serving size of foods.

Among the studies undertaken with polyols (mainly with maltitol, isomalt and sorbitol), the glycaemic response versus glucose is similar in people with normal and abnormal carbohydrate metabolism, as exemplified by type-1 and type-2 diabetes patients, provided insulin-dependent participants receive insulin via an artificial pancreas.[7] This is as experienced with carbohydrate foods generally.[67] This similarity of GI between disease states, however, may not extend to similarity in the insulinaemic index between states.[94]

In addition to having lower glycaemic responses, polyols, low-digestible sugars and bulking agents can reduce the glycaemic response to other carbohydrates. The magnitude of this effect is not great, but is not insignificant either and is in the order of 10–15%.[7,95–97] However, no such effect is reported with fructans.[98,99] The important conclusion here is that the low-glycaemic character intrinsic to these carbohydrates is not lost when they are consumed with other carbohydrates (or other macronutrients).

A further reduction in acute post-prandial glycaemia can occur when fat is included in the meal. This is common to both digestible and non-digestible carbohydrate whether used as sweeteners or not.[7] It is accompanied by an elevation of insulinaemia via an incretin response. A common view is that fats reduce glycaemia via stomach emptying, however, this would not explain the elevated insulin response; thus, both stomach emptying and a gastrointestinal incretin response contribute to the lower glycaemia. The implications of the elevated insulinaemia in such a circumstance remain to be researched. However, sugar–fat mixtures (and more generally, high-glycaemic carbohydrate–fat mixtures) are not viewed as beneficial and too high an insulin response may contribute to the development of obesity[100] and coronary heart disease.[101] Hence, it may be particularly important to reduce the glycaemic response of fatty foods (and the fat content of high-glycaemic foods). In this respect, this author notes the beneficial impact of low-glycaemic foods on long-term glycaemic control in diabetes patients appears greater among consumers of moderate- (35–40%) rather than low-fat (25%) diets.

A question arises as to whether sweeteners affect the cephalic phase insulin response (i.e. do sweeteners cause an elevation of insulin and so lowering of glycaemia, reflexively via the brain?). This appears not to happen to a significant extent with the sweeteners aspartame or saccharine.[92,102] Likewise, in diabetic patients, sweetness is reported to have no impact on food intake and macronutrient composition other than perhaps for a lowering of sucrose ingestion.[103,104]

1.8 LONG-TERM GLYCAEMIC CONTROL WITH SWEETENERS AND BULKING AGENTS

Fructosamine and glycated haemoglobin (HbA_{1c}) in blood are medium and long-term markers of day-long exposure to elevated blood glucose concentrations. Non-, pre- or undiagnosed diabetic individuals as well as diabetes patients with elevated HbA_{1c} are at increased risk of coronary heart disease, stroke and all cause mortality.[105–107] In diabetes patients, the elevation of HbA_{1c} is associated also with higher risk of retinopathy, nephropathy, perivascular disease, limb amputation and perivascular deaths.[105,107,108] While markers of risk for cardiovascular disease in interventional studies have usually been limited to lipid markers, it is recognised that good glycaemic control is of first importance in the control of diabetic hyperlipidaemia.[109] Possibly, glycated protein markers are underutilised as a tool to assess risk to health in both epidemiological and interventional studies. Mechanisms of increased risk are discussed elsewhere,[110] and indicate greater risk whenever anti-oxidant defences are low.

Of all the risk markers used often in intervention studies with diabetes patients, only fasting blood glucose, fructosamine and HbA_{1c} show a consistent improvement, either in direction alone or in both direction and statistical significance due to replacement of high- with low-glycaemic carbohydrate foods.[10,45] It should not go unrecognised that replacement of saturated fat with high-fibre high-carbohydrate diets also improves fasting blood glucose (total cholesterol and triglycerides) in type-2 diabetes patients, and a role for non-digestible carbohydrate in this response is evident.[43]

This risk to glycaemic control from high-glycaemic carbohydrate in type-2 diabetes is reduced by the use of several substrates in place of carbohydrate, including protein,[111] the polyol isomalt[7] and fructose[112,113] as well as low-glycaemic carbohydrates foods.[10,45] The implication is that it is the size of the overall glycaemic exposure in response to foods that associates with risk; this is more than simply explainable by GL, and GL more than GI, as noted previously.[4] This hierarchy requires careful understanding. Here, high (saturated) fat diets elevate the glycaemic response to foods chronically via deterioration in insulin sensitivity and beta-cell function, so amplifying the glycaemic response to carbohydrate foods rather than by supplying fuel for blood glucose formation. In essence, both saturated fat and high-glycaemic carbohydrate each pose a risk to glycaemic control and health. Additionally, both add energy to the diet, so potentially contributing directly to the obesogenic environment.

A further consideration is that when study participants already have good glycaemic control, then only two outcomes are possible by change of diet, either no effect or deterioration, which may take years before overt disease emerges. A third possibility arises when glycaemic control is poor; it may improve within weeks and months. This is evident (in the author's assessment) for mixed groups of type-1 and type-2 diabetes patients, for whom intervention with low-glycaemic carbohydrate diets seem most effective in people with poor glycaemic control. In such studies, poor glycaemic control associates in the first instance with moderate- rather than low-fat ingestion.

Based on the available evidence from intervention studies, low-glycaemic diets will correct about 30% of the deterioration in glycaemic response (author, unpublished), which implies an approximately 30% reduced risk of diabetic complications and heart disease.[114] Interestingly, when looking at initially healthy people via prospective epidemiological studies, high- versus low-GL diets appear to explain about 30–40% of the relative risk for type-2 diabetes[74,115,116] and perhaps more of cardiovascular disease, in women especially.[14,117] Interestingly too, use of an inhibitor of carbohydrate digestion in a pre-diabetic state can reduce the incidence of coronary heart disease by up to 50%.[118] How much more effective life-long exposure to low-glycaemic diets would be in current inactive societies remains uncertain – though prospective studies of 20 years duration suggest possibility of greater benefit than those of 5–6 years duration.[119,120] On the basis of such data, there would appear to be a significant public health benefit from minimising high-glycaemic carbohydrate consumption. It is reasonable, therefore, for food manufacturers to begin or continue to consider how they can either replace or minimise either high-glycaemic carbohydrate and saturated fats or both in foods, and in this sugars and sweeteners have a role.

Few recent studies have examined the impact of polyols. One study examined the influence of isomalt on both fasting and post-prandial plasma glucose and glycated haemoglobin in diabetes patients, showing improvement in all three parameters[121] as shown by the author's analysis of original tabulated data.[7] The isomalt was consumed in such a way that it was likely to have reduced the intake of sucrose. In healthy people, by contrast, isomalt appears to have no influence on fructosamine concentrations. Again, this is consistent with studies of other low-glycaemic carbohydrate foods, where severity of the disease condition impacts on the magnitude of effect.

Replacement of sucrose with fructo-oligosaccharides (<10 g per day) caused a relative reduction in fasting glucose in type-2 diabetes patients over 14 days by 10% in one study,[122] and by 6% in another when exchanging 20 g sucrose with fructo-oligosaccharides for 4 weeks. By contrast, no similar influence occurred in people without diabetes.[123–126] This is consistent with the lower glycaemic impact of fructans compared with sucrose. Such a result again mirrors an improvement in glycaemic control in diabetes patients but not persons with normal blood glucose concentrations when replacing high- with other low-glycaemic carbohydrate foods.[22] However, small shifts in glycaemic control away from the healthy normal appear partially correctable with low glycaemic carbohydrates that undergo fermentation (see Section 1.9).

The chronic effect of D-tagatose on blood glucose is unclear.[127] As might be predicted, no influence was seen on fasting blood glucose in normal individuals over 8 weeks. In eight diabetes patients, feeding supplemental tagatose had no effect on plasma glucose or HbA_{1c}, though the study was under-powered and it is unclear whether GL was significantly reduced. Supplementation with a similar amount of glucose would almost certainly have damaged glycaemic control.

Randomly bonded glucose (polydextrose) had no influence on glycaemic control (HbA_{1c}) in normal subjects,[128] but 20 g daily (resistant maltodextrin) reduced fasting glucose in type-2 diabetes patients,[129] again consistent with expectations with low-glycaemic carbohydrate foods. Carbohydrates such as these may well be corrective and preventive, neither of which can occur in fully healthy persons, and which in non-diabetic individuals (blood glucose >5 and <7 mmol/l) is hard to identify without combining information from many studies.[130]

Given that reduction of dietary (saturated) fat and high-glycaemic carbohydrates will each lower HbA_{1c}, it is not surprising that using a mixture of non-digestible carbohydrate and intense sweetener as a fat replacer should contribute to lower HbA_{1c} concentrations in diabetes patients.[131] Again, this observation plus those mentioned previously lead to the conclusion that improvement in glycaemic control arises from both a lower ingestion of saturated fat and the consumption of low-glycaemic carbohydrate including sweeteners and bulking agents in preference to high-glycaemic counterparts. However, intense sweeteners alone added to a diet may have little or no direct influence on long-term glycaemic control, as shown with aspartame.[93,132] Note, however, improvement in glycaemic control would be difficult to establish in individuals without diabetes within a period of a few months[133] or in diabetes patients in whom control was already well established, again within a period of a few months.[134] Benefits for individuals at risk of diabetes, hypertension and coronary heart disease may take years to develop as shown using pharmacological approaches to reducing post-prandial glycaemia.[135]

The observations to date support the view that sweeteners with a combination of low-glycaemic response and reduced energy value can contribute to an environment in which obesity, diabetes and potentially coronary heart disease and certain cancers are less likely to develop.

1.9 ARE LOW GLYCAEMIC CARBOHYDRATES OF BENEFIT IN HEALTHY PERSONS?

In short, the answer appears to be yes, and while it is unclear whether this means yes in all healthy persons or just a good proportion of healthy persons who would progress to metabolic disease and associated cancers remains to be established:

- Meta-analyses of prospective observational studies on populations of healthy persons suggest both low GI and low GL, especially when coupled with high unavailable carbohydrate content, lowers the risk of diabetes, cardiovascular disease, age-related macular degeneration and certain cancers in persons with no established disease at the outset of study.[16, 136]
- Meta-analysis of interventional studies in groups of persons with varied severity of glycaemic control shows both low GI or GL and high unavailable carbohydrate content partially corrects departures from normal of glycaemic control at all levels from near normal to moderate diabetes.[22]
- Meta-analysis of interventional studies that lower the acute glycaemic response of available carbohydrate, by way of introducing an unavailable carbohydrate to the diet for 3 months, can partially correct fasting blood glucose in healthy persons with fasting glucose >5 mmol/l and pre-diabetic as well as in diabetic individuals.[137]
- The potential relevance of low-glycaemic carbohydrate, as outlined by WHO/FAO[68] and ILSI NA[16] leads to the conclusion that low-glycaemic response assess either as GI or GL is now 'sufficiently mature' to apply in communications with the consumer.

1.10 GASTROINTESTINAL TOLERANCE IN RELATION TO THE GLYCAEMIC RESPONSE

The topic of gastrointestinal tolerance has been considered in detail elsewhere.[6, 138, 139] Here consideration is given to the impact of intolerance on the capacity for GL reduction by exchange of sugars, sweeteners and bulking agents, selecting examples to illustrate some key points. Thus the greater the amount of carbohydrate that is exchanged the greater the GL reduction that is possible; this is until people consuming the alternative carbohydrate turn away due to gastrointestinal intolerance, as indicated in Figure 1.6. For this purpose, intolerance is assessed from the proportion of people in sampled populations that experience mild watery stools after alternative carbohydrate ingestion. This proportion is summarised according to a model with binomial distribution that can be described by two parameters: one is the highest dose at which no response is observed by any individual in the population sample (threshold, D_0) when intake is in divided doses, and the other is the sharpness of the response (S) as the dose is increased further. In the realistic range of intakes in divided doses up to 50 g daily both polydextrose and isomalt are well tolerated in adults, allowing considerable advantage in GL reduction to be gained from carbohydrate exchange. Although maltitol is well tolerated, its glycaemic response is the highest so that it gains less in terms of GL reduction up to the threshold. Fructo-oligosaccharide, which has a low-glycaemic response, is evidently least well tolerated, and this limits its potential for reduction of GL. Data such as these are essential when examining risk and benefit of alternative carbohydrates. The outcomes illustrate, in addition to the potential for GL reduction, that polyols are not

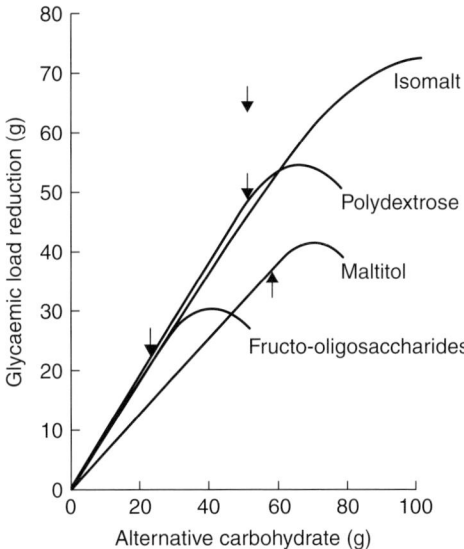

Fig. 1.6 This shows the balance of glycaemic load reduction and gastrointestinal intolerance for selected alternative carbohydrates. The curves are based on the author's unpublished analysis of information from the literature on tolerance[6, 138, 144, 145] and glycaemia (Table 1.2). The arrows indicate thresholds of tolerance.

inevitably more laxative than oligo- and polysaccharides. The latter suggests that laxation is dependent on the fermentation process as much as the amount of fermentable carbohydrate and its molecular weight. Erythritol, another polyol, is not without effect on gastrointestinal tolerance, though it is well tolerated owing to it being mostly absorbed. It is as effective at GL reduction as isomalt and polydextrose up to 50 g intake daily in divided doses. The magnitude of such reductions alone are of potential public health and clinical importance and would certainly contribute to reductions managed by carbohydrate exchange in the diet as a whole;[7] especially as some individuals are capable of tolerating considerable amounts of these carbohydrates, well above population threshold values.

1.11 CONCLUSION

While non-communicable diseases are set to overburden private and governmental health budgets, there is need of preventive methods to maintain health. Dietary change provides one such method and specifically reducing the glycaemic response to diet via food selection, food modification and ingredient choices and development is a valuable objective. There is now growing evidence from clinical data confirming the potential for reduced severity of disease, and from epidemiological data for reducing the risk of developing a variety of non-communicable diseases. Incorporation of the low-glycaemic approach as one component of a 'better diet' is likely to remain important. Strategies that combine reduced GI or GL, reduced saturated fat and reduced energy are likely to be most effective, and these attributes are found in alternative sugars and sweeteners.

Further, experimental evidence shows that even in foods where saturated fat cannot easily be lowered, the use of alternative sweeteners in place of higher glycaemic sugars and dextrins will elicit a diminished glycaemic and insulin response. Enhancement of the latter by dietary fat is hardly possible with very-low-glycaemic sugars and sweeteners, thus reducing a possibility of atherogenic insulinaemia. In such products, a reduced glycaemic and insulinaemic response owing to the use of low-glycaemic carbohydrates is a valuable objective, while use of additional saturated fats to reduce the acute-glycaemic response should be avoided.

Lastly, the potentiation of hyperglycaemia-induced overproduction of superoxides from the mitochondrial respiratory chain provides a possible mechanism of oxidative damage, aging, tumour formation, atherogenic endothelial damage and of diabetic complications.[110] Low-glycaemic carbohydrates may be especially valuable in avoiding these conditions, especially because some damage appears irreversible with contribution to a phenomenon of 'hyperglycaemic memory' in which some progressive damage occurs despite normalisation of glycaemic control. Health maintenance rather than therapeutic measures to restore health may thus prove to be the better option.

REFERENCES

1. J.D. Featherstone. The science and practice of caries prevention. *Journal of the American Dental Association* 2000, 131, 887–899.
2. T. Takatsuka. Influence of Palatinit(r) and xylitol on demineralisation/remineralisation of bovine enamel. *Cardiology Today* 2000, 1, 37–40.
3. G. Livesey. Energy values of dietary fibre and sugar alcohols for man. *Nutrition Research Reviews* 1992, 5, 61–84.
4. G. Livesey. Low-glycaemic diets and health: implications for obesity. *The Proceedings of the Nutrition Society* 2005, 64, 105–113.
5. W. Scheppach, H. Luehrs and T. Menzel. Beneficial health effects of low-digestible carbohydrate consumption. *British Journal of Nutrition* 2001, 85(Suppl. 1), S23–S30.
6. G. Livesey. Tolerance of low-digestible carbohydrates: a general view. *British Journal of Nutrition* 2001, 85(Suppl. 1), S7–S16.
7. G. Livesey. Health potential of polyols as sugar replacers, with emphasis on low glycaemic properties. *Nutrition Research Reviews* 2003, 16, 163–191.
8. P. Gourbeyre, S. Denery and M.J. Bodinier. Probiotics, prebiotics, and synbiotics: impact on the gut immune system and allergic reactions. *Journal of Leukocyte Biology*. 2011, 89, 685–695.
9. J.C. Miller and I. Lobbezoo. Replacing starch with sucrose in a high glycaemic index breakfast cereal lowers glycaemic and insulin responses. *European Journal of Clinical Nutrition* 1994, 48, 749–752.
10. J.C. Brand-Miller, S. Hayne, P. Petocz and S. Colagiuri. Low-glycemic index diets in the management of diabetes: a meta-analysis of randomized controlled trials. *Diabetes Care* 2003, 26, 2261–2267.
11. G. Livesey. Approaches to health via lowering of postprandial glycaemia. *British Journal of Nutrition* 2002, 88, 741–744.
12. J.C. Brand-Miller. Glycemic load and chronic disease. *Nutrition Reviews* 2003, 61, S49–S55.
13. P.C. Colombani. Glycemic index and load - dynamic dietary guidelines in the context of diseases. *Physiolology and Behaviour* 2004, 83, 603–610.
14. S. Liu and W.C. Willett. Dietary glycemic load and atherothrombotic risk. *Current Atherosclerosis Reports* 2002, 4, 454–461.
15. W. Kopp. Pathogenesis and etiology of essential hypertension: role of dietary carbohydrate. *Medical Hypotheses* 2005, 64, 782–787.
16. C.J. Chiu, S. Liu, W.C. Willett, T.M. Wolever, J.C. Brand-Miller, A.W. Barclay and A. Taylor. Informing food choices and health outcomes by use of the dietary glycemic index. *Nutrition Reviews* 2011, 69, 231–242.
17. J.F. Clapp. Maternal carbohydrate intake and pregnancy outcome. *Proceedings of the Nutrition Society* 2002, 61, 45–50.
18. J.C. Louie, J.C. Brand-Miller, T.P. Markovic, G.P. Ross and R.G. Moses. Glycemic index and pregnancy: a systematic literature review. *Journal of Nutrition and Metabolism* 2010, 2010, 282464.

19. M. Lampl and P. Jeanty. Exposure to maternal diabetes is associated with altered fetal growth patterns: a hypothesis regarding metabolic allocation to growth under hyperglycemic-hypoxemic conditions. *American Journal of Human Biology* 2004, 16, 237–263.
20. G.M. Shaw, T. Quach, V. Nelson, S.L. Carmichael, D.M. Schaffer, S. Selvin and W. Yang. Neural tube defects associated with maternal periconceptional dietary intake of simple sugars and glycemic index. *American Journal of Clinical Nutrition* 2003, 78, 972–978.
21. L. Pennell, C.M. Smith-Snyder, L.R. Hudson, G.B. Hamar and J. Westerfield. Practice changes in glycemic management and outcomes in coronary artery bypass surgery patients. *Journal Cardiovascular Nursing* 2005, 20, 26–34.
22. G. Livesey, R. Taylor, T. Hulshof and J. Howlett. Glycemic response and health a systematic review and meta-analysis: relations between dietary glycemic properties and health outcomes. *American Journal of Clinical Nutrition* 2008, 87, 258S–268S.
23. T.M. Larsen, S.-M. Dalskov, M. van Baak, S.A. Jebb, A. Papadaki, A.F.H. Pfeiffer, J.A. Martinez, T. Handjieva-Darlenska, M. Kunešová, M. Pihlsgård, S. Stender, C. Holst, W.H.M. Saris and A. Astrup. Diets with high or low protein content and glycemic index for weight-loss maintenance. *New England Journal of Medicine* 2010, 363, 2102–2113.
24. M.E.J. Lean and W.P.T. James. Prescription of diabetic diets in the 1980's. *The Lancet* 1986, 1, 723–725.
25. L.S. Lieberman. Dietary, evolutionary, and modernizing influences on the prevalence of type-2 diabetes. *Annual Review of Nutrition* 2003, 23, 345–377.
26. S.B. Eaton, M. Konner and M. Shostak. Stone agers in the fast lane: chronic degenerative diseases in evolutionary perspective. *American Journal of Medicine* 1988, 84, 739–749.
27. A.W. Thorburn, J.C. Brand and A.S. Truswell. Slowly digested and absorbed carbohydrate in traditional bushfoods: a protective factor against diabetes? *American Journal of Clinical Nutrition* 1987, 45, 98–106.
28. L. Cordain, J.B. Miller, S.B. Eaton, N. Mann, S.H. Holt and J.D. Speth. Slowly digested and absorbed carbohydrate in traditional bushfoods: a protective factor against diabetes? *American Journal of Clinical Nutrition* 2000, 71, 682–692.
29. S. Colagiuri and J.B. Miller. The carnivor connection - evolutionary aspects of insulin resistance. *European Journal of Clinical Nutrition* 2002, 56, 30–35.
30. L. Cordain, S.B. Eaton, A. Sebastian, N. Mann, S. Lindeberg, B.A. Watkins, J.H. O'Keefe and J. Brand-Miller. Origins and evolution of the Western diet: health implications for the 21st century. *American Journal of Clinical Nutrition* 2005, 81, 341–354.
31. K. O'Dea. Westernisation, insulin resistance and diabetes in Australian aborigines. *Medical Journal of Australia* 1991, 155, 258–264.
32. M. Toussaint-Samat. *History of Food*, Blackwell Publishing Ltd, Oxford, England; 2001.
33. B. Liebman. The changing American diet. *Nutrition Action* 1999, 26, 8–9.
34. L.S. Gross, L. Li, E.S. Ford and S. Liu. Increased consumption of refined carbohydrates and the epidemic of type 2 diabetes in the United States: an ecologic assessment. *American Journal of Clinical Nutrition* 2004, 79, 774–779.
35. O. Bermudez. Consumption of sweet drinks among American adults from the NHANES 1999-2000. *Experimental Biology (San Diego)* 2005, Abstract #839.5.
36. A. Kapur and K. Kapur. Relevance of glycemic index in the management of post-prandial glycemia. *Journal of the Association of Physicians of India* 2001, 49, 42–45.
37. ADA. Nutritional Recommendations and principles for people with diabetes mellitus. *Diabetes Care* 1997, 20, S14–S17.
38. J.A. Ingelfinger, P.H. Bennett, I.M. Liebow and M. Miller. Coronary heart disease in the Pima Indians. Electrocardiographic findings and postmortem evidence of myocardial infarction in a population with a high prevalence of diabetes mellitus. *Diabetes* 1976, 25, 561–565.
39. A.S. Krolewski, A. Czyzyk, J. Kopczynski and S. Rywik. Glycemic index, glycemic load, and cereal fiber intake and risk of type 2 diabetes in US black women. *Diabetologia* 1981, 21, 520–524.
40. A. Verrillo, A. de Teresa, P. Carandente Giarrusso, A. Scognamiglio, S. La Rocca and L. Lucibelli. Prevalence of impaired glucose tolerance, diabetes mellitus and ischemic heart disease in an Italian rural community. The Sanza Survey. *Bollettino Societa Italiana Biologia Sperimentale (Napoli)* 1984, 60, 485–491.
41. M.L. Morici, A. Di Marco, D. Sestito, R. Candore, C. Cangemi, F. Accardo, M. Donatelli, M.G. Cataldo and A. Lombardo. The impact of coexistent diabetes on the prevalence of coronary heart disease. *Journal of Diabetes Complications* 1997, 11, 268–273.

42. J.W. Anderson and K. Ward. Long-term effects of high-carbohydrate, high-fiber diets on glucose and lipid metabolism: a preliminary report on patients with diabetes. *Diabetes Care* 1978, 1, 77–82.
43. B.V. Howard, W.G. Abbott and B.A. Swinburn. Evaluation of metabolic effects of substitution of complex carbohydrates for saturated fat in individuals with obesity and NIDDM. *Diabetes Care* 1991, 14, 786–795.
44. H.C. Simpson, R.W. Simpson, S. Lousley, R.D. Carter, M. Geekie, T.D. Hockaday and J.I. Mann. A high carbohydrate leguminous fibre diet improves all aspects of diabetic control. *Lancet* 1981, 1, 1–5.
45. J.W. Anderson, K.M. Randles, C.W. Kendall and D.J. Jenkins. Carbohydrate and fiber recommendations for individuals with diabetes: a quantitative assessment and meta-analysis of the evidence. *Journal of The American College of Nutrition* 2004, 23, 5–17.
46. L. Story, J.W. Anderson, W.J. Chen, D. Karounos and B. Jefferson. Adherence to high-carbohydrate, high-fiber diets: long-term studies of non-obese diabetic men. *Journal of the American Dietetic Association* 1985, 85, 1105–1110.
47. B. Karlstrom, B. Vessby, N.G. Asp, M. Boberg, I.B. Gustafsson, H. Lithell and I. Werner. Effects of an increased content of cereal fibre in the diet of Type 2 (non-insulin-dependent) diabetic patients. *Diabetologia* 1984, 26, 272–277.
48. A.L. Kinmonth, R.M. Angus, P.A. Jenkins, M.A. Smith and J.D. Baum. Whole foods and increased dietary fibre improve blood glucose control in diabetic children *Archives of Diseases in Children* 1982, 57, 187–194.
49. Institute of Medicine. Dietary, functional, and total fiber, In: *Dietary Reference Intakes for Energy, Carbohydrate, Fiber, Fat, Fatty Acids, Cholesterol, Protein, and Amino Acids (Macronutrients)*, National Academic Press, Washington, DC; 2002, pp. 339–421.
50. A.R.P. Walker. Dietary fiber in health and disease: the South African experience. In: D. Kritchevsky and C. Bonfield (eds). *Dietary Fibre in Health and Disease*, Eagan Press, St. Paul, MN; 1995, pp. 11–25.
51. A.R.Walker. Health implications of fibre-depleted diets. *South African Medical Journal* 1977, 52, 767–770.
52. M. Cullen, J. Nolan, M. Moloney, J. Kearney, J. Lambe and M.J. Gibney. Effect of high levels of intense sweetener intake in insulin dependent diabetics on the ratio of dietary sugar to fat: a case-control study. *European Journal of Clinical Nutrition* 2004, 58, 1336–1341.
53. R.J. Stubbs, A.M. Prentice and W.P.T. James. Carbohydrate and energy balance. In: G.H. Anderson, B.J. Rolls and D. Steffen (eds). *Nutritional implications of macronutrient substitutes*, The New York Academy of Sciences, New York; 1997, pp. 44–69.
54. A.M. Prentice and S. Jebbs. Obesity in Britain: gluttony or sloth. *British Medical Journal* 1995, 311, 437–439.
55. A.M. Poynten, S.K. Gan, A.D. Kriketos, L.V. Campbell and D.J. Chisholm. Circulating fatty acids, non-high density lipoprotein cholesterol, and insulin-infused fat oxidation acutely influence whole body insulin sensitivity in nondiabetic men. *Journal of Clinical Endocrinology and Metabolism* 2005, 90, 1035–1040.
56. E.J. Mayer-Davis, J.H. Monaco, H.M. Hoen, S. Carmichael, M.Z. Vitolins, M.J. Rewers, S.M. Haffner, M.F. Ayad, R.N. Bergman and A.J. Karter. Dietary fat and insulin sensitivity in a triethnic population: the role of obesity. The Insulin Resistance Atherosclerosis Study (IRAS). *American Journal of Clinical Nutrition* 1997, 65, 79–87.
57. H.P. Himsworth. The dietetic factors determining the glucose tolerance and sensitivity to insulin of healthy man. *Clinical Science* 1935, 2, 67–94.
58. A.M. Opperman, C.S. Venter, W. Oosthuizen, R.L. Thompson and H.H. Vorster. Meta-analysis of the health effects of using the glycaemic index in meal-planning. *British Journal of Nutrition* 2004, 92, 367–381.
59. ADA. Evidence-Based Nutrition Principles and Recommendations for the Treatment and Prevention of Diabetes and Related Complications. *Diabetes Care* 2002, 25, 50S–60S.
60. G.L. Khor. Dietary fat quality: a nutritional epidemiologist's view. *Asia Pacific Journal of Clinical Nutrition* 2004, 13, S22.
61. T. Hung, J.L. Sievenpiper, A. Marchie, C.W. Kendall and D.J. Jenkins. Fat versus carbohydrate in insulin resistance, obesity, diabetes and cardiovascular disease. *Current Opinion in Clinical Nutrition and Metabolic Care* 2003, 6, 165–176.

62. D. Mozaffarian, A. Ascherio, F.B. Hu, M.J. Stampfer, W.C. Willett, D.S. Siscovick and E.B. Rimm. Interplay between different polyunsaturated fatty acids and risk of coronary heart disease in men. *Circulation* 2005, 111, 157–164.
63. J.H. Jr. O'Keefe and L. Cordain. Cardiovascular disease resulting from a diet and lifestyle at odds with our Paleolithic genome: how to become a 21st-century hunter-gatherer. *Mayo Clinical Proceedings* 2004, 79, 101–108.
64. Diabetes UK. The implementation of nutritional advice for people with diabetes. *Diabetes Medicine* 2003, 20, 786–807.
65. G. Reaven, K. Segal, J. Hauptman, M. Boldrin and C. Lucas. Effect of orlistat-assisted weight loss in decreasing coronary heart disease risk in patients with syndrome X. *American Journal of Cardiology* 2001, 87, 827–831.
66. R.B. Blacket, B. Leelarthaepin, C.A. McGilchrist, A.J. Palmer and J.M. Woodhill. The synergistic effect of weight loss and changes in dietary lipids on the serum cholesterol of obese men with hypercholesterolaemia: implications for prevention of coronary heart. *Australian and New Zealand Journal of Medicine* 1979, 9, 521–529.
67. K. Foster-Powell, S.H. Holt and J.C. Brand-Miller. International table of glycemic index and glycemic load values: 2002. *American Journal of Clinical Nutrition* 2002, 76, 5–56.
68. FAO/WHO. *Carbohydrates in Human Nutrition. (FAO Food and Nutrition Paper No. 66)*, Food and Agriculture Organisation, Rome, Italy; 1998.
69. K. Marsh, A. Barclay, S. Colagiuri and J. Brand-Miller. Glycemic index and glycemic load of carbohydrates in the diabetes diet. *Current Diabetes Reports* 2011, 11, 120–127.
70. , D.J. Jenkins, K. Srichaikul, C.W. Kendall, J.L. Sievenpiper, S. Abdulnour, A. Mirrahimi, C. Meneses, S. Nishi, X. He, S. Lee, Y.T. So, A. Esfahani, S. Mitchell, T.L. Parker, E. Vidgen, R.G. Josse and L.A. Leiter. The relation of low glycaemic index fruit consumption to glycaemic control and risk factors for coronary heart disease in type 2 diabetes/ *Diabetologia* 2010, 54, 271–279.
71. R.A. McCance and R.D. Lawrence. *Medical Research Council Special Report Series No 135*, HMSO, London; 1929.
72. D.A.T. Southgate. Determination of carbohydrates in foods. I. Available carbohydrate. *Journal of the Science of Food and Agriculture* 1969, 20, 326–330.
73. D.J. Jenkins, T.M. Wolever, R.H. Taylor, H. Barker, H. Fielden, J.M. Baldwin, A.C. Bowling, H.C. Newman, A.L. Jenkins and D.V. Goff. Glycemic index of foods: a physiological basis for carbohydrate exchange. *American Journal of Clinical Nutrition* 1981, 34, 362–366.
74. J. Salmeron, J.E. Manson, M.J. Stampfer, G.A. Colditz, A.L. Wing and W.C. Willett. Dietary fiber, glycemic load, and risk of non-insulin-dependent diabetes mellitus in women. *Journal of the American Medical Association* 1997, 277, 472–477.
75. J.C. Brand-Miller, M. Thomas, V. Swan, Z.I. Ahmad, P. Petocz and S. Colagiuri. Physiological validation of the concept of glycemic load in lean young adults. *Journal of Nutrition* 2003, 133, 2728–2732.
76. J.R. Anfinsen, T. Wolver, M. Solar, E. Hitcher and P.D. Wolff. Methods and systems for determining and controlling glycaemic responses. US Patent, US 0043106 A1; 2004.
77. J.A. Monro. Glycaemic glucose equivalent: combining carbohydrate content, quantity and glycaemic index of foods for precision in glycaemia management. *Asia Pacific Journal of Clinical Nutrition* 2002, 11, 217–224.
78. J.A. Monro. Expressing the glycaemic potency of foods. *Proceedings of the Nutrition Society* 2005, 64, 115–122.
79. F.S. Atkinson, K. Foster-Powell and J.C. Brand-Miller. International tables of glycemic index and glycemic load values: 2008. *Diabetes Care* 2008, 31, 2281–2283.
80. T.M. Wolever, H.H. Vorster, I. Bjorck, J. Brand-Miller, F. Brighenti, J.I. Mann, D.D. Ramdath, Y. Granfeldt, S. Holt, T.L. Perry, C. Venter and W. Xiaomei. Determination of the glycaemic index of foods: interlaboratory study. *European Journal of Clinical Nutrition* 2003, 57, 475–482.
81. F. Brouns, I. Bjorck, K.N. Frayn, A.L. Gibbs, V. Lang, G. Slama and T.H.M. Wolever. Glycaemic index methodology. *Nutrition Research Reviews* 2005, 18, 145–171.
82. T.M. Wolever, J.C. Brand-Miller, J. Abernethy, A. Astrup, F. Atkinson, M. Axelsen, I. Bjorck, F. Brighenti, R. Brown, A. Brynes, M.C. Casiraghi, M. Cazaubiel, L. Dahlqvist, E. Delport, G.S. Denyer, D. Erba, G. Frost, Y. Granfeldt, S. Hampton, V.A. Hart, K.A. Hatonen, C.J. Henry, S. Hertzler, S. Hull, J. Jerling, K.L. Johnston, H. Lightowler, M. Mann, L. Morgan, L.N. Panlasigui, C. Pelkman, T. Perry, A.F. Pfeiffer, M. Pieters, D.D. Ramdath, R.T. Ramsingh, S.D. Robert, C. Robinson, E. Sarkkinen,

F. Scazzina, D.C. Sison, B. Sloth, J.Staniforth, N. Tapola, L.M. Valsta, I.Verkooijen, M.O. Weickert, A.R. Weseler, P. Wilkie and J. Zhang. Measuring the glycemic index of foods: interlaboratory study. *American Journal of Clinical Nutrition* 2008, 87, 247S–257S.
83. Australian Standard. Glycaemic index of food. AS 4694-2007, Sydney, Australia; 2007.
84. International Standards Institute. Food products – determination of the glycaemic index (GI) and recommendation for food classification. ISO 26642, Geneva, Switzerland; 2010.
85. Canadian Food Inspection Agency. Nutrition Labelling Compliance Test http://www.inspection.gc.ca/english/fssa/labeti/nutricon/nutriconae.shtml#a11; 2010 (accessed 27 April 2011).
86. G. Livesey. Glycaemic responses and toleration. In: H. Mitchell (ed.). *Sweeteners and sugar alternatives in food technology*, Blackwell Publishing Ltd, Oxford, England; 2006, pp. 2–18.
87. A.J. Aziz, The glycemic index: methodological aspects related to the interpretation of health effects and to regulatory labelling. *Journal of AOAC International*, 2009, 92, 879–887.
88. J.W. DeVries. Glycemic index: the analytical perspective. *Cereal Foods World* 2007, 52, 45–49.
89. R.A. McCance and E.M. Widdowson. *The composition of Foods*, 5th edn, HMSO, London; 1991.
90. S. Gregersen, P.B. Jeppesen, J.J. Holst and K. Hermansen. Antihyperglycemic effects of stevioside in type 2 diabetic subjects. *Metabolism* 2004, 53, 73–76.
91. J. Rodin. Comparative effects of fructose, aspartame, glucose, and water preloads on calorie and macronutrient intake. *American Journal of Clinical Nutrition* 1990, 51, 428–435.
92. L. Abdallah, M. Chabert and J. Louis-Sylvestre. Cephalic phase responses to sweet taste. *American Journal of Clinical Nutrition* 1997, 65, 737–743.
93. N.H. Mezitis, C.A. Maggio, P. Koch, A. Quddoos, D.B. Allison and F.X. Pi-Sunyer. Glycemic effect of a single high oral dose of the novel sweetener sucralose in patients with diabetes. *Diabetes Care* 1996, 19, 1004–1005.
94. X. Lan-Pidhainy and T.M. Wolever. Are the glycemic and insulinemic index values of carbohydrate foods similar in healthy control, hyperinsulinemic and type 2 diabetic patients? *European Journal of Clinical Nutrition* 2011, 65, 727–734
95. Y. Ueda, S. Wakabayahi and A. Matsuoka. Effect of indigestible dextrin on blood glucose and urinary C-peptide levels following sucrose loading. *Journal of the Japan Diabetes Society* 1993, 36, 715–723 (in Japanese).
96. T. Wada, J. Sugatani, E. Terada, M. Ohguchi and M. Miwa. Physicochemical characterization and biological effects of inulin enzymatically synthesized from sucrose. *Journal of Agricultural Food Chemistry* 2005, 53, 1246–1253.
97. T.W. Donner, J.F. Wilber and D. Ostrowski. D-tagatose, a novel hexose: acute effects on carbohydrate tolerance in subjects with and without type 2 diabetes. *Diabetes Obesity and Metabolism*. 1999, 1, 285–291.
98. J.J. Rumessen, S. Bode, O. Hamberg and E. Gudmand-Hoyer. Fructans of Jerusalem artichokes: intestinal transport, absorption, fermentation, and influence on blood glucose, insulin, and C-peptide responses in healthy subjects. *American Journal of Clinical Nutrition* 1990, 52, 675–681.
99. F. Brighenti, M.C. Casiraghi, E. Canzi and A. Ferrari. Effect of consumption of a ready-to-eat breakfast cereal containing inulin on the intestinal milieu and blood lipids in healthy male volunteers. *European Journal of Clinical Nutrition* 1999, 53, 726–733.
100. S. Kudlacek and G. Schernthaner. The effect of insulin treatment on HbA1c, body weight and lipids in type 2 diabetic patients with secondary-failure to sulfonylureas. A five year follow-up study. *Hormone and Metabolic Research* 1992, 24, 478–483.
101. G.M. Reaven. Compensatory hyperinsulinemia and the development of an atherogenic lipoprotein profile: the price paid to maintain glucose homeostasis in insulin-resistant individuals. *Endocrinology and Metabolism Clinics of North America* 2005, 34, 49–62.
102. K.L. Teff, J. Devine and K. Engelman. Sweet taste and diet in type II diabetes. *Physiology and Behavior* 1995, 57, 1089–1095.
103. B.J. Tepper, L.M. Hartfiel and S.H. Schneider. Sweet taste and diet in type II diabetes. *Physiology and Behavior* 1996, 60, 13–18.
104. A.G. Renwick. Intense sweeteners, food intake and the weight of a body of evidence. *Physiology and Behavior* 1993, 55, 139–143.
105. I.M. Stratton, A.I. Adler, H.A. Neil, D.R. Matthews, S.E. Manley, C.A. Cull, D. Hadden, R.C. Turner and R.R. Holman. Association of glycaemia with macrovascular and microvascular complications of type 2 diabetes (UKPDS 35): prospective observational study. *BMJ* 2000, 321, 405–412.

106. K.T. Khaw, N. Wareham, R. Luben, S. Bingham, S. Oakes, A. Welch and N. Day. Glycated haemoglobin, diabetes, and mortality in men in Norfolk cohort of European prospective investigation of cancer and nutrition (EPIC-Norfolk). *BMJ* 2001, 322, 15–18.
107. J.A. Davidson. Treatment of the patient with diabetes: importance of maintaining target HbA(1c) levels *Current Medical Research Opinion* 2004, 20, 1919–1927.
108. DCCT Research Group. The Absence of a Glycemic Threshold for the Development of Long-Term Complications: The Perspective of the Diabetes Control and Complications Trial. *Diabetes* 1996, 45, 1289–1298.
109. CDA. Guidelines for the management of diabetes mellitus in the new millennium. *Canadian Journal of Diabetes Care* 2001, 23, 56–69.
110. M. Brownlee. Biochemistry and molecular cell biology of diabetic complications. *Nature* 2001, 414, 813–820.
111. M.C. Gannon, F.Q. Nuttall, A. Saeed, K. Jordan and H. Hoover. Effect of a high-protein, low-carbohydrate diet on blood glucose control in people with type 2 diabetes. *American Journal of Clinical Nutrition* 2003, 78, 734–741.
112. K. Osei, J. Falko, B.M. Bossetti and C.G. Holland. Metabolic effects of fructose as a natural sweetener in the physiologic meals of ambulatory obese patients with type II diabetes. *American Journal of Medicine* 1987, 83, 249–255.
113. J.E. Swanson, D.C. Laine, W. Thomas and J.P. Bantle. Metabolic effects of dietary fructose in healthy subjects. *American Journal of Clinical Nutrition* 1992, 55, 851–856.
114. G. Livesey, R. Taylor, T. Hulshof and J. Howlett. Glycemic response and health a systematic review and meta-analysis: relations between dietary glycemic properties and health outcomes. *American Journal of Clinical Nutrition* 2008, 87, 223S–236S.
115. J. Salmeron, A. Ascherio, E.B. Rimm, G.A. Colditz, D. Spiegelman, D.J. Jenkins, M.J. Stampfer, A.L. Wing and W.C. Willett. Dietary fiber, glycemic load, and risk of NIDDM in men. *Diabetes Care* 1997, 20, 545–550.
116. K.A. Meyer, L.H. Kushi, D.R.J. Jacobs, J. Slavin, T.A. Sellers and A.R. Folsom. Carbohydrates, dietary fiber, and incident type 2 diabetes in older women. *American Journal of Clinical Nutrition* 2000, 71, 921–930.
117. S. Liu, W.C. Willett, M.J. Stampfer, F.B. Hu, M. Franz, L. Sampson, C.H. Hennekens and J.E. Manson. A prospective study of dietary glycemic load, carbohydrate intake, and risk of coronary heart disease in US women. *American Journal of Clinical Nutrition* 2000, 71, 1455–1461.
118. J.L. Chiasson, R.G. Josse, R. Gomis, M. Hanefeld, A. Karasik, M. Laakso and SNTR Group. Acarbose treatment and the risk of cardiovascular disease and hypertension in patients with impaired glucose tolerance: the STOP-NIDDM trial. *Journal of the American Medical Association* 2003, 290, 486–494.
119. T.L. Halton, W.C. Willett, S. Liu, J.E. Manson, C.M. Albert, K. Rexrode and F.B. Hu. Low-carbohydrate-diet score and the risk of coronary heart disease in women. *New England Journal of Medicine* 2006, 355, 1991–2002.
120. T.L. Halton, S. Liu, J.E. Manson and F.B. Hu. Low-carbohydrate-diet score and risk of type 2 diabetes in women. *American Journal of Clinical Nutrition* 2008, 87, 339–346.
121. D. Pometta, C. Trabichet and M. Spengler. Effects of a 12-week administration of isomalt (Palatinit(R)) on metabolic control in Type-II-diabetics. *Aktuelle Ernährungsmedizin* 1985, 10, 174–177.
122. K. Yamashita, K. Kawai and M. Itakura. Effects of fructo-oligosaccharides on blood glucose and serum lipids in diabetic subjects. *Nutrition Research* 1984, 4, 961–966.
123. J. Luo, S.W. Rizkalla, M. Lerer-Metzger, J. Boillot, A. Ardeleanu, F. Bruzzo, A. Chevalier and G. Slama. A fructose-rich diet decreases insulin-stimulated glucose incorporation into lipids but not glucose transport in adipocytes of normal and diabetic rats. *Journal of Nutrition* 1995, 125, 164–171.
124. J. Luo, S.W. Rizkalla, C. Alamowitch, A. Boussairi, A. Blayo, J.L. Barry, A. Laffitte, F. Guyon, F.R.J. Bornet and G. Slama. Chronic consumption of short-chain fructooligosaccharides by healthy subjects decreased basal hepatic glucose production but had no effect on insulin-stimulated glucose metabolism. *American Journal of Clinical Nutrition* 1996, 63, 939–945.
125. B.R. Balacazar-Munoz, E. Martinez and M. Gonzalez-Ortiz. Effect of oral inulin administration on lipid profile and insulin sensitivity in dyslipidemic obese subjects. *Review of Medicine in Chile* 2003, 131, 597–604.
126. W. van Dokkum, B. Wezendonk, T.S. Srikumar and E.G. van den Heuvel. Effect of nondigestible oligosaccharides on large-bowel functions, blood lipid concentrations and glucose absorption in young healthy male subjects. *European Journal of Clinical Nutrition* 1999, 53, 1–7.

127. J.P. Saunders, T.W. Donner, J.H. Sadler, G.V. Levin and N.G. Makris. Effects of acute and repeated oral doses of D-tagatose on plasma uric acid in normal and diabetic humans. *Regulatory Toxicology and Pharmacology* 1999, 29, S57–S65.
128. Z. Jie, L. Bang-Yao, X. Ming-Jie, L. Hai-Wei, Z. Zu-Kang, W. Ting-Song and S.A. Craig. Studies on the effects of polydextrose intake on physiologic functions in Chinese people. *American Journal of Clinical Nutrition* 2000, 72, 1503–1509.
129. M. Nomura, Y. Nakajima and H. Abe. Effects of long-term administration of indigestible dextrin as soluble dietary fiber on lipid and glucose metabolism. *Nippon Eiyo Shokuryo Gakkaishi (Journal of the Japanese Society of Nutrition and Food Science)*, 1992, 45, 21–25 (in Japanese).
130. G. Livesey and H. Tagami. Interventions to lower the glycemic response to carbohydrate foods with a low-viscosity fiber (resistant maltodextrin): meta-analysis of randomized controlled trials. *American Journal of Clinical Nutrition* 2009, 89, 114–125.
131. N.Y. Reyna, C. Cano, V.J. Bermudez, M.T. Medina, A.J. Souki, M. Ambard, M. Nunez, M.A. Ferrer and G.E. Inglett. Sweeteners and beta-glucans improve metabolic and anthropometrics variables in well controlled type 2 diabetic patients. *American Journal of Therapeutics* 2003, 10, 438–443.
132. V.L. Grotz, R.R. Henry, J.B. McGill, M.J. Prince, H.Shamoon, J.R. Trout and F.X. Pi-Sunyer. Lack of effect of sucralose on glucose homeostasis in subjects with type 2 diabetes. *Journal of the American Dietetic Association* 2003, 103, 1607–1612.
133. R. Giacco, G. Clemente, D. Luongo, G. Lasorella, I. Fiume, F. Brouns, F. Bornet, L. Patti, P. Cipriano, A.A. Rivellese and G. Riccardi. Effects of short-chain fructo-oligosaccharides on glucose and lipid metabolism in mild hypercholesterolaemic individuals. *Clinical Nutrition* 2004, 23, 331–340.
134. S. Colagiuri, J.J. Miller and R.A. Edwards. Metabolic effects of adding sucrose and aspartame to the diet of subjects with noninsulin-dependent diabetes mellitus. *American Journal of Clinical Nutrition* 1989, 50, 474–478.
135. J.L. Chiasson, R.G. Josse, R. Gomis, M. Hanefeld, A. Karasik, M. Laakso. and SNTR Group. Acarbose for the prevention of Type 2 diabetes, hypertension and cardiovascular disease in subjects with impaired glucose tolerance: facts and interpretations concerning the critical analysis of the STOP-NIDDM Trial data. *Diabetologia*, 2004, 47, 969–975; Discussion 976–977.
136. A.W. Barclay, P. Petocz, J. McMillan-Price, V.M. Flood, T. rvan, P. Mitchell and J.C. Brand-Miller. Glycemic index, glycemic load, and chronic disease risk–a meta-analysis of observational studies. *American Journal of Clinical Nutrition* 2008, 87, 627–637.
137. G. Livesey and H. Tagami. The significant impact of a low viscous fibre on glycaemic response. In: J.W. van der Kamp, J.A. Jones, B. McClearly and D. Topping (eds). *Dietary Fibre: New Frontiers for Food and Health* Wageningen Academic Publishers, Wageningen, The Netherlands; 2010, pp. 475–491.
138. P. Marteau and B. Flourie. Tolerance to low-digestible carbohydrates: symptomatology and methods. *British Journal of Nutrition* 2001, 85(Suppl. 1), S17–S21.
139. J. Cummings, A. Lee and D. Storey. Workshop: physiology and tolerance of LDCs. *British Journal of Nutrition* 2001, 85(Suppl. 1), S59–S60.
140. International Organization for Standardization. Precision of test methods – Determination of repeatability and reproducibility for a standard test method by inter-laboratory tests (ISO 5725). International Organization for Standardization, Geneva, Switzerland; 1986.
141. R. Villegas, S. Liu, Y.T. Gao, G. Yang, H. Li, W. Zheng and X.O. Shu. Prospective study of dietary carbohydrates, glycemic index, glycemic load, and incidence of type 2 diabetes mellitus in middle-aged Chinese women. *Archives of Internal Medicine* 2007, 167, 2310–2316.
142. S.J. Janket, J.E. Manson, H. Sesso, J.E. Buring and S. Liu. A prospective study of sugar intake and risk of type 2 diabetes in women. *Diabetes Care* 2003, 26, 1008–1015.
143. G. Livesey. Amount and type of carbohydrate: quantity versus quality. Diabetes UK Annual Professionals Conference, Glasgow, Scotland. Diabetes UK, London; March 2008.
144. M.T. Flood, M.H. Auerbach and S.A. Craig. A review of the clinical toleration studies of polydextrose in food. *Food Chemistry and Toxicology* 2004, 42, 1531–1542.
145. A. Ruskone-Fourmestraux, A. Attar, D. Chassard, B. Coffin, F. Bornet and Y. Bouhnik. A digestive tolerance study of maltitol after occasional and regular consumption in healthy humans. *European Journal of Clinical Nutrition* 2003, 57, 26–30.
146. I. Macdonald and C.A. Williams. Effects of ingesting glucose and some of its polymers on serum glucose and insulin levels in men and women *Annals of Nutrition and Metabolism* 1988, 32, 23–29.
147. K. Okuma, I. Matsuda, Y. Katta and Y. Hanno. Pyrolysis of starch and its digestibility by enzymes - characterization of indigestible dextrins. *Denpun Kagaku* 1990, 37, 107–114 (in Japanese).

2 Dental Health

Anne Maguire

Centre for Oral Health Research, School of Dental Sciences, Newcastle University, UK

2.1 INTRODUCTION

Diet has an important influence on oral health and can affect teeth generally as they develop, during eruption into the mouth and locally after eruption.

Teeth are important for good nutrition; they enable an individual to enjoy a varied diet and prepare food for digestion. Teeth facilitate the development of appropriate speech patterns and good oral health, illustrated by a healthy smile, enhances facial appearance. Conversely, poor oral health can cause considerable pain, increase risk of other diseases, detract from a socially acceptable appearance reducing self-esteem and increase the financial burden of health care.

Dental caries has been labelled man's most common disease and is well recognised as an ongoing problem. Another dental condition, dental erosion, is recognised as a problem in individuals exposed to acids, other than those produced by bacteria in the mouth. This intra-oral exposure to acid may be from an extrinsic source through diet or environment, or an intrinsic source from vomiting, reflux or regurgitation of stomach acid. The main dietary sources of acid are fruit acids, ascorbic and phosphoric acid in frequently consumed acidic fruit juices, squashes and carbonated beverages (full and low calorie) as well as pickles, sauces, lactovegetarian foods and yoghurt. Further discussion with regard to dental erosion is beyond the remit of this section, but in view of the role of sweeteners in the manufacture of potentially acidic soft drinks, foods and confectionery products, the reader can find information on the aetiology, mechanisms and implications of dental erosion[1] and its management in relation to diet[2] elsewhere.

2.2 DENTAL CARIES

2.2.1 The problem

Dental caries is a common disease of affluence. Its prevalence rose in developed countries in parallel with the increased availability of sugar through refining and trade reaching its highest level between the 1950s and early 1970s.[3–5] Since the mid-1970s, caries prevalence has shown an overall decline in developed countries but it is a serious and growing problem

---

Sweeteners and Sugar Alternatives in Food Technology, Second Edition.
Edited by Dr Kay O'Donnell and Dr Malcolm W. Kearsley.
© 2012 John Wiley & Sons, Ltd. Published 2012 by John Wiley & Sons, Ltd.

in disadvantaged communities in both developing and industrialised societies[6–8] and is a major health problem, especially in economic terms.[9–11] Additionally, the evidence of caries decline among children and young people in a number of western societies seen in the 1970s and 1980s has been superseded by more recent studies showing that these reductions in caries prevalence may be ending or even reversing,[12–18] with a skewed distribution of disease, concentrated in lower socioeconomic strata.[8,19]

An individual's caries experience can be measured using an index that records the number of decayed missing and filled primary teeth (dmft) or permanent teeth (DMFT) in their mouth, depending on their age. In a caries-free mouth with no experience of dental caries, this value would be zero. Between 1999 and 2008, the average DMFT in 12-year olds in European countries ranged from 0.7 in Germany, Denmark and Great Britain to 3.4 in Latvia and 3.7 in Lithuania.[20,21] There was a similar pattern across the world in developed countries with similarly aged children having a DMFT of between 1.0 (Australia) and 1.7 (Japan), while for emerging BRIC nations it was between 2.8 (Brazil) and 3.9 (India).[22,23]

Almost certainly, fluoride, especially fluoride in toothpaste, has been responsible for the decline in caries prevalence, but dental caries persists as a major worldwide infectious disease, conditioned by diet.

2.2.2 Aetiology

The basic process of caries aetiology is well known. Dietary sugars are metabolised to acids by bacteria (particularly mutans streptococci) in dental plaque, a biofilm of microbial accumulations, which adheres closely to uncleaned surfaces of teeth.[24] The bacteria in this biofilm cooperate and compete by various mechanisms, resulting in changes in composition over time and as such could be considered in terms of a continuum; from those that are benign or less harmful to those that are virulent or more harmful.[25] The multi-factorial nature of dental caries makes it difficult to analyse the singular contribution of the biofilm to the caries process.[25] A shift in composition of the biofilm is both a cause and result of the tissue pathology in which high levels of fermentable carbohydrates result in an outgrowth of micro-organisms responsible for the pathology.[26,27] A caries-related biofilm is enriched in *Streptococcus mutans* as well as other bacterial species that are acidogenic and also thrive in an acidic environment.[28] At a 'critical pH' of 5.5, tooth enamel starts to lose calcium and phosphate ions, demineralising the enamel surface. If this outward flow of ions persists, the tooth surface becomes fragile and collapses to form a carious cavity. Before this stage (i.e. during the pre-cavitation stage), remineralisation can occur and the process is reversible; lesions can regress and heal if the tooth surface conditions are favourable to inward flow of calcium and phosphate ions. Favourable conditions could include dietary change resulting in less sugar being available for bacteria to metabolise to acids, allowing saliva (with adequate calcium and phosphate levels and a high pH) to encourage remineralisation. However, once a cavity is formed, the process is irreversible and caries will usually progress, requiring treatment involving removal of caries and restoration of the tooth.[29]

There is overwhelming evidence that fermentable carbohydrates, in particular sugars, are the main dietary factor in the aetiology of dental caries,[30] although sugars alone are not the sole determinant of whether food is cariogenic; other substances may influence cariogenicity, either hindering or enhancing the caries-promoting properties of sugars in the diet.[31] This evidence has accumulated over time from many types of studies, including human interventional and observational studies, animal experiments, plaque pH studies and enamel slab and incubation experiments. Comprehensive accounts of these studies can be found in texts by Rugg-Gunn[4] and Fejerskov and Kidd[29] as well as reviews by Sheiham[32] and Zero.[5]

Starchy foods can also cause caries, especially if heat treated and finely ground, but staple starch foods are not seen as a significant cause of caries. It is only when sugars are refined and added to the diet that dental caries occurs in significant amounts. Likewise, sugars in fresh fruit, vegetables and milk are not a threat to dental health – it is the non-milk extrinsic (NME) sugars in the diet that are the major cause of dental caries. NME sugars[33] when grouped together with concentrated sugars in honey, syrups and fruit juices are collectively described as 'free sugars'.[34] The WHO Report 'Diet, Nutrition and Prevention of Chronic Diseases', published in 2003,[35] recommended that they should not contribute more than 10% to total energy intake. In many economically developed countries, free sugars constitute a large proportion (up to 70%) of the total sugars in the diet[36–38] and contribute significantly more than the recommended proportion of food energy intake. As a result, a number of national reports have set targets for the contribution of free sugars to the diet, the average being 10% or less in terms of calories.[39,40]

Sreenby[41] described the pattern of the relationship between sugars supply and dental caries prevalence. At low levels of sugars intake of 10 kg per year (25 g per day), dental caries is low. At around 15 kg per year (40 g per day), dental caries prevalence shows a steep increase with the level of disease increasing with increasing sugars intake. At high levels of supply (35 kg per year or 93 g per day), caries prevalence plateaus at a high level as disease saturation occurs. Regular exposure to fluorides has raised the threshold of sugar intake at which caries progresses to cavitation, but fluoride has its limits and not all populations at risk of dental caries receive the dental health benefits of fluorides.

2.2.3 Control and prevention

Reduction in the sugars content of our diet has been a central strategy for disease control and prevention, not only for dental health but also for general health.[33,35,37,38,42–44] In terms of sugars control, there have been two main strategies for prevention; firstly, an overall reduction in the quantity of refined sugars in our diet, with more emphasis on the need for more fruit, vegetables and starchy foods. The second strategy has been to substitute free sugars with alternative sweeteners, particularly in foods, drinks and medicines likely to be consumed between meals, for example, snack foods and confectionery and those medicines with prolonged oral clearance likely to be used regularly and long-term by the chronically sick. The development and implementation of national strategies for dental health in relation to diet have been guided by government policy and guidelines from dental professional bodies.[45–49] For a full review of the roles of nutrition and diet in oral health and disease, the reader is referred to the textbooks by Rugg-Gunn and Nunn[2] and Fejerskov and Kidd.[29]

Since sugars have a predominant aetiological role in dental caries and the sweet taste of sugar is popular, it is not surprising that the substitution of alternative sweeteners for cariogenic sugars has been investigated enthusiastically. As a result, two main groups of alternative sweeteners have been identified for use in foods and drinks: reduced-calorie (nutritive) bulk sweeteners that have calories and low-calorie (non-nutritive) intense sweeteners with little or no calorific content.

This section describes and reviews the dental evidence concerning reduced and low-calorie sweeteners, and bulking agents. The evidence relating foods and drinks to dental disease comes from a variety of sources, therefore, a short description of cariogenicity and methods for its estimation or measurement is given first in the following text.

2.2.4 Determining cariogenicity

It is important that the terminology used to describe the dental effects of sweeteners is accurate and consistent. In simple terms, *cariogenicity* can be defined as the potential to produce dental caries, while *acidogenicity* describes the potential to produce acid.[50] Acidogenicity and fermentability are essentially terms used to describe findings from *in vitro* experiments and *in vivo* studies other than clinical trials, indicating possible effects of a sweetener on the development of dental caries, whereas cariogenicity, non-cariogenic and anti-cariogenic are clinical terms. Interpretation of these definitions varies, leading to some difficulties interpreting the findings of some studies. The terms 'cariostatic', 'anti-cariogenic' and 'anti-caries' have all been used when discussing dental therapeutic claims of some sweeteners, as have 'active' and 'passive' effects. Here, the properties of non-fermentability and non-cariogenicity are classed as passive effects while active caries-preventive (or caries-inhibitory) effects include the terms bacteriostatic and cariostatic. Only a reversal in the caries process, that is the remineralisation of a carious lesion, is described as a therapeutic or anti-cariogenic effect.

Human clinical trials provide the best method for assessing the effect of dietary items on dental disease; however, there are a number of obstacles to their use. Firstly, dental caries is a chronic disease; carious lesions sometimes take many years to develop, so that clinical trials need to last about 3 years and it is difficult to keep large numbers of human volunteers on the same diet for this length of time. Secondly, there is a wide variation in caries increments between individuals, necessitating large numbers of individuals to complete each trial. Ensuring compliance with a dietary test regime over long periods is also difficult and it is unethical to expect subjects to consume a diet that might be less healthy (e.g. contain more sugars) than their normal diet. In view of these difficulties, other methods of estimating the cariogenicity of foods have developed (Box 2.1).

2.2.4.1 Incubation experiments

These are 'test-tube experiments' and examine the ability of plaque micro-organisms to metabolise the test food to acid. Saliva, which contains oral micro-organisms, or pure cultures of oral micro-organisms have been substituted for dental plaque. Incubation experiments are a very stringent test since some acid may be formed from some foods but at such a slow rate as to be of little clinical relevance. Conversely, rapid acid production indicates that the food under test is potentially cariogenic.

Box 2.1 Methods of estimating cariogenicity.

- Incubation experiments
- Plaque pH experiments
- Animal experiments
- Enamel slab experiments (both *in vitro* and *in vivo*)

This type of experiment can be extended by adding dental enamel or hydroxyapatite to the incubation and measuring calcium and phosphate dissolution from the mineralised tissue.

2.2.4.2 Plaque pH experiments

These can be seen as incubation experiments *in vivo*. The pH in the dental plaque on the surface of teeth is monitored during and after the test food is eaten. Again, it is pH change or the acidogenicity of the food that is being measured, not cariogenicity.

There are three main methods of assessing plaque pH: (1) by micro-electrode probes (antimony or glass) inserted into the plaque; (2) by indwelling micro-electrodes (glass or iridium oxide) built into intra-oral appliances, which are worn for several days to allow plaque to accumulate and (3) by harvesting small samples of plaque from representative teeth and measuring the pH of these samples outside the mouth. The indwelling glass micro-electrode system, developed in Zürich, has been used to identify dentally safe snack foods, drinks and medicines since the Swiss Office of Health introduced legislation for the labelling of foods, drinks and medicines with regards to dental health in 1969.[50] Products could be labelled 'zahnschonend' ('ménage les dents', 'safe for teeth') only if they passed the cariogenicity test by not lowering plaque pH below 5.7 either during consumption or up to 30 minutes later, as well as having a low erosive potential.

2.2.4.3 Animal experiments

The main advantage of animal experiments (usually with rats) is that the dietary regime – type and amount of food and frequency of eating – can be carefully controlled. However, some polyols may cause severe intestinal disturbances and loss of weight at very high feeding levels in animals, while in other studies, high levels are well tolerated. Additionally, in some tests, the animals are super-infected with specific micro-organisms (e.g. mutans streptococci) to standardise experimental conditions. This can give a false indication of the relative cariogenicity of some foods, especially sugars.

2.2.4.4 Enamel slab experiments

To simulate caries attack in the mouth, intra-oral appliances, capable of holding slabs of dental enamel, are used. Plaque forms on the surface of the slabs that remain in the mouth for 1–4 weeks. On a number of occasions, throughout the day, the appliance is removed and the slabs put into solutions or suspensions of foods in order to simulate the eating of that food. The appliance is then re-inserted into the mouth. Caries development is measured quantitatively by examining the enamel slabs. Although different food or drink solutions can be compared in the same experiment, the test foods are not actually eaten, so the stimulatory effect of the food on salivary flow (which varies with different foods) is absent.

These types of studies are more short-term compared with clinical trials but they can provide important information relating to the mechanism of action of sweeteners, the oral flora and the protective role of saliva.

2.2.4.5 Clinical trials in human subjects

Clinical trials provide the strongest type of evidence since dental caries development is measured in human subjects and the cariogenicity of a food or drink under test can be measured, rather than estimated. The most superior type of clinical trial is a double blind randomised clinical trial (or RCT): subjects are randomly allocated to treatment groups, and they and the assessors do not know (are 'blind' to) the group identity of the subject. An estimate of the magnitude of any effect can be made in economic terms as well as percentage change in levels of disease, and the incidence of side effects can be monitored.

The evidence for the dental effects of sweeteners is described using published research, according to the type of study used to investigate the particular aspect of the sweetener. The terminology used in the original published research will be used, with reference to current terminology where relevant.

2.3 REDUCED-CALORIE BULK SWEETENERS

2.3.1 Erythritol

Erythritol is a sugar alcohol and theoretically, as such, could be considered to be refractory to biological decomposition by cariogenic organisms.[51] From the cariological point of view, it has been described as a promising sugar substitute, although the dental evidence is still very limited. In contrast to xylitol and most other polyols used as bulk sweeteners, erythritol does not have laxative properties[52] and therefore may have broader food applications in the future.

In an incubation experiment,[53] erythritol was neither utilised as a substrate for lactic acid production nor for plaque formation of mutans streptococci and certain oral micro-organisms. A rat experiment carried out by the same researchers showed significantly less caries (mean caries score = 3.1) in rats fed a diet containing 26% erythritol, compared with control rats fed with a diet containing 26% sucrose (mean caries score = 60.5).

Erythritol and xylitol saliva stimulants have been compared clinically in the control of dental plaque and mutans streptococci.[51,54] In these experiments, xylitol was found to be superior in reduction of mean weight of total plaque mass, plaque and salivary levels of mutans streptococci and plaque levels of total streptococci. However, more recent work by Makinen *et al.*[55] has shown erythritol to have more promise. Their study looked at the plaque and saliva levels of mutans streptococci following the daily use of erythritol, xylitol and sorbitol chewable tablets, complemented by twice daily use of a dentifrice containing the same polyols, and showed significant plaque reduction in subjects receiving xylitol and erythritol. Associated incubation experiments showed that xylitol, and especially erythritol, inhibited the growth of several strains of mutans streptococci, although no theories for the mechanism of growth inhibition by erythritol have so far been published.[52] In addition, erythritol has been shown to decrease the polysaccharide-mediated cell adherence, which contributes to dental plaque formation through a mechanism not dependent on growth inhibition.[52] Based on current evidence, erythritol is non-cariogenic.

2.3.2 Isomalt

Reports on the potential use of isomalt as a non-acidogenic sweetener date back to the 1970s, when incubation experiments demonstrated that isomalt was neither fermented nor used as a carbon source for growth by cariogenic streptococci.[56] Gehring and Karle[57] screened differing strains of mutans streptococci and lactobacilli with mixed human dental plaque *in vitro* for fermentation and growth with 1% isomalt. In contrast to sucrose, glucose or fructose, isomalt remained practically unfermented. Van der Hoeven[58] made similar observations while screening about 50 oral bacterial strains; the majority of the cultures tested did not ferment isomalt and only very few strains of streptococci and actinomyces showed a small pH drop. These findings were corroborated by Ziesenitz and Siebert[59] using a very sensitive method of determining acid production rates.

In a study of the acidogenic response of isomalt in human dental plaque,[60] while the intra-plaque pH dropped rapidly from pH 6.7 to around pH 4.6 after ingestion of sugars, the intra-plaque pH after isomalt ingestion remained high. The same year, Imfeld[50] tested the two components of isomalt separately and in combination, showing that all these compounds depressed plaque pH by insignificant amounts, which earned the 'safe for teeth' label for products made with isomalt.

More recently, Takasuka and co-workers[61,62] investigated the effects of isomalt on de- and remineralisation on bovine enamel lesions. They used an *in vitro* pH cycling model with 10% isomalt solutions, and an *in situ* study in which 12 volunteers used a 10% isomalt-containing toothpaste with an intra-oral appliance containing two demineralised enamel specimens. Treatment effects were assessed by chemical analysis of *in vitro* solutions and transversal microradiography. The researchers were able to conclude that 'Isomalt had a positive effect on the de/remineralisation balance when given under conditions relevant to practical use'.

Kaneko and co-workers,[63] in a study of plaque pH responses to isomaltulose and its hydrogenated derivative isomalt, showed no significant lowering of plaque pH and a low acidogenicity with isomalt.[64–66]

The potential dental effects of sugar-free sweets formulated with Lycasin® (also known as hydrogenated glucose syrup or maltitol syrup – see Section 2.3.4.1) or isomalt instead of sugars were evaluated in an incubation and plaque pH study.[67] The authors concluded that the extent of demineralisation of dental enamel was related not only to the fermentability of the sweets but also their acidity, highlighting the importance of reducing the levels of flavouring acids when re-formulating sweets with non-fermentable bulk sweeteners.

Karle and Gehring[68] reported on the non-cariogenic properties of isomalt in rats. The rats fed 30% sucrose had the highest caries score, but caries in isomalt-fed rats did not differ from the negative control rats fed the basic diet ($p < 0.30$), although caries was highly significantly reduced when compared with the sucrose-fed rats ($p < 0.001$). Similar trends of lower caries scores have been demonstrated in studies with desalivated rats[69] in which caries scores for xylitol-fed rats tended to be lower than those for isomalt, but not statistically significantly different. The non-cariogenic potential of isomalt has been corroborated by further experiments assessing the cariogenic potential of chocolates sweetened with sucrose, isomalt, sorbitol or xylitol in *S. mutans*-infected rats, which demonstrated the non-cariogenicity of the chocolate sweetened with isomalt as well as chocolate sweetened with xylitol.[57]

The question whether the oral flora might eventually adapt to isomalt was studied in rats fed *ad libitum* on diets containing either 16% sucrose or 16% isomalt for 14 weeks.[58] No striking differences were observed in the bacterial composition of the plaque between the sucrose- and isomalt-fed rats and there was no accumulation of isomalt-fermenting micro-organisms, demonstrating that no adaptation to isomalt had occurred over the 3-month period. This was confirmed by van der Hoeven[70] in a further rat experiment, but no human studies have been carried out to reinforce these findings.

Dental and other biological properties of isomalt have been reviewed by Schiweck and Ziesenitz[71] and Ziesenitz.[72] Summarising evidence from incubation experiments, rat studies and plaque pH studies, they concluded that isomalt is non-cariogenic.

2.3.3 Lactitol

Lactitol appears to be fermented by about as many types of plaque micro-organisms as sorbitol.[73] Metabolism was slow, but frequent sub-culturing of *S. mutans* increased its ability to ferment lactitol. Grenby *et al.*[74] concluded that lactitol had low cariogenic potential (greater than xylitol but less than other polyols) in incubation experiments.

Grenby and Desai[75] reported that dental plaque of volunteers who ate lactitol sweets for 3 days contained less soluble carbohydrate and more calcium and phosphorus compared with those eating sucrose sweets.

Two rat experiments[74,76] have shown lactitol to be of low cariogenicity, and plaque pH studies[50] have indicated only slight falls in plaque pH, designating lactitol as hypoacidogenic in humans. No clinical dental trials of lactitol have been reported.

2.3.4 Maltitol

A small number of studies have reported the dental effects of maltitol. Some of these studies[77–79] are difficult to interpret as 'maltitol-rich' sweeteners – containing about 93% maltitol – were tested.

Edwardson *et al.*[79] investigated the ability of a number of strains of oral bacteria to ferment maltitol, in comparison with Lycasin (maltitol syrup), sorbitol and xylitol. They reported:

> No streptococcal strain fermented maltitol, which agrees with findings of Naito[80] and Ikeda *et al.*;[81] Maltitol was only fermented by the lactobacilli (about two-thirds of the strains). However, the lactobacilli make up only a small part of the regular oral plaque flora... The results therefore do not indicate that maltitol is fermented to any great extent in dental plaque.

Other incubation studies by Ooshima *et al.*[82] led them to conclude:

> Fourteen strains of oral streptococci, including mutans streptococci, did not utilise the maltitol nor produce sufficient acid to demineralise tooth enamel.

This finding appeared to put maltitol into the same category as xylitol – although maybe not quite as non-fermentable as xylitol – and considerably less fermentable than sorbitol or mannitol. Imfeld[50] found that plaque pH did not fall below pH 6.0 when French and Japanese maltitols were tested by a pH telemetry system.

Rundegren *et al.*[83] reported enamel slab experiments in which four subjects dipped their appliances into 10% solutions of maltitol or sucrose (positive control) four times daily for 4 weeks; otherwise, the subjects consumed a normal diet. Some demineralisation was observed with maltitol but this was less severe than when sucrose was the substrate.

In the extensive rat caries studies of Ooshima *et al.*,[82] maltitol was substituted for either wheat flour or sucrose, in different groups. The authors concluded that 'maltitol did not induce dental caries in rats', and in addition, 'replacement of sucrose with maltitol resulted in a significant reduction in caries incidence'. It was possible to estimate the effect of substituting maltitol for wheat starch in a sucrose-containing diet (i.e. to test for any anti-cariogenic action of maltitol) in eight pairs of groups. In about half of these pairs, the group receiving maltitol and wheat starch developed less caries than wheat starch alone. Although not emphasised by the authors, this did point to a possible cariostatic action of maltitol. The rat experiments of Ooshima *et al.*[82] agree with results reported by Izumitani *et al.*,[78] which showed that a maltitol-rich mixture 'did not induce caries in rats'.

The effects of the frequent consumption of maltitol-containing lozenges on acid production and bacterial counts were studied by Birkhed et al.[77] The results showed lower ($p < 0.05$) pH values induced by a maltitol rinse and sorbitol rinse after a 3-month period of frequent consumption suggesting plaque adaptation. However, the maltitol tested was 93.5% D-maltitol, which was replaced by high purity crystalline D-maltitol in 1994. Plaque bacteria have been shown not to adapt to crystalline D-maltitol following exposure over 14 days in one study.[84] More recently, maltitol was shown to have similar favourable properties to xylitol in a school-based RCT of 'gummy bear snacks' in 8-year olds in the United States. Significant reductions in mutans streptococci in dental plaque were observed after 6 weeks consumption of gummy bears delivering either 11.7 g or 15.6 g of xylitol or a high dose (44.7 g) of maltitol daily via three exposures to the snack per day, but with no statistically significant difference between groups.[85] However, in a study of 6-month use of maltitol versus xylitol gum in adults, although there were statistically significant reductions in mutans streptococci in plaque and saliva with xylitol gum, levels at the end of maltitol gum use had increased significantly.[86] Although these results suggested that mutans streptococci may be able to adapt and grow in a maltitol-rich environment during longer term (>3 months) use in adults, the study had some weaknesses. It was a controlled clinical trial; interventions were not fully randomised, with subjects who expressed a preference for a particular flavour of gum given prioritised assignment to that gum group to enhance compliance. Other weaknesses of the study were that the average age of the maltitol group (34.5 years) was higher than the xylitol and control groups (28.0 years and 23.7 years, respectively), which had a negative association with plaque mutans streptococci levels. In addition, due to dropouts and exclusion due to reported use of xylitol gum, only 59% of the control group (no gum use) remained in the final analysis. In contrast, in a clinical trial conducted in 240 13–15-year olds in China, 4 weeks of chewing maltitol gum has been shown to lead to a similar statistically significant reduction of plaque growth and favourable salivary parameters as xylitol chewing gum when compared with chewing gum base in 13–15-year olds.[87–89]

In summary, although the studies are few, the possibility of fermentation of maltitol by plaque organisms seems small, and it can be classed as non-acidogenic. Rat experiments have shown maltitol to be non-cariogenic, and possibly caries inhibitory. Recent clinical trials have shown maltitol in chewing gum and in gummy bear snacks to possess similar properties to xylitol during short-term use, although further clinical trials with human volunteers are needed to confirm maltitol's properties, particularly during longer term use in adults.

2.3.4.1 Lycasin is the registered trade name of hydrogenated starch hydrolysates produced by Roquette

Swedish Lycasin (Lycasin 80/33), produced until the 1970s, was extensively tested in early animal experiments,[90] plaque pH experiments,[91] incubation experiments[92] and in one clinical trial[93] (the Roslagen study) of Lycasin-containing candies. It has generally been found to have favourable dental properties. When production moved to France, Lycasin 80/55 was developed and found to be superior.[94,95] Research since then has only been concerned with Lycasin 80/55. The dental effects of Lycasin have been extensively reviewed by Rugg-Gunn.[4,96]

At least nine reports of incubation experiments have shown that Lycasin can be fermented by plaque organisms, but that the rate is slow compared with sugars. Only some of the organisms capable of fermenting sugars can ferment Lycasin. Growth and metabolism of plaque micro-organisms has been found to be substantially less after use of Lycasin-sweetened lozenges compared with sucrose–glucose lozenges and in combination with xylitol, acid production is very slow.[97]

Lycasin has been subjected to a number of plaque pH studies,[60,94,95,98] which have shown its low acidogenicity, and the similarity of the dental effects of Lycasin and sorbitol, either as a 10% solution or as a boiled sweet or syrup, have been shown by Imfeld.[50] Under the Zurich indwelling electrode system for measuring plaque pH,[50] Lycasin is classed as 'safe for teeth'.

At least four animal experiments have indicated the low cariogenicity of Lycasin with the two fully reported experiments[99,100] concluding that Lycasin was non-cariogenic or virtually so, in rats.

The adaptation of mutans streptococci strains to ferment hydrogenated glucose syrup (Swedish Lycasin 80/33) was tested by frequent sub-culturing in a series of *in vitro* experiments.[73] A marked increase in fermentation of Lycasin by mutans streptococci was recorded but this property was lost when the adapted strain was sub-cultured once in glucose. This indicated that adaptation to ferment Lycasin was rather unstable and, since alternative sweeteners are often consumed in combination with other sugars and carbohydrate will always be present in the oral cavity, it was unlikely that gross adaptation would occur *in vivo*. The rat experiments carried out by Havenaar *et al.*[99] on French Lycasin reinforced this view, with no selection or adaptation of oral bacteria found when it was consumed frequently over a long period.

There has been one report of enamel slab experiments in human volunteers eating non-cariogenic sweets,[101] which concluded that Lycasin-containing confectionery could aid remineralisation of pre-cavitation carious lesions.

2.3.5 Sorbitol

Sorbitol has been tested extensively for acidogenicity and cariogenicity in several types of study (*in vitro* incubation experiments, plaque pH studies, animal experiments and human clinical trials). Comprehensive reviews have been published.[102,103] Sorbitol is fermented slowly by plaque organisms (mainly streptococci) but the rate is very much slower than that for glucose or sucrose.[104]

Sorbitol depresses plaque pH only slightly[50,91,96,105–108] and solutions (usually 10%) of sorbitol have often been used as a negative control in plaque pH studies because of this very slight depression.[109–111]

At least 11 studies have compared the effect of sorbitol on the development of dental caries in animals with the effect of sucrose.[103] Some dental caries developed in rats consuming sorbitol-containing diets, but the amount was much less than that which occurred with sugars. Most dental caries occurred in studies in which the rats were super-infected with mutans streptococci, micro-organisms that are particularly capable of metabolising sorbitol, with little dental caries occurring when the rats were not super-infected with these bacteria. Results of animal experiments need cautious interpretation, but indicate that dental caries could occur in those rare cases of individuals with very high *S. mutans* counts who consumed sorbitol to the exclusion of other dietary sugars.

In experiments in humans using the enamel slab technique,[112] sorbitol gave rise to 45% of the demineralisation (softening) of enamel compared with sucrose.

Most of the clinical trials evaluating the effects of sorbitol on dental health in humans have tested sorbitol in chewing gum although only a few clinical trials of chewing gum have been conducted specifically with sorbitol. The majority have been conducted with xylitol, sorbitol and mixtures of xylitol/sorbitol or sorbitol/maltitol. In none of these long-term studies, comprehensively reviewed by Birkhed and Bär[103] and Burt[113] and systematically reviewed by Lingstrom et al.,[114] Mickenautsch et al.[115] and Deshpande and Jadad[116] has there been any evidence of cariogenicity of sorbitol. The most recent systematic review included the results of 19 papers from 14 study populations that compared polyol-containing chewing gum with no gum. Using meta-analyses, the review showed that the use of xylitol, xylitol/sorbitol and sorbitol gums were associated with a mean preventive fraction (95% CI) for dental caries of 58.7% (35.4, 81.9), 52.8% (40.0, 66.0) and 20.0% (12.7, 27.3), respectively;[116] the caries-preventive effect of the combined sweeteners being between that of sorbitol gum and xylitol gum. These findings were in contrast to those of Lingstrom et al.[114] who had earlier concluded that the evidence for the use of sorbitol or xylitol in chewing gum was inconclusive and stressed the need for well-designed randomised clinical studies with high compliance to demonstrate the role of sorbitol and xylitol in caries prevention. This plea was re-iterated in the systematic review by Mickenautsch et al.,[115] although they did conclude that the evidence suggests that sugar-free chewing gum has a caries-reducing effect.

Three studies have evaluated the effectiveness of sorbitol in confectionery (other than chewing gum), and have failed to exhibit significant differences in caries incidence between test and control groups.[103]

Mutans streptococci, species of oral lactobacilli, enterococci and proprionibacteria do ferment sorbitol.[117] A few strains of actinomyces, as well as sanguis and mitior streptococci, also have this ability,[73,79] particularly in subjects consuming sorbitol-containing products for long periods of time,[77] since cariogenic micro-organisms can 'learn' to metabolise sorbitol when their sugar supply is restricted.[118] Dental plaque adaptation to sorbitol was comprehensively reviewed by Hogg and Rugg-Gunn,[119] who concluded that in the light of current evidence, frequent or long-term use of sorbitol was unlikely to represent any increased risk of dental caries in normal people, although it may present a small problem in people with low salivary flow who are high consumers of sorbitol. Birkhed et al.,[120] who looked at long-term (2–12 years) intake of sorbitol-containing products, did not find high caries activity among high consumers, and Birkhed and Bär[103] came to the same conclusion, stating that current evidence 'does not appear justified to generally question its (sorbitol's) use as a dentally safe bulk sugar substitute for sugar-free sweets'. However, these views were questioned by the results of a small study of five subjects by Waaler and co-workers,[121] which showed that plaque exposed to sorbitol for 12 weeks increased in its ability to metabolise sorbitol and that this effect persisted 12 weeks after the end of sorbitol exposure. The practical importance of this is unclear. Studies since then have shown dental plaque adaptation to sorbitol to occur in spite of the presence of xylitol,[122] although plaque's pH response to sucrose does not appear to be increased by prior sorbitol gum use.[123–125]

2.3.6 Mannitol

There is much less information on the dental effects of mannitol compared with sorbitol although it is being used increasingly in chewing gums, in combination with other polyols. Mannitol and sorbitol have commonly been tested in the same experiments and they show similar rates of fermentation[104,126] and similar plaque pH curves.[50] Three rat experiments[127–129] have also indicated that the cariogenicity of sorbitol and mannitol is similar, while the one enamel slab experiment to test mannitol[112] indicated a similar acidogenic potential to sorbitol. The recent systematic review of polyol-containing chewing gums showed a non-statistically significant mean preventive fraction (95% CI) of 10.7% (20.5, 41.9) for sorbitol/mannitol gums when compared with no gum use.[116] There have been no clinical trials of foods or chewing gum where mannitol is the sole sweetener.

2.3.7 D-tagatose

D-tagatose is a reduced-calorie bulk sugar. It is a ketohexose with a chemical structure similar to fructose and is nearly as sweet as, and has the bulk of, sucrose. However, only 15–20% of D-tagatose ingested is absorbed from the small intestine to provide almost zero (1.5 kcal/g) available energy.[130, 131]

In the United States, D-tagatose, has Generally Recognized As Safe (GRAS) status under US Food and Drug Administration (FDA) regulations, thereby permitting its use as a sweetener in foods and beverages, health foods and dietary supplements. It is included as a substance eligible for the dental caries health claim pertaining to sugar alcohols and dental caries.[38, 132] In some toothpastes and mouthrinses, D-tagatose has been used as an alternative to sorbitol, as a humectant and sweetener, maintaining moisture in the product as well as to improve taste. In addition, through its ability to reverse coaggregations of early and late colonisers of dental biofilms,[133] D-tagatose would appear to have some potential preventing and reducing plaque development and altering subgingival microbiota. This effect contrasts with sorbitol that showed little reversal effect and may offer the potential for conservative control of gingival and periodontal disease.[133] Apart from this incubation work, in terms of dental effects, the literature is sparse with only one report of plaque pH testing[50] in which a 10% solution of D-tagatose was found to be hypo-acidogenic, being only slowly fermented by oral bacteria. There are no reports of enamel slab or animal experiments, or clinical trials.

2.3.8 Xylitol

Ever since its discovery, many researchers have been impressed by xylitol's favourable dental properties and because of this, the dental literature is considerable.

It is well established that xylitol is non-cariogenic and attention has tended to focus on whether xylitol has caries-preventive (caries-inhibitory, cariostatic) or even anti-cariogenic (active therapeutic) properties.

A number of reviews have focussed on the value of xylitol in caries prevention.[134–145] Some reviewers have concluded that xylitol has a unique active role in caries prevention; others have been more cautious, saying the case is not yet proven. Areas for future xylitol studies to address current research gaps have been suggested,[144, 145] while the need to translate the existing evidence base into clear strategies for individual- and community-based caries prevention, particularly for high-risk groups, is a central theme in some reviews.[144, 146] Differentiation of the effects of chewing gum as opposed to any specific effects of polyols in chewing gums has been the subject of many reviews, most recently those by Burt[113] and Stookey.[147] In addition, two systematic reviews have been completed recently; one of sugar-free chewing gum[115] and one of polyol-containing chewing gums.[116] The results and conclusions of these systematic reviews are described in more detail in Section 2.3.5. Research gaps were identified that related to optimal dosing and relative polyol efficacy[116] and the need for confirmatory evidence based on well-designed randomised trials.[115] This conclusion contrasts with the earlier systematic review of dietary factors in the prevention of dental caries[114] in which 18 randomised or controlled clinical trials of polyol-based chewing gums met the inclusion criteria, which included at least 2 years follow-up and caries increment as a primary endpoint. This review concluded that the evidence for the use of sorbitol or xylitol in chewing gum was inconclusive but also highlighted the need for well-designed randomised clinical studies with adequate control groups and high compliance.

With regard to the evidence, the dental properties of xylitol have been investigated through a wide range of studies including *in vitro*, animal, *in situ* and *in vivo* experiments. Xylitol is not fermented by dental plaque as shown by incubation experiments,[79,104,148] and, in comparison with other polyols, xylitol is the least fermentable. There is also good evidence that acid production by plaque metabolising sugars is reduced by xylitol.[107,124,144,149–154] This seems to be adequately explained by a selective decrease in mutans streptococci in plaque exposed to xylitol and possibly by a decrease in plaque quantity. While it is clear that polyols in general reduce plaque and plaque adhesion by reducing the amount and type of polysaccharides formed by plaque bacteria, xylitol appears to be particularly effective, possibly by interfering with mechanisms of adhesion between plaque organisms, and between plaque organisms and the tooth surface.[52,66,135,144,155,156] Xylitol at a concentration of 1% and 3% has shown a clear inhibitory effect on experimental multi-species biofilms *in vitro*, preventing their formation, although the exact mechanism is still unclear.[157] Other evidence from well-controlled long-term clinical studies has indicated that xylitol decreases the growth of plaque, compared with sugars, and other polyols.[56,124,158–162] These studies included a trial of total sugar substitution (the Turku Study[158]) and trials of partial substitution (usually, chewing gum studies) as well as xylitol supplementation.[85,159,160,163–182]

The effect of xylitol and other sweeteners on tooth structure and the potential for remineralisation has been investigated *in vivo*,[183] in rat experiments[184–187] and *in situ*.[188] Remineralisation occurred in nearly all experiments where non-sugar sweeteners were used during the 'healing phase', but there was no clear indication that xylitol had a specific active effect compared with other non-sugar sweeteners when evaluated in these short-term studies. However, there is some evidence for a long-term preventive effect that appears to be due to persistent microbiological changes in the mouth, which are still evident 15 months[189] and 5 years[173] after xylitol use. This effect is in addition to the remineralising effects of calcium and phosphate in the saliva.

In some clinical trials, remineralisation has been measured by recording 'reversals' in pre-cavitation carious lesions.[158] The most dramatic results have been seen in chewing gum studies when xylitol gums have supplemented a sucrose-containing diet. The majority of these studies have shown substantial caries reductions being observed in those individuals consuming xylitol gum[138,166,167,170,173,190] although others have shown a lesser effect.[182,191] Significant reductions in caries development have also been observed when xylitol has been administered in confectionery form (other than chewing gum), either as a supplement to the diet,[192] or in partial substitution of dietary sugars.[163–165,169,193] Sugar-free gums are anti-cariogenic; they have an active therapeutic role in the remineralisation of early carious lesions. It has been suggested that both of the beneficial characteristics of chewing sugar-free gum – sugar substitution and saliva stimulation – could be responsible for this dental effect, rather than a specific polyol effect. This hypothesis has yet to be tested thoroughly as it requires a polyol-containing gum to be tested against a placebo gum rather than against no gum use; the option which currently predominates as a choice for a control in most chewing gum studies. Machiulskiene *et al.*[182] in their 3-year community intervention trial used a placebo gum along with polyol-containing gums and concluded that the caries preventive effect of chewing sugar-free gum related to the chewing process itself rather than being an effect of

> **Box 2.2** Specific effects of xylitol in reducing proportions of mutans streptococci in plaque.
>
> 1. Development of mutant xylitol-resistant strains, which may be less virulent in the oral environment.[137, 194–198]
> 2. Concentrations of ammonia and basic amino acids increase when plaque is exposed to xylitol, resulting in neutralisation of plaque acids.[199–201]
> 3. Xylitol can act in a bacteriostatic way: some strains of oral streptococci take up xylitol and convert it to xylitol-5-phosphate through the major route of sugar transport, the phosphoenolpyruvate phosphotransferase (PEP-PTS) system. The xylitol-5-phosphate inhibits glycolytic enzymes resulting in inhibition of cell growth and acid production.[137, 144, 202, 203]
> 4. Some streptococcal strains can take up xylitol, which participates in what is termed 'a futile xylitol-5-phosphate cycle' before being expelled from the bacterial cell.[203–207]

Source: Adapted from Maguire & Rugg-Gunn.[141]

gum sweeteners or additives, although the daily xylitol dose used was low at 2.9 g per day. Another chewing gum trial recently carried out in China has also used a placebo gum[87–89] and showed a specific and similar polyol effect for xylitol and maltitol. When xylitol and sorbitol chewing gums have been directly compared, xylitol gums appear to exhibit superior reductions in caries incidence and enhanced remineralisation compared with sorbitol gum groups. However, in a recent systematic review with meta-analyses,[116] a lack of high-quality head-to-head comparisons made it difficult to determine relative efficacy of different polyols in chewing gum.[116, 174, 175, 177, 178]

Where xylitol has been shown to have a dental effect, its properties appear to be better than other polyols. This is most probably due, in part, to both specific and non-specific effects of xylitol. The non-specific effect found with xylitol, as well as other polyols to a lesser extent, is a result of non-fermentability not encouraging bacterial growth. In addition, xylitol reduces proportions of mutans streptococci in plaque through specific effects (Box 2.2).

As well as the specific effects described in Box 2.2, the 'Mother and Child' study[208] together with its follow-up studies,[209, 210] and similar work by other researchers[211–214] have demonstrated an important caries-preventive property of xylitol. When consumed at an appropriate dosage by a mother at the critical period of mother–child transmission of oral flora, xylitol can reduce the transmission and colonisation of mutans streptococci to her child on a long-term basis. This action is mediated through xylitol's apparent specific effects on mutans streptococci and through the resultant reduction in caries development, xylitol can be described as acting in a caries-preventive manner.

There is ample evidence that the oral flora does not adapt to xylitol when tested over a prolonged period in humans.[77, 134, 162, 215, 216] An *in vitro* study has demonstrated that oral strains of lactobacilli can adapt to xylitol although the clinical significance of this is likely to be limited.[217] Any ability of a few organisms to ferment xylitol is negated by the inaction of other more numerous plaque organisms, so that no fall in plaque pH occurs on exposure to this polyol.[50, 95, 159, 218]

Another question that has arisen with regard to xylitol and other polyols concerns their possible additive effects in combination, as well as with other preventive agents. True synergy is an effect resulting from the combination of agents, which is greater than the simple addition of the effects of the agents alone. It is important that this definition is considered when the possible additive effects of polyols, including xylitol, used in combination with other therapeutic agents, is considered. There have been relatively few studies, the majority of these have studied possible synergistic effects between xylitol and sorbitol (almost exclusively using chewing gum as the vehicle) and a few studies have investigated synergy between xylitol and fluoride (mainly in toothpastes or mouthrinses) or xylitol and chlorhexidine (in mouthrinses or chewing gum). There is no clinical evidence for true synergy between xylitol and sorbitol, xylitol and fluoride, or xylitol and chlorhexidine. The combination of xylitol with sorbitol results in a diluted xylitol effect.[124, 159, 219–223] There is some evidence that xylitol and fluoride produce independent and additive effects,[187, 224–226] while more recent *in vitro* incubation experiments by Maehara *et al.*,[227] indicate that fluoride and xylitol together may have synergistic inhibitory effects on acid production by mutans streptococci. Several studies have evaluated fluoride toothpastes containing xylitol[160, 176, 200, 228–230] and shown an additive effect of these two ingredients, but no true synergy. From the limited number of studies of xylitol and chlorhexidine used in combination, xylitol appears to have a limited additive effect, but it is able to prolong chlorhexidine's suppressive effect on bacterial flora.[231–233] When used with triclosan in toothpaste, 10% xylitol showed significant reductions in plaque and saliva mutans levels at 6 months compared with a placebo, demonstrating an additive effect.[234] In an *in vitro* experiment, xylitol's use in a xylitol/chlorhexidine mouthrinse has recently been shown to have a statistically significant anti-vital effect on Streptococcus sanguis (an early biofilm coloniser), compared with xylitol and chlorhexidine alone.[235]

The debate continues on the xylitol dosage needed for a microbial and caries-preventive effect and any dose–response and frequency–response relationship.[143] To deliver a dose for a caries preventive effect, only partial substitution or supplementation of xylitol is required, and studies have shown that doses from less than 1 g per day in toothpaste, 5–20 g in confectionery and 2–10 g in chewing gums can produce positive dental benefits. Two studies, one an *in situ* study and the other a clinical chewing gum trial, have suggested a concentration-dependent response for xylitol.[125, 174, 177] The Belize chewing gum study[174, 177] concluded that chewing sugar-free gum five times per day was more effective than three times a day, suggesting a frequency and dose response as well as highlighting the importance of factors related to stimulated salivary secretion in the polyol-containing chewing gum effect. However, it is only relatively recently, as guidelines for the effective use of sugar substitutes (and xylitol in particular), to improve oral health have been developed in the United States and Europe,[45, 46, 49] that the issue of dose and frequency has started to be addressed. A series of prospective studies designed to determine the minimum effective amount and frequency of xylitol use and its dose- and frequency-response relationship with *S. mutans* or dental caries have been undertaken in the United States by Milgrom and co-workers in adults[236, 237] and children.[85] These studies have shown a linear reduction in mutans streptococci levels in plaque and saliva with increasing frequency of xylitol gum use at a constant daily dose of 10.32 g per day.[236] The dose-response relationship showed mutans streptococci levels decreased in plaque at 5 weeks and in plaque and unstimulated saliva at 6 months with increasing exposure to xylitol. A plateau or ceiling effect for a dose occurred between 6.88 g and 10.32 g per day. A xylitol dose exceeding 10.32 g per day was unlikely to increase effectiveness, while a dose of 3.44 g per day was unlikely to show changes in mutans streptococci levels. The school-based RCT[85] using polyol-based confectionery (gummy bear snacks) in 8-year olds observed reductions in *S. mutans*/sobrinus levels after 6-week consumption of gummy bear snacks containing xylitol (11.7 g per day or 15.6 g per day) or high-dose maltitol (44.7 g per day) and concluded that this dose form (ie. gummy bear snack) may be a suitable alternative to polyol-containing chewing gum in children. The same group have investigated the use of a paediatric topical oral syrup in 9–15 month infants to prevent dental caries in a high-risk group of children[192] and found a total daily dose of 8 g xylitol, administered two to three times per day during primary tooth eruption, was effective in preventing early childhood caries.

The range of vehicles for delivery of xylitol is now increasing as manufacturers respond to guidelines to promote oral health produced by government and professional bodies and clinicians start to include it in their clinical armamentaria to prevent disease.[143,238–240] The American Academy of Pediatric Dentistry's policy document adopted in 2006 and revised in 2010 entitled 'Policy on the Use of Xyltiol in Caries Prevention' is intended to assist oral health care professionals to make informed decisions about the use of xylitol-based products. There is also evidence that low-dose non-intentional xylitol exposure is occuring as it is added in small non-clinical amounts to various foods and children's vitamins,[237,241] although the implications of this are currently unclear. The range of uses for xylitol now encompasses substitution for free sugars in foods, drinks and confectionery, supplementation through chewing gum use, gummy snacks, lozenges and mints, as well as 'therapeutic' uses in toothpastes, mouthrinses, pacifiers, wafers and infant oral wipes or towlettes.[146] In view of the dental properties similar to xylitol's now being shown by other polyols, for example maltitol and erythritol, an ideal opportunity now exists to extend product development. Additional challenges include the need for novel and acceptable vehicles for populations with special or specific needs as well as products directed at a broader age range. The aim should be a comprehensive, cost-effective and dentally efficacious range of consumer- and professionally based products and applications. From the available evidence it can be concluded that: (1) xylitol is non-cariogenic; (2) xylitol in chewing gum is anti-cariogenic, as are other polyols in chewing gum; (3) the inhibition of mother/child transmission of cariogenic oral flora, leading to reduced caries development in young children, is caries preventive; (4) the dental properties of xylitol are superior to other polyols so far investigated thoroughly, it is likely to be due to a combination of several specific effects of xylitol as well as the general effects of polyols in sucrose substitution and saliva stimulation and (5) further dental research is necessary as some less well-researched polyols, for example maltitol and erythritol, appear to have similar dental properties to xylitol.

2.3.9 Key points from the dental evidence for reduced-calorie sweeteners and their use

1. There is a considerable amount of information on the effect of reduced-calorie sweeteners on dental health from various types of study – incubation experiments, plaque pH studies, enamel slab experiments, animal studies and human clinical trials. From these, it can be unequivocally concluded that polyols are non-cariogenic or virtually so.
2. Xylitol has been subjected to many studies of all types and can be classed as non-cariogenic. Xylitol is anti-cariogenic in chewing gum as are other polyols.
3. The dental properties of maltitol are more like xylitol than sorbitol. The little research that has been carried out on dental effects of erythritol suggests that it is non-cariogenic as is D-tagatose.
4. Concern has been expressed that the oral flora may adapt to polyols so that they lose their 'safe for teeth' property. The oral flora does not adapt to xylitol when tested over a 2-year period in humans. From the evidence available for the other polyols, frequent or

long-term use is unlikely to present any increased risk of dental caries in normal people. However, there is some evidence that frequent use of sorbitol may represent a small cariogenic risk in individuals with a very low salivary flow or those consuming high levels of sorbitol.
5. Rich sources of free sugars in the diet include confectionery, soft drinks, biscuits and cakes and substitution of these sugars by reduced-calorie sweeteners could reduce caries risk substantially. The caries preventive potential of sugars substitution has been demonstrated in smaller scale studies but not yet on an epidemiological scale.
6. The 'safe for teeth' concept is now thoroughly established with confectionery carrying a sugar-free, tooth-friendly logo found worldwide.
7. The growth of the use of sugar-free chewing gum has been a great success story. There seems little doubt that chewing polyol-containing sugarless gum benefits dental health positively, although the minimum daily dose and frequency required to obtain specific 'polyol effects' on mutans streptococci, dental plaque and caries incidence remains unclear.
8. Observational studies have shown an increased risk of developing dental caries in chronically sick children taking sugars-containing syrup medicines regularly and long term.[242–245] Polyol use has facilitated sugar substitution in these medicines and progress has been made towards sugar-free options for medicines prescribing and dispensing through sugar-free medicines campaigns.[246, 247]

2.4 HIGH-POTENCY (HIGH-INTENSITY) SWEETENERS

Little is known about the dental effects of high-potency sweeteners, but due to their composition, they are very unlikely to promote dental caries. As a consequence, the dental literature is sparse, and research has been directed towards the investigation of caries-inhibitory properties. However, since these low-calorie sweeteners provide a sweet taste with little volume, they are often combined with a bulking agent such as polydextrose, maltodextrin, resistant maltodextrin or other bulking polysaccharides, to improve their functional properties. It is important that the dental effects of these combinations are also considered. For example, sucralose-based sweeteners that contain bulking ingredients, which allow them to pour and measure more like granular sugar, have been found to have a cariogenic potential due to the presence of added fermentable carbohydrate in the form of maltodextrins.[248] Slow fermentation of maltodextrin has been shown to depress the pH of interdental plaque to 5.25 in a pH telemetry study of sweetening powder containing citrate, cyclamate, saccharin and maltodextrin.[50]

2.4.1 Acesulfame K

Acesulfame K has been reported to inhibit the growth of *S. mutans*,[249] to decrease acid formation from sucrose,[59] but not to affect the development of dental caries in rats,[250] although this latter study did show that while acesulfame K, saccharin and cyclamate did not inhibit dental caries in the rat separately, they did when used in combination. Further work by Brown and Best[251] showed a negative interaction of acesulfame K with hexitol metabolism by *S. mutans* as well as an apparent negative synergism between fluoride and acesulfame K. The mechanism for these synergistic negative effects was postulated as interference with the phosphoenolpyruvate (PEP)-based transport of sorbitol and mannitol into the bacterial cell.

2.4.2 Aspartame and Neotame

2.4.2.1 *Aspartame*

Aspartame is said to be non-fermentable, non-cariogenic in rats and to cause some inhibition of dental plaque organisms. The evidence for these dental effects is mainly from animal experiments[252–256] and one clinical trial of a sorbitol chewing gum, which also contained aspartame.[257]

Although the rat study by Tanzer and Slee[252] showed that aspartame did not inhibit dental caries in rats, Lout *et al.*[253] found that five times daily rinsing with 0.05% aspartame (at pH 3.0 – similar in pH and concentration to that found in carbonated beverages) did not potentiate caries activity in rats. In another rat study, sucralose, sorbitol and aspartame in drinking water induced little or no caries in super-infected desalivated rats, while, in contrast, sucrose and fructose induced extensive decay.[254] In a second experiment, where desalivated animals received Diet 2000 *ad libitum*, sucrose in solution promoted caries whereas sucralose, aspartame and saccharin were without effect.

Aspartame has been tested for its cariogenicity alone and in the presence of sucrose in young Sprague-Dawley rats inoculated with *S. mutans* and fed basal Diet 2000 with the additions of various combinations of sucrose and/or aspartame. The addition of 0.15% aspartame to 30% sucrose diet significantly reduced caries in comparison to rats fed only 30% sucrose diet, while in animals fed aspartame only, there was no caries.[255] In a further experiment,[256] rats receiving 0.15% aspartame and 0.30% aspartame alone developed no caries. Animals fed sucrose plus aspartame had statistically significantly lower caries than animals fed the same amounts of sucrose that led the authors to conclude that aspartame is non-cariogenic and anti-cariogenic, although this limited evidence suggests only a possible caries inhibition rather than a therapeutic effect and the clinical relevance of this requires corroboration.[258]

2.4.2.2 *Neotame*

Neotame's discovery and development is documented in the literature by Witt.[259] There are no studies of the dental effects of Neotame in the literature.

2.4.3 Cyclamate and saccharin

2.4.3.1 *Cyclamate*

There is one reference to cyclamate in the dental literature. This was a plaque pH study,[260] which compared the pH changes in dental plaque after rinsing with sugared (sucrose) or sugar-free (saccharin, cyclamate and sorbitol) versions of the same paediatric acetaminophen (paracetamol) solution. Plaque pH values showed a significant difference between groups after rinsing with mean pH values below 5.70 for 1 hour following rinsing with the sugared solution, whereas no mean pH value was below 5.80 with the sugar-free solution.

2.4.3.2 *Saccharin*

Saccharin has been reported to inhibit bacterial growth and metabolism of glucose-grown cariogenic and other streptococci *in vitro*.[249, 261–265] The mechanism for this is thought to be due to saccharin's inhibition of lactate dehydrogenase by competitive binding. Laboratory animal studies have shown less development of dental caries in rats fed a cariogenic diet supplemented with saccharin compared with controls,[252,262,266,267] although the studies by Siebert et al.[250] showed no caries inhibition with saccharin.

There have been two plaque pH experiments involving saccharin use in liquid oral medicines.[260,268] The study by Rekola[268] compared acid production from bacteria in dental plaque following rinsing with 10 syrupy medicines sweetened with sucrose, fructose, sorbitol, xylitol, saccharin or a combination of these. It concluded that xylitol, xylitol–saccharin and xylitol–sorbitol combinations in these medicines were non-acidogenic, sorbitol was hypo-acidogenic and sucrose and fructose were highly acidogenic. Mentes and co-workers, in another plaque pH study,[260] confirmed the hypo-acidogenicity of a sugar-free medicine containing saccharin, cyclamate and sorbitol, while Grobler et al.[269] found that diet drinks made with saccharin caused negligible falls in plaque pH.

An apparent negative synergism between the interaction of saccharin, acesulfame K and fluoride with hexitol metabolism by *S. mutans* has been described by Brown and Best,[251] based on their *in vitro* incubation experiments on the growth rate and growth yield of *S. mutans* NCTC 10449 in the presence of the hexitols sorbitol and mannitol. The mechanism postulated for this effect was an interference with the PEP (phosphoenolpyruvate)-dependent transport of sorbitol and mannitol into the bacterial cell.[260,269]

The clinical relevance of some these bacteriostatic and/or cariostatic properties ascribed to saccharin needs corroboration.[258]

2.4.4 Sucralose

A few studies have considered the dental effects of sucralose; a chlorinated derivative of sucrose. These studies have included incubation experiments, plaque pH studies and animal experiments.

In *in vitro* experiments,[270] sucralose, as a sole carbon source, was unable to support growth of ten strains of oral bacteria and dental plaque supporting the concept that it is non-cariogenic.

Sucralose has been shown to be non-cariogenic in experimental animals, with little or no caries developing in super-infected desalivated rats receiving sucralose through drinking water[271] or through their diet; Diet 2000 minus sucrose plus 93 mg% sucralose.[254] The severity of caries lesions in sucralose-fed rats after 35 days was significantly less than those in sucrose-fed rats. In further rat experiments, aimed at investigating the remineralising potential of various sweeteners,[187] all rats were exposed initially to a sucrose diet before being given a diet of sucralose, sorbitol, xylitol, sucrose or distilled water. Sucrose caused further caries progression, while removal of the cariogenic challenge allowed remineralisation to occur. The diets of sweeteners and distilled water led to remineralisation and the authors concluded that 'no (reduced or low-calorie) sweetening agent was superior to another in this respect'. Indeed, the follow-on diet of distilled water produced the lowest levels of caries.

Sucralose-based sweeteners that contain bulking ingredients, such as maltodextrins, which provide granular properties and allow them to pour and measure more like sucrose, do have cariogenic potential due to the presence of added fermentable carbohydrate.[248] These combination products (sucralose alone or in combination with maltodextrin or maltodextrin and dextrose) have been compared with sucrose both in aqueous solution[272] and in coffee,[273] using plaque pH testing *in vivo* following rinsing with these liquids. Rinsing with sucralose, or sucralose in combination with maltodextrin and/or dextrose (commercially available formulations, of sucralose) was less acidogenic than rinsing with a sucrose solution of equivalent sweetness, measured by mean area under the curve, although the inclusion of maltodextrin/dextrose with sucralose did produce sufficient acid to reduce the minimum plaque pH recorded close to that for sucrose. When sucralose in hot coffee was tested, the acidic nature of unsweetened coffee, which led to a modest pH depression, appeared to be blunted by sucralose. This study confirmed sucralose's non-acidogenicity and indicated that it may reduce the acidogenic potential of coffee, although the clinical relevance of this is unclear. Similar plaque pH experiments have been undertaken by Meyerowitz et al.[274] on sucralose (alone or bulked with maltodextrin or with maltodextrin/dextrose) compared with sucrose in iced tea. Rinsing with tea and sucrose resulted in significantly lower minimum pH, a greater drop in pH and larger area under the curve than rinsing with the solutions containing sucralose. A recent overview of the safety of sucralose and sucralose-mixture products[275] confirmed sucralose's lack of potential to induce dental caries but did not refer to the lack of dental data for sucralose-mixture products.

2.4.5 Other sweeteners

2.4.5.1 Neohesperidine DC

There are no reported studies of the dental effects of Neohesperidine DC, a modified glycoside, extracted from lemon and bitter orange peel.

2.4.5.2 Thaumatin

There is very little information on the dental effects of thaumatin. It has been shown to favour remineralisation of early caries in rats when added to starch to give a sweetness equivalent to sucrose.[276] In addition, Matsukubo and Takazoe,[277] in their review of sucrose substitutes, reported that mutans streptococci do not produce acid or insoluble glucan from thaumatin.[278] Otherwise, there have been no other studies on thaumatin's dental properties.

2.4.5.3 Carbohydrate sweeteners

Isomaltulose (Palatinose) is a reducing glucose–fructose disaccharide that has been investigated for its dental properties. The research has included plaque pH studies in which isomaltulose was compared with its hydrogenated derivative isomalt,[63] and with xylitol.[66] In incubation experiments it has been compared with sucrose, glucose, sorbitol and xylitol.[64,279] In rat experiments[280,281] caries scores were consistently lower in rats fed a palatinose diet compared with those fed a sucrose diet while, in a human study, significantly lower acid production was found following mouthrinsing with an isomaltulose solution compared with a glucose solution.[282] Overall, isomaltulose has been found to be hypoacidogenic compared with sucrose and glucose.

It has only been since 2008 that isomaltulose has been included as a substance eligible for the health claim 'non-cariogenic' in the United States although it has been used commercially in Japan since 1985 and is approved for use in the EU as a Novel Food.

2.4.6 Key points from the dental evidence for high-potency (high-intensity) sweeteners and their use

- The high-potency sweeteners can be considered as non-cariogenic and safe for teeth with regard to dental caries.
- High-potency sweeteners are used extensively in soft drinks, particularly carbonated soft drinks, excessive consumption of which can cause dental erosion.

As high-potency sweeteners are used at low concentrations, possible direct effects (inhibitory or otherwise) on bacterial metabolism are likely to be less important than indirect effects on caries through salivary stimulation.

2.5 BULKING AGENTS

2.5.1 Polydextrose

The general properties of polydextrose as a replacement for sugar and provider of dietary fibre in foods[283] have been reviewed by Murray,[284] Burdock and Flamm[285] and Mitchell et al.[286] Its dental properties have been investigated by Imfeld[50] and Setsu.[287] The evidence for the dental effects of polydextrose is confined to incubation experiments, rat experiments and plaque pH studies. In incubation experiments, polydextrose did not serve as a substrate for plaque formation and cellular aggregation of *S. mutans*, although some strains metabolised polydextrose at a low level.[50] In rat experiments polydextrose has been reported to be non-cariogenic, while plaque pH studies have shown it to be hypo-acidogenic.[288]

2.5.2 Fructose polymers

2.5.2.1 Short-chain fructo-oligosaccharides (sc-FOS)

Short-chain fructo-oligosaccharides (sc-FOS) are present naturally in plant foods such as banana, garlic, barley and honey. They consist of sucrose to which one, two or three additional fructose units are adjoined to produce ketose, nystose and fructofuranosylnystose respectively.

Most oligosaccharides are resistant to digestion in the upper gastro-intestinal tract, although a transport system involved in the uptake of oligosaccharides with 3–4 glucose units has been identified in *S. mutans*.[289,290] Many of the bacteria found in dental plaque are similar to species found in the large intestine, however, in view of the small amounts of naturally occurring oligosaccharides consumed and their intrinsic location within foods, it is very unlikely that, as consumed, they are harmful to teeth.[291]

In vitro studies have shown that sc-FOS are rapidly metabolised to acid following incubation with several strains of oral streptococci and they also induce plaque growth *in vitro* suggesting a potential cariogenicity. Indeed, Hartemink et al.[292] in a study of their degradation and fermentation, concluded that sc-FOS are potentially as cariogenic as sucrose. In studies that have investigated the acidogenicity of the sc-FOS nystose found in fructo-oligosccharide mixes,[59] and Neosugar, which contains nystose, the results have been equivocal, which may be due to between-strain differences in the ability of *S. mutans* to metabolise nystose.[291] Novel manufactured oligosaccharides are increasingly being included in foods designed to improve bone health[293] and improve absorption of trace elements[294] as well as being used as prebiotics in infant formula and fortified milks to improve intestinal flora, in particular the population of bifidobacteria in the colon.[295] Prebiotics are defined as 'non-digestible food ingredients that beneficially affect the host by selectively stimulating the growth and/or activity of one or a limited number of bacteria in the colon, and thus improve host health'.[296] Many of the species of colonic bacteria are also present in dental biofilms and in view of the increasing use of prebiotic and functional foods, further studies to determine the dental effects of fructo-oligosaccharides would appear to be necessary, especially for those products likely to be consumed frequently or between meals.

2.5.2.2 Inulins (longer chain fructo-oligosaccharides >20 units)

There are no studies on the dental properties of inulins, which belong to the fructans class of carbohydrates, although a review of the applications of inulin and oligofructose in health and nutrition recently reported 'fructans are non-cariogenic as they are not used by *S. mutans* to form acid and glucans that are responsible for dental caries'.[297]

2.5.2.3 Maltodextrins

Maltodextrins are short-chain glucose polymers with a dextrose equivalent of less than 20. They are frequently used in infant drinks and infant dried foods as an anti-caking agent.[298] Human plaque pH studies have shown maltodextrins to be as acidogenic as sucrose and therefore potentially cariogenic.[299]

One plaque pH study,[272] compared rinsing with an aqueous solution of sucralose (alone or in combination with maltodextrin or maltodextrin and dextrose) with an aqueous solution of sucrose. The mean pH minimum for the sucralose rinse was significantly higher than the sucralose/maltodextrin (SM), sucralose/maltodextrin/dextrose (SMD) and sucrose rinses. The mean difference between resting and minimum pH for the sucralose rinse was significantly lower when compared to the SM, SMD and sucrose rinses, although the differences for these parameters seen for the SM and SMD groups versus the sucrose group were also statistically significantly lower. Rinsing with aqueous solutions of sucralose, or sucralose in combination with maltodextrin and/or dextrose (commercially available formulations, of sucralose) was less acidogenic than rinsing with a sucrose solution of equivalent sweetness. In another plaque pH study in which 3 different maltodextrins (Dextrose Equivalents of 5.5, 14.0 and 18.5) made up as 10% solutions and three commercially available maltodextrin-containing childrens drinks were investigated, it was concluded that although maltodextrins appeared to be significantly less acidogenic than 10% sucrose, they could lead to a substantial drop in plaque pH and therefore may have a potential to demineralise enamel.[300]

2.5.2.4 Resistant maltodextrin

There are no studies on the dental properties of resistant maltodextrin.

2.5.3 Key points from the dental evidence for bulking agents

In view of the increasing use of these bulking agents, more research on their dental effects is necessary, since the existing literature is limited in both the number of studies and their scope.

2.6 SUMMARY

Reduced-calorie sweeteners can make an important contribution to oral health and this now being recognised by government, regulators and consumers as well as manufacturers. Most reduced-calorie sweeteners are virtually non-cariogenic, xylitol is non-cariogenic and erythritol as well as maltitol appear to have properties similar to xylitol. When used in sugarless chewing gums, polyols can be classed as anti-cariogenic, being non-cariogenic salivary stimulants, and encouraging the acid neutralising and remineralising effects of fast flowing alkaline saliva. Incorporation of reduced-calorie sweeteners into 'safe for teeth' confectionery products has provided a tooth-friendly alternative to sugar-containing confectionery products (a significant source of free sugars), and reinforced the dental health message regarding dietary sugars intake.

There has been good progress in the reformulation of sugar-containing medicines with prolonged oral clearance; substitution of cariogenic free sugars with reduced and low-calorie sweeteners has helped to reduce caries risk of individuals who require this form of medication regularly and long term. This remains especially important in those chronically sick individuals for whom dental caries or its treatment may lead to increased morbidity. Low-calorie sweeteners are non-cariogenic, however, their low bulk in food, drink and medicine products is often compensated by the inclusion of potentially cariogenic bulking agents, and it is important that any potential dental health risk is recognised by manufacturers, regulators and consumers.

Investigation of the dental properties of sweeteners and sugar alternatives needs to continue and should be encouraged for all potentially favourable new sweetening and bulking agents, especially those likely to be used in food, drink and medicine products likely to be consumed between meals.

REFERENCES

1. A. Lussi (ed). *Dental Erosion. From Diagnosis to Therapy*, Karger, Basel, Switzerland; 2006.
2. A. Rugg-Gunn, J. Nunn. *Nutrition, Diet and Oral Health*, 1st ed, Oxford University Press, Oxford, England; 1999.
3. E. Newbrun. Sucrose, the arch criminal of dental caries. *ASDC Journal of Dentistry for Children* 1969, 36, 239–248.
4. A.J. Rugg-Gunn. *Nutrition and Dental Health*. Oxford University Press, Oxford, England; 1993.
5. D.T. Zero. Sugars - The arch criminal? *Caries Research* 2004, 38(3), 277–285.
6. E. Al Hosani and A.J. Rugg-Gunn. Combination of low parental educational attainment and high parental income related to high caries experience in pre-school children in Abu Dhabi. *Community Dentistry and Oral Epidemiology* 2000, 26, 31–36.
7. G.N. Davies. Early childhood caries: a synopsis. *Community Dentistry and Oral Epidemiology* 1998, 26(Suppl. 1), 106–116.
8. T.M. Marthaler. Changes in dental caries 1953–2003. *Caries Research* 2004, 38, 173–181.
9. D.E. Barmes. Epidemiology of dental caries. *Journal of Clinical Periodontology*. 1977, 4, 80–93.
10. World Health Organization. *Dental caries levels at 12 years (Mimeograph of data from global data bank)*. World Health Organization, Geneva, Switzerland; 1989.
11. P.E. Petersen. Global policy for improvement of oral health in the 21st century–implications to oral health research of World Health Assembly 2007, World Health Organization. *Community Dentistry and Oral Epidemiology* 2009, 37(1), 1–8.
12. K.W. Stephen. Caries in young populations - worldwide. In: W.H. Bowen and L.A. Tabak (eds). *Cariology for the Nineties*, University Rochester Press, Rochester, NY; 1993, pp. 37–50.
13. M. O'Brien. *Childrens Dental Health in the United Kingdom 1993*. HMSO, London; 1994.

14. O. Fejerskov and V. Baelum. Changes in prevalence and incidence of the major oral diseases. In: B. Guggenheim and S. Shapiro (eds). *Oral Biology at the Turn of the Century*, Karger, Basel, Switzerland; 1998, pp. 1–11.
15. P. Petersen. Changing oral health profiles of children in Central and Eastern Europe - Challenges for the 21st century. *IC Digest (International College of Dentists - European Section)* 2003, 2, 12–13.
16. C. Stecksén-Blicks. K. Sunnegårdh and E. Borssén. Caries experience and background factors in 4-year-old children: time trends 1967–2002. *Caries Research* 2004, 38, 149–155.
17. N.B. Pitts, J. Boyles, Z.J. Nugent, N. Thomas and C.M. Pine. The dental caries experience of 5-year-old children in Great Britain (2005/6). Surveys co-ordinated by the British Association for the Study of Community Dentistry. *Community Dental Health* 2007, 24, 59–63.
18. N.B. Pitts and D.J. Evans. The dental caries experience of 5 year old children in the United Kingdom. Surveys co-ordinated by the British Association for the Study of Community Dentistry in 1995/96. *Community Dental Health* 1997, 14, 47–52.
19. US Department of Health and Human services. Healthy People 2010 Midcourse Review. Focus Area 21; Oral Health. http://www.healthindicators.gov/Indicators/Initiative_Healthy-People-2020_4/Selection; 2010.
20. T.M. Marthaler, J. Brunelle, M.C. Downer and G.J. Truin. The prevalence of dental caries in Europe 1990-1995. *Caries Research* 1996, 30(4), 237–255.
21. World Health Organization. WHO Oral Health Country/Area Profile Programme. Chosen Region: Europe - 'EURO'. http://www.mah.se/CAPP/Country-Oral-Health-Profiles/EURO/; 2007.
22. World Health Organization. WHO Oral Health Country/Area Profile Programme. Chosen Region: The Americas – 'AMRO'. http://www.mah.se/CAPP/Country-Oral-Health-Profiles/AMRO/; 2007.
23. World Health Organization. WHO Oral Health Country/Area Profile Programme. Chosen Region: Western Pacific - 'WPRO'. http://www.mah.se/CAPP/Country-Oral-Health-Profiles/WPRO/; 2007.
24. J.M. ten Cate. Biofilms, a new approach to the microbiology of dental plaque. *Odontology/The Society of the Nippon Dental University* 2006, 94(1), 1–9.
25. S.M Adair and Q. Xie. Antibacterial and probiotic approaches to caries management. *Advances in Dental Research* 2009, 21(1), 87–89.
26. P.D. Marsh. Microbial ecology of dental plaque and its significance in health and disease. *Advances in Dental Research* 1994, 8, 263–271.
27. P.D. Marsh. Are dental diseases examples of ecological catastrophes? *Microbiology* 2003, 149, 279–294.
28. D. Beighton. The complex oral microflora of high-risk individuals and groups and its role in the caries process. *Community Dentistry and Oral Epidemiology* 2005, 33, 248–255.
29. O. Fejerskov and E.A. Kidd (eds). *Dental Caries. The Disease and its Clinical Management*, 2nd edn, Blackwell Publishing Ltd., Oxford, England; 2008.
30. U. Arens. *Oral Health, Diet and other Factors*, Elsevier, Amsterdam, The Netherlands; 1998.
31. W.H. Bowen. Food components and caries. *Advances in Dental Research* 1994, 8(2), 215–220.
32. A. Sheiham. Dietary effects on dental diseases. *Public Heath Nutrition* 2001, 4(2B): 569–591.
33. UK Department of Health. *Report of Panel on Dietary Sugars. No. 37. Dietary sugars and human disease*, HMSO, London; 1989.
34. J. Mann. Sugar revisited – again. *Bulletin of the World Health Organization* 2003, 81(8), 552.
35. World Health Organization and the Food Agriculture Organization of the United Nations. *Diet, nutrition and the prevention of chronic diseases. Report of a joint WHO/FAO Expert Consultation, 28 January–1 February 2002. WHO Technical Report Series No. 916*. World Health Organization, Geneva, Switzerland; 2003.
36. A. Walker, *et al*. National Diet and Nutrition Survey: Young people aged 4 to 18 years. Volume 2: Report of the Oral Health Survey. London, TSO; 2000.
37. National Academy of Science and Institute of Medicine. *Dietary Reference Intakes for Energy, Carbohydrate, Fiber, Fat, Fatty Acids, Cholesterol, Protein and Amino Acids*, N.A.Press, Washington, DC; 2002.
38. American Dietetic Association. Position of the American Dietetic Association: use of nutritive and nonnutritive sweeteners. *Journal of the American Dietetic Association* 2004, 104(2), 255–275; erratum *Journal of the American Dietetic Association* 2004, 104(6), 1013.
39. M.D.C.M. Freire, G. Cannon and A. Sheiham. *Sugar and Health - An analysis of the recommendations on sugars and health in 115 authoritative scientific reports on food, nutrition and public health published throughout the world in thirty years between 1961-1991. Department Monograph Series, No. 1*. Department of Epidemiology and Public Health, University College, London; 1992.

40. N. Steyn, N. Myburgh and J. Nel. Evidence to support a food-based dietary guideline on sugar consumption in South Africa. *Bulletin of the World Health Organization* 2003, 81(8), 599–608.
41. L.M. Sreenby. Sugar availability, sugar consumption and dental caries. *Community Dentistry Oral Epidemiology* 1982, 10, 1–7.
42. UK Department of Health. *Dietary reference values for food energy and nutrients for the United Kingdom. Committee on Medical Aspects of Food Policy, COMA. Report No. 41.* HMSO, London; 1991.
43. World Health Organization. *Global Strategy on Diet, Physical Activity and Health*, World Health Organization, Geneva, Switzerland; 2004.
44. UK Department of Health. *Choosing Better Oral Health: An Oral Health Plan for England*, Department of Health; Dental and Opthalmic Services Division, London; 2005.
45. Scottish Intercollegiate Guidelines Network(SIGN). Preventing dental caries in children at high caries risk. A national clinical guideline. Publication Number 47; 2000.
46. Scottish Intercollegiate Guidelines Network(SIGN). Prevention and management of dental decay in the pre-school child. A national clinical guideline. Publication Number 83; 2005.
47. UK Department of Health and British Association for the Study of Community Dentistry. *Delivering Better Oral Health. An Evidence-based Toolkit for Prevention*, Department of Health and the British Association for the Study of Community Dentistry, London; 2007.
48. American Academy of Pediatric Dentistry. Guideline on caries-risk assessment and management for infants, children and adolescents. *Pediatric Dentistry* 2010, 32, 101–106.
49. American Academy of Pediatric Dentistry. Policy on the use of xylitol in caries prevention. *Pediatric Dentistry* 2010, 32(Special Issue), 36–38.
50. T. Imfeld. *Identificaton of Low Caries Risk Dietary Components*, Karger, Basel, Switzerland; 1983.
51. K.K. Makinen, K.P. Isotupa, T. Kivilompolo, P.L. Makinen, J. Toivanen and E. Soderling. Comparison of erythritol and xylitol saliva stimulants in the control of dental plaque and mutans streptococci. *Caries Research* 2001, 35(2), 129–135.
52. E.M. Soderling and A-M. Hietala-Lenkkeri. Xylitol and erythritol decrease adherence of polysaccharide-producing oral streptococci. *Current Microbiology* 2010, 60(1), 25–29.
53. J. Kawanabe, M. Hirasawa, T. Takeuchi, T. Oda and T. Ikeda. Noncariogenicity of erythritol as a substrate. *Caries Research* 1992, 26(5), 358–362.
54. K.K. Makinen, *et al.* The effect of polyol-combinant saliva stimulants on S. mutans levels in plaque and saliva of patients with mental retardation. *Special Care In Dentistry* 2002, 22(5), 187–193.
55. K. Makinen, *et al.* Similarity of the effects of erythritol and xylitol on some risk factors of dental caries. *Caries Research* 2005, 39(3), 207–215.
56. F. Gehring. Formation of acids by cariogenically important streptococci from sugars and sugar alcohols with special reference to isomaltitol and isomaltulose. *Zeitschrift Fur Ernahrungswissenschaft. Journal of Nutritional Sciences. Supplementa*, 1973, 15, 16–27.
57. F. Gehring and E.J. Karle. Sweetening agent, Palatinit under specific consideration as to microbiological and caries-prophylactic aspects. *Zeitschrift Fur Ernahrungswissenschaft* 1981, 20(2), 96–106.
58. J.S. van der Hoeven. Influence of disaccharide alcohols on the oral microflora. *Caries Research* 1979, 13, 301–306.
59. S.C. Ziesenitz and G. Siebert. The metabolism and utilization of polyols and other bulk sweeteners compared with sugar. In: T.H. Grenby (ed). *Developments in Sweeteners – 3*, Elsevier Applied Science, London; 1987, pp. 109–149.
60. F. Gehring and H-D. Hufnagel. Intra-und extraorale pH-Messungen an Zahnplaque des Menschen nach Spulungen mit einigen Zucker-und Saccharoseaustausch-stoff-Losungen. *Oralprophylaxe* 1983, 5, 13–19.
61. T. Takatsuka. Influence of Palatinit® and xylitol on demineralisation/remineralisation on bovine enamel. *Cariology Today* 2000, 1, 37–40.
62. T. Takatsuka, R.A. Exterkate and J.M. ten Cate. Effects of Isomalt on enamel de- and remineralization, a combined in vitro pH-cycling model and in situ study. *Clinical Oral Investigations* 2008, 12(2), 173–177.
63. T. Kaneko, T. Matsukubo, T. Yatake, Y. Muramatsu and Y. Takaesu. Evaluation of acidogenicity of commercial isomaltooligosaccharides mixture and its hydrogenated derivative by measurement of pH response under human dental plaque. *Bioscience, Biotechnology, and Biochemistry* 1995, 59(3), 372–377.

64. Y. Maki, *et al.* Acid production from isomaltulose, sucrose, sorbitol, and xylitol in suspensions of human dental plaque. *Caries Research* 1983, 17(4), 335–339.
65. N. Sasaki, K. Okuda, V. Topitsoglou and G. Frostell. Inhibitory effect of xylitol on the acid production activity from sorbitol by Streptococcus mutans and human dental plaque. *Bulletin of Tokyo Dental College* 1987, 28(1), 13–18.
66. P. Lingstrom, F. Lundgren, D. Birkhed, I. Takazoe and G. Frostell. Effects of frequent mouthrinses with palatinose and xylitol on dental plaque. *European Journal of Oral Sciences* 1997, 105(2), 162–169.
67. T.H. Grenby and M. Mistry. Laboratory studies of sweets re-formulated to improve their dental properties. *Oral Diseases* 1996, 2(1), 32–40.
68. E.J. Karle and F. Gehring. Palitinit- ein neuer Zuckeraustauschstoff und seine kariesprophylaktische Beuteilung. *Deutsche Zahnarztliche Zeitschrift* 1978, 3, 189–191.
69. E.J. Karle and F. Gehring. Studies on cariogenicity of sugar substitutes in xerostomized rats. *Deutsche Zahnarztliche Zeitschrift* 1979, 34(7), 551–554.
70. J.S. van der Hoeven. Cariogenicity of dissacharide alcohols in rats. *Caries Research* 1980, 61–66.
71. H. Schiweck and S.C. Ziesenitz. Physiological properties of polyols in comparison with easily metabolisable saccharides. In: T.H. Grenby (ed). *Advances in Sweeteners*, Blackie Academic and Professional, London; 1996.
72. S.C. Ziesenitz. Chapter 6, Basic structure and metabolism of isomalt. In: T.H. Grenby (ed). *Advances in Sweeteners*, Blackie Academic and Professional, London; 1996, pp. 109–133.
73. R. Havenaar, J.H. Huis in 't Veld, O.B. Dirks and J.D. de Stoppelaar. Some bacteriological aspects of sugar substitutes. In: B. Guggenheim (ed). *Health and Sugar Substitutes*, Karger, Basel, Switzerland; 1979.
74. T.H. Grenby and A. Phillips. Dental and metabolic effects of lactitol in the diet of laboratory rats. *British Journal of Nutrition* 1989, 61(1), 17–24.
75. T.H. Grenby and T. Desai. A trial of lactitol in sweets and its effects on human dental plaque. *British Dental Journal* 1988, 164, 383–387.
76. J.S. van der Hoeven. Cariogenicity of lactitol in program-fed rats. *Caries Research* 1986, 20, 441.
77. D. Birkhed, S. Edwardsson, M.L. Ahlden and G. Frostell. Effects of 3 months frequent consumption of hydrogenated starch hydrolysate (Lycasin), maltitol, sorbitol and xylitol on human dental plaque. *Acta Odontologica Scandinavica* 1979, 37(2), 103–115.
78. A. Izumitani, N. Sumi, Y. Kusamura, T. Ooshima and S. Sobue. Caries-inducing activity of maltitol-rich sweetener in experimental dental caries of rats. *Japanese Journal of Pedodontics* 1985, 23, 56–61.
79. S. Edwardsson, D. Birkhed and B. Mejare. Acid production from Lycasin, maltitol, sorbitol and xylitol by oral streptococci and lactobacilli. *Acta Odontologica Scandinavica* 1977, 35(5), 257–263.
80. F. Naito. The properties and application of Maltbit. *New Food Industry (Japan)* 1971, 13(9), 1–21.
81. T. Ikeda, M. Imai, Y. Sumitani and K. Narisawa. Maltitol fermentability and dental plaque formability of Strep. Mutans. *The 15th general meeting of the basic dentistry society,Tokyo, Japan*, 25–26 September 1973.
82. T. Ooshima, *et al.* Noncariogenicity of maltitol in specific pathogen-free rats infected with mutans streptococci. *Caries Research* 1992, 26(1), 33–37.
83. J. Rundegren, T. Koulourides and T. Ericson. Contribution of maltitol and Lycasin to experimental demineralisation in the human mouth. *Caries Research* 1980, 14, 67–74.
84. A. Maguire, A.J. Rugg-Gunn and W.G. Wright. Adaptation of dental plaque to metabolise maltitol compared with other sweeteners. *Journal of Dentistry* 2000, 28(1), 51–59; erratum *Journal of Dentistry* 2000, 28(7), 537 (Note: corrected to A.J. Rugg-Gunn and W.G. Wright).
85. K.A. Ly, C.A. Riedy, P. Milgrom, M. Rothen, M.C. Roberts and L. Zhou. Xylitol gummy bear snacks: a school-based randomized clinical trial. *BMC Oral Health* 2008, 8, 20.
86. S. Haresaku, T. Hanioka, A. Tsutsui, M. Yamamoto, T. Chou and Y. Gunjishima. Long-term effect of xylitol gum use on mutans streptococci in adults. *Caries Research* 2007, 41(3), 198–203.
87. V. Macioce, *et al.* Effect of maltitol or xylitol sugar-free chewing gums on salivary parameters related to dental caries development. *Caries Research* 2010, 44, 232.
88. C. Thabuis, *et al.* Effect of maltitol or xylitol sugar-free chewing gums on plaque parameters related to dental caries development. *Caries Research* 2010, 44, 211.
89. X.J. Li, B. Zhong, H.X. Xu, M. Yi and X.P. Wang. Comparative effects of the maltitol chewing gums on reducing plaque. *Hua Xi Kou Qiang Yi Xue Za Zhi* 2010, 28(5), 502–504.
90. G. Frostell, P.H. Keyes and R.H. Larson. Effect of various sugars and sugar substitutes on dental caries in hamsters and rats. *Journal of Nutrition* 1967, 93(1), 65–76.

91. G. Frostell. Effects of mouth rinses with sucrose, glucose, fructose, lactose, sorbitol and Lycasin on the pH of dental plaque. *Odontologisk Revy* 1973, 24(3), 217–226.
92. F. Bramstedt and K. Trautner. Sugar substitutes and biochemistry of dental plaque. *Deutsche Zahnarztliche Zeitschrift* 1971, 26, 1135–1144.
93. G. Frostell, *et al.* Substitution of Sucrose by Lycasin in Candy - Roslagen Study. *Acta Odontologica Scandinavica* 1974, 32(4), 235–254.
94. G. Frostell and D. Birkhed. Acid production from Swedish Lycasin (candy quality) and French Lycasin (80/55) in human dental plaques. *Caries Research* 1978, 12(5), 256–263.
95. D. Birkhed and S. Edwardsson. Acid production from sucrose substitutes in human dental plaque. In: B. Guggenheim (ed). *Health and Sugar Substitutes*, Karger, Basel, Switzerland; 1979, pp. 211–217.
96. A.J. Rugg-Gunn. Effect of Lycasin upon plaque pH when taken as a syrup or as a boiled sweet. *Caries Research* 1988, 22(6), 375–376.
97. T.H. Grenby. Dental properties of antiseptic throat lozenges formulated with sugars or Lycasin. *Journal of Clinical Pharmacy and Therapeutics* 1995, 20(4), 235–241.
98. G. Frostell. Lycasin as a sugar substitute. *Deutsche Zahnarztliche Zeitschrift* 1977, 32(5 Suppl. 1), S71–S75.
99. R. Havenaar, J.S. Drost, J.D. de Stoppelaar, J.H. Huis in't Veld and O.B. Dirks. Potential cariogenicity of Lycasin 80/55 in comparison to starch, sucrose, xylitol, sorbitol and L-sorbose in rats. *Caries Research* 1984, 18(4), 375–384.
100. T.H. Grenby. Dental effects of Lycasin in the diet of laboratory rats. *Caries Research* 1988, 22(5), 288–296.
101. S.A. Leach, J.A. Speechley, M.J. White and J.J. Abbott. Remineralisation in vivo by stimulating salivary flow with Lycasin: a pilot study. In: S.A. Leach (ed). *Factors Relating to Demineralisation and Remineralisation of The Teeth*. IRL Press, Oxford, England; 1986, pp. 69–79.
102. D. Birkhed, S. Edwardsson, S. Kalfas and G. Svensater. Cariogenicity of sorbitol. *Swedish Dental Journal* 1984, 8(3), 147–154.
103. D. Birkhed and A. Bar. Sorbitol and dental caries. *World Review of Nutrition and Dietetics* 1991, 65, 1–37.
104. M.L. Hayes and K.R. Roberts. The breakdown of glucose, xylitol and other sugar alcohols by human dental plaque bacteria. *Archives of Oral Biology* 1978, 23(6), 445–451.
105. L.S. Fosdick, H.R. Englander, K.C. Hoerman, I. Kesel and R.G. Kesel. A comparison of pH values of in vivo dental plaque after sucrose and sorbitol mouthwashes. *Journal of The American Dental Association* 1957, 55, 191–195.
106. T. Imfeld. Evaluation of the cariogenicity of confectionery by intra-oral wire-telemetry. *Helvetica Odontologica Acta* 1977, 21, 1–28.
107. O. Aguirre-Zero, D.T. Zero and H.M. Proskin. Effect of chewing xylitol chewing gum on salivary flow rate and the acidogenic potential of dental plaque. *Caries Research* 1993, 27(1), 55–59.
108. K.K. Park, B.R. Schemehorn, G.K. Stookey, H.H. Butchko and P.G. Sanders. Acidogenicity of high-intensity sweeteners and polyols. *American Journal of Dentistry* 1995, 8(1), 23–26.
109. M.A. Pollard. Potential cariogenicity of starches and fruits as assessed by the plaque-sampling method and an intraoral cariogenicity test. *Caries Research* 1995, 29(1), 68–74.
110. M.E. Curzon and M.A. Pollard. Integration of methods for determining the acido/cariogenic potential of foods: a comparison of several different methods. *Caries Research* 1996, 30(2), 126–131.
111. M.A. Pollard, *et al.* Acidogenic potential and total salivary carbohydrate content of expectorants following the consumption of some cereal-based foods and fruits. *Caries Research* 1996, 30(2), 132–137.
112. T. Koulourides, R. Bodden, S. Keller, L. Manson-Hing, J. Lastra and T. Housch. Cariogenicity of nine sugars tested with an inter-oral device in man. *Caries Research* 1976, 10, 427–441.
113. B.A. Burt. The use of sorbitol- and xylitol-sweetened chewing gum in caries control. *Journal of The American Dental Association* 2006, 137(2), 190–196.
114. P. Lingstrom, *et al.* Dietary factors in the prevention of dental caries: a systematic review. *Acta Odontologica Scandinavica* 2003, 61(6), 331–340.
115. S. Mickenautsch, S.C. Leal, V. Yengopal, A.C. Bezerra and V. Cruvinel. Sugar-free chewing gum and dental caries: a systematic review. *Journal of Applied Oral Science* 2007, 15(2), 83–88.
116. A. Deshpande and A.R. Jadad. The impact of polyol-containing chewing gums on dental caries: a systematic review of original randomized controlled trials and observational studies. *Journal of The American Dental Association* 2008, 139(12), 1602–1614.

117. W.M. Edgar. Sugar substitutes, chewing gum and dental caries–a review. *British Dental Journal* 1998, 184(1), 29–32.
118. A.R. Firestone and J.M. Navia. In vivo measurements of sulcal plaque pH in rats after topical applications of xylitol, sorbitol, glucose, sucrose, and sucrose plus 53 mM sodium fluoride. *Journal of Dental Research* 1986, 65(1), 44–48.
119. S.D. Hogg and A.J. Rugg-Gunn. Can the oral flora adapt to sorbitol? *Journal of Dentistry* 1991, 19(5), 263–271.
120. D. Birkhed, G. Svensater and S. Edwardsson. Cariological studies of individuals with long-term sorbitol consumption. *Caries Research* 1990, 24(3), 220–223.
121. S.M. Waaler, G. Rolla and S. Assev. Adaptation of dental plaque to sorbitol after 3 months' exposure to chewing gum. *Scandinavian Journal of Dental Research* 1993, 101(2), 84–86.
122. S. Assev and G. Rolla. Does the presence of xylitol in a sorbitol-containing chewing gum affect the adaptation to sorbitol by dental plaque? *Scandinavian Journal of Dental Research* 1994, 102(5), 281–283.
123. W.J. Loesche, N.S. Grossman, R. Earnest and R. Corpron. The effect of chewing xylitol gum on the plaque and saliva levels of Streptococcus mutans. *Journal of The American Dental Association* 1984, 108(4), 587–592.
124. E. Soderling, K.K. Makinen, C.Y. Chen, H.R. Pape Jr, W. Loesche and P.L. Makinen. Effect of sorbitol, xylitol, and xylitol/sorbitol chewing gums on dental plaque. *Caries Research* 1989, 23(5), 378–384.
125. K. Wennerholm, J. Arends, D. Birkhed, J. Ruben, C.G. Emilson and A.G. Dijkman. Effect of xylitol and sorbitol in chewing-gums on mutans streptococci, plaque pH and mineral loss of enamel. *Caries Research* 1994, 28(1), 48–54.
126. T.H. Grenby. Nutritive sucrose substitutes and dental health. In: T.H. Grenby, K.J. Parker and M.G. Lindley (eds). *Developments in Sweeteners*, Applied Science Publishers, London; 1983, pp. 51–58.
127. J.H. Shaw. Inability of low levels of sorbitol and mannitol to support caries activity in rats. *Journal of Dental Research* 1976, 55(3), 376–382.
128. K.W. Shyu and M.Y. Hsu. The cariogenicity of xylitol, mannitol, sorbitol and sucrose. *Proceedings of the National Science Council, Republic of China* 1980, 4, 21–26.
129. T.H. Grenby and J. Colley. Dental effects of xylitol compared with other carbohydrates and polyols in the diet of laboratory rats. *Archives of Oral Biology* 1983, 28(8), 745–758.
130. G.V. Levin, L.R. Zehner, J.P. Saunders and J.R. Beadle. Sugar substitutes: their energy values, bulk characteristics, and potential health benefits. *American Journal of Clinical Nutrition* 1995, 62(Suppl. 5), 1161S–1168S.
131. H. Bertelsen, B.B. Jensen and B. Buemann. D-tagatose–a novel low-calorie bulk sweetener with prebiotic properties. *World Review of Nutrition and Dietetics* 1999, 85, 98–109.
132. G.V. Levin. Tagatose, the new GRAS sweetener and health product. *Journal of Medicinal Food* 2002, 5(1), 23–36.
133. Y. Lu and G.V. Levin. Removal and prevention of dental plaque with D-tagatose. *International Journal of Cosmetic Science* 2002, 24(4), 225–234.
134. A. Bar. Caries prevention with xylitol. A review of the scientific evidence. *World Review of Nutrition and Dietetics* 1988, 55, 183–209.
135. E. Soderling and A. Scheinin. Perspectives on xylitol-induced oral effects. *Proceedings of The Finnish Dental Society* 1991, 87(2), 217–229.
136. T. Imfeld. Clinical caries studies with polyalcohols. A literature review. *Schweiz Monatsschr Zahnmed* 1994, 104, 941–945.
137. L. Trahan. Xylitol: a review of its action on mutans streptococci and dental plaque–its clinical significance. *International Dental Journal* 1995, 45(1 Suppl 1), 77–92.
138. A.A. Scheie amd O.B. Fejerskov. Xylitol in caries prevention: what is the evidence for clinical efficacy? *Oral Diseases* 1998, 4(4), 268–278.
139. K.K. Makinen. Xylitol-based caries prevention: is there enough evidence for the existence of a specific xylitol effect?. *Oral Diseases* 1998, 4(4), 226–230.
140. K.K. Makinen. The rocky road of xylitol to its clinical application. *Journal of Dental Research* 2000, 79(6), 1352–1355.
141. A. Maguire and A.J. Rugg-Gunn. Xylitol and caries prevention–is it a magic bullet? *British Dental Journal* 2003, 194(8), 429–436.
142. C. Van Loveren. Sugar alcohols: what is the evidence for caries-preventive and caries-therapeutic effects? *Caries Research* 2004, 38(3), 286–293.

143. K.A. Ly, P. Milgrom and M. Rothen. Xylitol, sweeteners, and dental caries. *Pediatric Dentistry* 2006, 28(2), 154–163; discussion 92–98.
144. E.M. Soderling. Xylitol, Mutans Streptococci and Dental Plaque. *Advances in Dental Research* 2009, 21, 74–78.
145. S. Twetman. Current controversies–is there merit? *Advances in Dental Research* 2009, 21(1), 48–52.
146. D.H. Kitchens. Xylitol in the prevention of oral diseases. *Special Care in Dentistry* 2005, 25(3), 140–144.
147. G.K. Stookey. The effect of saliva on dental caries. *Journal of The American Dental Association* 2008, 139(Suppl), 11S–17S.
148. D. Drucker and J. Verran. Comparative effects of the substance-sweeteners glucose, sorbitol, sucrose, xylitol and trichlorosucrose on lowering of pH by two oral Streptococcus mutans strains in vitro. *Archives of Oral Biology* 1980, 24, 965–970.
149. S. Twetman and C. Stecksen-Blicks. Effect of xylitol-containing chewing gums on lactic acid production in dental plaque from caries active pre-school children. *Oral Health and Preventive Dentistry* 2003, 1(3), 195–199.
150. C. Stecksen-Blicks, *et al.* Effect of xylitol on mutans streptococci and lactic acid formation in saliva and plaque from adolescents and young adults with fixed orthodontic appliances. *European Journal of Oral Sciences* 2004, 112(3), 244–248.
151. A. Sengun, Z. Sari, S.I. Ramoglu, S. Malkoc and I. Duran. Evaluation of the dental plaque pH recovery effect of a xylitol lozenge on patients with fixed orthodontic appliances. *Angle Orthodontist* 2004, 74(2), 240–244.
152. P. Lif Holgerson, I. Sjostrom, C. Stecksen-Blicks and S. Twetman. Dental plaque formation and salivary mutans streptococci in schoolchildren after use of xylitol-containing chewing gum. *International Journal of Paediatric Dentistry* 2007, 17(2), 79–85.
153. C.H. Splieth, M. Alkilzy, J. Schmitt, C. Berndt and A. Welk. Effect of xylitol and sorbitol on plaque acidogenesis. *Quintessence International* 2009, 40(4), 279–285.
154. G. Campus, M.G. Cagetti, G. Sacco, G. Solinas, S. Mastroberardino and P. Lingstrom. Six months of daily high-dose xylitol in high-risk schoolchildren: a randomized clinical trial on plaque pH and salivary mutans streptococci. *Caries Research* 2009, 43(6), 455–461.
155. M. Rekola. Comparative effects of xylitol- and sucrose-sweetened chew tablets and chewing gums on plaque quantity. *Scandinavian Journal of Dental Research* 1981, 89(5), 393–399.
156. E. Soderling, L. Alaraisanen, A. Scheinin and K.K. Makinen. Effect of xylitol and sorbitol on polysaccharide production by and adhesive properties of streptococcus mutans. *Caries Research* 1987, 21, 109–116.
157. C. Badet, A. Furiga and N. Thebaud. Effect of xylitol on an *in vitro* model of oral biofilm. *Oral Health and Preventive Dentistry* 2008, 6(4), 337–341.
158. A. Scheinin and K.K. Makinen. Turku Sugar Studies I-XXI. *Acta Odontologica Scandinavica* 1975, 33(70), 1–349.
159. D. Birkhed, S. Edwardsson, U. Wikesjo, M.L. Ahlden and J. Ainamo. Effect of 4 days consumption of chewing gum containing sorbitol or a mixture of sorbitol and xylitol on dental plaque and saliva. *Caries Research* 1983, 17(1), 76–88.
160. L.G. Petersson, D. Birkhed, A. Gleerup, M. Johansson and G. Jonsson. Caries-preventive effect of dentifrices containing various types and concentrations of fluorides and sugar alcohols. *Caries Research* 1991, 25(1), 74–79.
161. T.H. Grenby abd A.H. Bashaarat. A clinical trial to compare the effects of xylitol and sucrose chewing-gums on dental plaque growth. *British Dental Journal* 1982, 152(10), 339–343.
162. V. Topitsoglou, D. Birkhed, L.A. Larsson and G. Frostell. Effect of chewing gums containing xylitol, sorbitol or a mixture of xylitol and sorbitol on plaque formation, pH changes and acid production in human dental plaque. *Caries Research* 1983, 17(4), 369–378.
163. P. Alanen, P. Isokangas and K. Gutmann. Xylitol candies in caries prevention: results of a field study in Estonian children. *Community Dentistry and Oral Epidemiology* 2000, 28(3), 218–224.
164. A. Scheinin and J. Banoczy. Collaborative WHO xylitol field studies in Hungary. An overview. *Acta Odontologica Scandinavica* 1985, 43(6), 321–325.
165. A. Scheinin *et al.* Collaborative WHO xylitol field studies in Hungary. I. Three year caries activity in institutionalised children. *Acta Odontologica Scandinavica* 1985, 43, 327–347.
166. P. Isokangas. Xylitol chewing gum in caries prevention. A longitudinal study on Finnish school children. *Proceedings of The Finnish Dental Society* 1987, 83(Suppl. 1), 1–117.

167. D. Kandelman and G. Gagnon. Clinical results after 12 months from a study of the incidence and progression of dental caries in relation to consumption of chewing-gum containing xylitol in school preventive programs. *Journal of Dental Research* 1987, 66(8), 1407–1411.
168. P. Isokangas, P. Alanen, J. Tiekso and K.K. Makinen. Xylitol chewing gum in caries prevention: a field study in children. *Journal of The American Dental Association* 1988, 117(2), 315–320.
169. D. Kandelman, A. Bar and A. Hefti. Collaborative WHO xylitol field study in French Polynesia. I. Baseline prevalence and 32-month caries increment. *Caries Research* 1988, 22(1), 55–62.
170. P. Isokangas, J. Tiekso, P. Alanen and K.K. Makinen. Long term effect of xylitol chewing gum on dental caries. *Community Dentistry and Oral Epidemiology* 1989, 17, 200–203.
171. D. Kandelman and G. Gagnon. A 24-month clinical study of the incidence and progression of dental caries in relation to consumption of chewing gum containing xylitol in school preventive programs. *Journal of Dental Research* 1990, 69(11), 1771–1775.
172. L.M. Steinberg, F. Odusola and I.D. Mandel. Remineralizing potential, antiplaque and antigingivitis effects of xylitol and sorbitol sweetened chewing gum. *Clinical Preventive Dentistry* 1992, 14(5), 31–34.
173. P. Isogangas, K.K. Makinen, J. Tiekso and P. Alanen. Long-term effect of xylitol chewing gum in the prevention of dental caries: a follow-up 5 years after termination of a prevention program. *Caries Research* 1993, 27(6), 495–498.
174. K.K. Makinen, *et al.* Xylitol chewing gums and caries rates: a 40-month cohort study. *Journal of Dental Research* 1995, 74(12), 1904–1913.
175. K.K. Makinen, *et al.* Stabilisation of rampant caries: polyol gums and arrest of dentine caries in two long-term cohort studies in young subjects. *International Dental Journal* 1995, 45(1 Suppl. 1), 93–107.
176. J.L. Sintes, *et al.* Enhanced anticaries efficacy of a 0.243% sodium fluoride/10% xylitol/silica dentifrice: 3-year clinical results. *American Journal of Dentistry* 1995, 8(5), 231–235.
177. K.K. Makinen, P.P. Hujoel, C.A. Bennett, K.P. Isotupa, P.L. Makinen and P. Allen. Polyol chewing gums and caries rates in primary dentition: a 24-month cohort study. *Caries Research* 1996, 30(6), 408–417.
178. K.K. Makinen, *et al.* Conclusion and review of the Michigan Xylitol Programme (1986-1995) for the prevention of dental caries. *International Dental Journal* 1996, 46(1), 22–34.
179. K.K. Makinen, *et al.* Properties of whole saliva and dental plaque in relation to 40-month consumption of chewing gums containing xylitol, sorbitol of sucrose. *Caries Research* 1996, 30(3), 180–188.
180. K.K. Makinen, *et al.* Polyol-combinant saliva stimulants: a 4-month pilot study in young adults. *Acta Odontologica Scandinavica* 1998, 56(2), 90–94.
181. P.P. Hujoel, *et al.* The optimum time to initiate habitual xylitol gum-chewing for obtaining long-term caries prevention. *Journal of Dental Research* 1999, 78(3), 797–803.
182. V. Machiulskiene, B. Nyvad and V. Baelum. Caries preventive effect of sugar-substituted chewing gum. *Community Dentistry and Oral Epidemiology* 2001, 29(4), 278–288.
183. A. Scheinin, E. Soderling, U. Scheinin, R.L. Glass and M.L. Kallio. Xylitol-induced changes of enamel microhardness paralleled by microradiographic observations. *Acta Odontologica Scandinavica* 1993, 51(4), 241–246.
184. S.A. Leach and R.M. Green. Effect of xylitol-supplemented diets on the progression and regression of fissure caries in the albino rat. *Caries Research* 1980, 14(1), 16–23.
185. S.A. Leach and R.M. Green. Reversal of fissure caries in the albino rat by sweetening agents. *Caries Research* 1981, 15, 508–511.
186. R. Havenaar, J.H. Huis in 't Veld, J.D. de Stoppelaar and O.B. Dirks. Anti-cariogenic and remineralizing properties of xylitol in combination with sucrose in rats inoculated with Streptococcus mutans. *Caries Research* 1984, 18(3), 269–277.
187. W.H. Bowen and S.K. Pearson. The effects of sucralose, xylitol, and sorbitol on remineralization of caries lesions in rats. *Journal of Dental Research* 1992, 71(5), 1166–1168.
188. R.H. Manning, W.M. Edgar and E.A. Agalamanyi. Effects of chewing gums sweetened with sorbitol or a sorbitol/xylitol mixture on the remineralisation of human enamel lesions in situ. *Caries Research* 1992, 26(2), 104–109.
189. K.K. Makinen, *et al.* Thirty-nine-month xylitol chewing-gum programme in initially 8-year-old school children: a feasibility study focusing on mutans streptococci and lactobacilli. *International Dental Journal* 2008, 58(1), 41–50.
190. J.I. Virtanen, R.S. Bloigu and M.A. Larmas. Timing of first restorations before, during, and after a preventive xylitol trial. *Acta Odontologica Scandinavica* 1996, 54(4), 211–216.

191. H. Kovari, K. Pienihakkinen and P. Alanen. Use of xylitol chewing gum in daycare centers: a follow-up study in Savonlinna, Finland. *Acta Odontologica Scandinavica* 2003, 61(6), 367–370.
192. P. Milgrom, *et al.* Xylitol pediatric topical oral syrup to prevent dental caries: a double-blind randomized clinical trial of efficacy. *Archives of Pediatrics & Adolescent Medicine* 2009, 163(7), 601–607.
193. E. Honkala, S. Honkala, M. Shyama and S.A. Al-Mutawa. Field trial on caries prevention with xylitol candies among disabled school students. *Caries Research* 2006, 40(6), 508–513.
194. H.J. Beckers. Influence of xylitol on growth, establishment, and cariogenicity of Streptococcus mutans in dental plaque of rats. *Caries Research* 1988, 22(3), 166–173.
195. L. Trahan, E. Soderling, M.F. Drean, M.C. Chevrier and P. Isokangas. Effect of xylitol consumption on the plaque-saliva distribution of mutans streptococci and the occurrence and long-term survival of xylitol-resistant strains. *Journal of Dental Research* 1992, 71(11), 1785–1791; erratum *Journal of Dental Research* 1993, 72(1), 87–88.
196. L. Trahan, G. Bourgeau and R. Breton. Emergence of multiple xylitol resistant (fructose PTS-) mutants from human isolates of mutans streptococci during growth on dietary sugars in the presence of xylitol. *Journal of Dental Research* 1996, 75, 1892–1900.
197. E. Soderling, L. Trahan, T. Tammiala-Salonen and L. Hakkinen. Effects of xylitol, xylitol-sorbitol, and placebo chewing gums on the plaque of habitual xylitol consumers. *European Journal of Oral Sciences* 1997, 105(2), 170–177.
198. J.M. Tanzer, A. Thompson, Z.T. Wen and R.A. Burne. Streptococcus mutans: fructose transport, xylitol resistance, and virulence. *Journal of Dental Research* 2006, 85(4), 369–373.
199. K.K. Makinen and A. Scheinin. Turku sugar studies. VII. Principal biochemical findings on whole saliva and plaque. *Acta Odontologica Scandinavica* 1976, 34(5), 241–283.
200. K.K. Makinen, E. Soderling, H. Hurttia, O.P. Lehtonen and E. Luukkala. Biochemical, microbiologic, and clinical comparisons between two dentifrices that contain different mixtures of sugar alcohols. *Journal of The American Dental Association* 1985, 111(5), 745–751.
201. E. Soderling, J. Talonpoika and K.K. Makinen. Effect of xylitol-containing carbohydrate mixtures on acid and ammonia production in suspensions of salivary sediment. *Scandinavian Journal of Dental Research* 1987, 95(5), 405–410.
202. L. Trahan, M. Bareil, L. Gauthier and C. Vadeboncoeur. Transport and phosphorylation of xylitol by a fructose phoshotransferase system in Streptococcus mutans. *Caries Research* 1985, 19, 53–63.
203. H. Miyasawa-Hori, S. Aizawa and N. Takahashi. Difference in the xylitol sensitivity of acid production among Streptococcus mutans strains and the biochemical mechanism. *Oral Microbiology and Immunology* 2006, 21(4), 201–205.
204. A. Pihlanto-Leppala, E. Soderling and K.K. Makinen. Expulsion mechanism of xylitol 5-phosphate in Streptococcus mutans. *Scandinavian Journal of Dental Research* 1990, 98, 112–9.
205. E. Soderling and A. Pihlanto-Leppala. Uptake and expulsion of 14C-xylitol by xylitol-cultured Streptococcus mutans ATCC 25175 in vitro. *Scandinavian Journal of Dental Research* 1989, 97, 511–519.
206. A.H. Rogers, K.A. Pilowsky, P.S. Zilm and N.J. Gully. Effects of pulsing with xylitol on mixed continuous cultures of oral streptococci. *Australian Dental Journal* 1991, 36(3), 231–235.
207. L. Trahan, S. Neron and M. Bareil. Intracellular xylitol-phosphate hydrolysis and efflux of xylitol in Streptococcus sobrinus. *Oral Microbiology and Immunology* 1991, 6(1), 41–50.
208. E. Soderling, P. Isokangas, K. Pienihakkinen and J. Tenovuo. Influence of maternal xylitol consumption on acquisition of mutans streptococci by infants. *Journal of Dental Research* 2000, 79(3), 882–887.
209. P. Isokangas, E. Soderling, K. Pienihakkinen and P. Alanen. Occurrence of dental decay in children after maternal consumption of xylitol chewing gum, a follow-up from 0 to 5 years of age. *Journal of Dental Research* 2000, 79(11), 1885–1889.
210. E. Soderling, P. Isokangas, K. Pienihakkinen, J. Tenovuo and P. Alanen. Influence of maternal xylitol consumption on mother-child transmission of mutans streptococci: 6-year follow-up. *Caries Research* 2001, 35(3), 173–177.
211. I. Thorild, B. Lindau and S. Twetman. Salivary mutans streptococci and dental caries in three-year-old children after maternal exposure to chewing gums containing combinations of xylitol, sorbitol, chlorhexidine, and fluoride. *Acta Odontologica Scandinavica* 2004, 62(5), 245–250.
212. I. Thorild, B. Lindau and S. Twetman. Caries in 4-year-old children after maternal chewing of gums containing combinations of xylitol, sorbitol, chlorhexidine and fluoride. *European Archives of Paediatric Dentistry* 2006, 7(4), 241–245.

213. P. Oscarson, P. Lif Holgerson, I. Sjostrom, S. Twetman and C. Stecksen-Blicks. Influence of a low xylitol-dose on mutans streptococci colonisation and caries development in preschool children. *European Archives of Paediatric Dentistry* 2006, 7(3), 142–147.
214. Y. Nakai, C. Shinga-Ishihara, M. Kaji, K. Moriya, K. Murakami-Yamanaka and M. Takimura. Xylitol gum and maternal transmission of mutans streptococci. *Journal of Dental Research* 2010, 89(1), 56–60.
215. M. Larmas, A. Scheinin, F. Gehring and K.K. Makinen. Turku sugar studies XX. Microbiological findings and plaque index values in relation to 1-year use of xylitol chewing gum. *Acta Odontologica Scandinavica* 1976, 34(6), 381–396.
216. K.K. Makinen, E. Soderling, M. Hamalainen and P. Antonen. Effect of long-term use of xylitol on dental plaque. *Proceedings of The Finnish Dental Society* 1985, 81(1), 28–35.
217. C. Badet, B. Richard, M. Castaing-Debat, P.M. de Flaujac and G. Dorignac. Adaptation of salivary Lactobacillus strains to xylitol. *Archives of Oral Biology* 2004, 49(2), 161–4.
218. W.M. Edgar and M.W. Dodds. The effect of sweeteners on acid production in plaque. *International Dental Journal* 1985, 35(1), 18–22.
219. G. Frostell. Interaction between xylitol and sorbitol in plaque metabolism. *Swedish Dental Journal* 1984, 8(3), 137–146.
220. S. Assev and G. Rolla. Further studies on the growth inhibition of Streptococcus mutans OMZ 176 by xylitol. *Acta pathologica, microbiologica, et immunologica Scandinavica. Section B, Microbiology* 1986, 94, 97–102.
221. S. Assev and G. Rolla. Effects of xylitol/sorbitol combinations on bacterial growth and metabolism in Streptococcus sobrinus OMZ 176. *Apmis* 1993, 101(12), 933–938.
222. S. Assev and G. Rolla. Effect of xylitol-containing chewing gum on sorbitol metabolism in dental plaque. *European Journal of Oral Sciences* 1995, 103(2 (Pt 1)), 103–105.
223. S.M. Waler and G. Rolla. Xylitol, mechanisms of action and uses. *Norske Tannlaegeforenings Tidende* 1990, 100(4), 140–143.
224. A.H. Rogers. Effects of xylitol and fluoride on the response to glucose pulses of Streptococcus mutans T8 growing in continuous culture. *Oral Microbiology and Immunology* 1992, 7, 124–126.
225. B.T. Amaechi, S.M. Higham and W.M. Edgar. The influence of xylitol and fluoride on dental erosion in vitro. *Archives of Oral Biology* 1998, 43(2), 157–161.
226. B.T. Amaechi, S.M. Higham and W.M. Edgar. Caries inhibiting and remineralizing effect of xylitol in vitro. *Journal of Oral Science* 1999, 41(2), 71–76.
227. H. Maehara, Y. Iwami, H. Mayanagi and N. Takahashi. Synergistic inhibition by combination of fluoride and xylitol on glycolysis by mutans streptococci and its biological mechanism *Caries Research* 2005, 39, 521–528.
228. M.T. Smits and J. Arends. Influence of xylitol- and/or fluoride-containing toothpastes on the remineralization of surface softened enamel defects in vivo. *Caries Research* 1985, 19(6), 528–535.
229. J. Arends, M. Smits, J.L. Ruben and J. Christoffersen. Combined effect of xylitol and fluoride on enamel demineralization in vitro. *Caries Research* 1990, 24(4), 256–257.
230. J.L. Sintes, A. Elias-Boneta, B. Stewart, A.R. Volpe and J. Lovett. Anticaries efficacy of a sodium monofluorophosphate dentifrice containing xylitol in a dicalcium phosphate dihydrate base. A 30-month caries clinical study in Costa Rica. *American Journal of Dentistry* 2002, 15(4), 215–219.
231. T. Nuuja, J.H. Meurman, H. Murtomaa, S. Kortelainen and J. Metteri. The effect of a combination of chlorhexidine diacetate, sodium fluoride and xylitol on plaque wet weight and periodontal index scores in military academy cadets refraining from mechanical tooth cleaning for 7-day experimental periods. *Journal of Clinical Periodontology* 1992, 19(2), 73–76.
232. T. Nuuja, J.H. Meurman and H. Torkko. Xylitol and the bactericidal effect of chlorhexidine and fluoride on Streprococcus mutans and Streptococcus sanguis. *Acta Odontologica Scandinavica* 1993, 51, 109–114.
233. G.H. Hildebrandt and B.S. Sparks. Maintaining mutans streptococci suppression: with xylitol chewing gum. *The Journal of the American Dental Association*, 2000, 131(7), 909.
234. L. Jannesson, S. Renvert, P. Kjellsdotter, A. Gaffar, N. Nabi and D. Birkhed. Effect of a triclosan-containing toothpaste supplemented with 10% xylitol on mutans streptococci in saliva and dental plaque. A 6-month clinical study. *Caries Research* 2002, 36(1), 36–39.
235. E-M. Decker, G. Maier, D. Axmann, M. Brecx and C. von Ohle. Effect of xylitol/chlorhexidine versus xylitol or chlorhexidine as single rinses on initial biofilm formation of cariogenic streptococci. *Quintessence International* 2008, 39(1), 17–22.

236. K.A. Ly, P. Milgrom, M.C. Roberts, D.K. Yamaguchi, M. Rothen and G. Mueller. Linear response of mutans streptococci to increasing frequency of xylitol chewing gum use: a randomized controlled trial [ISRCTN43479664]. *BMC Oral Health* 2006, 6, 6.
237. P. Milgrom, K.A. Ly, M.C. Roberts, M. Rothen, G. Mueller and D.K. Yamaguchi. Mutans streptococci dose response to xylitol chewing gum. *Journal of Dental Research* 2006, 85(2), 177–181.
238. US Department of Health and Human Services. *Oral Health in America: A Report of the Surgeon General*.US Department of Health and Human Services, National Institute of Dental and Craniofacial Research, National Institutes of Health, Rockville, MD; 2000.
239. H. Lynch and P. Milgrom. Xylitol and dental caries: an overview for clinicians. *Journal of the California Dental Association* 2003, 31(3), 205–209.
240. P. Milgrom, D.T. Zero and J.M. Tanzer. An examination of the advances in science and technology of prevention of tooth decay in young children since the Surgeon General's Report on Oral Health. *Academic Pediatrics*, 2009, 9(6), 404–409.
241. M.C. Roberts, *et al.* How xylitol-containing products affect cariogenic bacteria. *Journal of The American Dental Association* 2002, 2002(133), 435–441.
242. I.F. Roberts and G.J. Roberts. Dental disease in chronically sick children. *Journal of Dentistry for Children* 1981, 48, 346–351.
243. D.J. Kenny and P. Somaya. Sugar load of oral liquid medications on chronically sick children. *Journal of the Canadian Dental Association* 1989, 55, 43–46.
244. R. Feigal, M.C. Gleeson, T.M. Beckman and M.E. Greenwood. Dental caries related to liquid medication intake in young cardiac patients. *Journal of Dentistry for Children* 1984, 51, 360–362.
245. A. Maguire, A.J. Rugg-Gunn and T.J. Butler. Dental health of children taking antimicrobial and non-antimicrobial liqiuid oral medication long-term. *Caries Research* 1996, 30, 16–21.
246. I.C. Mackie. Helping the consumer to make better choices of medicines. In: A.J. Rugg-Gunn (ed). *Sugarless - Towards the Year 2000*, Royal Society of Chemistry, Newcastle upon Tyne, England; 1993, pp. 104–111.
247. D.J. Evans, D. Howe, A. Maguire and A.J. Rugg-Gunn. The development and evaluation of a campaign to facilitate the prescribing and dispensing of sugar-free medicines among general medical practitioners and pharmacists in the north-east of England. *Community Dent Health* 1999, 16, 131–137.
248. I.D. Mandel and V.L. Grotz. Dental considerations in sucralose use. *Journal of Clinical Dentistry* 2002, 13(3), 116–118.
249. A.T. Brown, L.C. Breeding and W.C. Grantham. Interaction of saccharin and acesulfam with *Streptococcus Mutans*. *Journal of Dental Research* 1982, 61(Special Issue A), 191.
250. G. Siebert, S.C. Ziesenitz and J. Lotter. Marked caries inhibition in the sucrose-challenged rat by a mixture of non-nutritive sweeteners. *Caries Research* 1987, 21, 141–148.
251. A.T Brown and G.M. Best. Apparent synergism between the interaction of saccharin, acesulfame K, and fluoride with hexitol metabolism by Streptococcus mutans. *Caries Research* 1988, 22(1), 2–6.
252. J.M. Tanzer and A.M. Slee. Saccharin inhibits tooth decay in laboratory models. *Journal of The American Dental Association* 1983, 106(3), 331–3.
253. R.K. Lout, L.B. Messer, A. Soberay, K. Kajander and J. Rudney. Cariogenicity of frequent aspartame and sorbitol rinsing in laboratory rats. *Caries Research* 1988, 22(4), 237–241.
254. W.H. Bowen, S.K. Pearson and J.L. Falany. Influence of sweetening agents in solution on dental caries in desalivated rats. *Archives of Oral Biology* 1990, 35(10), 839–844.
255. S. Das, A.K. Das, R.A. Murphy and R. Worawongvasu. Aspartame and dental caries in the rat. *Pediatric Dentistry* 1991, 13(4), 217–220.
256. S. Das, A.K. Das, R.A. Murphy and S. Warty. Cariostatic effect of aspartame in rats. *Caries Research* 1997, 31(1), 78–83.
257. B.B. Beiswanger, A.E. Boneta, M.S. Mau, B.P. Katz, H.M. Proskin and G.K Stookey. The effect of chewing sugar-free gum after meals on clinical caries incidence. *Journal of The American Dental Association* 1998, 129(11), 1623–6.
258. T. Imfeld. Efficacy of sweeteners and sugar substitutes in caries prevention. *Caries Research* 1993, 27(Suppl. 1), 50–5.
259. J. Witt. Discovery and development of neotame. *World Review of Nutrition and Dietetics* 1999, 85, 52–57.
260. A. Mentes. pH changes in dental plaque after using sugar-free pediatric medicine. *Journal of Clinical Pediatric Dentistry* 2001, 25(4), 307–312.

261. H.A. Linke. Growth inhibition of glucose-grown cariogenic and other streptococci by saccharin in vitro. *Zeitschrift Fur Naturforschung. Section C. Journal of Biosciences* 1977, 32(9–10), 839–843.
262. H.A. Linke. Inhibition of dental caries in the inbred hamster by saccharin. *Annals of Dentistry* 1980, 39(4), 71–74.
263. T.H. Grenby and J.M. Bull. Action of saccharin on oral bacteria. *Caries Research* 1979, 13, 89.
264. H.A. Linke and J.S. Kohn. Inhibitory effect of saccharin on glycolytic enzymes in cell-free extracts of *Steptococcus mutans*. *Caries Research*, 1984, 18, 12.
265. T.H. Grenby and M.G. Saldanha. Studies of the inhibitory action of of intense sweeteners on oral micro-organisms relating to dental health. *Caries Research* 1986, 20, 7–16.
266. J.M. Tanzer, L. Grant and J. Ciarcia. Effects of bicarbonate-based dental powder, fluoride, and saccharin on dental caries and on Streptococcus sobrinus recoveries in rats. *Journal of Dental Research* 1987, 66(3), 791–794.
267. J.M. Tanzer, L. Grant and T. McMahon. Bicarbonate-based dental powder, fluoride, and saccharin inhibition of dental caries associated with Streptococcus mutans infection of rats. *Journal of Dental Research* 1988, 67(6), 969–972.
268. M. Rekola. In vivo acid production from medicines in syrup form. *Caries Research* 1989, 23(6), 412–416.
269. S.R. Grobler, G.N. Jenkins and D. Kotze. The effects of the composition and method of drinking of soft drinks on plaque pH. *British Dental Journal* 1985, 158, 293–296.
270. D.A. Young and W.H. Bowen. The influence of sucralose on bacterial metabolism. *Journal of Dental Research* 1990, 69(8), 1480–1484.
271. W.H. Bowen, D.A. Young and S.K. Pearson. The effects of sucralose on coronal and root-surface caries. *Journal of Dental Research* 1990, 69(8), 1485–1487.
272. L.M. Steinberg, F. Odusola, J. Yip and I.D. Mandel. Effect of aqueous solutions of sucralose on plaque pH. *American Journal of Dentistry* 1995, 8(4), 209–211.
273. L.M. Steinberg, F. Odusola and I.D. Mandel. Effect of sucralose in coffee on plaque pH in human subjects. *Caries Research* 1996, 30(2), 138–142.
274. C. Meyerowitz, E.P. Syrrakou and R.F. Raubertas. Effect of sucralose–alone or bulked with maltodextrin and/or dextrose–on plaque pH in humans. *Caries Research* 1996, 30(6), 439–444.
275. V.L. Grotz and I.C. Munro. An overview of the safety of sucralose. *Regulatory Toxicology and Pharmacology* 2009, 55(1), 1–5.
276. S.A. Leach, E.A. Agalamanyi and R.M. Green. Remineralisation of teeth by dietary means. In: S.A. Leach SA and W.M. Edgar (eds). *Demineralisation and Remineralisation of the Teeth*, IRL Press, Oxford, England; 1983, pp. 51–73.
277. T. Matsukubo and I. Takazoe. Sucrose substitutes and their role in caries prevention. *International Dental Journal* 2006, 56(3), 119–130.
278. T. Ikeda. Sugar substitutes: reasons and indications for their use. *International Dental Journal* 1982, 32(1), 33–43.
279. K. Ohta and I. Takazoe. Effect of isomaltulose on acid production and insoluble glucan synthesis by *Streptococcus mutans*. *Bulletin of Tokyo Dental College* 1983, 24, 1–11.
280. T. Ooshima, *et al*. Non-cariogenicity of the dissacharide palatinose in experimental caries of rats. *Infection and Immunity* 1983, 39, 43–49.
281. N. Sasaki, *et al*. Cariogenicity of isomaltulose (palatinose), sucrose and mixture of these sugars in rats infected with Streptococcus mutans E-49. *Swedish Dental Journal* 1985, 9, 149–155.
282. V. Topitsoglou, *et al*. Effect of frequent rinses with isomaltulose (Palatinose) solution on acid production in human dental plaque. *Caries Research* 1984, 18, 47–51.
283. S.A. Craig, J.F. Holden and M.Y. Khaled. Determination of polydextrose as dietary fiber in foods. *Journal of Aoac International* 2000, 83(4), 1006–1012.
284. P.R. Murray PR. Polydextrose. Polydextrose. In: G.G. Birch and M.G. Lindley (eds). *Low-calorie products*. Elsevier Applied Science, London; 1988, pp. 83–100.
285. G.A. Burdock and W.G. Flamm. A review of the studies of the safety of polydextrose in food. *Food and Chemical Toxicology* 1999, 37(2–3), 233–264.
286. H. Mitchell, M.H. Auerbach and F.K. Moppett. Polydextrose. In: L.O. Nabors (ed). *Alternative Sweeteners*, 3rd edn, Marcel Dekker, New York; 2001, pp. 499–518.
287. E. Setsu. Cariogenicity of polydextrose and refined polydextrose as a substrate. *Nichidai Koko Kagaku* 1989, 15(1), 1–11.

288. H.R. Muhlemann Polydextrose- ein kalorienarmer Zuckerersatzstoff. Zahnmedizinische Prufungen. *Swiss Dentistry* 1980, 1(3), 29.
289. R.R.B. Russell, J. Aduse-Opoku, I.C. Sutcliffe, A. Tao and J.J. Land Ferretti. A binding protein dependant transport system in *Streptococcus mutans* responsible for multiple sugar metabolism. *Journal of Biological Chemistry* 1992, 267, 4631–4637.
290. L. Tao, I.C. Sutcliffe, R.R.B. Russell and J.J. Ferretti. Cloning and expression of the multiple sugar metabolism (msm) operon of *Streptococcus mutans* in heterologous streptococcal hosts. *Infection and Immunity* 1993, 61, 1121–1125.
291. P.J. Moynihan. Update on the nomenclature of carbohydrates and their dental effects. *Journal of Dentistry* 1998, 26(3), 209–218.
292. R. Hartemink, M.C. Quataert, K.M. van Laere, M.J. Nout and F.M. Rombouts. Degradation and fermentation of fructo-oligosaccharides by oral streptococci. *Journal of Applied Bacteriology* 1995, 79(5), 551–557.
293. A. Nzeusseu, D. Dienst, V. Haufroid, G. Depresseux, J.P. Devogelaer and D.H. Manicourt. Inulin and fructo-oligosaccharides differ in their ability to enhance the density of cancellous and cortical bone in the axial and peripheral skeleton of growing rats. *Bone* 2006, 38(3), 394–399.
294. V. Ducros, *et al.* Influence of short-chain fructo-oligosaccharides (sc-FOS) on absorption of Cu, Zn, and Se in healthy postmenopausal women. *Journal of The American College Of Nutrition* 2005, 24(1), 30–37.
295. G. Moro, *et al.* Dosage-related bifidogenic effects of galacto- and fructooligosaccharides in formula-fed term infants *Journal of Pediatric Gastroenterology and Nutrition* 2002, 34, 291–295.
296. G.R. Gibson and M.B. Roberfroid. Dietary modulation of the human colonic micro-biota: introducing the concept of prebiotics. *Journal of Nutrition* 1995, 125, 1401–1412.
297. N. Kaur and A.K. Gupta. Applications of inulin and oligofructose in health and nutrition. *Journal of Biosciences* 2002, 27, 703–714.
298. J.F. Tahmassebi, M.S. Duggal, G. Malik-Kotrou and M.E. Curzon. Soft drinks and dental health: A review of the current literature. *Journal Of Dentistry* 2006, 34(1), 2–11.
299. P.J. Moynihan, M.E.L. Gould and N. Huntley. Effect of glucose polymers in water, milk and a milk substitute (Calogen) on plaque pH *in vitro*. *International Journal of Paediatric Dentistry* 1996, 6, 19–24.
300. G.R. Al-Khatib, M.S. Duggal and K.J. Toumba. An evaluation of the acidogenic potential of maltodextrins in vivo. *Journal Of Dentistry* 2001, 29(6), 409–414.

3 Digestive Health

Henna Röytiö, Kirsti Tiihonen and Arthur C. Ouwehand

Active Nutrition, DuPont Nutrition & Health, Kantvik, Finland

3.1 INTRODUCTION; PREBIOTICS, SWEETENERS AND GUT HEALTH

This chapter will discuss the influence of specific substances, prebiotics, on our health and well being and in particular, their influence on digestive health. Although prebiotics have varying degrees of sweetness, their main function in a product is usually to provide a health benefit, and any contribution they make to the overall sweetness of a product will be discussed elsewhere in this book.

3.2 INTESTINAL MICROBIOTA

The microbiota (formerly microflora) inhabiting our intestine, mainly the colon, plays a major role in our health and well being. We are often only aware of this when our microbiota is in disarray, for example during diarrhoea. However, the microbiota that inhabit the gastrointestinal tract and its activity significantly influence human health.

The intestinal microbiota consists of an estimated 400 culturable species,[1] however, molecular studies suggest the true number may well be over 1000 species as only a fraction of the colonic microbiota can be cultured with current techniques.[2] Although each of us has our own microbiota (which is as unique as a finger print), some species are common to virtually all humans and are considered to the core microbiota. Recently, it has been suggested that three core enterotypes exist and people are colonised predominantly by *Bacteroides* (enterotype 1), *Prevotella* (enterotype 2) or *Ruminococcus* (enterotype 3).[3] In an adult, the colonic microbiota may weigh as much as 1 kg and consist of 10^{14} microbes.[4] The microbiota provides us with a protective mechanism against incoming (pathogenic) microbes, competitive exclusion[5] and modulate our immune response.[6] The enormous variety and number of microbes provide a wide array of metabolic activity and some microbes may have protective functions in the gastrointestinal tract. The use of non-digestible carbohydrates (prebiotics) selectively stimulates the growth and/or activity of one or a limited number of bacteria in the microbiota. An effective prebiotic increases the overall well being of the gastrointestinal tract by positive changes in the microbiota, by reducing pH and decreasing the risk of gastrointestinal diseases.

3.3 GUT HEALTH

Although there is no clear definition of 'gut health',[7] in general it will include the maintenance of bodily health and the prevention of disease rather than curing existing gastrointestinal disease. The composition, but especially the activity, of the intestinal microbiota is of paramount importance in this respect. Several dietary strategies have, therefore, been devised to modulate the composition and activity of the microbiota, mainly by functional foods and dietary supplements containing prebiotics or probiotics or a combination of the two, and synbiotics.[8] Probiotics are 'live micro-organisms that when administered in adequate amounts confer a health benefit on the host'[9] and have long been used to modify the intestinal microbiota and otherwise provide health benefits. More recently, prebiotics have been introduced. The concept of prebiotics has been defined as, 'the selective stimulation of growth and/or activity(ies) of one or a limited number of microbial genus(era)/species in the gut microbiota that confer(s) health benefits to the host'.[10] Until recently, the modulation of the composition of the intestinal microbiota has received most attention, as it is relatively easy to determine. Microbiota activity in the intestine is more difficult to assess but is probably more important as this is responsible for the actual health effects and gives a mechanistic understanding of the actual health benefits; the ultimate aim.

3.4 PREBIOTICS VERSUS FIBRE

Most prebiotics are oligosaccharides, although sugar alcohols and other modified carbohydrates may also fulfil a similar function. Dietary fibre has many properties in common with prebiotics and the two terms are often used interchangeably. Prebiotics and fibre are, however, not the same. Both are not digested by human digestive enzymes, but prebiotics are fermented selectively in the colon and exert their health effects via the colonic microbiota. Dietary fibre, on the other hand, may not be fermented at all or is fermented by a wider range of colonic microbes and may exert health benefits in other ways, for example improved bowel function. Other physiological effects that are associated with increased consumption of dietary fibre include increased fermentation by colonic microbiota, immunomodulation, attenuated blood glucose and cholesterol levels. The current definition of dietary fibre is based on its chemical properties, and additionally, at least one of the previously mentioned beneficial physiological effects is needed to fulfil the definition.[11]

3.5 ENDOGENOUS PREBIOTICS

In addition to exogenous prebiotics, the body produces a large quantity of 'endogenous prebiotics', mainly in the form of mucins. Human milk also includes oligosaccharides that have an effect on the intestinal microbiota of the infant and thus, can be considered as the first and original prebiotic.[8]

3.5.1 Milk oligosaccharides

Breast-fed infants, unlike bottle-fed infants, have an intestinal ecosystem characterised by a strong prevalence of bifidobacteria, lactobacilli, *Enterococcus* and *Streptococcus*.[12,13] Owing to the characteristic differences between the microbiota of breast-fed and bottle-fed babies, the research of bifidogenic substrates in human milk has traditionally been substantial. Breast milk has been shown to contain at least 130 different human milk oligosaccharides (HMOs), of which galacto-oligosaccharides (GOS) are the most prominent and these substrates contribute to the development of the infant's intestinal microbiota.[8,14] HMOs are resistant to digestion in the upper gastrointestinal tract and reach the colon where they have bifidogenic properties. A considerable portion of HMOs is excreted in faeces, thus representing the model for prebiotics.[15] There is a general consensus that the intestinal microbiota that develops in breast-fed infants, through the bifidogenic properties of HMOs, is beneficial for the development of the immune system, in infancy and also later in life; the higher prevalence of bifidobacteria has been linked to a lower risk of gastrointestinal and respiratory tract infections in the breast-fed infants.[16] Human milk also contains other active constituents and protective nutrients. These include secreted antibodies, fatty acids, lactoferrin and homologues of host cell surface glycoconjugates that inhibit the growth and colonisation of the pathogenic microbes to epithelial surfaces and may prevent infection by acting as decoy receptors for pathogens.[8,17]

3.5.2 Secreted substrates in the gut

The intestine and associated organs secrete a wide array of compounds such as mucins, bile and secretory immuoglobulin A (SIgA). The mucus layer in the gut forms the first line of defence against invading substances and is formed of mucins secreted by goblet cells. Mucins are a rich source of carbohydrates and nitrogen and are metabolised by a variety of intestinal bacteria, therefore, possibly affecting the intestinal microbial ecosystem. Also, IgA and bile are degraded and/or hydrolysed by the intestinal microbiota. In general, this does not lead to release of energy, but is more likely to be a protective reaction by the microbiota. Conjugated bile acids have a strong antimicrobial activity; deconjugation has been suggested as a means of reducing antimicrobial activity. IgA is thought to prevent innate immune reactions against commensal bacteria and a large part of the faecal microbes are coated with IgA.[18]

3.6 PREBIOTICS

Prebiotics are usually carbohydrate compounds that are not absorbed or digested in the small intestine and reach the colon intact where they are fermented into short chain fatty acids (SCFA) and gas. Prebiotics have the ability to specifically stimulate the beneficial endogenous populations in the colon, such as lactobacilli and bifidobacteria, and they often decrease the colonic pH through increased production of SCFA (see Figure 3.1). Through the modification of the endogenous microbiota composition and/or activity, they should confer health benefits to the consumer.[10]

3.7 CURRENT PREBIOTICS

Most of the currently available prebiotics are extracted from natural sources, some of which are subsequently modified. Fructo-oligosaccharides (FOS) and inulin are the most studied prebiotics and other prebiotics with established clinical efficacy are GOS and lactulose.[19]

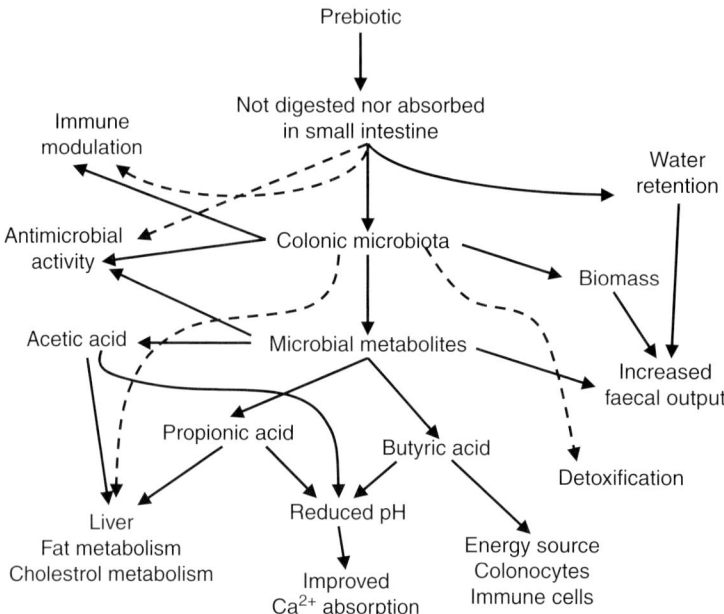

Fig. 3.1 Prebiotic pathways. A schematic representation of the direct and indirect interactions of prebiotic on different health benefits. Solid lines indicate well-established interactions, while dashed lines indicate possible interactions.

FOS and inulin are fructose polymers; large molecules with a degree of polymerisation (DP) greater than 20 are referred as inulin. Inulin is synthesised from sucrose or extracted from chicory roots. FOS can be produced by limited enzymatic hydrolysis of inulin into smaller molecules. FOS and inulin selectively stimulate the growth of bifidobacteria; this has been confirmed in numerous *in vitro* and *in vivo* studies. FOS is easily degraded in the proximal colon by gut microbes producing high levels of SCFAs.[8,20] The necessary dose for a bifidogenic effect for FOS is between 4 and 10 g per day[21,22] and for inulin is between 5 and 8 g per day,[23,24] although the magnitude of the increased proportion of bifidobacteria depends on the initial population levels.[23]

GOS (sometimes referred to as trans-galacto-oligosaccharides (TOS)) are naturally present in human and cow's milk, but can also be synthesised from lactose. *In vitro* studies have shown that GOS are utilised by bifidobacteria and lactobacilli and can decrease the growth of *Bacteroides* spp. and *Clostridium difficile*.[25,26] The same bifidogenic effects, with doses ranging from 3.6 to 10 g, have been seen in several human feeding trials, together with other beneficial modifications of the microbiota and microbial enzyme activities. Increased numbers of lactobacilli in the colon and decreased growth of, for example *Bacteroides* and *Clostridium histolyticum* group members have been demonstrated in healthy human volunteers.[27–29]

Lactulose is a semi-synthetic disaccharide composed of galactose and fructose. It is easily fermented by the gut micro-organisms yielding high concentrations of SCFA leading to a reduced pH[30] and is used medically for the treatment of constipation.[31] Lactulose has been shown to increase the growth of bifidobacteria *in vivo* and *in vitro*.[32–34] *In vitro*, lactulose also decreases *Clostridium* spp. growth[35] and the number of *Bacteroides* spp., and microbial enzyme activities and faecal toxic metabolites.[8] In humans,[36] a significant bifidogenic effect has been observed with a daily dose of 10 g lactulose. In another human intervention trial, the same dose was also shown to be bifidogenic, and furthermore, to decrease clostridial numbers.[37] A concomitant decrease in potentially pro-carcinogenic faecal enzymes and increased levels of SCFA has also been found together with beneficial modifications of the microbiota owing to lactulose consumption.[38]

New emerging prebiotic candidates also exist and include, for example, polydextrose (PDX), xylo-oligosaccharides (XOS), soy-oligosaccharides (SOS), lactitol and isomalto-oligosaccharides (IMO). There is preliminary supporting scientific evidence on the beneficial effects of these compounds.[39–41]

PDX is a randomly bonded glucose and sorbitol polymer that is slowly fermented by the colonic microbiota and thus available both in the proximal and distal colon, offering sustained carbohydrate fermentation.[42] In humans, fermentation of PDX leads to the growth of favourable microbiota and diminished putrefactive microbes, as well as enhanced production of SCFAs, lower faecal pH and suppressed production of carcinogenic metabolites (e.g. indole and *p*-cresol).[43]

XOS, xylobiose, xylotriose and xylo-tetraose are made up of xylose molecules.[39] XOS are fermented by bifidobacteria and the bifidogenic properties have been demonstrated *in vitro* and human studies.[8]

The main sugars in SOS, extracted from soy meal, are raffinose and the tetrasaccharide stachyose. Both have been shown to reach the colon and have bifidogenic properties. *In vitro* studies and human feeding studies have demonstrated an increase in number of faecal bifidobacteria and lactobacilli after consumption of SOS. Concomitant decreases in *Bacteroides* spp., clostridia and potentially toxic microbial metabolites have also been reported.[8]

IMO are a mixture of glucose oligomers, which can be produced from cornstarch.[24] All IMOs, isomaltose, isomaltotriose and panose, have been shown to support the growth of most species of bifidobacteria in pure culture and in the presence of human microbiota are selectively fermented by bifidobacteria. Fermentation results in a significant increase in lactic acid and acetate concentration.[8]

3.8 HEALTH BENEFITS

Prebiotic health benefits in the gastrointestinal tract can be mediated in several ways. Simple competition for space and nutrients by resident bacteria utilising prebiotics, and thus creating competitive advantage over the opportunistic pathogens, has an impact in resistance against intestinal infections. Such an effect is most probably greatest in the small intestine, where bacterial numbers are smaller and less species occupy the mucosa. Therefore, prebiotics that favour growth of lactobacilli, one of the main acid-resistant occupants of the small intestine,[44] should be a good choice for reducing the risk of infections. Metabolites produced by the resident bacteria may also reduce the risk of infections. Simple lowering of pH, especially by lactic acid producing bacteria, favours the more adapted and 'sturdy' residential microbes as opposed to opportunistic pathogens. In addition to decreasing susceptibility to infections, acidification of the lumen positively affects mineral availability.[45] Some indications of the role of FOS and inulin on improved mineral absorption (calcium and magnesium) from the colon and a consequently improved bone turnover rate exist,[46–48] although the evidence is mainly experimental and human studies have shown contradictory results.[49,50]

Prebiotics can also have a direct effect on reducing infection susceptibility by acting as receptor analogues for pathogens in the intestinal lumen. Manno-oligosaccharides are known to inhibit attachment of *Escherichia coli* to mucus; however, the prebiotic functions of manno-oligosaccharides have not been studied in detail.

Moving down the gastrointestinal tract to the colon, prebiotics can, by balancing the composition of the microbial community, influence development of inflammatory bowel disease (IBD), such as Crohn's disease or ulcerative colitis. IBD is thought to result from a combination of genetic predisposition, immunological factors and imbalanced gut microbiota.[51,52] However, the sequence of events resulting in the disease is unclear. Therefore, the suggested link between, for example, sulphate reducing bacteria (cause or consequence) and inflammatory processes in IBD patients remains unconfirmed. Some studies have reported impaired SCFA production in patients with IBD. Thus, treatment of patients with prebiotics may reduce the risk of disease and increase the duration of remission. For dietary treatment of patients at risk for ulcerative colitis, prebiotics that are complex in structure and can reach the distal colon unchanged appear promising.[53] For example, in patients with IBD, some evidence exists of the effectiveness of FOS and inulin to improve colonic inflammation and related symptoms[54] with a concomitant increase in intestinal *Bifidobacterium* numbers.[55] A non-fatal but nonetheless disabling condition, irritable bowel syndrome (IBS), is a functional disorder without clear underlying food intolerance or organic bowel disease that is characterised by diarrhoea, constipation or a combination of these.[84] The use of prebiotics to improve IBS may be controversial since dietary fibre may be ineffective or even sometimes worsen the situation, and furthermore, the reduction of rapid colonic bacterial fermentation may alleviate symptoms.[85] Prebiotics supporting controlled fermentation rate may, however, prove useful for patients with IBS. For example, GOS (in amounts between 3.5 g and 7.0 g per day) has been shown to help alleviate the symptoms of IBS by reducing flatulence and bloating while improving stool consistency in IBS patients during a 12-week study period, thus having significant therapeutic value for the IBS patients ($n = 44$).[56] Some evidence also exists of the preventive effects of GOS on the incidence and duration of traveller's diarrhoea[57] and in constipation in elderly people.[58]

Recent evidence suggests that inulin-type fructans might be able to modulate the release of hormones associated with appetite control (plasma GLP-1 and peptide YY), thereby affecting the gut–brain axis of appetite regulation.[59] Increased post-prandial satiety, lower hunger rates and a subsequently decreased total energy intake during meals following prebiotic consumption have been demonstrated,[60] and furthermore, satiety increasing effects were evident in both a short-term 2-week intervention[61] and in a long-term 12-month study in adolescent girls and boys ($n = 100$).[62]

The establishment of high levels of bifidobacteria in the infant gut has been attributed to the presence of galactose-containing oligosaccharides in human milk; thus, the inclusion of GOS to infant formulas has attracted considerable commercial interest[63] and these prebiotic supplemented formulas (GOS and FOS, 9:1) have been shown to modify the microbiota of the formula-fed babies in such a way that it resembles that of breast-fed babies.[64] Furthermore, the inclusion of prebiotics to infant formulas has been shown to reduce the cumulative incidence of intestinal and respiratory infections[65,66] and the cumulative incidence of atopic dermatitis in high-risk populations during a 6-month intervention[67] and the protective effects prevailed until at least the end of a 2-year follow-up period.[68] However, these studies have been criticised on the grounds for containing several weaknesses[69] and thus, the evidence of GOS (and FOS) on enhancing the immunity of infants remains inconclusive. A positive correlation between the GOS-induced bifidogenicity and improved activity of the immune system (phagocytosis, natural killer cell activity and production of anti-inflammatory and inflammatory cytokines) has also been demonstrated in healthy elderly volunteers.[29] Whether these changes in the immune system relate to actual improvements in immunity, that is reducing the risk and/or duration of an illness, remains to be elucidated.

PDX has been shown to be fermented slowly, probably owing to its complex structure with both $\alpha(1-4)$ and $\alpha(1-6)$ linkages requiring cooperation of several bacterial species for fermentation.[70] In healthy but constipated volunteers, PDX has been shown to shorten the orofaecal transit time and to improve stool consistency, which may relate to the increased SCFA production in the whole length of the colon.[71] The management of risk for colon cancer also requires prebiotics that can deliver benefits to the whole colon. PDX has been shown in challenge models to decrease tumour formation and balance deviated COX-2 gene expression, especially in the distal parts of colon.[72,73] COX-2 expression is necessary in inflammatory processes but chronic overexpression of COX-2 is not beneficial. Pharmaceutical treatment of chronic inflammatory diseases and colon cancer with specific COX-2 inhibitors are being developed, but serious systemic side effects of COX-2 inhibition have been described.[74] Prebiotic intervention may offer a local means to manage risk for colon diseases without tampering with the systemic balance of COX-2 activities. Induced butyrate production has been proposed as one of the possible mediating mechanisms. Butyrate is the main energy source for epithelial cell and can regulate differentiation processes.[75,76] Other SCFAs can also play a role, not forgetting that the vast bacterial metabolome (i.e. the total of metabolites that can be produced by the microbiota) most certainly includes other components that affect the host cell metabolism. Complex structure may not be a prerequisite since inulin-type fructans appear efficacious against cancer development.[77] Possibly, efficient stimulation of bifidobacteria is sufficient.[78] Clinical intervention studies in search for cancer risk decreasing dietary treatments, especially combinations of prebiotics and probiotics will shortly be carried out.[79] In addition to gastrointestinal effects, PDX has shown to attenuate post-prandial blood glucose and lipid responses in humans.[43,80,81] Also, inulin has shown to attenuate lipid responses in animal models.[82] PDX can help reduce caloric intake especially when it is used as a between-meal appetite suppressor.[83] The mechanism is thought to be related to prolonged gastric emptying but the role of microbial fermentation is also been studied.

More studies are needed to understand the structure–function differences between different prebiotics. In the future, prebiotic documentation of bifidogenic effects will be developed into more detailed understanding of the prebiotic mode of action and substrate requirement for a balanced complex gut microbial ecosystem allowing targeted use of prebiotics for desired health benefits.

Although bifidogenicity, that is the ability to stimulate the faecal *Bifidobacterium* levels, has often been used as an indicator for probiotic efficacy, it is not a valid biomarker for health benefits. Increases in bifidobacteria and lactobacilli levels fulfil part of the prebiotic definition but the actual health benefit for the consumer of the prebiotic still needs to be documented.[18]

3.9 SYNBIOTICS

The prebiotic concept aims to stimulate the growth and/or activity of endogenous beneficial microbes in the colon. An alternative approach is to consume exogenous beneficial microbes, that is probiotics. This will introduce microbes into the gastrointestinal tract with known

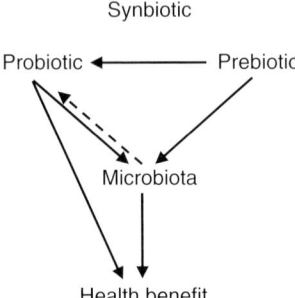

Fig. 3.2 A schematic representation of interactions between the components of a synbiotic (probiotic + prebiotic). Solid lines indicate well-established interactions, while dashed lines indicate possible interactions.

properties and health benefits. Obviously, one can combine both concepts and this will create a synbiotic environment.[10]

Strictly speaking, in this combination, the prebiotic part will modulate the composition and/or activity of the intestinal microbiota, which in turn will help the probiotic's functionality. The probiotic will exert its beneficial effects, either through the microbiota or directly on the host. The two components would not need to interact directly (see Figure 3.2). However, currently, synbiotics are often interpreted such that the prebiotic is a substrate for the probiotic and will facilitate the growth and activity of the probiotic.[86] (see Figure 3.2). Although the interpretation of providing a ready energy source makes sense from the probiotics point of view, it may have a negative impact on the functionality of the prebiotic when part of it has been utilised by the probiotic microbe. To date, it is not known to what extent a probiotic microbe is actually metabolically active in the intestine. It is, however, known that microbes in a different metabolic state have different influences on the host[87] and the combination of a probiotic strain and an appropriate prebiotic may give rise to a different response by the host.

It may thus be important to understand whether a synbiotic is a product where the prebiotic directly supports the probiotic or whether a synbiotic is a product where the prebiotic indirectly aids the probiotic through interaction with the intestinal microbiota (see Figure 3.2). In the former, one has to select for a substrate that suits the probiotic while in the latter, one is freer to select any combination of pre- and probiotics.

3.10 SAFETY CONSIDERATIONS

Most commonly, available prebiotics are naturally present in food or have been consumed for decades as part of the normal diet. As such, they are usually well tolerated. However, in purified form or through supplementation, potential overconsumption may cause undesirable side effects, such as abdominal pain, excessive flatus and, in extreme cases, diarrhoea.

Laxative effects of non-digestible carbohydrates include increased stool weight, water content and shortened colonic transit time (see Figure 3.1). While these mild laxative properties are often desirable, some side effects of this laxation may be unpleasant such as abdominal distension, flatus and diarrhoea. Intestinal tolerance of non-digestible carbohydrates differs from person to person and also depends on adaptation to the products. Adaptation may occur through our daily diet as most foods contain several fermentable carbohydrates such as inulin. Typically, toleration is better if the daily dose of supplemented prebiotic is divided into several portions.

Intestinal toleration of non-digestible compounds depends on their osmotic and fermentation effects. As the non-digestible substance pass from the small intestine to the colon, water absorption from the digesta in the lumen to the gut tissue is reduced, resulting in a higher moisture content in the faeces. Increased luminal volume stimulates the gut motility and the digesta transit. In the colon, the fermentation of the prebiotics decreases the luminal osmotic pressure by producing SCFAs that are partly absorbed by the host. However, if the fermentation is faster than absorption, the build-up of the fermentation products can also increase the luminal osmotic pressure. By liberating gases, fermentation can also cause flatus. Osmotic pressure is proportional to the number of molecules and thus osmotic diarrhoea is more probable with low molecular weight substances. Inulin, for example, is better tolerated than its partial hydrolysate FOS.[88] Recommended formulation doses for highly fermentable oligosaccharides are typically less than 10 g per portion. However, for an unknown reason, a small minority of the general population seems to be very sensitive to fermentable carbohydrates.[89] GOS are well tolerated at a daily dose of less than 10 g.[58,90] The intestinal side effects will increase at a daily dose of 15 g,[91] but a daily dose of 20 g is still acceptable.[92] Inulin and FOS have also been shown to be tolerated up to 20 g per day.[93] PDX is fermented slowly and incompletely[42] and can be tolerated up to 50 g per day.[94,95]

Although some prebiotics are modified forms of naturally existing substances most are extracted from natural sources. As such, they have been part of the human diet for many years, without known detrimental effects. Current prebiotics are, therefore, considered safe for consumption. Recent investigations have, however, suggested that inulin may enhance tumour formation in a mouse model for intestinal cancer,[96] although this is in contrast to earlier findings in the same model[97] and has been suggested to relate to the basal diet used in the former study.[9] In another report, FOS has been suggested to promote the colonisation and translocation of *Salmonella enterica* Typhimurium in rats.[98] The mechanism behind this may be the increased intestinal permeability that was observed in the rats consuming FOS.[99] The relevance of these observations for humans is not known, particularly the relationship between increased colonisation and a low-calcium diet. It indicates, however, that safety issues should not be neglected even where prebiotics form part of the normal diet.

3.11 CONCLUSION

The science behind prebiotic food ingredients has dramatically increased in the 15 years since the concept was introduced. Bifidogenicity and other beneficial changes in the intestinal microbiota are well established for many prebiotics. Now, clear health benefits in relation to these microbiota changes are also emerging. To understand the mechanisms behind these, continued investigations into the composition and activity of the intestinal microbiota as well as its interactions with the host tissues remain essential.

For the future, prebiotics with a greater selectivity may be possible, enabling the tailored design of prebiotics for different groups of individuals, so that optimal prebiotic efficacy can be achieved. Also other activities may be combined, for example, inhibition of pathogen adhesion and immune modulation.

There are still many opportunities for research in the field of prebiotics.

ACKNOWLEDGEMENTS

Partially funded by the National Technology Agency of Finland (TEKES).

REFERENCES

1. W.E.C. Moore and L.V. Holdeman. Human fecal flora: the normal flora of 20 Japanese-Hawaiians. *Applied Microbiology* 1974, 27, 961–979.
2. H. Hayashi, M. Sakamoto and Y. Benno. Phylogenetic analysis of the human gut microbiota using 16S rDNA clone libraries and strictly anaerobic culture-based methods. *Microbiology and Immunology* 2002, 46, 535–548.
3. M. Arumugam, *et al.* Enterotypes of the human gut microbiome. *Nature* 2011, 473, 174–180.
4. D.C. Savage. Microbial ecology of the gastrointestinal tract. *Annual Review of Microbiology* 1977, 31, 107–133.
5. I. Adlerberth, M. Cerquetti, I. Poilane, A. Wold and A. Collignon. Mechanisms of colonisation and colonisation resistance of the digestive tract. *Microbial Ecology in Health and Disease* 2000, 11(Suppl. 2), 223–239.
6. A. Ouwehand, E. Isolauri and S. Salminen. The role of the intestinal microflora for the development of the immune system in early childhood. *European Journal of Nutrition* 2002, 41(Suppl. 1), I/32–I/37.
7. O. Adolfsson, S.N. Meydani and R.M. Russel. Yogurt and gut function. *American Journal of Clinical Nutrition* 2004, 80, 245–456.
8. K.M. Tuohy, G.C.M. Rouzaud, W.M. Bruck and G.R. Gibson. Modulation of human gut microflora towards improved health using prebiotics – assessment of efficacy. *Current Pharmaceutical Design* 2005, 11, 75–90.
9. FAO/WHO. http://www.who.int/foodsafety/fs_management/en/probiotic_guidelines.pdf; 2002.
10. M. Roberfroid, *et al.* Prebiotic effects: metabolic and health benefits. *British Journal of Nutrition* 2010, 104(Suppl. 2), S1–S63.
11. K. Raninen, J. Lappi, H. Mykkänen and K. Poutanen. Dietary fiber type reflects physiological functionality: comparison of grain fiber, inulin and polydextrose. *Nutrition Foundation* 2011, 69(1), 9–21.
12. D. Mariat, O. Firmesse, F. Levenez, V. Guimaraes, H. Sokol, J. Dore, G. Corthier and J.P. Furet. The Firmicutes/Bacteroidetes ratio of the human microbiota changes with age. *BMC Microbiology* 2009, 9, 123.
13. G. Solis, C.G. de Los Reyes-Gavilan, N. Fernandez, A. Margolles and M. Gueimonde. Establishment and development of lactic acid bacteria and bifidobacteria microbiota in breast-milk and the infant gut. *Anaerobe*, 2010, 16, 307–310.
14. G. Boehm, J. Jelinik, B. Stahl, K. van Laere, J. Knol, S. Fanaro, G. Moro and V. Vigi. Prebiotics in infant formulas. *Journal of Clinical Gastroenterology* 2004, 38(Suppl. 2), S76–S79.
15. G.V. Coppa, S. Bruni, L. Morelli, S. Soldi and O. Gabrielli. The first prebiotics in humans: 'human milk oligosaccharides'. *Journal of Clinical Gastroenterology* 2004, 38(Suppl. 2), S80–S83.
16. D.S. Newburg and W.A. Walker. Protection of the neonate by the innate immune system of developing gut and of human milk. *Pediatric Research* 2007, 61, 2–8.
17. D.S. Newburg. Oligosaccharides in human milk and bacterial colonization. *Journal of Pediatric Gastroenterology And Nutrition* 2000, 30(Suppl. 2), S8–S17.
18. A. Ouwehand, M. Derrien, W. de Vos, K. Tiihonen and N. Rautonen. Prebiotics and other microbial substances for gut functionality. *Current Opinion in Biotechnology* 2005, 16, 212–217.
19. J.A.E. van Loo. Prebiotics promote good health. The basis, the potential, and the emerging evidence. *Journal of Clinical Gastroenterology* 2004, 38(Suppl. 2), S70–S75.
20. T.S. Manning and G.R. Gibson. Microbial-gut interactions in health and disease. *Best Practice and Research. Clinical Gastroenterology* 2004, 18, 287–298.
21. A. Costabile, S. Kolida, A. Klinder, E. Gietl, M. Bauerlein, C. Frohberg, V. Landschutze and G.R. Gibson. A double-blind, placebo-controlled, cross-over study to establish the bifidogenic effect of a very-long-chain inulin extracted from globe artichoke in healthy human subjects. *British Journal of Nutrition* 2010, 104, 1007–1017.

22. R.K. Buddington, C.H. Williams, S.C. Chen and S.A. Witherly. Dietary supplement of neosugar alters the fecal flora and decreases activities of some reductive enzymes in human subjects. *American Journal of Clinical Nutrition* 1996, 63, 709–716.
23. S. Kolida and G.R. Gibson. Prebiotic capacity of inulin-type fructans. *Journal of Nutrition* 2007, 137, 2503S–2506S.
24. S. Kolida, D. Meyer and G.R. Gibson. A double-blind placebo-controlled study to establish the bifidogenic dose of inulin in healthy humans. *European Journal of Clinical Nutrition* 2007, 61, 1189–1195.
25. R.J. Palfaram, G.R. Gibson and R.A. Rastall. Effect of pH and dose on the growth of gut bacteria on prebiotic carbohydrates *in vitro*. *Anaerobe* 2002, 8, 287–292.
26. M.J. Hopkins and G.T. Macfarlane. Nondigestible oligosaccharides enhance bacterial colonization resistance against *Clostridia difficile in vitro*. *Applied and Environmental Microbiology* 2003, 69, 1920–1927.
27. K.M. Tuohy, G.C. Rouzaud, W.M. Bruck and G.R. Gibson. Modulation of the human gut microflora towards improved health using prebiotics–assessment of efficacy. *Current Pharmaceutical Design* 2005, 11, 75–90.
28. F. Depeint, G. Tzortzis, J. Vulevic, K. I'anson and G.R. Gibson. Prebiotic evaluation of a novel galactooligosaccharide mixture produced by the enzymatic activity of Bifidobacterium bifidum NCIMB 41171, in healthy humans: a randomized, double-blind, crossover, placebo-controlled intervention study. *American Journal of Clinical Nutrition* 2008, 87, 785–791.
29. J. Vulevic, A. Drakoularakou, P. Yaqoob, G. Tzortzis and G.R. Gibson. Modulation of the fecal microflora profile and immune function by a novel trans-galactooligosaccharide mixture (B-GOS) in healthy elderly volunteers. *American Journal of Clinical Nutrition* 2008, 88, 1438–1446.
30. K.M. Khan and C.A. Edwards. Effect of substrate concentration on short chain fatty acid production in *in vitro* cultures of human faeces with lactulose, a rapidly fermented carbohydrate. *Microbial Ecology in Health and Disease* 2002, 14, 160–164.
31. M. Saarela, K. Hallamaa, T. Mattila-Sandholm and J. Mättö. The effect of lactose derivates lactulose, lactitol and lactobionic acid on the functional and technological properties of potentially probiotic *Lactobacillus* strains. *International Dairy Federation* 2003, 13, 291–302.
32. Y. Bouhnik, A. Attar, F.A. Joly, M. Riottot, F. Dyard and B. Flourié. Lactulose ingestion increases faecal bifidobacterial counts: a randomized double-blind study in healthy humans. *European Journal of Clinical Nutrition* 2004, 8, 462–466.
33. K.M. Tuohy, C.J. Ziemer, A. Klinder, Y. Knöbel, B.L. Pool-Zobel and G.R. Gibson. A human volunteer study to determine the prebiotic effects of lactulose powder on human colonic microbiota. *Microbial Ecology in Health and Disease* 2002, 14, 165–173.
34. M. Bielecka, E. Biedrzycka, A. Majkowska, J. Juskiewitcz and M. Wróblewska. Effect of non-digestible oligosaccharides on gut microecosystem in rats. *Food research international* 2002, 35, 139–144.
35. C.E. Rycroft, M.R. Jones, G.R. Gibson and R.A. Rastall. A comparative *in vitro* evaluation of the fermentation properties of prebiotic oligosaccharides. *Journal of Applied Microbiology* 2001, 91, 878–887.
36. Y. Bouhnik, A. Attar, F.A. Joly, M. Riottot, F. Dyard and B. Flourie. Lactulose ingestion increases faecal bifidobacterial counts: a randomised double-blind study in healthy humans. *European Journal of Clinical Nutrition* 2004, 58, 462–466.
37. K. Tuohy, C.J. Ziemer, A. Klinder, Y. Knöbel, B.L. Pool-Zobel and G.R. Gibson. A human volunteer study to determine the prebiotic effects of lactulose powder on human colonic microbiota. *Microbial Ecology in Health and Disease* 2002, 14, 165–173.
38. J. Ballongue, C. Schumann and P. Quignon. Effects of lactulose and lactitol on colonic microflora and enzymatic activity. *Scandinavian Journal of Gastroenterology. Supplement* 1997, 222, 41–44.
39. G.T. Macfarlane and J.H. Cummings. Probiotics and prebiotics: can regulating the activities of intestinal bacteria benefit health? *British Medical Journal* 1999, 318, 999–1003.
40. H. Mäkeläinen, S. Forssten, M. Saarinen, N. Rautonen and A.C. Ouwehand. Xylo-oligosaccharides enhance the growth of bifidobacteria and *Bifidobacterium lactis* in a simulated colon model. *Beneficial Microbes*, 2010, 1, 81–91.
41. H. Mäkeläinen, J. Stowell, A.C. Ouwehand and N. Rautonen. Xylo-oligosaccharides and lactitol promote the growth of *Bifidobacterium lactis* and Lactobacillus species in pure cultures. *Beneficial Microbes* 2010, 1, 139–146.
42. S.J. Lahtinen, K. Knoblock, A. Drakoularakou, M. Jacob, J. Stowell, G. Gibson and A.C. Ouwehand. Effect of molecule branching and glycosidic linkage on the degradation of polydextrose by gut microbiota. *Bioscience, Biotechnology, and Biochemistry* 2010, 74, 2016–2021.

43. Z. Jie, L. Bang-yao, X. Ming-jie, L. Hai-wei, Z. Zu-kang, W. Ting-song and S.A.S. Craig. Studies on the effects on polydextrose intake on physiologic functions in Chinese people. *American Journal of Clinical Nutrition* 2000, 72, 1503–9.
44. W.L. Hao and Y.K. Lee. Microflora of the gastrointestinal tract: a review. *Methods in Molecular Biology* 2004, 268, 491–502.
45. T.P. Trinidad, T.M. Wolever and L.U. Thompson. Effect of acetate and propionate on calcium absorption from rectum and distal colon of humans. *American Journal of Clinical Nutrition* 1996, 63, 574–578.
46. E.G. van den Heuvel, T. Muys, W. van Dokkum and G. Schaafsma. Oligofructose stimulates calcium absorption in adolescents. *American Journal of Clinical Nutrition* 1996, 69, 544–548.
47. H. Younes, C. Coudray, J. Bellanger, C. Demigne, Y. Rayssiguier and C. Remesy. Effects of two fermentable carbohydrates (inulin and resistant starch) and their combination on calcium and magnesium balance in rats. *British Journal of Nutrition* 2001, 86, 479–485.
48. L. Holloway, S. Moynihan, S.A. Abrams, K. Kent, A.R. Hsu and A.L. Friedlander. Effects of oligofructose-enriched inulin on intestinal absorption of calcium and magnesium and bone turnover markers in postmenopausal women. *British Journal of Nutrition* 2007, 97, 365–372.
49. M. Tahiri, J.C. Tressol, J. Arnaud, F.R. Bornet, C. Bouteloup-Demange, C. Feillet-Coudray, M. Brandolini, V. Ducros, D. Pepin, F. Brouns, A.M. Roussel, Y. Rayssiguier and C. Coudray. Effect of short-chain fructo-oligosaccharides on intestinal calcium absorption and calcium status in postmenopausal women: a stable-isotope study. *British Journal of Nutrition* 2003, 77, 449–457.
50. E.G. van den Heuvel, T. Muijs, F. Brouns and H.F. Hendriks. Short-chain fructo-oligosaccharides improve magnesium absorption in adolescent girls with a low calcium intake. *Nutrition Research* 2009, 29, 229–237.
51. P. Chandran, S. Satthaporn, A. Robins and O. Eremin. Inflammatory bowel disease: dysfunction of GALT and gut bacterial flora (II). *Surgeon* 2003, 1, 125–136.
52. L.E. Oostenbrug, H.M. van Dullemen, G.J. te Meerman and P.L. Jansen. IBD and genetics: new developments. *Scandinavian Journal of Gastroenterology* 2003, 239, 63–68.
53. O. Kanauchi, Y. Matsumoto, M. Matsumura, M. Fukuoka and T. Bamba. The beneficial effects of microflora, especially obligate anaerobes, and their products on the colonic environment in inflammatory bowel disease. *Current Pharmaceutical Design* 2005, 11, 1047–1053.
54. T. Hussey, R. Issenman, R. Persad, A. Otley and B. Christensen. Nutrition therapy in pediatric Crohn's diseases patients improves nutrition status and decreases inflammation. *Journal of Pediatric Gastroenterology and Nutrition* 2003, 37, 338–342, poster 45.
55. J.O. Lindsay, K. Whelan, A.J. Stagg, P. Gobin, H.O. Al Hassi, N. Rayment, M.A. Kamm, S.C. Knight and A. Forbes. Clinical, microbiological and immunological effects of fructo-oligosaccharide in patients with Crohn's disease. *Gut* 2006, 55, 348–355.
56. D.B. Silk, A. Davis, J. Vulevic, G. Tzortzis and G.R. Gibson. Clinical trial: the effects of a trans-galactooligosaccharide prebiotic on faecal microbiota and symptoms in irritable bowel syndrome. *Alimentary Pharmacology and Therapeutics* 2009, 29, 508–518.
57. A. Drakoularakou, G. Tzortis, R.A. Rastall and G.R. Gibson. A double-blind, placebo-controlled, randomized human study assessing the capacity of a novel galacto-oligosaccharide mixture in reducing travellers' diarrhea. *European Journal of Clinical Nutrition* 2010, 64, 146–52.
58. U. Teuri and R. Korpela. Galacto-oligosaccharides relieve constipation in elderly people. *Annals of Nutrition and Metabolism* 1998, 42, 319–327.
59. D. Bosscher and J. Van Loo. Oligofructose-enriched inulin. Keeping optimal body weight. *Nutrafoods* 2008, 7, 21–25.
60. P.D. Cani, E. Joly, Y. Horsmans and N.M. Delzenne. Oligofructose promotes satiety in healthy human: a pilot study. *European Journal of Clinical Nutrition* 2006, 60, 567–572.
61. P.D. Cani, E. Lecourt, E.M. Dewulf, F.M. Sohet, B.D. Pachikian, D. Naslain, F. De Backer, A.M. Neyrinck and N.M. Delzenne. Gut microbiota fermentation of prebiotics increases satietogenic and incretin gut peptide production with consequences for appetite sensation and glucose response after a meal. *American Journal of Clinical Nutrition* 2009, 90, 1236–1243.
62. S.A. Abrams, I.J. Griffin, K.M. Hawthorne and K.J. Ellis. Effect of prebiotic supplementation and calcium intake on body mass index. *Journal of Pediatrics* 2007, 151, 293–298.
63. R. Crittenden and M.J. Playne. Production, properties and applications of food-grade oligosaccharides. *Trends in Food Science and Technology* 1996, 7, 353–361.

64. M.M Rinne, M. Gueimonde, M. Kalliomäki, U. Hoppu, S.J. Salminen and E. Isolauri. Similar bifidogenic effects of prebiotic-supplemented partially hydrolyzed infant formula and breastfeeding on infant gut microbiota. *Fems Immunology and Medical Microbiology* 2005, 43, 59–65.
65. S. Arslanoglu, G. Moro and G. Boehm. Early supplementation of prebiotic oligosaccharides protects formula-fed infants against infections during the first 6 months of life. *Journal of Nutrition* 2007, 137, 2420–2424.
66. E. Bruzzese, M. Volpicelli, V. Squeglia, D. Bruzzese, F. Salvini, M. Bisceglia, P. Lionetti, M. Cinquetti, G. Iacono, S. Amarri and A.A. Guarino. Formula containing galacto- and fructo-oligosaccharides prevents intestinal and extra-intestinal infections: an observational study. *Clinical Nutrition* 2009, 28, 156–161.
67. G. Moro, S. Arslanoglu, B. Stahl, J. Jelinek, U. Wahn and G.A. Boehm. Mixture of prebiotic oligosaccharides reduces the incidence of atopic dermatitis during the first six months of age. *Archives of Disease in Childhood* 2006, 91, 814–819.
68. S. Arslanoglu, G.E. Moro, J. Schmitt, L. Tandoi, S. Rizzardi and G. Boehm. Early dietary intervention with a mixture of prebiotic oligosaccharides reduces the incidence of allergic manifestations and infections during the first two years of life. *Journal of Nutrition* 2008, 138, 1091–1095.
69. EFSA Panel on Dietetic Products, Nutrition and Allergies (NDA). Scientific opinion on the substantiation of a health claim related to Immunofortis and strengthening of the baby's immune system. *The EFSA Journal* 2010, 8, 1430.
70. S.A.S Craig, J.F. Holden, M.H. Auerbach and H.I. Frier. Polydextrose as a soluble fiber: physiological and analytical aspects. *Cereal Foods World* 1998, 43, 370–376.
71. C. Hengst, S. Ptok, A. Roessler, A. Fechner and G. Jahreis. Effects of polydextrose supplementation on different faecal parameters in healthy volunteers. *International Journal of Food Sciences and Nutrition* 2008, 60, 96–105.
72. S. Ishizuka, T. Nagai and H. Hara. Reduction of aberrant crypt foci by ingestion of polydextrose in the rat colorectum. *Nutrition Research* 2003, 23, 117–121.
73. H. Mäkivuokko, J. Nurmi, P. Nurminen, J. Stowell and N. Rautonen. In vitro effects on polydextrose by colonic bacteria and Caco-2 cell cyclooxygenase gene expression. *Nutrition and Cancer* 2005, 52, 94–104.
74. J.M. Dogne, C.T. Supuran and D. Pratico. Adverse cardiovascular effects of the coxibs. *Journal of Medicinal Chemistry* 2005, 48, 2251–2257.
75. M. Bugaut and M. Bentejac. Biological effects of short-chain fatty acids in non-ruminant mammals. *Annual Review of Nutrition* 1993, 13, 217–241.
76. Y.S. Kim, O.D. Tsa, A. Morita and A. Bella. Effect of sodium butyrate and three human colorectal adenocarcinoma cell lines in culture. *Falk Symposium* 1982, 31, 317–323.
77. B.L. Pool-Zobel. Inulin-type fructnas and reduction in colon cancer risk: review of experimental and human data. *British Journal of Nutrition* 2005, 93(Suppl. 1), S73–S90.
78. B.L. Pool-Zobel, et al. *Lactobacillus* and *Bifidobacterium* mediated antigenotoxicity in the colon of rats. *Nutrition and Cancer* 1996, 26, 365–380.
79. J. van Loo, Y. Clune, M. Bennett and J.K. Collins. The SYNCAN project: goals, set-up, first results and settings of the human intervention study. *British Journal of Nutrition* 2005, 93(Suppl. 1), S91–S98.
80. Y. Shimomura, et al. Attenuated response of the serum triglyceride concentration to ingestion of a chocolate containing polydextrose and lactitol in place of sugar. *Bioscience, Biotechnology, and Biochemistry* 2005, 69(10), 1819–1823.
81. T.J. Vasankari and M. Ahotupa. Supplementation of a polydextrose reduced hamburger meal induced postprandial hypertriglyceridemia. *AHA Congress Proceedings*, 2005, 112(17), II-833.
82. N.M. Delzenne. The hypolipidaemic effect of inulin: when animal studies help to approach the human problem. *British Journal of Nutrition* 1999, 82(1), 3–4.
83. N.A. King, S.A. Craig, T. Pepper and J.E. Blundell. Evaluation of the independent and combined effects of xylitol and polydextrose consumed as a snack on hunger and energy intake over 10 d. *British Journal of Nutrition* 2005, 93(6), 911–5.
84. F. Cremonini and N.J. Talley. Diagnostic and therapeutic strategies in the irritable bowel syndrome. *Minerva Medica* 2004, 95, 427–441.
85. K.L. Dear, M. Elia and J.O. Hunter. Do interventions which reduce colonic bacterial fermentation improve symptoms of irritable bowel syndrome? *Digestive Diseases and Sciences* 2005, 50, 758–766.

86. R. Van der Meulen, L. Avonts and L. deVuyst. Short fractions of oligofructose are preferentially metabolized by *Bifidobacterium animalis* DN-173 010. *Applied and Environmental Microbiology* 2004, 70, 1923–1930.
87. Van Baarlen, *et al.* Differential NF-κB pathways induction by *Lactobacillus plantarum* in the duodenum of healthy humans correlating with immune tolerance Differential NF-κB pathways induction by *Lactobacillus plantarum* in the duodenum of healthy humans correlating with immune tolerance. *Proceedings of the National Academy of Sciences*, 2009, 106, 2371–2376.
88. P. Coussement. Inulin and oligofuctose: safe intakes and legal status. *Journal of Nutrition* 1999, 129, 1412S–1417S.
89. J. Absolonne, M. Jossart, P. Coussment and M. Roberfroid. Digestive acceptability of oligofructose. *Proceedings of the First Orafti Research Conference, Orafti, Tienen, Belgium;* 1995; pp. 151–161.
90. M. Ito, Y. Deguchi, A. Miyamori, K. Matsumoto, H. Kikuchi, K. Matsumoto, Y. Kobayashi, T. Yajima and T. Kann. Effects of administration of galactooligisaccharides on the human faecal microflora, stool weight and abdominal sensation. *Microbial Ecology in Health and Disease* 1990, 3, 285–292.
91. U. Teuri, R. Korpela, M. Saxelin, L. Montonen and S. Salminen. Increased fecal frequency and gastrointestinal symptoms following ingestion of galacto-oligosaccharide-containing yogurt. *Journal of Nutritional Science and Vitaminology* 1998, 44, 465–471.
92. T. Sako, K. Matsumoto and R. Tanaka. Recent progress on research and applications of non-digestible galacto-oligosaccharides. *International Dairy Federation* 1999, 9, 69–80.
93. I.G. Carabin and W.G. Flamm. Evaluation of safety of inulin and oligofructose as dietary fiber. *Regulatory Toxicology and Pharmacology* 1999, 30, 268–282.
94. M.T. Flood, M.H. Auerbach and S.A.S. Craig. A review of the clinical toleration studies of polydextrose in food. *Food and Chemical Toxicology* 2004, 42, 1531–1542.
95. P. Marteau and P. Seksik. Tolerance of probiotics and prebiotics. *Journal of Clinical Gastroenterology* 2004, 38(Suppl. 2), S67–S69.
96. A.M. Pajari, J. Rajakangas, E. Päivärinta, V.M. Kosma, J. Rafter and M. Mutanen. Promotion of intestinal tumor formation by inulin is associated with an accumulation of cytosolic b-catenin in Min mice. *International Journal of Cancer* 2003, 106, 653–660.
97. F. Pierre, *et al.* Short-chain fructooligosaccharides reduce the occurrence of colon tumors and develop gut-associated lymphoid tissue in Min mice. *Cancer Research* 1997, 57, 225–228.
98. S.J.M. ten Bruggencate, I.M.J. Bovee-Oudenhoven, M.L.G. Lettink-Wissink, M.B. Katan and R. van der Meer. Dietary fructooligosaccharides and inulin decrease resistance of rats to *Salmonella*: protective role of calcium. *Gut* 2004, 53, 530–535.
99. S.J.M. ten Bruggencate, I.M.J. Bodee-oudenhoven, M.L.G. Lettink-Wissink and R. van der Meer. Dietary fructooliglosaccharides increase intestinal permeability in rats. *Journal of Nutrition* 2005, 135, 837–842.

4 Calorie Control and Weight Management

Michele Sadler[1] and Julian D. Stowell[2]

[1] MJSR Associates, Kent, UK
[2] Active Nutrition, DuPont Nutrition & Health, Reigate, UK

4.1 INTRODUCTION

Non-nutritive, intense sweeteners have a useful role in the diet providing sweetness without calories. Using intense sweeteners to replace sugar can result in energy reduction in foods and beverages, and hence potentially in energy intake. However, this depends on the nutritional composition of particular foods and drinks, and also on the physiological, psychological and behavioural effects of consuming foods with sweeteners in contrast to those with sugar. This chapter will look at the scientific evidence behind these complex issues.

4.2 CALORIC CONTRIBUTION OF SUGARS IN THE DIET

It is difficult to establish with any degree of accuracy the caloric contribution of sugars in our diet. Food balance data, based on food available for consumption, does not take account of home production or wastage and various methods that have been developed to assess individual intakes typically result in under-reporting. A comprehensive assessment of carbohydrate intake has been the subject of a joint Food and Agriculture Organization of the United Nations (FAO) and World Health Organization (WHO) consultation in Rome.[1] This consultation concluded that, as a percentage of energy, total carbohydrate intake ranges from 40% to over 80%, with the developed countries such as those in North America, Western Europe and Australia at the low end, and developing countries in Asia and Africa at the high end of the range. Sugars, defined as mono- and disaccharides, account for 9–27% of energy intake and the consultation reported an inverse trend between total carbohydrate and sugar intake.[1]

4.3 CALORIE CONTROL AND ITS IMPORTANCE IN WEIGHT MANAGEMENT

Energy balance, whereby intake of calories equals energy expenditure, is essential for weight maintenance. For weight loss, a negative energy balance is necessary while a positive energy balance will, without exception, result in weight gain. Current opinion is that all calories are

---

Sweeteners and Sugar Alternatives in Food Technology, Second Edition.
Edited by Dr Kay O'Donnell and Dr Malcolm W. Kearsley.
© 2012 John Wiley & Sons, Ltd. Published 2012 by John Wiley & Sons, Ltd.

equal with regard to energy balance whatever their macronutrient source, and energy density (kcal/g) is increasingly viewed as a key factor in the regulation of energy intake.[2–4] Though intake of sugars is popularly linked with increasing prevalence of obesity, for example Bray *et al.* have suggested that increased use of high-fructose corn syrup in beverages may play a role in the increasing prevalence of obesity,[5] scientific evidence for this hypothesis is lacking. The European Food Safety Authority (EFSA) has concluded that epidemiological studies do not show a positive association between total sugar intake and obesity, and that a cause and effect relationship has not been established between total sugar intake and body weight gain.[6]

That any single nutrient explains the obesity epidemic is in any case unrealistic.[2] Scientific research has identified very few nutrient- or food-based determinants of obesity, with the strongest evidence relating to diets that are high in fat or low in fibre.[7] Jenkins has noted a general increase in caloric intake of 500 kcal per day in the United States since 1980,[8] and this combined with a trend towards sedentary lifestyles is more likely to explain the increasing prevalence of obesity than a change in the intake, or a high intake, of any single dietary component.

From the late 1980s to the mid-1990s, a low-fat, high-carbohydrate diet was advocated for weight loss.[9] This concept was turned on its head with resurrection of the Atkins diet and the advent of other low-carbohydrate diets. Evidence-based reviews support the use of low-calorie or energy-deficit diets and low-fat diets as those most likely to achieve modest weight loss.[10,11] However, few dietary approaches stand the test of time very well with regard to weight maintenance, though many studies suggest that increasing physical activity can be helpful for weight maintenance.[10,12]

Hence, reducing total calorie intake and increasing physical activity should be key targets in avoiding a positive energy balance over time. Dr James O. Hill of the University of Colorado has raised awareness of the energy balance concept via the 'America on the Move' programme (www.americaonthemove.org). The basis of this programme is that an additional 2000 steps per day (or equivalent) combined with cutting 100 kcal per day from the usual diet can keep people from gaining an average of 1–2 lb (0.45–0.91 kg) annually. The Calorie Control Council is an international association of manufacturers of low-calorie and reduced-fat foods and beverages. This group has achieved remarkable success with its online weight loss and maintenance programme (www.caloriescount.com). This programme promotes the benefits of balance of macro- and micronutrient intakes, of energy balance and of a balanced approach to physical activity. In this way, thousands of consumers have been helped to achieve and sustain an appropriate body weight. In the United Kingdom, the Department of Health has a public health programme 'Change4Life', with a focus on increased activity and easy ways for consumers to reduce the calories they eat and drink (www.nhs.uk/Change4Life).

4.4 CALORIE REDUCTION IN FOODS

Traditional carbohydrates such as sucrose (sugar), glucose and other mono- and disaccharides (collectively known as sugars) and starch are important components of foods. Their technological, physiological and psychological benefits are very well documented and between them they contribute sweetness, bulk, texture, viscosity, browning ability and calories to foods. As well as helping to extend shelf life, they also impart distinctive, pleasant flavours. However, certain patterns (high frequency) of carbohydrate consumption have been implicated in tooth decay, and the general issue that dietary guidelines for sugars may be exceeded maintains a focus on the intake of sugars in relation to the obesity debate. In the

Table 4.1 Caloric value and relative sweetness of some sugar alternatives.

	Caloric value (kcal/g)		Relative sweetness (sucrose = 1)
	Europe	United States	
Erythritol	a	0.2	0.7
Isomalt	2.4	2	0.4
Lactitol	2.4	2	0.4
Maltitol	2.4	2.1	0.9
Mannitol	2.4	1.6	0.5
Sorbitol	2.4	2.6	0.5
Xylitol	2.4	2.4	0.95
Inulin	1	a	<0.1
Polydextrose	1	1	<0.1
Tagatose	a	1.5	0.9

aNo current legislation. High-intensity sweeteners do not provide a source of energy in use; they are used in such small quantities that they are all effectively non-nutritive. All the reduced-calorie sweeteners and bulking agents shown here have a low glycaemic response compared to sucrose and glucose.

United Kingdom, food manufacturers are being encouraged by policy makers to reduce the sugar content and portion size of products, two proposed strategies to assist consumers to eat fewer calories. Similarly, the 2010 Dietary Guidelines for Americans suggest consumers should consume fewer foods with added sugars as one way to balance calories to manage body weight (www.health.gov/dietaryguidelines/2010.asp).

If fully caloric sugars (e.g. sucrose or glucose) providing 4 kcal/g are replaced with non-nutritive, intense sweeteners, as shown in Table 4.1, this can result in valuable caloric savings. While some intense sweeteners such as aspartame do have a caloric value, they are used in such small quantities that they are effectively non-nutritive. In practice, calorie reduction is dependent on the nutritional composition of the food matrix. Calorie savings are most easily achieved in foods where sugar is the main energy source; for example, in sugar-sweetened beverages, close to zero energy density can be achieved. In more complex food matrices where sugars are one of many energy-providing components, calorie savings can be harder to achieve. This is dependent on what is being used to replace the sugars, and if for example, reduced sugar results in a higher proportion of fat in the food, the energy density per unit weight will increase. Nevertheless non-nutritive sweeteners and reduced-calorie sugar alternatives, when used in combination in certain foods, can maintain the sweetness and bulk normally provided by sugars. Figure 4.1 shows how speciality carbohydrates (such as Litesse® polydextrose), fructose and polyols (such as xylitol and lactitol) can be used to reduce the caloric density and glycaemic response of typical, commercially available, cereal-based snacks.

Table 4.2 shows how intense sweeteners, low-calorie bulking agents and other speciality carbohydrates can be integrated into a daily diet to achieve substantial caloric savings. In the example given, the combined use of intense sweeteners, polyols and bulking agents facilitates a reduction in caloric intake of 443 kcal per day. The amounts of these ingredients used lies well within their documented safety and toleration limits (37 g for polyols and bulking agents).

Achieving calorie reductions in foods is only one part of the quest to help consumers maintain energy balance, as the question then arises whether consumption of such foods contributes to an overall energy reduction in the diet, and thus to weight loss or avoidance of weight gain. Scientifically, this is complex to determine. Research has focussed on differences

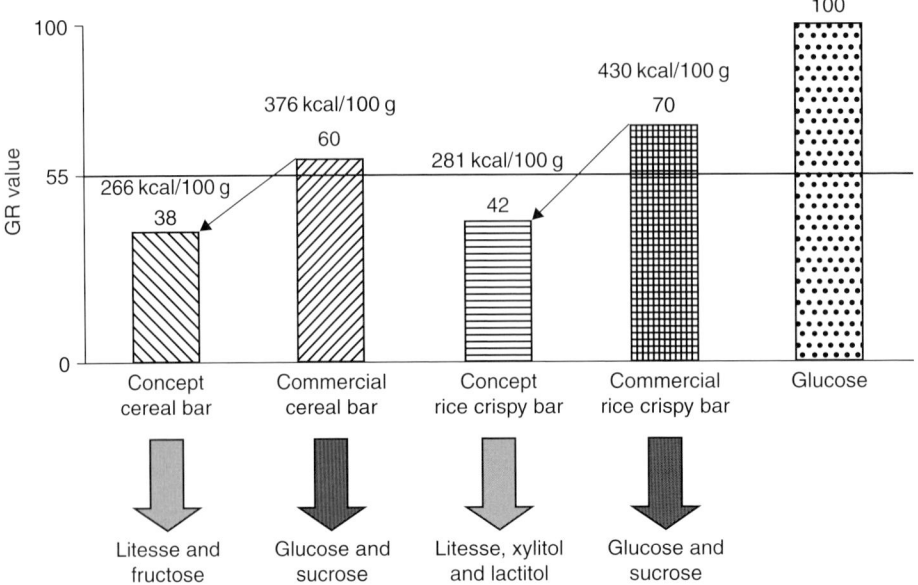

Fig. 4.1 The role of speciality carbohydrates in cereal-based snacks showing the average GR values for four snack bars. The relative GR of these snack bars was measured using portions of food that represented 25 g of total carbohydrate (including polyols, not Litesse) versus 25 g glucose (GR, glycaemic response relative to glucose).

in brain response to sugars and non-nutritive sweeteners, and differences in dietary behaviour, for example through effects on appetite and satiety. Study outcomes include ratings of appetite and satiety, energy intake and energy compensation, changes in body weight and maintenance of body weight. These aspects are discussed in the following text.

4.5 APPETITE AND SATIETY RESEARCH

It has been proposed that sweetness unaccompanied by calories confuses the body's regulatory mechanisms leading to a loss of control over appetite, which may induce overconsumption, though this has been refuted.[13] It has also been hypothesised that low-calorie 'diet' products may lead to a false energy saving, as the energy deficit is compensated for at the next meal or later during the day.[14] Studies to investigate such effects have included short-term effects on appetite, satiety and energy intake. Satiety refers to the effects of a food or a meal after eating has ended, whereas satiation refers to the processes involved in the termination of a meal.[15] In evaluating the role of non-nutritive and bulk sweeteners in these processes, the primary consideration is whether sweeteners encourage consumers to eat less, more or the same number of calories compared with sugar-sweetened foods.

The usual methodology to study the acute or short-term regulation of food intake is the preload-test meal paradigm. The ingested preload beverage, snack or meal may vary in volume, energy, energy density or nutrient composition, and is assumed to influence later energy intake. Subjective measurements of hunger and thirst, satiety and desire to eat are thus taken before and at predetermined time intervals after the preload and test meal, and food intake for the remainder of the day may also be recorded. The validity of this methodology and the important aspects of study design have recently been reviewed.[16]

Table 4.2 The use of intense sweeteners, polyols and bulking agents to achieve caloric reduction in a typical diet.

	Full calorie (kcal)	Reduced calorie (kcal)	Calories saved (kcal)	Typical quantities of polyols and bulking agents (g)
Breakfast Two slices of toast with butter (74 g)	175	175	–	–
Tea/coffee	29	29	–	–
Snack Two chocolate chip cookies (30 g)	150	120	30	4.5 and 4.5 (HIS)
Tea/coffee	29	29	–	–
Lunch Ham salad baguette	325	325	–	–
Fruit yoghurt	125	80	45	HIS
Tea/coffee	29	29	–	–
Snack Chocolate bar (50 g)	260	150	110	3 and 15 (HIS)
Carbonated beverage (1 can)	139	1	138	HIS
Evening meal Steak (200 g)	400	400	–	–
Jacket potato (180 g)	245	245	–	–
Carrots (60 g)	13	13	–	–
Peas (60 g)	32	32	–	–
Ice cream (50 g)	160	40	120	5 and 5 (HIS)
Tea/coffee	29	29	–	–
Totals	2140	1697	443	12.5 and 24.5 (HIS)

Source: Full-calorie values: www.weightlossresources.co.uk; reduced-calorie values: Danisco Sweeteners Database. HIS, high-intensity sweeteners.

In addition, imaging techniques are being developed to estimate the effect of dietary manipulation on specific areas of brain activity known to be involved in hunger and satiety.[17] Using magnetic resonance imaging, investigations with one intense sweetener have shown that it does not trigger the prolonged decrease in activity in the upper hypothalamus that is observed following the ingestion of sugar, suggesting that aspartame does not trigger sugar-like responses in the brain.[18] This preliminary evidence suggests that further research is warranted to explore whether the brain shows the same response to caloric sugars and to non-nutritive sweeteners.

4.6 SWEETENERS AND SATIETY, ENERGY INTAKES AND BODY WEIGHT

A number of scientific reviews have addressed different aspects of the role of intense sweeteners in satiety, energy intake and body weight control, though not all have reached the same conclusions.

4.6.1 Satiety and energy intake

Many acute, short-term studies have investigated hunger, satiety and energy intakes during a single meal or snack after consumption of a sweet stimulus. A comprehensive review of the area[4] highlighted that such studies have differed in the format (e.g. liquid and solid foods), energy density and palatability of the test foods, as well as the preload volumes, energy differentials, the time interval to consumption of the test meal and the dependent measures. Hence, it is difficult to draw firm conclusions from such studies. However, studies designed to compare the effect of sugar-sweetened foods and drinks with those containing intense sweeteners on hunger ratings have generally found no significant differences.[4]

Since liquids are known to produce shorter sensory stimulation than solid foods, a currently debated issue is whether sweetened beverages have a lower satiety value than solid foods, and the implications of this.[16,19]

In their wide ranging review, Bellisle and Drewnowski[4] also concluded that more recent studies have generally failed to show that non-nutritive sweeteners stimulate appetite relative to unsweetened test items or water, and that there are no significant effects on energy compensation when sugar-containing foods or drinks are replaced by those containing sweeteners. For people of ideal weight, an earlier review concluded that there is evidence of compensation, in men rather than women and in children rather than adults, such that the consumption of low-energy foods is followed by increased energy intake at a later time.[20] Blundell et al., in their review of the methodology, highlighted that the outcomes observed in lean people may not be replicated in obese people.[16]

Regarding energy intake over periods of at least 24 hours, a meta-analysis[21] considered evidence from 15 randomised controlled trials for the effect of aspartame alone, or aspartame with other intense sweeteners conducted in healthy adults. The authors concluded that energy intakes were reduced with the aspartame condition compared with all types of control except non-sucrose controls such as water. A mean reduction of about 10% of energy intake was reported, representing a deficit of 222 kcal per day. If sustained, this could result in a predicted weight loss of around 0.2 kg per week (confidence intervals 50% above and below this estimate). Regarding compensation for the missing energy in the sweetener condition, the authors calculated from short-term studies that the average energy compensation was 32% overall, but when considering only soft drinks, energy compensation was on average 15.5%.

Though few clinical studies have explored the effects of long-term use of intense sweeteners on energy intake, some show reduced energy intake with sweetener use,[22,23] but overall the results are equivocal.[4] Epidemiological studies have not provided evidence that sweetener use is associated with a reduced sugar intake. Though such studies provide evidence that subjects with higher body mass index (BMI) have greater use of sweeteners, this is consistent with overweight people using sweeteners for weight control.[4]

A systematic review by the Canadian Diabetes Association National Nutrition Committee of the benefits and risks associated with non-nutritive intense sweeteners in diabetes management[25] concluded that for non-diabetic individuals, the effects of consuming intense sweeteners on appetite, hunger and food intake do not differ from those of water, and that the effects of sweeteners on the energy content of a subsequent meal do not differ from those of sugar. A review by Benton[20] also concluded that sweeteners do not increase energy intake or hunger ratings.

A review of earlier data on sweeteners and body weight made the case that non-nutritive sweeteners have a role in reducing between-meal caloric intake but their scope may be modest due to the difficulties in replacing traditional ingredients.[24] However, major developments in food science during the last 20 years have meant that where reduced-calorie versions of foods can be produced this is often without compromise in taste.

4.6.2 Body weight management

It is important to acknowledge that the impact of food intake on weight management is a complex area of research and it is necessary to review the totality of available data before reaching conclusions. The most important outcome measure for weight loss is to show a reduction in body fat, though not all studies have measured this endpoint. A number of reviews have addressed different aspects of long-term studies of sweetener use and body weight in adults, and again have not all reached the same conclusions.

An earlier review by Drewnowski[2] concluded from a study investigating aspartame, that while the sweetener did not promote rapid weight loss, its inclusion in a multi-disciplinary weight management programme may promote the long-term maintenance of reduced body weight – though a key unanswered question was whether a more energy-dilute diet was consumed in the aspartame group.

The review by the Canadian Diabetes Association also considered weight control, and concluded that intense sweeteners used by obese people as an adjunct to multi-disciplinary programmes may improve weight loss and weight control, particularly if they are used to replace energy-dense foods.[25] Benton also investigated whether sweeteners are useful while dieting and to prevent weight gain. It was concluded that, possibly because of methodological issues, there is no body of evidence that long-term use of non-nutritive sweeteners helps to prevent weight gain.[20]

De la Hunty et al.[21] investigated the effect of aspartame alone or aspartame with other intense sweeteners on weight loss and weight maintenance in healthy adults. A significant reduction in body weight was reported, which was established to correspond to approximately a 3% reduction in body weight (2.3 kg for a 75 kg adult). Over the average study length of 12 weeks, this equated to an estimated rate of weight loss of approximately 0.2 kg per week for a 75 kg adult, which was consistent with their findings for energy intake.

Use of sweeteners may be useful for managing weight in overweight children, as shown by a 6-month intervention programme in families with at least one overweight child. The effects of an intense sweetener plus exercise group instructed to reduce sugar intake by 100 kcal per day and to increase physical activity by 2000 steps per day were compared with a control group asked not to change their diet or physical activity level.[26] In the sweetener/exercise group, the overweight children achieved their goal of cutting out 100 kcal of sugar per day on 78% of study days. The goal was to reduce the overweight children's rate of body weight increase and to prevent an increase or produce a decrease in BMI for age, rather than weight loss *per se*. Significantly more children in the sweetener/exercise group met these goals than in the control group. In addition, the parents showed no significant weight gain during the 6-month intervention period. These results suggest a useful role for non-nutritive sweeteners in weight management as an adjunct to healthy lifestyle measures.

While a more recent review concluded that there is insufficient evidence that an exchange of sugar for non-sugar carbohydrates as part of a reduced-fat or energy-restricted diet results in lower body weight, the authors also concluded that observational studies have suggested a possible relationship between consumption of sugar-sweetened beverages and body weight.[27] An interesting study comparing dietary strategies in a group of weight loss maintainers compared with a group of normal weight subjects concluded that weight loss maintainers used more dietary strategies to remain at a stable weight.[28] These included use of sugar-modified foods, reduced intake of sugar-sweetened beverages, increased consumption of beverages with intense sweeteners, as well as greater restriction of fat intake and use of fat-modified foods, suggesting a useful role for foods and drinks made with non-nutritive sweeteners.

4.7 RELEVANCE OF ENERGY DENSITY AND GLYCAEMIC RESPONSE

4.7.1 Energy density

The lower the caloric density of a food, the fewer the calories that are likely to be consumed in an *ad libitum* situation. Barbara Rolls and colleagues working at Penn State University have devoted many years to the study of the relationship between caloric density and caloric intake. Their overwhelming conclusion is that overconsumption is linked to both increasing portion sizes and the trend towards energy-dense foods. For example, a study in 39 women, investigating the combined effects of energy density and portion size on energy intake, found that subjects consumed 56% more energy (925 kJ) when served the largest portion of a higher energy dense entrée compared with the smallest portion of a lower energy dense entrée. Subjects did not compensate for the differing energy intakes at a subsequent meal.[29] Hence, manipulation of food composition to produce lower energy density may be a useful strategy for reducing energy intake.

In their review, Bellisle and Drewnowski[4] addressed the question whether reducing the energy density of sweet foods and drinks by the use of sweeteners is a useful option for the control of body weight. They concluded that the usefulness of such foods and drinks depends on imprecise physiological control systems and the absence of powerful compensatory mechanisms. Though many studies are compatible with this, since it is not always possible to achieve reduced energy density in solid or semi-solid foods, reduced-calorie beverages may represent the optimal use of intense sweeteners in the context of a weight control strategy.

4.7.2 Glycaemic response

A further factor having an impact on satiation and satiety that is relevant to bulk sweeteners and bulking agents is the possible inverse relationship between glycaemic response and satiety. A growing number of studies suggest that reducing the glycaemic impact of the diet may help consumers to eat fewer calories. However, not all investigators and reviews have reached the same conclusion. Studies have included short-term effects on appetite ratings and *ad libitum* food intake, and long-term effects focussing on weight loss.

A systematic review of short-term studies (1 day or less) concluded that lower glycaemic foods or meals have a greater satiating effect than higher glycaemic foods or meals.[30] Confounding factors that may influence both satiety and glycaemic impact were accounted for in the analysis, and it was concluded that the mechanisms may include specific effects of blood glucose on satiety (the glucostatic theory) and other stimuli (e.g. peptides) involved in appetite control.

The main body of information linking glycaemic response to weight-control mandates in favour of a diet reduced in glycaemic response compared with our current norm (see Chapter 1 for further information). Medium-term clinical trials generally show less weight loss on diets with a high glycaemic index (GI) or high glycaemic load (GL) compared with low-GI/GL diets. In a comprehensive meta-analysis of 23 studies measuring changes in body weight after switching to low-GI from high-GI diets,[31] a reduction in GL (a measure that takes account of the amount of the food eaten) was associated with reduced body weight, and vice versa, suggesting a positive role for low-GI foods in appetite control under free living conditions. Reductions in body weight were also found with lower GI diets *per se*, and lower intakes of available carbohydrate and metabolisable energy. A separate meta-analysis also investigated the effect of low-GI or low-GL diets on changes in body weight and concluded that overweight or obese people following low-GI diets lost more weight than those consuming control diets, and that low-GI diets resulted in significantly greater reductions in body fat.[32] However, more research with longer follow-up is needed to explore whether this improvement continues in the longer term.

A study assessing the role of low-GI diets on weight regain, the Diogenes 8 European Country Dietary Intervention Study, found that a difference in GI of >10 units between diets showed benefits for weight maintenance, though a stronger effect was reported for a higher protein diet.[33]

4.8 LEGISLATION RELEVANT TO REDUCED CALORIE FOODS

A favourable legislative environment is essential if consumers are to take full advantage of positive developments in food science. It is inevitable that food-labelling laws will need to be updated to take account of new innovations. The interactions between food and ingredient companies, consumer groups and governments are ongoing around the world.

In Japan, nutritional labelling is regulated by the Health Promotion Act (2 August 2002; Law No. 103).[34] This Act has been amended eight times to date and contains provisions for nutrient labelling and nutrient claims. Nutrients covered in the Act are protein, fat, carbohydrate, sodium, minerals and vitamins. The mandatory requirements for nutrient labelling include energy, protein, fat, carbohydrate and sodium. Nutrient content claims (as of 13 July 2011) are allowed as follows:

- *Low calorie*: Equal to or less than 40 kcal per 100 g, or 20 kcal per 100 mL.
- *Reduced-calorie*: Equal to or greater than 40 kcal per 100 g, or 20 kcal per 100 mL.
- *Calorie free*: Equal to or less than 5 kcal per 100 g, or 5 kcal per 100 mL.
- *Low sugar*: Equal to or less than 5 g per 100 g, or 2.5 g per 100 mL.
- *Reduced sugar*: Equal to or greater than 5 g per 100 g, or 2.5 g per 100 mL.
- *Sugar free*: Equal to or less than 0.5 g per 100 g, or 0.5 g per 100 mL.

In Europe, relevant legislation includes Regulation (EC) 1333/2008 on Food Additives[35] that recently replaced the Sweeteners Directive (94/35/EC) and the Directive on foodstuffs for particular nutritional uses or PARNUT (2009/39/EC)[36] that includes provisions on slimming products. This Directive is about to be replaced by a Regulation on Specialised Food.[37]

The Food Additive Regulation lists all currently approved intense and bulk sweeteners, together with their conditions of use. Specification of the additives was recently published in Regulation 231/2012. All food additives will also be subject to re-authorisation and sweeteners shall be completed before 31 December 2020.[38]

The current PARNUT Directive[39] includes foods intended for use in energy-restricted diets for weight reduction. However, the intention of the future Regulation on Specialised Food will not cover provisions on energy-restricted diets. Instead, foods that claim to have effects on weight reduction/maintenance will need to comply with the Nutrition and Health Claims Regulation 1924/2008. For example, a food product could bear a nutrition claim, such as 'low-energy diet', under the condition that the following criteria for the nutrition claim mentioned in the Annex of Regulation 1924/2008 are fulfilled:

- *Low energy*: The product does not contain more than 40 kcal (170 kJ) per 100 g for solids or more than 20 kcal (80 kJ) per 100 mL for liquids.
- *Energy reduced*: The energy value is reduced by at least 30%, with an indication of the characteristic(s) that make(s) the food reduced in its total energy value. A proposal is currently discussed that would change this claim to 'at least 25%', that would harmonise this claim with Codex.
- *Energy free*: The product does not contain more than 4 kcal (17 kJ) per 100 mL. For table-top sweeteners, the limit of 0.4 kcal (1.7 kJ) per portion, with equivalent sweetening properties to 6 g of sucrose (approximately 1 teaspoon of sucrose), applies.
- *Low sugars*: The product contains no more than 5 g of sugars per 100 g for solids or 2.5 g of sugars per 100 mL for liquids.
- *Sugar free*: The product contains no more than 0.5 g of sugars per 100 g or 100 mL.
- *With no added sugar*: The product does not contain any added mono- or disaccharides or any other food used for its sweetening properties. If sugars are naturally present in the food, the following indication should also appear on the label: CONTAINS NATURALLY OCCURRING SUGARS.

Nutrition claims under discussion in the European Union and likely to be adopted and published early in 2012 include the following*:

- *Reduced sugars*: If energy is equal to or less than the original product.
- *Now contains X% less energy and/or sugars*: If the reduction in content is at least 15%. The claim shall be followed by a statement indicating the content of the nutrient or energy for which the claim is made, prior to reformulation, expressed per 100 g or 100 mL. A claim may be used for a maximum of 1 year following placing the reformulated product on the market. Products that have been placed on the market and labelled before the end of this period may continue to be sold until stocks are exhausted.

In the United States, the Nutrition Labeling and Education Act of 1990 established the principles and format for the nutrition labelling of foods. This Act is complemented by a number of additional entries in the Code of Federal Regulations (CFR)[40] relating to nutrient content claims. For example:

- *Reduced-calorie*: Defined as 25% fewer calories than a full-calorie equivalent product (21CFR101.60).
- *Low calorie*: Defined as foods that contribute <40 kcal per reference amount customarily consumed (RACC) or <120 kcal per 100 g food.
- *Calorie free*: Defined as foods that contain less than 5 kcal per RACC.

---

* This proposed amendment to the reduced sugars claim and the new *X%* less nutrition claim were not approved by the European Parliament.

- *Light*: Defined as follows (21CFR101.56):
  - Foods that derive over 50% of their calories from fat; the fat content is reduced by at least 50% compared to a full-fat equivalent product.
  - Foods that derive less than 50% of their calories from fat; the fat content is reduced by at least 50% and the calorie content is reduced by at least one-third.

In 2004, the Calorie Control Council (www.caloriecontrol.org) entered into dialogue with the US Food and Drug Administration (FDA) on the subject of permissible claims relating to low-calorie foods. The FDA confirmed acceptance of the following: *Low-calorie [name of food] may be useful in weight control. Obesity increases the risk of developing diabetes, heart disease and certain cancers.*

Codex Alimentarius has guidelines for the use of nutrition and health claims (CAC/GL 23-1997, Rev. 1-2004).[40] Nutrition claims should be consistent with national nutrition policy and support that policy. Only nutrition claims that support national nutrition policy should be allowed. The only nutrition claims permitted shall be those relating to energy, protein, carbohydrate, fat and components thereof, fibre, sodium and vitamins and minerals, for which Nutrient Reference Values (NRVs) have been laid down in the Codex Guidelines for Nutrition Labelling (CAC/GL 2-1985, Adopted 1985. Revision 1993. Amendment 2003, 2006, 2009 and 2010). Examples of nutrition claims are as follows:

- *Low energy*: 40 kcal (170 kJ) per 100 g (solids) or 20 kcal (80 kJ) per 100 mL (liquids).
- *Energy free*: 4 kcal per 100 mL (liquids).
- *Sugar free*: 0.5 g per 100 g (solids) or 0.5 g per 100 mL (liquids).
- *Energy reduced*: At least 25% relative difference to a different version of the same or similar food.

Clearly, the labelling of foods with regard to the use of sweeteners as ingredients and to nutrition and health claims is a complex subject. Many countries now have their own nutrition-labelling regulations and it is important whenever contemplating a launch of a new food product to obtain local expert opinions as to the labelling implications.

4.9 CONCLUSIONS

Overall, evidence suggests that foods formulated with non-nutritive intense sweeteners, reduced-calorie bulk sweeteners and bulking agents can play an interesting and useful role in helping consumers to improve the energy balance of their diets and as part of weight reducing diets. Contrary to some reports, the main body of published data shows that these ingredients, when incorporated into foods with lower caloric density and/or reduced glycaemic impact, can help consumers to eat less calories overall. A balanced approach to weight loss and maintenance is essential for long-term success.

ACKNOWLEDGEMENT

We are pleased to acknowledge the expert contribution of Paul Tenning who updated the section on legislation relevant to reduced-calorie foods.

REFERENCES

1. FAO. Carbohydrates in human nutrition. FAO Food and Nutrition Paper No. 66, Report of a Joint FAO/WHO Expert Consultation, 1998.
2. A. Drewnowski. Intense sweeteners and energy density of foods: implications for weight control. *European Journal of Clinical Nutrition* 1999, 53(10), 757–763.
3. B.J. Rolls, *et al.* Energy density but not fat content of foods affected energy intake in lean and obese women. *American Journal of Clinical Nutrition* 1999, 69(5), 863–871.
4. F. Bellisle and A. Drewnowski. Intense sweeteners, energy intake and the control of body weight. *European Journal of Clinical Nutrition* 2007, 61(6), 691–700.
5. G.A. Bray, S.J. Nielsen and B.M. Popkin. Consumption of high-fructose corn syrup in beverages may play a role in the epidemic of obesity. *American Journal of Clinical Nutrition* 2004, 79(4), 537–543.
6. EFSA. Scientific opinion on the substantiation of health claims related to intense sweeteners. *European Food Safety Authority Journal* 2011, 9(6), 2229.
7. S.A. Jebb. Dietary determinants of obesity. *Obesity Reviews* 2007, 8(Suppl 1), 93–97.
8. D.J. Jenkins, *et al.* Too much sugar, too much carbohydrate, or just too much? *American Journal of Clinical Nutrition* 2004, 79(5), 711–712.
9. B. Rolls and J.O. Hill. *Carbohydrates and weight management, ILSI North America Monograph*, ILSI Press, Washington, DC; 1998.
10. J.P. Wilding. Treatment strategies for obesity. *Obesity Reviews* 2007, 8(Suppl 1), 137–144.
11. A. Avenell, *et al.* What are the long-term benefits of weight reducing diets in adults? A systematic review of randomized controlled trials. *Journal of Human Nutrition and Dietetics* 2004, 17(4), 317–335.
12. D.R. Bensimhon, W.E. Kraus and M.P. Donahue. Obesity and physical activity: a review. *American Heart Journal* 2006, 151(3), 598–603.
13. B. Rolls. Reply to PJ Rogers and JE Blundell. *American Journal of Clinical Nutrition* 1993, 58(1), 121–122.
14. P.J. Rogers and J.E. Blundell. Intense sweeteners and appetite. *American Journal of Clinical Nutrition* 1993, 58(1), 120–122.
15. J.E. Blundell and P.J. Rogers. Hunger, hedonics and the control of satiation and satiety. In: M.I. Friedman, M.G. Tordoff and M.R. Kare (eds). *Chemical Senses Volume 4: Appetite and Nutrition*, Marcel Dekker, New York; 1991, pp. 127–148.
16. J.E. Blundell, *et al.* Appetite control: methodological aspects of the evaluation of foods. *Obesity Reviews* 2010, 11(3), 251–270.
17. N. Delzenne, *et al.* Gastrointestinal targets of appetite regulation in humans. *Obesity Reviews* 2010, 11(3), 234–250.
18. P.A. Smeets, *et al.* Functional magnetic resonance imaging of human hypothalamic responses to sweet taste and calories. *American Journal of Clinical Nutrition* 2005, 82(5), 1011–1016.
19. R. Hammersley, M. Reid and M. Duffy. How may refined carbohydrates affect satiety and mood? *Nutrition Bulletin* 2007, 32(Suppl 1), 61–70.
20. D. Benton. Can artificial sweeteners help control body weight and prevent obesity? *Nutrition Research Reviews* 2005, 18(1), 63–76.
21. A. de la Hunty, S. Gibson and M. Ashwell. A review of the effectiveness of aspartame in helping with weight control. *Nutrition Bulletin* 2006, 31, 115–128.
22. M.G. Tordoff and A.M. Alleva. Effect of drinking soda sweetened with aspartame or high-fructose corn syrup on food intake and body weight. *American Journal of Clinical Nutrition* 1990, 51(6), 963–969.
23. A. Raben, *et al.* Sucrose compared with artificial sweeteners: different effects on ad libitum food intake and body weight after 10 wk of supplementation in overweight subjects. *American Journal of Clinical Nutrition* 2002, 76(4), 721–729.
24. D. Booth. Sweeteners and body weight. In: S. Marie and T.R. Piggot (eds). *Handbook of Sweeteners*, Blackie and Son, Glasgow; 1991.
25. R. Gougeon, M. Spidel, K. Lee and C.J. Field. Canadian Diabetes Association National Nutrition Committee Technical Review: non-nutritive intense sweeteners in diabetes management. *Canadian Journal of Diabetes* 2004, 28, 385–399.
26. S.J. Rodearmel, *et al.* Small changes in dietary sugar and physical activity as an approach to preventing excessive weight gain: the America on the Move family study. *Pediatrics* 2007, 120(4), 869–879.
27. M.A. van Baak and A. Astrup, Consumption of sugars and body weight. *Obesity Reviews* 2009, 10 (Suppl 1) 9–23.

28. S. Phelan, *et al.* Use of artificial sweeteners and fat-modified foods in weight loss maintainers and always-normal weight individuals. *International Journal of Obesity* 2009, 33(10), 1183–1190.
29. T.V. Kral, L.S. Roe and B.J. Rolls. Combined effects of energy density and portion size on energy intake in women. *American Journal of Clinical Nutrition* 2004, 79(6), 962–968.
30. F.R. Bornet, *et al.* Glycaemic response to foods: impact on satiety and long-term weight regulation. *Appetite* 2007, 49(3), 535–553.
31. G. Livesey, *et al.* Glycemic response and health—a systematic review and meta-analysis: relations between dietary glycemic properties and health outcomes. *American Journal of Clinical Nutrition* 2008, 87(1), 258S–268S.
32. D.E. Thomas, E.J. Elliott and L. Baur. Low glycaemic index or low glycaemic load diets for overweight and obesity. *Cochrane Database of Systematic Reviews* 2007 (3), CD005105.
33. T.M. Larsen, *et al.* The Diet, Obesity and Genes (Diogenes) Dietary Study in eight European countries – a comprehensive design for long-term intervention. *Obesity Reviews* 2010, 11(1), 76–91
34. Labeling System for Nutrient. http://www.mhlw.go.jp/english/topics/foodsafety/fhc/04.html.
35. Official Journal of the European Union. http://eur-lex.europa.eu/LexUriServ/LexUriServ.do?uri=CELEX:32008R1333:EN:NOT; 2007.
36. Official Journal of the European Union. http://eur-lex.europa.eu/Result.do?arg0=953%2F2009&arg1=&arg2=&titre=titre&chlang=en&RechType=RECH_mot&Submit=Search; 2009.
37. European Commission – Press Release. http://ec.europa.eu/prelex/detail_dossier_real.cfm?CL=en&DosId=200612; 2011.
38. Official Journal of the European Union. http://eur-lex.europa.eu/Result.do?arg0=231%2F2012&arg1=&arg2=&titre=titre&chlang=en&RechType=RECH_mot&Submit=Search; 2010.
39. e-CFR. http://eur-lex.europa.eu/Result.do?arg0=953%2F2009&arg1=&arg2=&titre=titre&chlang=en&RechType=RECH_mot&Submit=Search; 2012.
40. Guidelines for Use of Nutrition and Health Claims. http://faculty.ksu.edu.sa/18869/Food%20Standards/Nutrition%20and%20health%20claims.pdf; 2004.

Part Two
High-Potency Sweeteners

5 Acesulfame K

Christian Klug[1] and Gert-Wolfhard von Rymon Lipinski[2]

[1]*Nutrinova Nutrition Specialties and Food Ingredients GmbH, Sulzbach, Germany*
[2]*MK Food Management Consulting GmbH, Bad Vilbel, Germany*

5.1 INTRODUCTION AND HISTORY

Acesulfame K, the potassium salt of acesulfame, was discovered accidentally in 1967 by Clauß and Jensen.[1] Acesulfame K belongs to the class of dihydro-oxathiazinone dioxides, which were then synthesised for the first time. Today, it is one of the most important intense sweeteners (see Figure 5.1).

A systematic investigation of this new class of substances was subsequently carried out and this showed that many substances belonging to this class were also sweet. Introduction of different substituents at positions 5 and 6 of the ring system significantly influenced intensity and quality of the sweet taste. All of these substances were, however, sweet, and the strongest sweetness was found for compounds with short side chains in position 6 of the ring system. Additionally, a series of substances with similar ring structure were synthesised to investigate the potential of related classes of substances. It was, in contrast, found that every change within the ring system resulted in a complete loss of sweetness.

Investigations into the sensory properties showed that 6-methyl-1.2.3-oxathiazin-4(3H)-one-2.2-dioxide had better sensory properties than the other compounds. Several salts of this compound were also studied and among these the potassium and calcium salts were rated more favourably than the sodium salt. Considerations about ease of production also showed that acesulfame was the substance most easily manufactured and among the salts, the potassium salt was easier to purify than the calcium salt. Therefore, the substance known today, as acesulfame K, was selected for systematic development and investigation for use as an intense sweetener.

The original process for the commercial production of acesulfame K was based on a starting material that was difficult to handle.[1] Therefore, an alternative process based on sulfamic acid and diketene was developed. These react to form acetoacetamide-*N*-sulphonic acid, which, in turn, cyclises in the presence of sulphur trioxide to the acesulfame ring system.[2] As the free acesulfame behaves like an acid, it forms a potassium salt, acesulfame K, when reacted with potassium hydroxide (see Figure 5.2). Several other, rather hypothetical, syntheses are described in the scientific literature and patent applications. None of these, however, have gained practical importance. Some of the proposed production routes do not even produce any or, at least, any significant amounts of acesulfame K.[3]

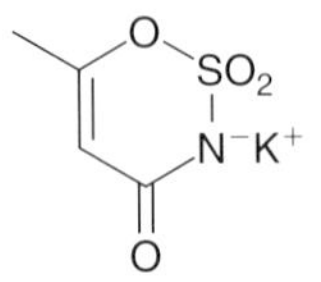

Fig. 5.1 Chemical structure of acesulfame K.

Acesulfame K was discovered and developed by Hoechst AG and is now marketed worldwide by the former food ingredients business of this company, now Nutrinova Nutrition Specialties & Food Ingredients GmbH under the trademark Sunett®.

5.2 ORGANOLEPTIC PROPERTIES

5.2.1 Acesulfame K as the single sweetener

Acesulfame K is approximately 200 times sweeter than sucrose when used at moderate sweetness levels. As with all intense sweeteners, the sweetness intensity depends on the use level and decreases with increasing sweetness to values in the range of 130–100 times sweeter than sucrose. Acesulfame K is approximately one-third as sweet as sucralose, half as sweet as sodium saccharin, about as sweet as aspartame and four to five times sweeter than sodium cyclamate.[4]

The sweetness of acesulfame K is perceived quickly and without any unpleasant delay (especially in comparison with aspartame and sucralose). The sweetness does not, as a rule, linger and normally does not persist longer than the intrinsic taste of the food in which it is used.[5]

As with other sweeteners, the organoleptic properties of acesulfame K are influenced by the food in which it is consumed. At low concentrations, the sweet taste is pure, while a side taste may become occasionally perceptible at elevated levels.[6] In practice, this is of rather limited importance, as normally, blends of acesulfame with other sweeteners are used for high sweetness levels rather than the single sweetener.

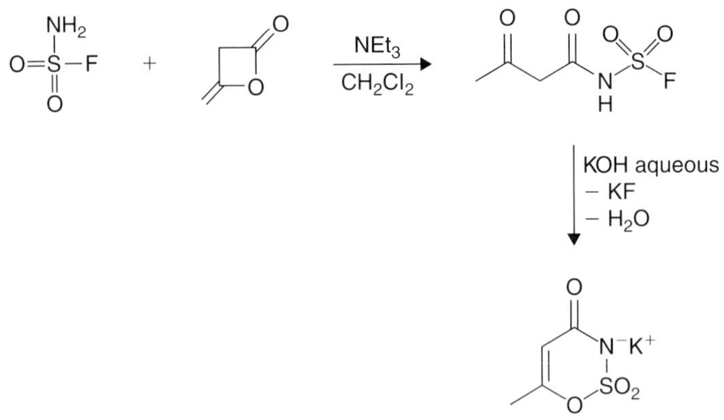

Fig. 5.2 Synthesis of acesulfame K according to the 'sulphur trioxide process'.[2]

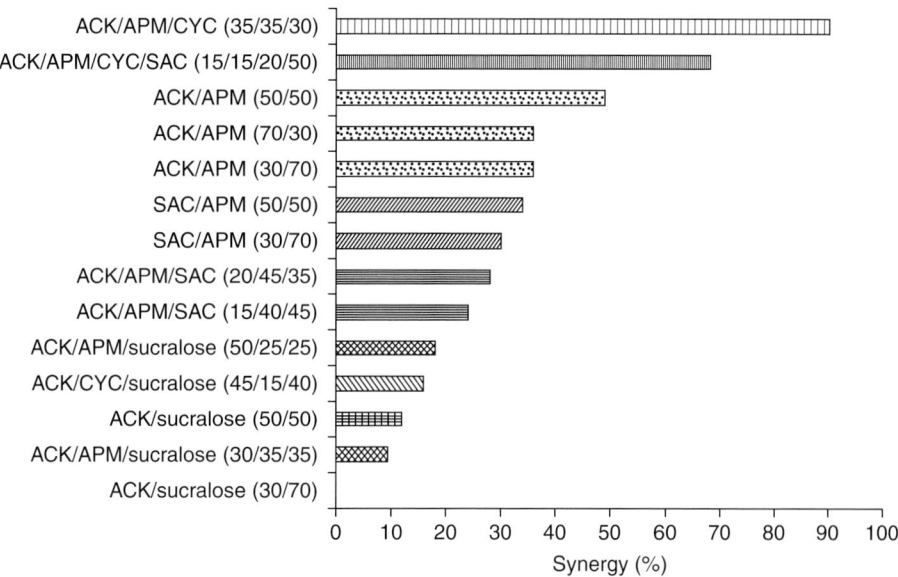

Fig. 5.3 Synergy of binary, ternary and quaternary sweetener blends in a carbonated soft drink at 10% sugar equivalence (all blend ratios given as sweetness contribution) (ACK, acesulfame K; APM, aspartame; CYC, cyclamate; SAC, saccharin).

It has been reported that with increasing temperature the sweetness intensity of acesulfame K decreased less than the sweetness of other intense sweeteners.[7]

5.2.2 Blends of acesulfame K with other sweetening agents

In blends of acesulfame K with aspartame, sodium cyclamate and sucralose, synergistic enhancement of the sweetness intensity and improvement of the taste quality are observed and these are often significant. In ternary blends of acesulfame K with aspartame and cyclamate, the quantitative sweetness enhancement may be as high as 90%, and in binary blends of acesulfame K and aspartame, it can still be in the range of 40–50% (see Figure 5.3). In contrast, the synergism in sweetener blends with sucralose is comparably low and blends of acesulfame K and saccharin do not produce any sweetness enhancement.

The synergistic sweetness enhancement, often called quantitative synergism, has been studied in detail in blends of acesulfame K with aspartame. A blend of equal parts of both sweeteners has 300 times the sweetness of sucrose, even at elevated concentrations (see Figure 5.4). In contrast, the single sweeteners are only 200 times as sweet as sucrose, even at lower use levels. Synergistic sweetness enhancement is not an isolated phenomenon, but is observed over a broad range of concentrations. The maximum is found in the range of 40–50% of acesulfame K and 50–60% of aspartame, and decreases only if one of the sweeteners prevails.[8,9]

The second beneficial effect with blends of acesulfame K and other intense sweeteners is an improvement of sweetness quality, often called qualitative synergism. It is explained by overlap of the different time–intensity profiles of the sweeteners that brings the resulting profile closer to sugar. In blends of acesulfame K and aspartame or sucralose, the lasting sweetness of these sweeteners is reduced (see Figure 5.5). Blends of acesulfame K and

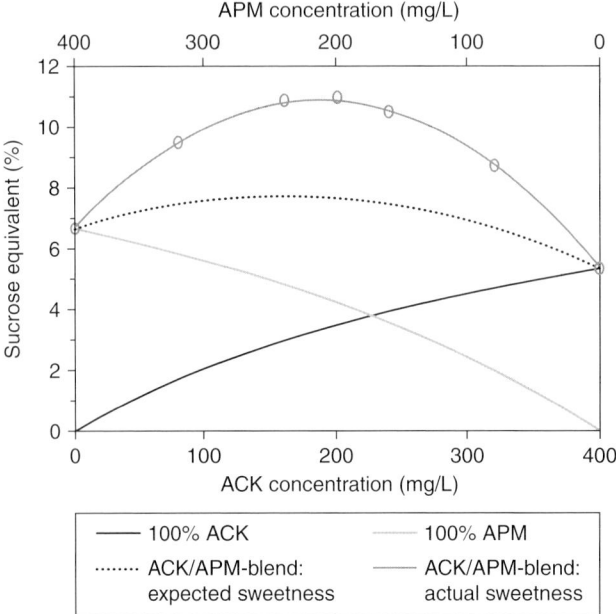

Fig. 5.4 Sweetness intensities of acesulfame K (ACK) and aspartame (APM) and their blend (50/50).[9]

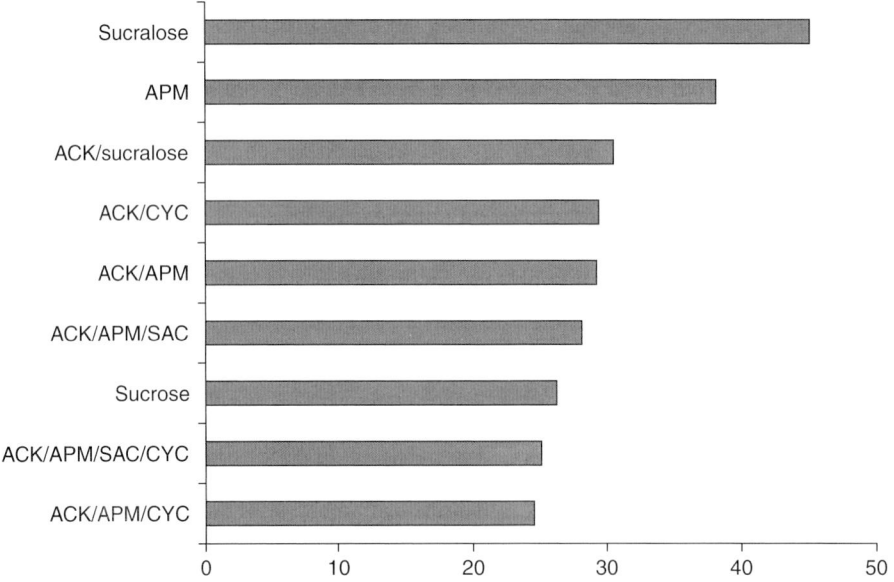

Fig. 5.5 Sweet aftertaste of different high-intensity sweeteners and sweetener blends versus sucrose (4% sucrose equivalence, aqueous solutions)[11] (ACK, acesulfame K; APM, aspartame; CYC, cyclamate; SAC, saccharin).

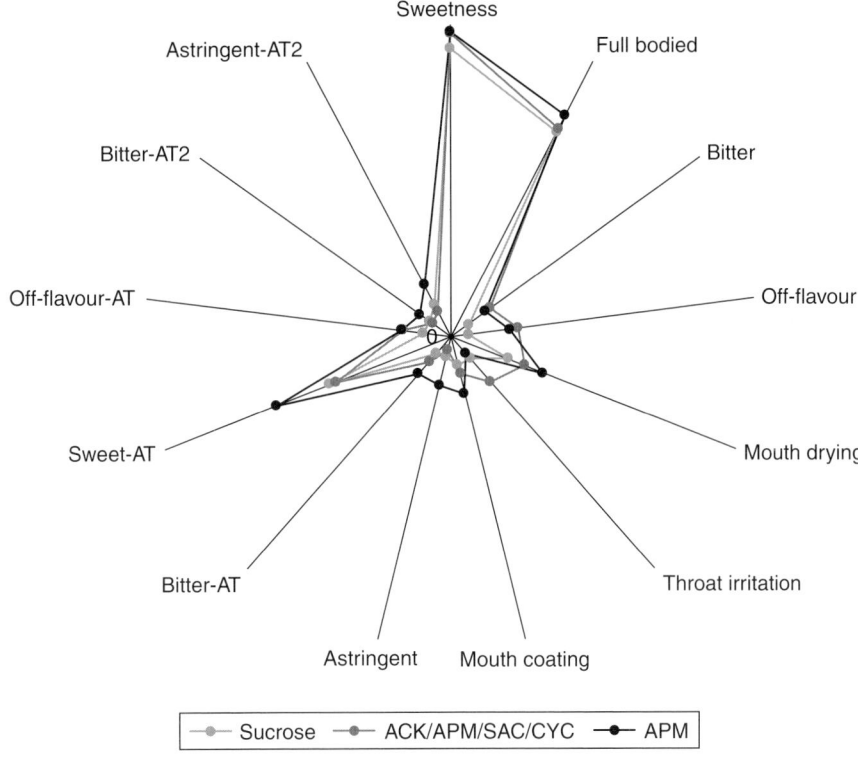

Fig. 5.6 Taste profile of aspartame and ACK/APM/SAC/CYC versus sucrose (4% sucrose equivalence, aqueous solutions)[11] (ACK, acesulfame K; APM, aspartame; CYC, cyclamate; SAC, saccharin).

aspartame; acesulfame K, aspartame and sodium cyclamate or acesulfame K, aspartame, sodium cyclamate and sodium saccharin come especially close to the sweetness profile of sugar (see Figure 5.6).[10–12]

The lingering sweetness of aspartame and sucralose was substantially reduced when blending either of the sweeteners with acesulfame K. Blends consisting of two or especially three different sweeteners on repeated presentation exhibited less reduction in sweetness intensity over four repeated sips than a single sweetener at an equivalent sweetness level. These findings indicate that the decline in sweetness intensity observed after repeated exposure to a sweet stimulus could be reduced by the blending of sweeteners.[13]

The rapid sweetness perception of acesulfame K combines advantageously with the taste profiles of sugar alcohols. The resultant blends have a full and rounded sweetness and are normally rated better than the sweetness of sugar alcohols alone, which are sometimes perceived as 'flat'. Dosages of sugar alcohols correspond to those necessary to replace the 'bulk' of sucrose in a formulation. Acesulfame K is used in quantities necessary to adjust and round the sweetness to the required sweetness level.

As with sugar alcohols, acesulfame K can also be combined with sweet carbohydrates. Blends with glucose or glucose–fructose syrups can be produced that meet the sucrose sweetness profile of sugar using either acesulfame K alone or with suitable sweetener blends containing acesulfame K.[14] Combined applications with complex carbohydrates like oligofructose or inulin are especially interesting. In these applications, positive effects on sweetness quality and mouthfeel are obtained in addition to a strong synergistic sweetness enhancement for blends of acesulfame K and aspartame, which go far beyond the quantitative synergy of the two high-intensity sweeteners.[15]

5.2.3 Compatibility with flavours

New investigations into taste interaction of intense sweeteners with certain flavours showed that the quality of different sweetener blends can be substantially different from flavour to flavour. Comprehensive data are available for acesulfame K and its blends, for example, for cola, orange, peach, strawberry, lemon-lime, tea and coffee,[16–20] and are of practical importance when optimising the sweetness of food products.

5.3 PHYSICAL AND CHEMICAL PROPERTIES

5.3.1 Appearance

Acesulfame K forms white to colourless monoclinic crystals that are non-hygroscopic. Its bulk density is 1.81 g/cm^3. The molar mass is 201.2.

5.3.2 Solubility

Acesulfame K is freely soluble in water and also in aqueous alcoholic solutions with a high water content. Its solubility is greater than required for practical application in both aqueous systems and in syrups of sorbitol, glucose or glucose–fructose (see Table 5.1). In addition to its use in a solid form, acesulfame K can also be used in the form of a concentrated stock solution. In contrast, it is sparingly soluble in most organic solvents.[21]

Table 5.1 Physicochemical properties and solubility of acesulfame K.

Appearance	Solid, crystalline (monocline), non-hygroscopic
Colour	Colourless to white
Colour in solution	Colourless
Dry matter	>99%
Melting point	>200°C (under decomposition)
Bulk density	1.81 g/cm^3
Equilibrium humidity	>80% RH
pH value (10% solution)	5.5–7.5
Solubility in water	~150 g/L at 0°C
	~270 g/L at 20°C
	~1300 g/L at 100°C
Solubility in ethanol	~220 g/L in 15% aqueous ethanol
	~100 g/L in 50% aqueous ethanol
Solubility in sugar syrups (at 20°C)	≥100 g/L for sucrose syrup (62.5% dry matter)
	≥160 g/L for invert sugar syrup (62.5% dry matter)
	~150 g/L for fructose syrup (50% dry matter)
Solubility in sugar alcohol syrups (at 20°C)	≥75 g/L for sorbitol syrup (70% dry matter)
	≥100 g/L for maltitol syrup (80% dry matter)
	≥250 g/L for isomalt syrup (25% dry matter)

5.3.3 Stability

5.3.3.1 Storage stability

Acesulfame K can be stored for many years in solid form without visible or analytically detectable changes.

Acesulfame K is, accordingly, stable in dry preparations like powdered beverages or desserts and tablets, but also in products having a low-water content like hard candy or chewing gum.

5.3.3.2 Stability in solution

Under the normal storage and processing conditions of liquid foods and beverages, acesulfame K is stable and at the common pH levels and customary storage conditions of beverages, the sweetness of acesulfame K remains unchanged. At pH 3.0 and storage at 30°C, even after several months, no perceptible loss in sweetness was observed. Within the common 'best before' period, sweetener loss is well below 10% of the original concentration and sweetness loss is not perceivable. Even storage at 40°C results in perceptible losses only after several months.[21,22] Only at very low pH levels and continuing high storage temperatures can slight losses be detected analytically after prolonged storage.

Investigation of reaction kinetics at 20°C showed a half-life of 11.5 years at pH 5.77 and 6.95 years at pH 3.22.[23]

5.3.3.3 Temperature stability

Dry, solid acesulfame K is stable at high temperatures. The decomposition limit depends on the rate of heating and is around 225°C under conditions of melting point determination (rapid heating). Slow heating may result in decomposition at slightly lower temperatures.

Acesulfame K is stable under the normal heating conditions used in the processing of foods. Pasteurisation or ultra-high temperature (UHT) treatment used for dairy products do not result in any loss of acesulfame K. Drying processes like spray drying, foam-mat drying or drying in a fluidised bed are also possible without sweetener loss.[24]

Similarly, canning and sterilisation, as used for fruit and vegetables, do not result in acesulfame K losses, when carried out under normal conditions.

Temperatures in a baking process may be as high as 180°C in the crust or outer parts of the product but only around 100°C in the crumb or inner parts of the product. Accordingly, they remain far below the decomposition limit of acesulfame K. In several studies on the baking stability of acesulfame K, no loss on baking was found, even after excess heating that rendered the product very dark, and therefore, organoleptically unacceptable.[25]

Microwave treatment of products containing acesulfame K is possible without any loss of sweetness.[26]

For sweets and confections, acesulfame K can be added in the usual way before cooking if the acids are added after cooking.

Kinetic investigations have shown that a decrease in sweetness is not anticipated even under extreme processing conditions and acesulfame K has been found to be 'extremely stable'.[23]

5.3.3.4 Stability in contact with food constituents

Reactions of acesulfame K with food constituents have not been observed. In model systems in which excess concentrations of acesulfame K and several typical food constituents were used to increase sensitivity and in which solutions were stored at elevated temperatures, no indication of reactions of acesulfame K with these compounds were found. Reactions with flavour compounds have also not been observed.[27]

5.4 PHYSIOLOGICAL PROPERTIES

In the human body, acesulfame K is absorbed, and similarly excreted, rapidly. After a single dose, excretion is virtually complete within 24 hours and it is excreted completely unchanged in the urine. Therefore, it is non-caloric. No metabolism was observed in humans or other animal species.[28]

Acesulfame K is not metabolised by bacteria in the oral cavity or in the gut. It cannot therefore be transformed into acids that attack the tooth enamel. Ziesenitz and Siebert showed that *Streptococcus mutans*, a bacteria species contributing to the formation of caries, was inhibited by acesulfame K at high concentrations.[29] In blends of acesulfame K with several other sweeteners[30] and also in blends of acesulfame K, saccharin and fluoride,[31] synergistic inhibitory effects were found.

In the human body, acesulfame K is inert. In a screening programme for potential pharmacological effects, no substance-specific influences on the functions of the body were found. Only effects attributable to potassium were detected.[32] These included absence of influence on secretion of insulin and the blood glucose level at normal intake levels. Under conditions equivalent to a real food consumption situation, levels of insulin and blood glucose remained equivalent to the control group while within a group consuming sugar, insulin was secreted and the blood glucose level increased as expected. Subsequently, in the group receiving sugar, the blood glucose level fell below the starting level.[33,34] As a decrease in blood glucose is, among others factors, held responsible for the feeling of hunger, such secondary effects would not be expected after consumption of acesulfame K. Subjective higher hunger ratings and increased food intake that were reported in a test situation after intake of sweetener solutions including acesulfame K seem questionable.[35] In this study, the control group received a glucose solution 1 hour before a test meal but glucose calories were not taken into account in the interpretation of the results. The total intake in calories from a solution of sweetening agent and test meal were higher for glucose than for the sweeteners studied.

No allergic reactions and no substantiated claims of adverse reactions with acesulfame K are known.

5.5 APPLICATIONS

Owing to its favourable range of characteristics, acesulfame K can be used in a wide variety of foods and beverages.

5.5.1 Beverages

Low-calorie and calorie-reduced beverages are a very important field of application for acesulfame K. It has good stability at the usual low pH levels of carbonated beverages. The taste quality of blends with acesulfame K with other sweeteners is superior to single sweeteners even at the fairly high sweetness level of these beverages. Consequently, such blends are used almost exclusively and beverages benefit especially from synergistic sweetness enhancement and quality improvement. When intense sweeteners are combined appropriately, the sweetness profile of sugar can be matched. Should differences be perceptible, they are usually based on the higher sensation of 'body' and the mouthfeel normally provided by sugar. Synergistic sweetness enhancement may result in substantial savings in the quantity of sweetener used. Therefore, use of blends may also be very economic.

When acesulfame K is blended with other sweeteners for beverage use, it may be reasonable to deviate from blend ratios that provide the highest synergistic sweetness enhancement. For example, in blends of acesulfame K and aspartame, variation of blend ratios allows modification of the time–intensity profile of sweetness and adaption to flavour profiles.

Detailed analyses of the influence of sweeteners on flavour perception have shown that sweetener blends are normally superior to single sweeteners. In highly complex cola flavours, depending on the sweetener blend, several top notes like 'spicy' can be enhanced. Particularly good results for the general cola perception are obtained when the notes 'spicy' and 'lemon' are rated as balanced or the note 'spicy' is stronger than 'lemon-lime'. Blends of acesulfame K and aspartame (30/70), acesulfame K and sugar (0.23/99.77), acesulfame K, aspartame and cyclamate (15/15/70) or acesulfame K and sucralose (40/60) were found to support a good perception of the cola flavour.[17]

For lemon-lime-flavoured beverages, blends of acesulfame K and aspartame (30/70 or 50/50) and acesulfame K, aspartame and cyclamate (15/15/70) come closest to the sweetness profile of sugar than other blends (see Figure 5.7). The blend of acesulfame K and aspartame (30/70) is the only sugar-free blend that provides fruitiness equivalent to sugar.[19] Blends of acesulfame K and aspartame (30/70) also resulted in time–intensity profiles for sweetness and flavour comparable to sugar in orange-flavoured beverages.

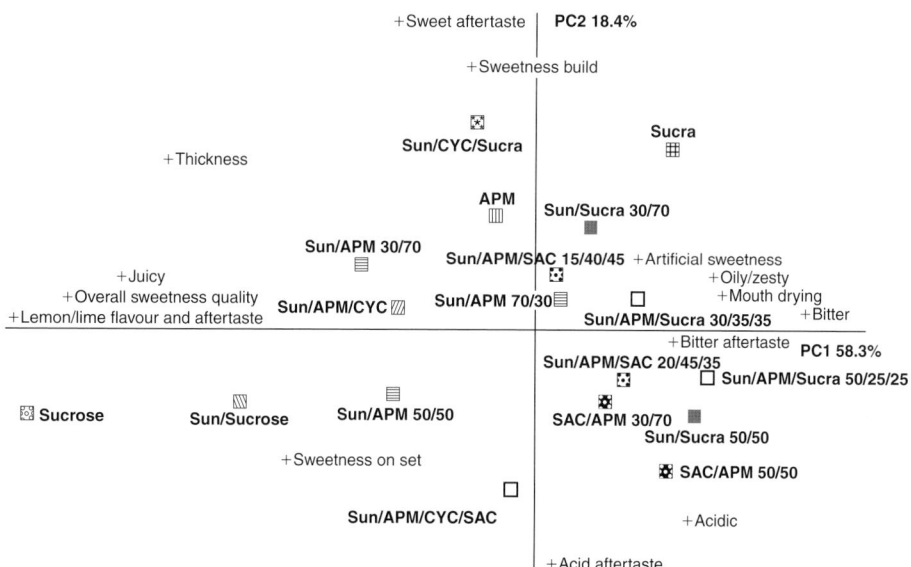

Fig. 5.7 Principle component analysis plot of selected sweetener blends (ratios given as sweetness contribution) in lemon-lime-flavoured drinks (APM, aspartame; CYC, cyclamate; SAC, saccharin; sucra, sucralose; SUN Sunett (R) acesulfame K).

Table 5.2 Flavour-specific recipe guidelines for use of acesulfame K in sugar-free beverages at 10% sucrose equivalence.

Flavour	Acesulfame K (mg/L)	Aspartame (mg/L)	Sucralose (mg/L)	Sodium cyclamate (mg/L)	Sodium saccharin (mg/L)
Lemon-lime	110	257	–	–	–
	168	168	–	–	–
	175	–	100	–	–
	115	125	55	–	–
	100	100	–	280	–
	65	65	–	280	40
Cola	110	260	–	–	–
	97	–	146	–	–
	194	97	36	–	–
	100	100	–	280	–
Orange	109	258	–	–	–
	175	–	100	–	–
	110	121	53	–	–
	100	100	–	280	–
Strawberry	179	179	–	–	–
	175	–	100	–	–

For natural raspberry flavour, blends of acesulfame K and aspartame ranging from 40/60 to 25/75 were rated especially favourably. In contrast, for artificial raspberry flavour, the ratios between 50/50 and 20/80 gave good results.

It is generally advantageous for beverages, if fruit flavours are enhanced and sweeteners are used that do not have side-tastes or off-tastes. Blends of acesulfame K with aspartame, acesulfame K, aspartame and cyclamate and well-balanced blend ratios of acesulfame K and sucralose have these advantages.[36]

Flavour-specific recipe guidelines for use of acesulfame K and its blends in sugar-free beverages are shown in Table 5.2.

In addition to low-calorie beverages, reduced-calorie beverages are becoming increasingly popular. In these products, sugar is only partially replaced by intense sweeteners like acesulfame K or its blends. Blends of acesulfame K with fructose, glucose, glucose–fructose syrups and also sugar are suitable for this purpose. All these carbohydrates combine well with acesulfame K and the resulting beverages show slightly higher viscosity than those containing only intense sweeteners and are therefore perceived as having more body. Very good results were obtained in these beverages with blends of acesulfame K and aspartame, which allow reduction of the carbohydrate content by up to 80% (see Table 5.3). If

Table 5.3 Flavour-specific recipe guidelines for use of acesulfame K in sugar-reduced beverages at 10% sucrose equivalence.

Sugar replacement ratio (%)	50	50	65	65	80	80
Sucrose (g/L)	50	–	35	–	20	–
HFCS 55 (solids) (g/L)	–	50	–	35	–	20
Acesulfame K (g/L)	0.080	0.076	0.098	0.088	0.115	0.108
Aspartame (g/L)	0.080	0.076	0.098	0.088	0.115	0.108

HFCS, high-fructose corn syrup.

acesulfame K and aspartame are combined with glucose–fructose syrups or high-fructose corn syrups, slightly higher synergies occur so that concentrations of high-intensity sweeteners can be reduced. In parts of the world in which sugar is expensive or not of consistent quality, such combinations can be used to stabilise beverage quality and to provide an acceptable cost basis.[37]

In fruit nectars, carbohydrates in the juice provide the basic sweetness. Therefore, lower use levels than for carbonated beverages are necessary, for example up to 200 mg/L for acesulfame K, approximately 100 mg/L of acesulfame K and 50 mg/L of aspartame or 60–80 mg/L of both acesulfame K and aspartame.[38]

For the common pH levels and storage periods of beverages, the sweetness of acesulfame K does not change and remains stable. Therefore, the taste of beverages containing acesulfame K also remains stable unless other constituents or ingredients cause stability problems.

For sugar-free powdered beverages, the same use levels of intense sweeteners can be used as for liquid beverages, calculated on the ready-to-drink beverage. Suitable bulking ingredients are maltodextrins or non-hygroscopic sugar alcohols, and for reduced-sugar powders, sucrose or glucose can be used. In liquid beverage concentrates, acesulfame K can be used without problems owing to its good solubility. For concentrates, blends are used rather than single sweeteners.

While less important than for non-alcoholic beverages, intense sweeteners are used in shandies, coolers and mixers with low alcohol content, the so-called alcopops or flavoured alcoholic beverages (FAB). In lemon-flavoured shandies, blends of acesulfame K and aspartame (50/50), acesulfame K and sucralose (50/50) or acesulfame K, aspartame and cyclamate (15/15/70) give good taste results and are virtually indistinguishable from sugar-containing products. For cola-flavoured shandies, blends of acesulfame K and aspartame or sucralose in the range of 40/60 give optimum results. Maximum levels approved in the European Union (EU) favour the use of sweetener blends as most single sweeteners do not give adequate sweetness. The additional benefit of these blends is the synergistic sweetness enhancement.[39]

5.5.2 Dairy products and edible ices

Flavoured dairy products are normally sweetened. Where intense sweeteners are used, a distinction should be made between products that are not heated and those that are heat-treated. Non-fermented dairy products, like flavoured milk or cocoa-based beverages, often undergo UHT treatment to improve their shelf life. Under these conditions, acesulfame K may be used as the single sweetener or in blends with other heat-stable sweeteners like cyclamate or sucralose.[40]

Fermented dairy products like yoghurt or white cheese are, generally, either not heated or only pasteurised. Blends of acesulfame K and aspartame in most cases provide a well-balanced sweetness. Stirred-style yoghurt or white cheese can be sweetened without problems, as intense sweeteners are normally added with the fruit preparations.[41] Application in sundae-style yoghurt and white cheese is not as easy as fruit preparations, containing intense sweeteners, as they have a lower density than those containing sugar. Therefore, there is a tendency for the layers to mix. Suitable recipes containing thickeners that form intermediate layers with increased viscosity prevent such undesirable mixing.

In functional dairy products like probiotic or prebiotic yoghurts, the strong quantitative synergism occurring particularly in blends of equal parts of acesulfame K and aspartame with the prebiotic bulking ingredients, inulin or oligofructose, is a particular advantage.[15]

Acesulfame K is not degraded by lactic acid bacteria and other micro-organisms used in the production of fermented milk products.[42]

For heat-processed desserts, the sweetener systems suggested for flavoured milk, for powdered products and for fruit-flavoured yoghurts are recommended.

Fruit-flavoured water-based edible ice can be sweetened with intense sweeteners and without bulking ingredients. However, the good taste compatibility of acesulfame K with a sugar alcohol bulking agent allows production of sugar-free ice cream of equivalent texture and taste to the full sugar version. The pasteurisation of ice cream mixes does not harm acesulfame K.

Recipe guidelines for use of acesulfame K and its blends in dairy products and ice cream are shown in Tables 5.4 and 5.5

5.5.3 Bakery products and cereals

Fine bakery wares normally require water-soluble bulking ingredients for the adjustment of texture. In conventional products, sugar is a bulking ingredient as well as a sweetener. Sugar-free bakery products of acceptable texture can therefore only be produced when sugar alcohols replace the sugar functions. Therefore, acesulfame K can be combined with sugar alcohols and/or other bulking agents in almost all such bakery products. The necessary use levels are normally about or below 1000 mg/kg.

Acesulfame K has no influence on the texture or appearance of bakery products. Taste and texture of bakery products with sugar alcohols or blends of sugar alcohols, polydextrose and acesulfame K are rated good when recipes are designed appropriately. It should, however, be noted that neither sugar alcohols nor acesulfame K undergo browning reactions. Therefore, bakery products with no-added sugar are paler than comparable conventional products. Browning and the resulting slightly darker colour can be achieved by addition of small quantities of fructose, if desired.

In cookies, cakes and cheese or fruit fillings, there is no analytical loss of acesulfame K on baking. This holds true even for prolonged baking with dark discolouration of cookies as well as for too high oven temperature, which results in black spots on the surface and therefore unacceptable products.[21]

Edible paper should have a sweet taste but not show any discolouration during baking. Use of sugar as the sweetening agent is problematic, as browning reactions occurring during baking yield darker products than desirable. Acesulfame K allows production of edible paper having light colour and pleasant sweetness, as it is stable on baking and does not cause browning.

For the production of sugar-free or reduced sugar breakfast cereals and nutrition bars, acesulfame K can be blended with the other ingredients before extrusion and is not degraded during the extrusion process. Owing to the low use levels that are around 1000 mg/kg for the single sweetener and in blends with sucralose (175–250 mg/kg) in the range 500–750 mg/kg, acesulfame K does not have the binding function of sugar as seen in conventional products. Pre-sweetened cereal flakes can be produced by spraying a solution of acesulfame K onto the product with subsequent drying.

5.5.4 Sweets and chewing gum

Sugar alcohols determine the properties of sugar-free sweets and confections as they form by far the largest proportion of the ingredients. Selection of the appropriate sugar alcohol(s)

Table 5.4 Recipe guidelines for use of acesulfame K in dairy products.

Application	Acesulfame K (mg/kg)	Aspartame (mg/kg)	Sodium cyclamate (mg/kg)	Sucralose (mg/kg)	Sucrose (g/kg)	Fructose (g/kg)	Inulin/oligo-fructose (g/kg)
Yoghurt	200	200	–	–	–	–	–
	160	160	–	–	–	10	–
	140	140	–	–	–	–	20
	110	–	–	100	–	–	–
	130	–	–	85	–	–	10
Drinkable yoghurt	135	135	–	–	–	–	20
	160	160	–	–	–	–	–
Milk and cocoa drinks	280	–	220	–	–	–	–
	330	110	–	–	–	–	–
	220	–	–	–	50	–	–
	75	75	–	–	50	–	–
	300	–	–	–	–	–	20

Table 5.5 Recipe guidelines for use of acesulfame K in ice cream.

Application	Acesulfame K (mg/L)	Aspartame (mg/L)	Isomalt (g/L)	Polydextrose (g/L)	Lactitol (g/L)
Vanilla ice cream	500	–	50	50	50
Chocolate soft ice cream	300	300	–	70	–

is therefore responsible for the processing characteristics, texture and storage stability of the products. As use levels of acesulfame K are low and normally in the range of few hundred mg/kg, it does not have any significant influence on these properties or on the storage characteristics.

The sweetness intensity of the commonly used sugar alcohols in confectionery products, isomalt and sorbitol is well below the sweetness of the blends of sugar and starch syrups that are used in conventional products. As the sweetness intensity of acesulfame K is perceived quickly, it is especially suitable to round off the sweetness of sugar alcohols in confectionery products and increase it to the customary higher level. Such combinations are especially compatible with fruit flavours and the results are often better than in conventional products.[43,44] In micro breath mints, the rapidly perceptible sweetness of acesulfame K is rated favourably and in addition, longer perception of sweetness and mint flavour is achieved.

Sugar-free confectionery products are produced in a similar way as sugar-based products, but often under slightly different cooking conditions. When acids are added before cooking, it is advisable to add acesulfame K together with the other ingredients (like flavours and colours) normally added after cooking. Otherwise, high cooking temperatures and low pH levels may cause slight sweetness losses. Whenever acids are added after cooking or the cooking mass is buffered adequately, acesulfame K can be mixed with the other ingredients before cooking, and no decomposition is likely.

Sugar alcohols and bulking agents are necessary to adjust the texture of some sugar-free hard and soft candies, gelatine-, gum-, jelly-based and similar products. They have the important function of keeping the products soft on chewing and even during storage. Acesulfame K, again, serves to round and increase the sweetness. Although the pH level of these products is often low, acesulfame K is sufficiently stable and does not change its sweetness during storage.

Sugar-free chocolate also requires water-soluble bulking ingredients, preferably non-hygroscopic disaccharide polyols. As for the other products of this group, acesulfame K is used to round the sweetness and increase it to the usual level. The easiest way to add acesulfame K is before grinding and rolling of the other ingredients so the desired fine granulation is achieved. During grinding, rolling and conching, acesulfame K remains completely stable. It has no influence on the texture and flavour development characteristics of chocolate and storage characteristics, and it does not support formation of fat bloom or crystallisation of other ingredients.[45]

The availability of disaccharide alcohols has resulted in improved options for the production of sugar-free marzipan. They allow production of sugar-free products coming close to conventional marzipan and having the same sweetness when acesulfame K is added.

Sugar-free chewing gum also requires water-soluble bulking ingredients and in this application sorbitol, xylitol and maltitol are normally used. While sorbitol is less sweet than sugar, xylitol has similar sweetness to sugar. Both sugar alcohols can be combined advantageously with acesulfame K. Owing to its high water solubility, acesulfame K is quickly dissolved from the mass of the chewing gum and therefore provides the important, initial sweetness impact. The high solubility, however, results in a rapid decrease in the sweetness of the product. This rapid loss can be reduced by using encapsulated acesulfame K giving a delayed release of the sweetener or by combining the acesulfame K with other sweeteners giving a delayed sweetness onset but a longer lasting sweetness.

Fine crystals of acesulfame K can be blended into the chewing gum mass directly. For recipes using sorbitol syrup, it can also be dissolved in this syrup and the two added together.[46]

As in other confectionery products, acesulfame K is stable in chewing gum and no reactions between acesulfame K and flavours have been observed.

In addition to its function as a sweetener, acesulfame K has flavour-enhancing properties, especially in mint-flavoured chewing gum but also in fruit-flavoured products like bubblegum. This can also be used to advantage in sugar-based products in which addition of only a few hundred mg/kg of acesulfame K are sufficient to prolong the flavour perception.

5.5.5 Jams, marmalades, preserves and canned fruit

It is difficult to produce jams and marmalades using only intense sweeteners. Texture and shelf stability have to be provided by bulking ingredients. Replacement of sucrose by the same amount of sugar alcohols results in sugar-free products, but not necessarily products with significant calorie reduction. In these products, acesulfame K can adjust the sweetness to the usual level. Small-scale production using a combination of acesulfame K and sorbitol gave products of excellent taste quality. Such products have, however, not gained significant importance in the market place to date.

Use of gelling agents like amidated pectins allows production of preserves with a reduced level of dry matter. Stability is, however, lower than for conventional products, as flavours change more rapidly over time and there is also a higher tendency for synerisis to occur. Owing to their higher water activity, these products are subject to microbial attack so addition of preservatives, preferably sorbates, is advisable.[47]

Production of canned fruit with intense sweeteners gives products with a low viscosity covering syrup. Acesulfame K can be used for products with higher pH levels that are sterilised at 121°C as well as for products with a lower pH level that are normally sterilised at lower temperatures. Blends of acesulfame K and sucralose are especially suitable for the production of sugar-free and calorie-reduced canned fruit.[48]

5.5.6 Delicatessen products

In delicatessen products, sweetening agents of different types are used to mask the taste of the commonly used vinegar, at least in part, and therefore to balance the overall taste. This applies especially for pickles, salad dressings, delicatessen salads and marinated fish. The tendency to produce products free from preservatives requires lower pH levels than for preserved products. Therefore, acceptability of these products decreases unless the taste is balanced by use of sweetening agents. In several countries these products have become sweeter over the last few decades to match consumer preferences.

Many delicatessen products are not free from micro-organisms but do not provide the necessary nutrients for growth of spoilage-causing organisms. They are therefore preferably sweetened with products that cannot be metabolised by micro-organisms and therefore do not promote microbial growth. Acesulfame K is not metabolised by micro-organisms and freely soluble, so it is well suited for use in these products. The taste of vinegar can be masked with a few hundred mg/kg.

From a technical point of view, the use of intense sweeteners instead of sugar may be necessary for high-fat emulsions in which solubility of sweet carbohydrates in the small quantity of aqueous phase is not sufficient to achieve the desired sweetness.

Up to 80% of the calories of tomato ketchup or spicy sauces are provided by carbohydrates, mostly as added sugar. Use of intense sweeteners like acesulfame K, especially in blends with sucralose, allows production of calorie-reduced alternatives with a well-balanced sweetness. Bulking ingredients like polydextrose, modified starches or other thickeners are necessary to increase the viscosity of the product and bring it to the expected level.[49]

5.5.7 Table-top sweeteners

Table-top sweeteners are one of the most important fields of application for intense sweeteners. They are marketed in the form of tablets, granular powders, solutions and a variety of other specialities.

Sweetener tablets are either produced as effervescent tablets or with addition of disintegrating agents to achieve fast dissolution in beverages. For effervescent tablets, lactose is commonly used as the carrier. Carbon dioxide is provided by sodium bicarbonate with tartaric acid serving to release the gas. To avoid too early a release of carbon dioxide, further inert carriers can also be added, including gelatine or carboxymethyl cellulose.[21]

Direct tabletting of acesulfame K is not easy. For production of the non-effervescent tablets, binding agents like carboxymethyl cellulose or polyvinyl pyrrolidone can be used. These also serve as disintegrating agents and favour dissolution of the tablets. Tabletting becomes easier if the ingredients are granulated.

Depending on the intended use of table-top sweeteners, for example coffee or tea, it can be advisable to develop specific sweetener blends. For tea, the taste is more rounded when acesulfame K and aspartame are used in the range of 30/70 or 50/50, or with additional cyclamate in the range of 15/15/70. Blends of acesulfame K and sucralose in the range of 55/45 are well suited to sweeten coffee.[20]

Details of the composition of sweetener tablets with acesulfame K are shown in Table 5.6.

For granular sweeteners, a bulking ingredient is required that provides the necessary volume to render acesulfame K easy to handle for consumers. If a granular sweetener free from calories is desired, such bulking ingredients can be salts of organic acids like citrates or tartrates.

Reduced-calorie granular sweeteners can be produced using lactose, sugar or glucose as bulking agents or, if sugar-free products are intended, sugar alcohols. A blend of 99.7% sorbitol and 0.3% acesulfame K is approximately equivalent to sugar in sweetness intensity

Table 5.6 Examples for application of acesulfame K in table-top sweeteners for hot beverages.

	Tea (mg/tablet)	Coffee (mg/tablet)
Acesulfame K	4.6	6.6
Aspartame	10.6	6.6
Lactose	34	34
Carboxymethylcellulose	3.6	3.6
Leucine	4.0	4.0

and also has a pleasant taste. Whenever isomalt is used instead of sorbitol, the handling characteristics are improved, as it is non-hygroscopic. This blend should, however, be set to 99.6% isomalt and 0.4% acesulfame K to achieve the sweetness intensity of sugar. Such products are produced by simple blending of the dry, solid ingredients or by granulation to reduce the risk of separation of the ingredients.

Spoon-for-spoon products are a speciality that can be dosed in equal volumes as sugar. Suitable carriers for these products are maltodextrins, which form bubbles during spray drying or foam-mat drying and therefore allow production of low-density products. These maltodextrins are, however, hygroscopic so humidity-proof packages are required.

In some European countries like Germany, liquid table-top sweeteners are popular. Owing to its good solubility, acesulfame K allows easy production of such products. When these solutions are adjusted to a pH range of 5–6 with citrate buffers, they can be stored for many years without discolouration. It is advisable to add small quantities of potassium sorbate to prevent growth of micro-organisms that could metabolise the buffer substance.

Another type of speciality table-top sweetener is sugar cubes with reduced-sugar content. A special process allows manufacture of a lower density product compared with a conventional sugar cube. The sugar content per sugar cube can therefore be significantly reduced even if the size of the cubes is only slightly smaller than normal. The same sweetness as in conventional cubes is achieved either by adding acesulfame K alone or by adding blends of acesulfame K and aspartame.

5.5.8 Pharmaceuticals

The active substances in pharmaceuticals often have a poor taste. Owing to its quickly perceptible sweetness, acesulfame K can mask such undesirable taste components or at least temper these, especially in the case of bitter-tasting products. It can be used in lozenges, chewing tablets or effervescent tablets in exactly the same way as in foods. Use in pharmaceutical syrups based on sorbitol is a particularly interesting application, and the solubility of acesulfame K is much higher than necessary for these products, thus avoiding any solubility issues.

An interesting, though not yet exploited, possibility would be development of salts of cationic pharmaceuticals with acesulfame as the anion. This could not only modify the taste but also potentially influence the solubility and absorption in the body.

5.5.9 Cosmetics

Oral hygiene products like toothpaste and mouthwash are commonly sweetened to render them acceptable. As fermentable carbohydrates are potentially cariogenic, they are not suitable for these products.

For toothpaste, humectants are required that prevent the paste from drying and hardening. Glycerol and sorbitol are mainly used for this purpose. While glycerol is only moderately sweet, sorbitol provides some sweetness, although this is normally not considered sufficient. Acesulfame K can increase the sweetness and simultaneously mask the less desirable taste of certain surfactants. It does not have any negative influence on the properties of toothpaste and especially does not affect the texture. Slow degradation of acesulfame K was observed only if calcium carbonate was used as the abrasive or if prolonged storage of such alkaline pastes at elevated temperatures (well above room temperature) was carried out. Acesulfame

K remained stable, however, in these pastes when stored at room temperature as well as in pastes containing other abrasives, especially when the common abrasive dicalcium phosphate was used. Use levels are a few hundred mg/kg for recipes for the European taste and up to approximately 5 g/kg in very sweet American recipes.[50]

Acesulfame K can be used in mouthwash without problems as it is not only soluble in water but also in mixtures of alcohol and water as used in mouthwash. Its solubility is normally sufficient even for the production of concentrates. It harmonises well with common flavours like mint and wintergreen.

Use of acesulfame K in lipsticks and lip balm is also possible. As these are non-aqueous products, a very fine powder grade should be used for these applications.

5.5.10 Tobacco products

Two key applications of acesulfame K are possible in tobacco. It can be used to sweeten cigarette tips, which are often slightly sweet, and it is also suitable for use in chewing tobacco.

5.5.11 Technical applications

No technical applications of acesulfame K are known. It was, however, reported that acesulfame allows production of ionic liquids. Ionic liquids are discussed as a replacement for organic solvents, as they may have better environmental properties. Acesulfame anions would be advantageous compared with the presently used fluoro-organic anions as they are derived from an extensively studied, safe food additive.[51] It is unclear whether these liquids may be used in the future, however.

5.6 SAFETY AND ANALYTICAL METHODS

5.6.1 Pharmacology

In the human body, acesulfame K is inert. No substance-specific effects were found in the screening programme for potential pharmacological effects.[31] These included secretion of insulin and the level of blood glucose.

5.6.2 Toxicology

Acesulfame K was evaluated for its safety in a comprehensive evaluation programme.

In a study on the acute oral toxicity, the LD_{50} was determined to be 7431 mg/kg.[52]

In long-term studies, acesulfame K was well tolerated. In a study on subchronic toxicity, up to 10% and in studies on chronic toxicity up to 3% in the diet were administered. Non-specific effects observed at the highest dose level in the subchronic study were reversible after termination of feeding and therefore obviously caused by the high potassium content of the diet and osmotic effects. None of the studies indicated any negative health effects of acesulfame K consumption.[52–55]

Several *in vitro* and *in vivo* studies for genotoxicity did not show genotoxic effects of acesulfame K. Among these were Ames tests with and without metabolic activation, studies on chromosome aberration, malignant transformation, DNA binding, unscheduled DNA synthesis (UDS), hypoxanthine guanine phosphoribosyltransferase (HGPRT) and dominant lethal tests.[56] Dose-related effects claimed from a study in mice on chromosome aberration[57] could neither be confirmed by examination of the slides from the study nor by repetition under the same conditions.[58–60]

Investigations on reproductive toxicity in several species gave no indications of effects on reproduction. No differences between test and control groups, in fertility in general, number of animals per litter, body weights, growth and mortality, were found.[61–63]

5.6.3 Safety assessments and acceptable daily intake

Safety data for acesulfame K were evaluated by international agencies and national health authorities. The Joint Expert Committee for Food Additives of the WHO and FAO (JECFA),[64] the former Scientific Committee for Food of the EU (SCF),[65] the US Food and Drug Administration (US FDA)[66] and health authorities, for example, of Japan, Canada and Australia, unanimously concluded that acesulfame K was safe for use in food.

Part of the safety evaluation was assessment of hydrolysis products, which may be formed in traces and under extreme conditions like low pH level and prolonged storage at high temperatures outside the common range for foods. They could only be detected in spiking studies with exaggerated concentrations much higher than those used in food. They were assessed as causing no safety problems at levels to be expected in food.[66]

An acceptable daily intake (ADI) of 0–15 mg/kg was allocated by all these agencies and authorities except for the SCF of the EU, which has not yet raised the former internationally valid value of 0–9 mg/kg. Allocation of an ADI concludes that consumption by children and pregnant women is also acceptable. The ADI of acesulfame K is equivalent to the sweetness of approximately 180 g sugar per day for the 60 kg adult. When blends with other sweeteners are used, this may increase to more than 250 g per day as a result of the synergistic sweetness enhancement of some blends. Therefore, sugar equivalents to the ADI of acesulfame K are much higher than the sugar consumption.

5.6.4 Analytical methods

The assay of acesulfame K is carried out by non-aqueous titration with 0.1 N perchloric acid in glacial acetic acid, preferably with potentiometric indication.[67]

Standard method for the detection and quantification of acesulfame K in foods and beverages is reverse-phase liquid chromatography with UV detection at 227 nm or using diode array detectors. It is often simple as beverages or aqueous extracts from food can be injected directly into the columns after filtration and appropriate dilution. Products containing some protein or larger amounts of carbohydrates should be clarified by adding zinc sulphate and potassium hexacyanoferrate.[68] Other UV-active sweeteners like aspartame and saccharin as well as some preservatives may be determined simultaneously by such methods.[68,69]

Other analytical methods like ion chromatography, capillary electrophoresis or isotachophoresis have also been developed but seem to be of no practical importance.[70–72]

5.7 REGULATORY STATUS

5.7.1 Approvals

On the basis of favourable safety assessments, acesulfame K is approved worldwide for food use, normally for a wide variety of foods and especially beverages, table-top sweeteners, dairy products and chewing gum. The EU approval includes more than 30 product categories.[73] In the United States, acesulfame K is approved as a 'general purpose sweetener and flavour enhancer' for use under good manufacturing practice.[66] This means it can be used without formal limits in all non-standardised products and in standardised products when it is listed as such or by general reference to intense sweeteners or sweetening agents in the list of ingredients. The US approval only excludes meat and poultry.

The General Standard for Food Additives of the Codex Alimentarius also lists acesulfame K for a wide variety of products.[74]

While sometimes intake above the ADI is claimed, reliable studies show intakes well below the ADI in all groups of the population.[75–77]

5.7.2 Purity criteria

Purity criteria for acesulfame K are laid down in the EU directive on special purity criteria for sweeteners,[78] the European Pharmacopoeia,[79] the Food Chemicals Codex,[80] the US Pharmacopoeia/National Formulary,[81] the monograph on acesulfame K published by JECFA[67] and a series of national regulations. They normally require an assay of at least 99% and limit several heavy metals.

For UV-active organic impurities, and especially 5-chloro acesulfame K, a limit of 20 mg/kg is generally set. This undesirable compound, which may be present in the material of some suppliers, can also be analysed in foods and beverages with a highly sensitive chromatographic method.

REFERENCES

1. K. Clauß and H. Jensen. Oxathiazinone dioxides – a new group of sweetening agents. *Angewandte Chemie* 1973, 85, 965.
2. A. Linkies and D. Reuschling. Ein neues Verfahren zur Herstellung von 6-Methyl-1.2.3-oxathiazin-2.2-dioxid Kaliumsalz (Acesulfame K). *Synthesis* 1990, 1990(5), 405.
3. M. Boehshar and A. Burgard. 5-chloro acesulfame K – a characteristic indicator for application of the "sulfur trioxide" process in the manufacture of acesulfame K. *Research Disclosure Journal* 2004, RD 477036.
4. K. Hoppe. Neue Vergeichstabellen zur Süßeintensität von 16 Süßungsmitteln. *Lebensmittelindustrie* 1991, 38(1), 13.
5. K. Paulus and M. Braun. Süßkraft und Geschmacksprofil von Süßstoffen. *Ernährungsumschau* 1988, 35, 384.
6. K. Hoppe and B. Gassmann. Bestimmung der Missgeschmacksschwellen von Saccharin, Cyclamat, Acesulfam und Aspartam. *Nahrung* 1985, 29, 417.
7. K. Hoppe and B. Gassmann. Zur Süßungsfähigkeit von Saccharin und Cyclamat in Heißgetränken. *Nahrung* 1979, 23, 319.
8. R.A. Frank, S.J. Mize and R. Carter. An assessment of binary mixture interactions for nine sweeteners. *Chemical Senses* 1989, 17, 621.

9. G.E. Du Bois, D.E. Walters, S.S. Schiffmann, Z.S. Warwick, B.J. Booth, S.D. Pecore, K. Gibes, B.T. Carr and L.M. Brands. Concentration-response relationships of sweeteners. A systematic study. In: D.E. Walters, F.T. Orthoefer and G.E. Du Bois (eds). *Sweeteners Discovery, Molecular Design and Chemoreception.* ACS Symposium Series No. 450. American Chemical Society, Washington, DC; 1991, p. 261.
10. N. Ayya and H.T. Lawless. Quantitative and qualitative evaluation of high-intensity sweeteners and sweetener mixtures. *Chemical Senses* 1992, 17, 245.
11. L.Y. Hanger, A. Lotz and S. Lepeniotis. Descriptive profiles of selected high intensity sweeteners (HIS) blends and sucrose. *Journal of Food Science* 1996, 61, 456.
12. A.C. Noble, N.L. Matysiak and S. Bonnans. Comparison of temporal perception of fruitiness in model systems sweetened with aspartame, an aspartame and acesulfame K blend , or sucrose. *Food Technology* 1991, 45(11), 121.
13. S.S. Schiffman, E.A. Sattely-Miller, B.G. Graham, J. Zervakis, H.H. Butchko and W.W. Stargel. Effect of repeated presentation on sweetness intensity of binary and ternary mixtures of sweeteners. *Chemical Senses* 2003, 28(3), 219.
14. S. Rathjen. Nutrinova GmbH. US Patent Application 2005/0037121, 17 February 2005.
15. M. Wiedmann and M. Jager. Innovative sweetening systems: synergies, functional & health benefits. *Food Ingredients and Analysis International* 1997, November/December, 51.
16. S. Meyer. Taste interactions-Adjusting sweeteners to flavours. *World of Food Ingredients* 2000, December, 42.
17. S. Meyer. Sweeteners in cola – optimisation through taste interactions. *Soft Drinks International* 2002, February, 27.
18. S. Meyer. Custom-tailored sweetness for fruit flavours. *Soft Drinks International* 2001, September, 38.
19. S. Rathjen. Fine tuning sweetener blends in lemon-lime drinks. *International Food Ingredients* 2003, April/May, 26.
20. K. Sälzer. New customised sweetening systems for tea and coffee. *Innovations in Food Technology* 2004, February, 2.
21. G.-W. von Rymon Lipinski. Properties and applications of acesulfame-K. In: D.G. Mayer and F.H. Kemper (eds). *Acesulfame K*, Marcel Dekker, New York; 1991, p. 209.
22. G.-W. von Rymon Lipinski. Stability and Synergism – Important characteristics for the Application of Sunett (R) Food Ingredients Conference Proceedings, Maarssen, The Netherlands;1989; p. 249.
23. C. Coiffard, L.J.M. Coiffard and Y. de Roeck-Holtzhauer. Influence of pH on the thermodegradation of acesulfame K in aqueous diluted solutions. *STP Pharma Sciences* 1997, 7(5), 382.
24. A. Lotz, C. Klug and G.-W. von Rymon Lipinski. Stability of acesulfame K during high temperature processing under conditions relevant for dairy products. *Zeitschrift für Lebensmitteltechnologie und Verfahrenstechnik* 1992, 43(5), EFS 21.
25. C. Klug, G.-W. von Rymon Lipinski and D. Böttger. Baking stability of acesulfame K. *Zeitschrift für Lebensmitteluntersuchung und Forschung* 1992, 194(5), 476.
26. M. Korb, B. Kniel and E. Meyer. Mikrowellenstabilität der Süßstoffe Acesulfam und Aspartam. *ZFL International Journal of Food Technology and Food Process Engineering* 1992, 43(9), 494.
27. K. Clauß. Unpublished results.
28. M. Volz, O. Christ, H. Eckert, J. Herok, H.-M. Kellner and W. Rupp. In: D.G. Mayer and F.H. Kemper (eds). *Acesulfame K.*, Marcel Dekker, New York; 1991, p. 7.
29. S.C. Ziesenitz and G. Siebert. Non-nutritive sweeteners as inhibitors of acid formation by oral microorganisms. *Caries Research* 1986, 20, 498.
30. G. Siebert, S.C. Ziesenitz and J. Lotter. Marked caries inhibition in the sucrose-challenged rat by a mixture of non-nutritive sweeteners. *Caries Research* 1987, 21, 141.
31. A.T. Brown and G.M. Best. Apparent synergism between the interaction of saccharin, acesulfame K and fluoride with hexitol metabolism by Streptococcus mutans. *Caries Research* 1988, 22, 2.
32. H.G. Alpermann, E. Granzer, J. Kaiser and R. Muschaweck. Pharmacological studies with acesulfame-K. In: D.G. Mayer and F.H. Kemper (eds). *Acesulfame K*, Marcel Dekker, New York; 1991, p. 137.
33. B. Härtel, H.-J. Graubaum and B. Schneider. Einfluss von Süßstoff-Lösungen auf die Insulinsekretion und den Blutglucosespiegel. *Ernährungsumschau* 1993, 40, 152.
34. J. Steiniger, H.-J. Graubaum, H.-D. Steglich, A. Schneider and C. Metzner. Gewichtsreduktion mit saccharose- und süßstoffhaltiger Reduktionsdiät. *Ernährungsumschau* 1995, 42, 430.

35. J.P. Rogers, J. Carlyle, A.J. Hill and J.E. Blundell. Uncoupling sweet taste and calories: compensation of the effects of glucose and three intense sweeteners on hunger and food intake. *Physiology and Behavior* 1988, 39, 561.
36. R.F. Baron and L.Y. Hanger. Using acid level, acesulfame potassium/aspartame blend ratio and flavour type to determine optimum flavour profiles of fruit flavoured beverages. *Journal of Sensory Studies*, 1998 13(3), 269.
37. S. Meyer and A. Derieth. Partial sugar replacement: breaking the boundaries. *Innovations in Food Technology* 2002, May, 14.
38. M. Abou-Zaid, A. El-Said and A. Askar. Süßstoffmischungen für kalorienarmen Aprikosennektar. *Flüssiges Obst* 1991, 58(4), 193.
39. G. Ritter. Poppig, alkoholhaltig, kalorienarm. Einsatz von Süßstoffen bei der herstellung von Alcopops. *Getränkeindustrie* 1997, 10, 706.
40. G.-W. von Rymon Lipinski. Sunett® - ein Süßstoff für Milcherzeugnisse. *Deutsche Molkerei Zeitung dmz* 1990, 111(6), 176.
41. G.-W. von Rymon Lipinski. Süßstoffe für Fruchtzubereitungen. *Deutsche Molkerei Zeitung dmz* 1991, 112, 574.
42. A. Lotz, C. Klug and K. Kreuder. Stabilität von Acesulfam und Aspartam bei der Herstellung von fermentierten Milcherzeugnissen. *Zeitschrift für Lebensmitteltechnologie und Verfahrenstechnik* 1991, 42(3), EFS 7.
43. L. Hanger, G.-W. von Rymon Lipinski and Z. Nakhost. Acesulfame K: applications in hard candy production. *The Manufacturing Confectioner* 1995, November, 75.
44. G.-W. von Rymon Lipinski and E. Gorstelle. The application of Sunett® in confectionery. *Confectionery Production* 1989, 55(9), 597.
45. G.-W. von Rymon Lipinski and E. Klein. Sunett® - Anwendungen in Süßwaren. *Süßwaren* 1988, 32(9), 336.
46. G.-W. von Rymon Lipinski and E. Lück. Hoechst AG. German Patent Application, DE-3120857, 5 January 1983.
47. L. Hörlein, A. Lotz and K. Gierschner. Herstellung von brennwertverminderter diabetikergeeigneter Erdbeerkonfitüre unter Verwendung der beiden Süßstoffe Acesulfam-K und Aspartam. *Industrielle Obst- und Gemüseverwertung* 1995, 80(1), 3.
48. K. Sälzer. The world of fruits in cans. *Fruit Processing* 2005, May/June, 169.
49. K. Sälzer. More ketchup, less calories – Sunett® presents new opportunities for low calorie sauces. *Innovations in Food Technology*, 2005, February, 46.
50. R. Schmidt, E. Janssen, O. Häussler, X. Duriez and R.F. Baron. Eine hochwertige Süßungsalternative für Zahnpasten. *SÖFW Journal* 1998, 124(3), 148.
51. E. Carter, S. Culver, P. Fox, R. Goode, I. Ntai, M. Tickell, R. Traylor, N. Hoffman and J. Davis Jr. Sweet success: ionic liquids derived from non-nutritive sweeteners. *Chemical Communication* 2004, 630–631.
52. E.J. Sinkeldam, H.P. Til, A.P. Groot, M.I. Willems, R. Kreiling and D.G. Mayer. Toxicity studies of acesulfame-K: a new high-intensity sweetener. In: D.G. Mayer and F.H. Kemper (eds). *Acesulfame K.*, Marcel Dekker, New York; 1991, p. 27.
53. R.B. Beems, H.P. Til, J. Newman and D.G. Mayer. Carcinogenicity study with acesulfame-K in mice. In: D.G. Mayer and F.H. Kemper (eds). *Acesulfame K.*, Marcel Dekker, New York; 1991, p. 59.
54. P.G.J. Reuzel and C.A. van der Heijden. Long-term oral toxicity study with acesulfame-K in beagles. In: D.G. Mayer and F.H. Kemper (eds). *Acesulfame K.*, Marcel Dekker, New York; 1991, p. 71.
55. E.J. Sinkeldam, C.F. Kuper, R.B. Beems, A.J. Newman and V.J. Feron. Combined chronic toxicity and carcinogenicity study with acesulfame-K in rats. In: D.G.Mayer and F.H. Kemper (eds). *Acesulfame K.*, Marcel Dekker, New York; 1991, p. 43.
56. R. Jung, R. Kreiling and D.G. Mayer. Acesulfame-k: studies for genotoxic effects. In: D.G. Mayer and F.H. Kemper (eds). *Acesulfame K.*, Marcel Dekker, New York; 1991, p. 87.
57. A. Mukherjee and J. Chakrabarti. In vivo cytogenetic studies on mice exposed to acesulfame K – a non-nutritive sweetener. *Food and Chemical Toxicology* 1997, 35, 1177.
58. E. Selzer Rasmussen. Evaluation of the clastogenicity of acesulfame K. *Pharmacology & Toxicology* 1999, 85(Suppl. 1), 60.
59. W. Völkner. RCC-CCR Project 621100, 10 December 1998.
60. W. Völkner. RCC-CCR Project 609900, 25 June 1998.

61. C. Baeder, G. Horstmann, W. Weigand and M. Kramer. Oral embryotoxicity study of acesulfame-K in rabbits. In: D.G. Mayer and F.H. Kemper (eds). *Acesulfame K.*, Marcel Dekker, New York; 1991, p. 115.
62. H.B.W.M. Koëter. Effect of acesulfame-K on pregnancy of the rat. In: D.G. Mayer and F.H. Kemper (eds). *Acesulfame K.*, Marcel Dekker, New York; 1991, p. 105.
63. E.J. Sinkeldam, H.B.W.M. Koëter, H.R. Immel and C.A. van der Heijden. Multigeneration study with acesulfame-K in rats. In: D.G. Mayer and F.H. Kemper (eds). *Acesulfame K.*, Marcel Dekker, New York; 1991, p. 121.
64. Joint FAO/WHO Expert Committee on Food Additives (JECFA). In: *Toxicological Evaluation of Certain Food Additives and Contaminants. WHO Food Additives Series No. 28.* WHO, Geneva, Switzerland; 1991, p. 183.
65. Scientific Committee for Food. Sweeteners, Report No. 16. Commission of the European Communities, Luxembourg; 1986.
66. Food and Drug Administration. Docket No. 2002F-0220. Federal Register; 2003, 68, 75411.
67. Joint FAO/WHO Expert Committee on Food Additives (JECFA). In: *Compendium of Food Additives Specifications, 57th Session, FAO Food and Nutrition Paper No. 52, Addendum 9*, FAO, Rome, Italy; 5 June to 14 June 2001, p. 1.
68. U. Hagenauer-Hener, C. Frank, U. Hener and A. Mosandl. Bestimmung von Acesulfam, Aspartam, Saccharin, Coffein, Sorbinsäure und Benzoesäure in Lebensmitteln. *Deutsche Lebensmittelrundschau* 1990, 86(11), 348.
69. Comité Européen de Normalisation (CEN). Standard 12856; 1999.
70. T.A. Biemer. Analysis of saccharin, acesulfame K and sodium cyclamate by high-performance ion chromatography. *Journal of Chromatography* 1989, 463(2), 463.
71. Q.-C. Chen and J. Wang. Simultaneous determination of artificial sweeteners, preservatives, caffeine, theobromine and theophylline in food and pharmaceutical preparations by ion chromatography. *Journal of Chromatography* 2001, 937(1/2), 57.
72. R.A. Frazier, E.L. Inns, N. Dossi, J.M. Ames and H.E. Nursten. Development of a capillary electrophoresis method for the simultaneous analysis of artificial sweeteners, preservatives and colours in soft drinks. *Journal of Chromatography* 2000, 876(1/2), 213.
73. Directive 94/35/EC on sweeteners for use in foodstuffs, Official Journal of the European Communities, 1994, L 237, 3; as amended by Directive 96/83/EC, *Official Journal of the European Communities*, 1996, L 48, 16.
74. Codex Alimentarius General Standard for Food Additives. Codex Stan 192. http://www.codexalimentarius.net/web/more_info.jsp?id_sta=4
75. L. A. Wilson, K. Wilkinson, H. M. Crews, A. M. Davies, C. S. Dick and V. L. Dumsday. Urinary monitoring of saccharin and acesulfame-K as biomarkers for exposure to these additives. *Food Additives and Contaminants* 1999. 16, 227.
76. A. Renwick. The intake of intense sweeteners – an update review. *Food Additives and Contaminants* 2006. 23(4), 327.
77. K. Huvaere K. and J. van Loco. Scientific Institute of Public Health. Report D/2010/0000/00. http://www.wiv-isp.be/pdf/verslag_zoetstoffen.pdf; 2010.
78. Directive 95/31/EC laying down specific purity criteria concerning sweeteners for use in foodstuffs, *Official Journal of the European Communities*, 1995, L 178, 1; as amended by Directive 2001/52/EC, *Official Journal of the European Communities*, 2001, L 190, 18.
79. Directorate for the Quality of Medicines of the Council of Europe (EDQM). *European Pharmacopoeia*, 5th edn, Council of Europe, Strasbourg, France; 2004, p. 911.
80. Committee on Food Chemical Codex. *Food Chemical Codex*, 4th edn, National Academy Press, Washington, DC; 1996, p. 10.
81. USP28-NF23. *The United States Pharmacopeia/The National Formulary*. United States Pharmacopeial Convention, Rockville, MD; 2004, p. 2947.

6 Aspartame, Neotame and Advantame

Kay O'Donnell
Weybridge, UK

6.1 ASPARTAME

6.1.1 Introduction

Aspartame is a nutritive intense sweetener produced by combining the amino acids L-phenylalanine and L-aspartic acid by a methyl-ester link (Figure 6.1).

Its full chemical name is (3S)-3-amino-N [alpha-S-alpha methoxy carbonylphenythyl] succinic acid but it is often described by its synonym L-aspartyl-L-phenylalanine methyl ester (Figure 6.1). In Europe, it is assigned the E number 951.

Aspartame was discovered by Schlatter in 1965 in the laboratories of G.D. Searle and there followed a period of rigorous testing before it first appeared in the US market in 1981 under the brand name of *NutraSweet*. The brand was heavily promoted and contributed to the phenomenal commercial success of aspartame as a sucrose replacement in the 1980s and 1990s.

6.1.2 Synthesis

Although G.D. Searle discovered and recognised the potential of aspartame as a sweetener, the Ajinomoto Company of Japan developed and patented many of the processes for its commercial production.[1,2] Many further patents refining the original process have subsequently been granted to Ajinomoto and other companies.

In the late 1970s, Toyo Soda filed patents utilising enzymes to link N-protected aspartic acid to β-phenylalanine methyl ester followed by crystallisation and purification steps. This biocatalyst method is claimed to have greater specificity and therefore better yields.[3]

The method used by Holland Sweetener Company (HSC) was based on the Toyo Soda patents. HSC withdrew from the aspartame business at the end of 2006 citing oversupply and price erosion in the aspartame market. Today, key suppliers in the market include the Ajinomoto, The NutraSweet Company and several Chinese, Korean and Indian suppliers.

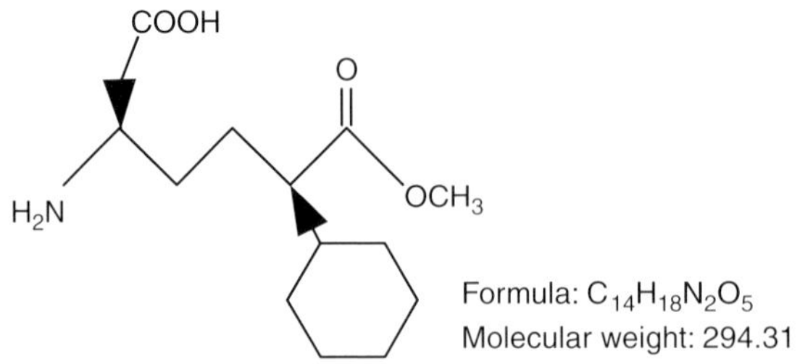

Fig. 6.1 Aspartame structure.
Source: Ajinomoto.

Raw materials for the production of aspartame are the two amino acids L-phenylalanine (produced by fermentation) and aspartic acid. The Toyo Soda process can use DL-phenylalanine, which is chemically synthesised and can offer cost benefits over the use of the purer optically active 'L' form. Aspartic acid is produced by chemical synthesis.

The reactive groups of the amino acids are first protected, with the exception of the groups that will form the methyl ester link.[2,3]

The amino acids are then coupled either chemically or enzymatically and the reactive groups removed. This is followed by a series of crystallisation steps to remove impurities.[2]

Crystallisation methods vary from static (Ajinomoto, NutraSweet) to stirred (HSC), and the technique employed will influence the type, size, shape and other properties of the crystal formed. This can influence the physical attributes of the product and therefore its performance in some applications, for example in dry mixes and encapsulated products.

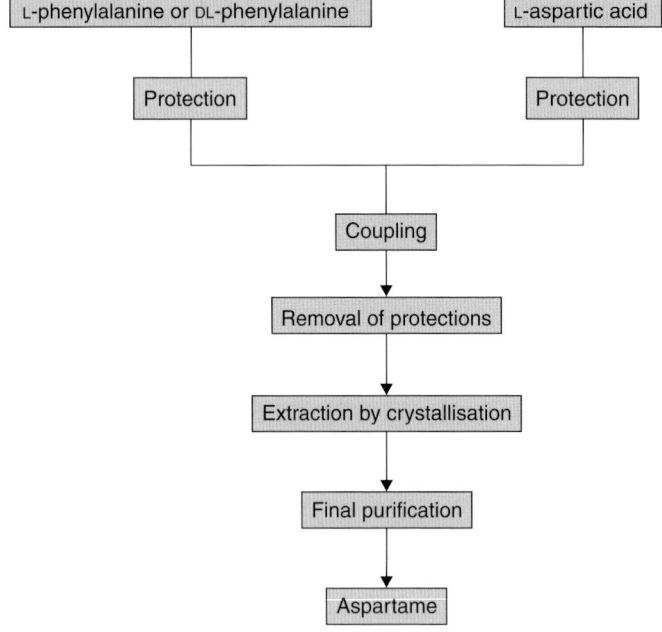

Fig. 6.2 Production of aspartame.
Source: Ajinomoto.

6.1.3 Sensory properties

Aspartame has a clean sweet taste and has approximately 180–200 times the sweetness of sucrose.[4] Unlike many other intense sweeteners, its taste profile is good enough and its maximum sweetness intensity of 13–14%[4] sucrose equivalence (SE), high enough for it to be used as the sole sweetener in most applications. At the time of its approval in the United Kingdom and United States in the early 1980s,[5] it offered a very significant benefit when compared with other intense sweeteners – its sweetener profile being much more similar to sucrose than any other permitted intense sweetener (Figure 6.3).

Relative sweetness (RS) of aspartame varies with concentration, as with all intense sweeteners. At threshold sweetness (0.34%)[6] in water RS is 400 (where sucrose = 1), while at 10% SE it falls to 130.[6,7] RS can also be affected by other factors including pH, temperature and application and is, therefore, difficult to predict. The generally quoted RS is 180–200 and this is a useful starting point for most formulations. Sweetness profile for aspartame is similar to sucrose – onset time is very slightly longer than sucrose, and in some applications it has a slightly lingering 'tail'. Where this is not desirable, it may be modified by blending with other intense sweeteners, sugars or other ingredients, for example naringin or aluminium potassium sulphate.[8]

6.1.3.1 Synergy and flavour enhancement

Aspartame is synergistic with many bulk and intense sweeteners with the levels of synergy being dependent on concentration and blend constituents. Synergy has been reported with glucose,[6] sucrose,[6,9,10] fructose,[6,10] polyols, saccharin, cyclamate,[11] acesulfame K[12] and stevia.[11]

It is interesting that combinations of aspartame and sucralose do not appear to be synergistic with respect to sweetness intensity and sweetness suppression has been reported with combinations of aspartame and sucrose in neutral solutions. This suppression changes to synergy when the pH is decreased.[9]

Aspartame is frequently blended with acesulfame K or saccharin. In these instances, sweetness intensity synergy is noted (up to 30%[13]). Equally important in these blends are

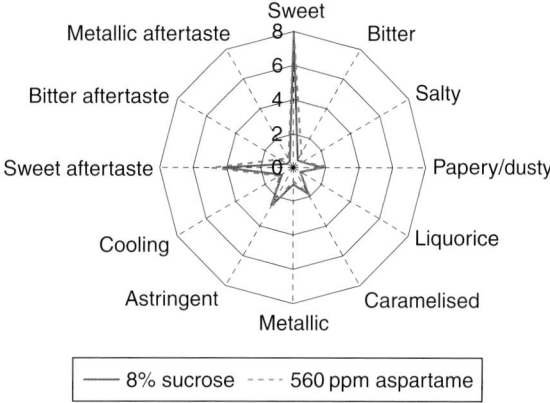

Fig. 6.3 Descriptive taste profile in water.
Source: The NutraSweet Company.

the taste modifying effects of aspartame on the bitter and metallic side tastes of acesulfame K and saccharin.[13]

Aspartame also has some flavour-enhancing properties and in particular soft fruit flavours can be enhanced.[14]

6.1.3.2 Aspartame salt

An aspartame–acesulfame salt is now commercially available. The two molecules are linked through an ionic bond and the salt has a RS of about 350. The product is composed of 64:36 aspartame:acesulfame K on a weight basis and in solution it dissociates to aspartame and acesulfame K.[15] The solid product is non-hygroscopic, dissolves more quickly than aspartame and is said to have stability benefits[15] in applications such as chewing gum. It was developed by HSC and branded as *TwinSweet*.

6.1.4 Physicochemical properties

Aspartame is a white, colourless, crystalline substance, is regarded as ecologically safe and is a biodegradable non-regulated material.[7]

6.1.4.1 Solubility (Figure 6.4)

The solubility of aspartame in water (pH 6–7) at 25°C is approximately 1%.[7] This can be increased by elevating the temperature and/or increasing the acidity. The use of some hydrocolloids, for example CMC, has also been reported to significantly increase solubility, even at low temperature.[16] The lowest solubility is at its isoelectric point at pH 5.4.[4]

Aspartame is sparingly soluble in other solvents.[7]

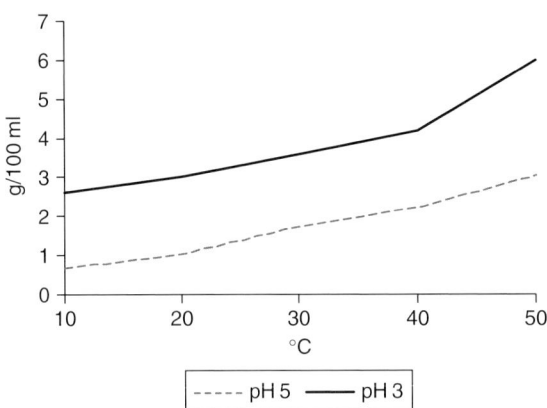

Fig. 6.4 Aspartame solubility.
Source: Ajinomoto.

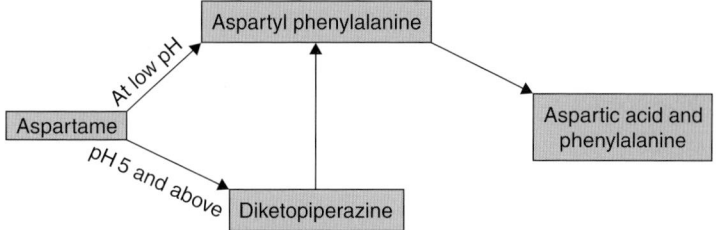

Fig. 6.5 Principal hydrolysis products of aspartame.

6.1.4.2 Stability

Dry

In solid form at ambient temperature and even at elevated temperature, aspartame has excellent stability (more than 5 years). In foods, optimum stability is achieved in systems with lower than 8% moisture.[6]

Stability decreases with increasing temperature and available moisture. It has been reported that addition of isomaltulose can improve the stability of aspartame over time.[17]

Liquid

Aspartame is less stable in liquid systems and the stability is primarily a function of pH, temperature and time.

The aspartame molecule slowly hydrolyses at low pH to produce methanol and the tasteless molecule aspartyl-phenylalanine (AP). An alternative route at pH 5 and above is that aspartame may cyclise to form its diketo-piperazine (DKP) with the elimination of methanol.[14] These conversion products can be subsequently hydrolysed to the individual amino acids – aspartic acid and phenylalanine[12, 18, 19] (see Figure 6.5).

In liquid systems, stability of aspartame at different pH follows a bell-shaped curve (see Figure 6.6). It is most stable in the range of pH 3–5 (with an optimum pH for stability at pH 4.2), which, fortunately, is the pH range of many food systems. Stability decreases with increasing temperature. It should be noted from Figure 6.6 that within the pH range

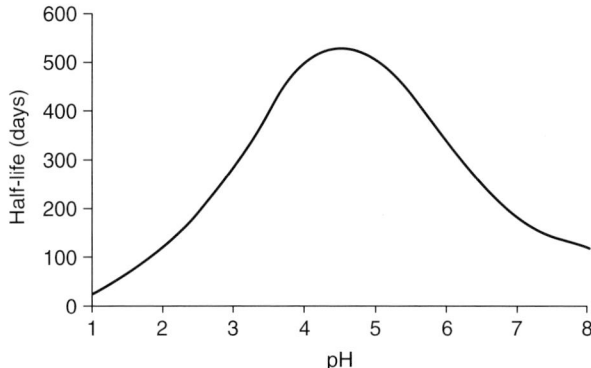

Fig. 6.6 Aspartame stability changes with pH at 20°C.
Source: Ajinomoto.

Table 6.1 Aspartame losses under different processing conditions.

Process	pH	Aspartame lost (%)
HTST (30 s/75°C)	3.5	<1
	6	1
UHT (15 s/136°C)	3.5	1
	6	42
Canning (fruit core temperature)	3.8–4	2–4

Source: Reproduced with permission from Ajinomoto.
HTST, high-temperature short-time; UHT, ultra-high temperature.

of 2.5–5.5 small changes in pH towards the optimum of pH 4.2 can have a significant and positive impact on stability.

Aspartame degradation in liquids during heat treatment is limited providing the pH is in the optimum range (see Table 6.1). High-temperature short-time (HTST) systems are the preferred method to minimise aspartame losses.

Stability during storage of beverages containing aspartame follows first-order kinetics and, to an extent, is predictable if parameters of pH and temperature are known.[6] Expected results can, however, be confounded by other interactions with ingredients that can have a positive or negative impact on stability and also perceived sweetness level. For example, synergy with other sweeteners can have an impact on perceived sweetness of beverages over time.

Figure 6.7 shows a synergy curve for aspartame and acesulfame K. Maximum synergy occurs when a 50:50 blend of aspartame of acesulfame K is used. On storage, the level of aspartame will decrease over time and as the level of aspartame decreases, the level of

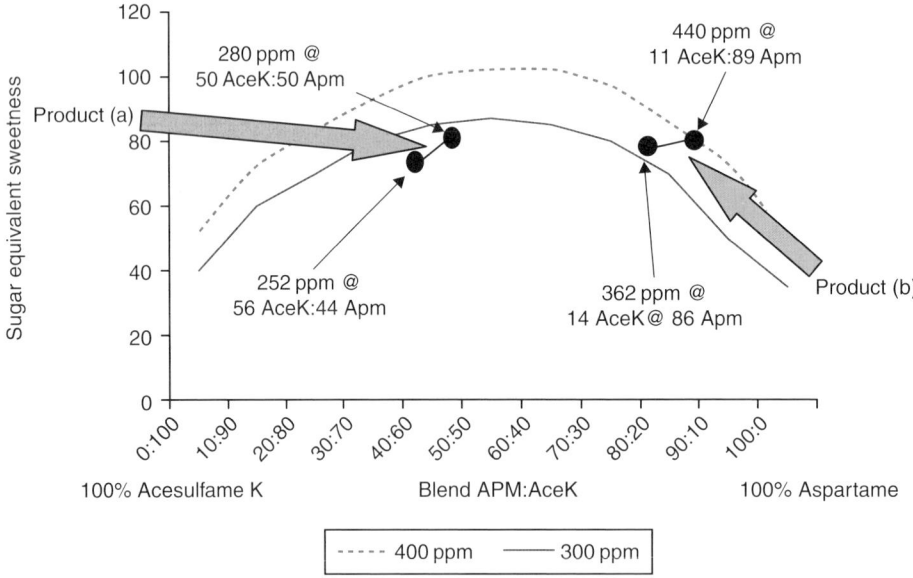

Fig. 6.7 Optimisation of an aspartame/acesulfame K blend. (AceK, acesulfame K; Apm, aspartame).

synergy also declines. So, in effect, the loss of aspartame on storage is amplified by the additional impact of lower synergy.

If, on the other hand, the beverage is formulated using a blend on the right-hand side of the curve, over time, aspartame will also hydrolyse, but the synergy level between the two sweeteners will increase, compensating in part for the loss of aspartame. So, the perceived sweetness of the beverage will reduce to a much lesser extent over time. Therefore, to maintain the sweetness level over time, it is recommended to formulate on the right-hand side of the curve.

Aspartame stability can also be affected by some flavour compounds. For example, cinnamon and some terpene compounds appear to accelerate degradation.[14]

Other compounds, for example polymerised polyphenols, have been shown to confer additional stability in aspartame-sweetened beverages.[20]

Since aspartame is made up of amino acids it can, under certain conditions, take part in Maillard reactions and form brown-coloured compounds. Use of isomaltulose has been shown to stabilise aspartame, reduce sweetness loss and reduce Maillard browning in some systems.[17]

Various factors can impact the stability and perceived sweetness of aspartame containing products and due to synergy, there is not always a direct relationship between the level of aspartame in a product and its acceptability. It is therefore advisable to perform full storage and acceptability trials on products containing aspartame.

6.1.5 Physiological properties

6.1.5.1 General

Aspartame is a nutritive sweetener, in that it is broken down in the body to its two constituents: amino acids and methanol. The amino acid metabolites follow the normal routes of digestion and utilisation in the body as they would if generated from other food sources (Figure 6.8).

Aspartic acid makes up approximately 40% of the aspartame molecule. It is absorbed in the intestinal lumen and plays an important role in nitrogen and energy metabolism, in the mitochondria.[1]

Phenylalanine makes up over half the aspartame molecule. It is an essential amino acid (i.e. the body cannot synthesise it) and it must, therefore, be obtained from foods for normal growth to be maintained. Adults require a minimum of 1–2 g of L-phenylalanine per day. The requirement for L-phenylalanine is highest in infants (where growth rate is rapid) and it reduces as growth rate declines.[21,22]

Methanol is a potentially harmful metabolite. Various studies have shown that even at abuse levels, the methanol levels produced by aspartame consumption are barely measurable, insignificant and do not represent a health risk.[22–24]

Toxic levels of methanol are around 200–500 mg/kg body weight (BW).[23] One 330 mL can of soft drink sweetened with aspartame at 550 mg/L would, in theory, generate 18.3 mg of methanol – that is approximately 0.26 mg/kg BW methanol in a 70 kg person. For comparison, a 220 mL glass of tomato juice would generate 47 mg methanol or 0.67 mg/kg BW in a 70 kg person.[7]

The small amount of methanol produced is likewise excreted from the body in an identical fashion to methanol produced from other food sources (e.g. banana, tomato juice).

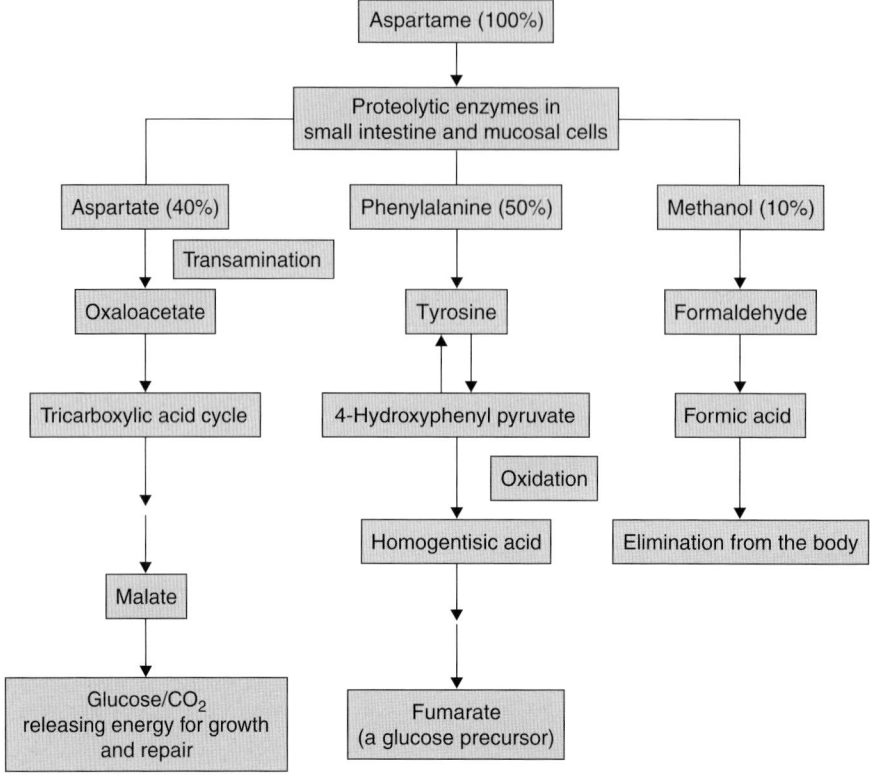

Fig. 6.8 Generalised and simplified scheme of aspartame metabolites and their metabolic fate.

6.1.5.2 Phenylketonuria (PKU)

The presence of phenylalanine as one of the breakdown products of aspartame is relevant for consumers with PKU. These people are unable to convert the essential amino acid phenylalanine into tyrosine, usually because of a defect in, or a deficiency of, the enzyme phenylalanine hydroxylase. In normal individuals, this enzyme converts phenylalanine to tyrosine, which is in turn catabolised. A deficiency in the enzyme leads to elevated levels of phenylalanine in tissues and accumulation of the end products of the transamination pathway of phenylalanine, which leads to brain damage and mental retardation.[21] The treatment for children born with this inborn error metabolism is a rigidly controlled diet to restrict phenylalanine intake and, therefore, prevent elevated levels of L-phenylalanine in the blood until adolescence or adulthood. For this reason, it is a requirement in many markets for products using aspartame, particularly soft drinks that would normally be classed as phenylalanine free, to have a note on the pack that the product contains a source of phenylalanine.

6.1.5.3 Oral health

Aspartame is not fermented by tooth plaque bacteria and is to be considered tooth-friendly.[26]

6.1.5.4 *Effect on blood glucose*

Aspartame does not affect blood glucose levels and is, therefore, suitable for use in foods for diabetics.[27-29]

6.1.5.5 *Effect on weight loss and weight maintenance*

There has been long debate on the use of aspartame and other intense sweeteners and their impact on weight control.[26-28] A review of a number of studies concluded that the use of aspartame as a substitute for sugar in the diet should allow a reduction of overall calorie intake. This may be partially compensated for, but overall use of aspartame is an effective way to lose weight. The rate of weight loss that can be achieved is low but meaningful and more than counterbalances the current average rate of weight gain of 0.007 kg per week in the population. It is therefore a useful adjunct to other weight loss regimes.[30,31]

6.1.6 Applications

Aspartame is used in many applications in the food and pharmaceutical industry. Current estimates suggest that it accounts for 32% of the global high-intensity sweetener market[32] and the major markets for the product are soft drinks and table-top sweeteners. However, it is also used in confectionery, pharmaceutical tablets and dry syrups, dairy products, dry mix products and bars. It is not used to any great extent in baked goods, products that undergo extensive and prolonged heat treatment or liquid products at or near neutral pH that require a long shelf life. It would not give the required stability in these products unless it was protected in some way, for example by encapsulation.

Use levels of aspartame vary with application and are also dependent on its use as a sole sweetener or in combination with other sweeteners.

6.1.6.1 *Use in manufacturing*

Aspartame is available in a number of physical forms from ultra-fine powder to granular product to aid dissolution, flow properties and granulation/direct compression properties.

6.1.6.2 *Soft drinks*

In soft drink applications, aspartame can be added, in dry form, directly to the syrup tank after the acid (to aid rapid dissolution). Alternatively, a more concentrated premix of aspartame suspended in water can be added to prevent clumping and adhesion of the powder to the mixing tank.[33]

6.1.6.3 *Yoghurt*

In stirred yoghurts, aspartame is usually incorporated with the fruit preparation after fermentation.

In set yoghurts, aspartame is added after pasteurisation, but before fermentation. Aspartame can be degraded by yoghurt cultures and, therefore, in processes where it is added before fermentation, it is important to increase the aspartame addition level by 15–30% to compensate for losses during fermentation.[7] Alternatively, careful selection of the yoghurt culture can significantly reduce these losses.[34]

6.1.6.4 *Confectionery*

In high-boiled confectionery, aspartame should be added at the end of the boiling process along with the flavours and other heat sensitive ingredients, to minimise process losses.

In chocolate, ultra-fine aspartame should be used and added at the end of the conching process.

In chewing gum, aspartame should be added towards the end of the kneading process, when the last part of the bulking agent is added.[6]

In many gum products aspartame is encapsulated before use to extend the sweetness profile and give 'long-lasting flavour' gum.

6.1.7 Analysis

The usual method of aspartame analysis is HPLC.[34–36] Perchloric acid titration may also be used.[37]

6.1.8 Safety

Aspartame is probably the most rigorously tested food ingredient to date. Prior to its approval in major markets, aspartame and its breakdown products underwent extensive tests to validate its safety.[38]

Its commercial success in the market during the 1980s and 1990s focussed attention on the product. Despite the fact that aspartame's metabolites bring nothing new to the diet and are handled by the body in the same way as they would be from other foods (Section 6.1.5), a number of reports linking aspartame to adverse safety effects appeared. These allegations, which linked aspartame to a variety of adverse effects including weight gain, neurotoxic effects, cancer and epilepsy, were based on a series of anecdotal reports and poorly controlled studies, and gained publicity via the Internet and popular press.

Scientifically controlled peer reviewed studies have consistently failed to produce evidence of a causal effect between aspartame consumption and adverse health events.

Additional reviews of aspartame safety covering studies done pre- and post-approval have been carried out by regulatory bodies. These have confirmed that aspartame is a safe and thoroughly tested food ingredient.[38,40]

A 2002 review of over 500 studies conducted by the EU SCF (EU Scientific Committee for Food) concluded that 'intake of its (aspartame's) component parts can be compared with intakes of the same substances from natural food'.[39]

A recent evaluation of the safety of aspartame at current use levels based on a review of toxicological and epidemiological studies concluded that the studies failed to find a link between aspartame and its decomposition products and have consistently found no adverse effects with aspartame doses of up to 400 mg/kg BW. It further concluded that there was no credible evidence that aspartame is a carcinogen and that the weight of evidence suggests that it is safe at current use levels as a non-nutritive sweetener.[40]

Table 6.2 Maximum EU use levels for aspartame in various applications.

Application	EU aspartame maximum use level
Soft drinks	600 mg/L
Confectionery	1000–6000 mg/kg
Baked goods	600 mg/kg
Milk drinks	600 mg/L

The European Food Safety Association (EFSA) issued a statement in 2011 concluding that a review of recent studies does not give reason to reconsider previous safety assessments of aspartame. Further it 'reaffirmed that any possible risks from aspartame have been considered by scientific bodies worldwide and current acceptable daily intake (ADI) ensures consumers are protected'.[41,42] Later in 2011, EFSA announced a full re-evaluation and risk assessment of all the evidence (published and unpublished), on aspartame in response to a request from the European Commission. The re-evaluation is due to be completed by July 2012.[42]

JECFA (Joint FAO/WHO Expert Committee on Food Additives) have assigned aspartame an ADI of 40 mg/kg BW. Unlike most other intense sweeteners, this level is sufficiently high for aspartame to be used as a sole sweetener in applications such as soft drinks.

FDA have assigned aspartame an ADI of 50 mg/kg BW.

EFSA have assigned aspartame an ADI of 40 mg/kg BW.

6.1.9 Regulatory status

Aspartame is permitted in all major markets. The maximum use level is dependent on application and EU maximum use levels for some major application areas are detailed in Table 6.2.

6.2 NEOTAME

Although in the 1980s and 1990s aspartame was an impressive commercial success, a number of research groups around the world looked for the next-generation sweeteners that had greater sweetness, improved stability, good sweetness profile and could be produced at lower cost.

Neotame, as it became branded, was a product of a research collaboration started in the mid-1980s between Tinti and Nofre at Claude Barnard University France and The NutraSweet Company.[43,44]

The French research group found a number of different compounds, which were many thousands of times sweeter than sucrose,[44–46] and the NutraSweet Company selected compounds to take to the next stage of development and licenced the technology to produce them. In 1991, initial details of some of these compounds, including the one that would later be called Neotame, were released.[46]

Neotame is N-[N-(3,3-dimethylbutyl)-L-aspartyl]-L-phenylalanine-1-methyl ester[43] (Figure 6.9).

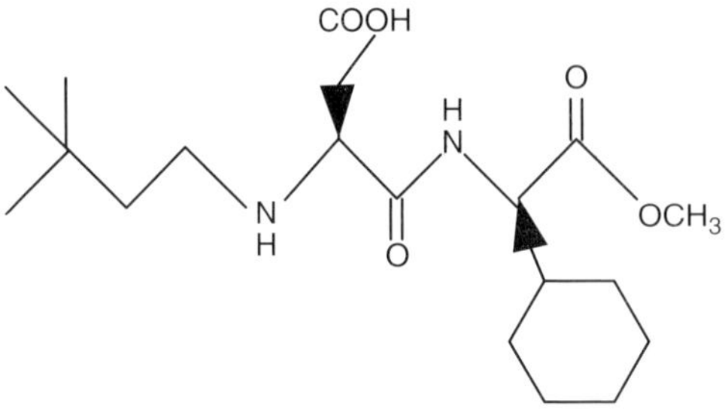

128 Sweeteners and Sugar Alternatives in Food Technology

Formula: $C_{20}H_{30}N_5O_5 \cdot H_2O$
Molecular weight: 378.47

Fig. 6.9 Neotame structure.

6.2.1 Neotame structure and synthesis

Neotame is a derivative of aspartame. Manufacture is from aspartame and 3,3-dimethylbutyraldehyde via reduction alkylation followed by purification, drying and milling[43] (Figure 6.10).

6.2.2 Sensory properties

6.2.2.1 *Taste*

Neotame is approximately 8000 times as sweet as sucrose[47] and has a clean sweet taste similar to sucrose. It has a liquorice side-taste that is increasingly noticeable, as sweetness tails off and also as its concentration increases. It does not have bitter or metallic side-tastes.

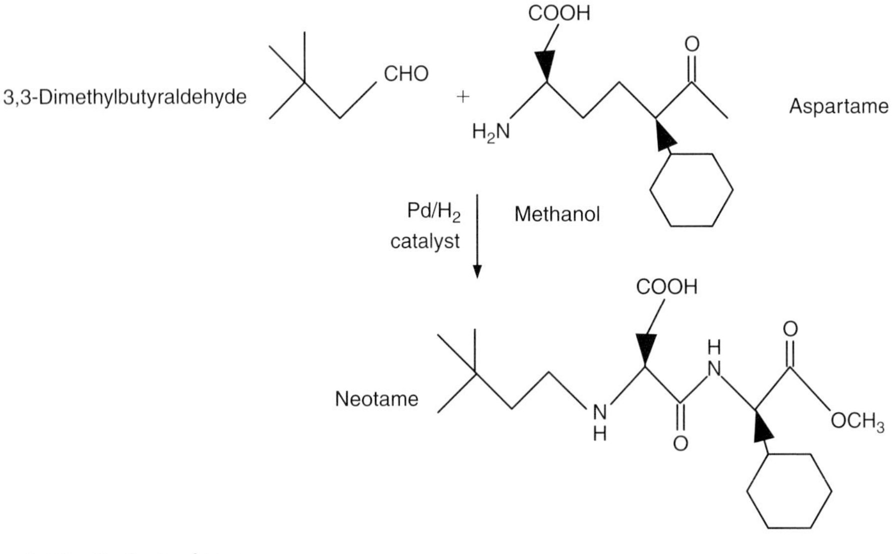

Fig. 6.10 Synthesis of Neotame.

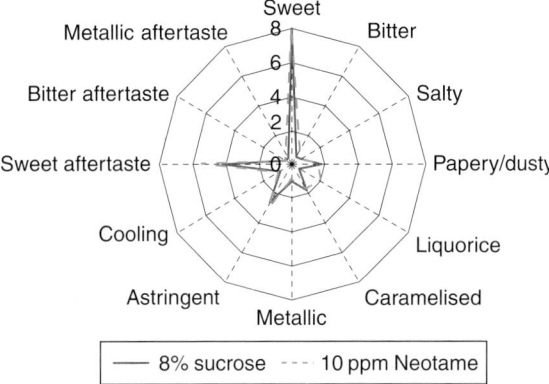

Fig. 6.11 Descriptive taste profile in water.
Source: The NutraSweet Company.

Maximum sweetness intensity occurs at 15.1% SE[47] and it can therefore, be used as the sole sweetener in some applications. However, in many applications, the liquorice side-taste would preclude this.

Figure 6.11 shows comparative flavour profiles of Neotame compared with sucrose and aspartame.

6.2.2.2 Temporal properties

Neotame has a slightly slower onset time than aspartame and sucrose and, in a similar manner to aspartame, Neotame has a slightly lingering sweetness.

The onset time and lingering sweetness of Neotame has been reported to be modified by the addition of flavours, hydroxyl amino acids (e.g. tyrosine, serine) and emulsifiers.[48]

6.2.2.3 Synergy and flavour enhancement

Sweetness synergy with Neotame is somewhat limited. It has some synergy with saccharine (14–25%)[49] but very little with other sweeteners.[48,49] The liquorice back-taste of Neotame is often reduced when it is blended with other bulk and intense sweeteners.[49,50]

Neotame has some interesting flavour-enhancing/modification properties. It is reported to enhance fruit, mint, cinnamic aldehyde and vanilla flavours (allowing a reduction in the amount of flavour used), and also maintain perceived acid levels in products, again allowing a reduction in the level used.[47]

Neotame is also reported to mask off-tastes, even at sub-threshold use levels, associated with soy, vitamin and mineral fortification.

This flavour modification effect of Neotame should be taken into account when formulating a product, as it is likely that this will result in changes to the flavour system being required.

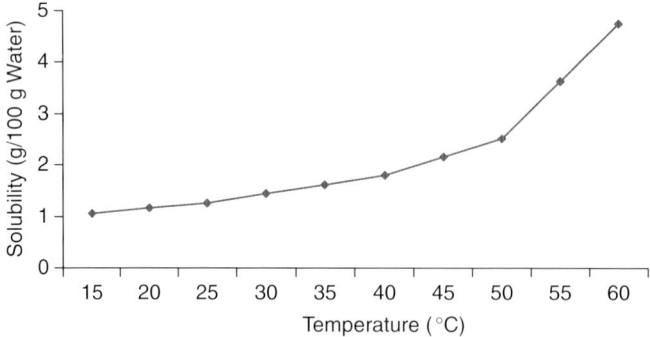

Fig. 6.12 Solubility of Neotame in water.

6.2.3 Physiochemical properties

Neotame is a white/off-white crystalline powder with a melting point of 80.9–83.4°C.[50] (Figure 6.9)

It can form both basic and acid salts and metal chelates.[50]

6.2.3.1 Solubility

Neotame is slightly more soluble than aspartame in water and significantly more soluble than aspartame in some solvents (ethanol). Solubility in water at 25°C is 1.3% w/w and this increases with rising temperature[50] (Figure 6.12). Formation of Neotame salts will also increase the solubility.[51] Dissolution rate in water at common use levels is quite rapid, for example 50 mg in 900 mL at 37°C dissolves in 5 minutes.[50]

6.2.3.2 Stability

Dry stability of Neotame is good and in excess of 5 years. In dry conditions, the major degradation product is de-esterified Neotame (formed by the hydrolysis of the methyl ester group).[51]

Fluorescent lighting and polyethylene packaging have no effect on stability.[47] Neotame has similar stability to aspartame in many products and is more stable in neutral pH conditions (i.e. in baked and dairy products).[50]

Stability, over time, is dependent on three main variables: pH, temperature and moisture. Stability decreases with increasing temperature. The pH stability follows a bell-shaped curve similar to aspartame and the optimum pH for maximum stability is 4.5.[51] Addition of divalent or trivalent cations[52] and beta-cyclodextrin[53] have been reported to improve stability.

In aqueous solutions (pH 2–8), the main degradation product from Neotame is de-esterified Neotame[49] produced by hydrolysis of the methyl ester: N-[N-(3,3-dimethylbutyl)-L-α aspartyl]-L-phenylalanine (see Figure 6.13).

Neotame degradation does not produce DKP or phenylalanine.

Stability in soft drinks utilising Neotame should be managed in a similar way to aspartame beverages. Degradation can be minimised by formulation as close to the optimum pH for stability (pH 4.5) as possible and by maintaining a low temperature. As an indication, in a trial formulation at pH 3.2 and 20°C, 89% Neotame remained after 8 weeks.[47]

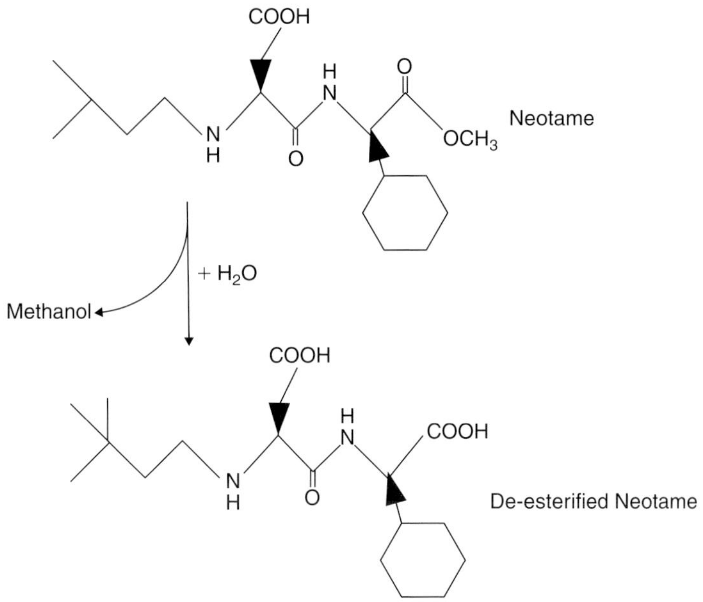

Fig. 6.13 Degradation of Neotame.

Neotame does not react with reducing sugars to generate Maillard reaction products.[47]

6.2.4 Physiological properties

In vivo, as in *in vitro*, the main degradation products from Neotame are de-esterified Neotame (see Figure 6.13) and methanol in equimolar quantities.[47]

The use level of Neotame is very small and, therefore, the amount of methanol produced is very small and well below levels of methanol considered to be safe.

Absorbed Neotame and de-esterified Neotame are excreted in the urine and faeces.[47]

Phenylalanine is not produced as a result of Neotame hydrolysis and, therefore, phenylalanine statements for PKU patients are not required on products containing Neotame.[47]

6.2.4.1 Oral health

Neotame is not metabolised by oral bacteria and is considered to be non-cariogenic.[50]

6.2.4.2 Blood glucose effects

Neotame does not affect glycaemic control in patients with non-insulin-dependent diabetes.[50]

6.2.5 Applications

Neotame is used in a number of applications in markets where it is approved.

Many products (particularly, soft drinks), use combinations of Neotame and sugar or high fructose glucose syrup (HFGS). It is also used in combination with other intense sweeteners (e.g. saccharin and sucralose).

In soft drinks, the use level in a 'typical' carbonated soft drink (9–12% sucrose) would be 0.045–0.07%. A 30% sugar reduction on a full sugar product can be obtained using 6.75–9.0% sugar and 0.002–0.003% of Neotame.[50]

Use of Neotame in chewing gum is usually in combination with other sweeteners, which are often encapsulated. Use of free Neotame in the gum base is claimed to give long-lasting sweetness in gum.[54]

Other applications that currently utilise Neotame include dairy drinks, ambient sauces, confectionery bars, chewing gum, fresh breath capsules and savoury snacks.

Neotame can be prepared in a number of different forms – agglomerated, encapsulated,[47] co-crystallised with sugar,[51] acid or basic salts, metal complexes or cyclodextrin complexes.[55] These different preparations can make for easier handling of a product used at very low levels.

6.2.6 Safety

Neotame has been the subject of numerous safety studies according to the general principles in US FDA and other international guidelines.[47]

The projected consumption level in the population is estimated to be 0.05 mg/kg BW per day at the 90th percentile level.[47] The no observed effect level (NOEL) for chronic toxicity and carcinogenicity is estimated to be at least 1000 mg/kg BW per day in rats and 800 mg/kg BW per day in dogs.[47] Both Neotame and its main degradation product, de-esterified Neotame, have been shown to be non-mutagenic and well tolerated above projected chronic consumption levels.[47]

JECFA has assigned an ADI of 0–2 mg/kg BW.[56]

6.2.7 Regulatory

Neotame's ISN number (E number) is 961.

Neotame is currently approved in many markets including Argentina Australia, Belarus, Brazil, Chile, China, EU, Georgia, India, Japan New Zealand, Nigeria, Norway, Russia, South Africa, Turkey, United Arab Emirates and the United States.

6.3 ADVANTAME

While the NutraSweet Company was developing Neotame, Ajinomoto was also looking to develop the next generation of intense sweeteners. Ajinomoto is a major producer of aspartame and has an active sweetener discovery and development programme.

The first product delivered by this programme was Advantame and, like Neotame, it was discovered when Ajinomoto re-evaluated the lead compounds discovered by Nofre and also the Coca Cola Company.[57] Use of SAR, computer modelling followed by synthesis, screening for potency, taste profile, ease of manufacture, physicochemical properties and likely metabolic fate, identified a compound initially designated ANS9801 and later named Advantame.

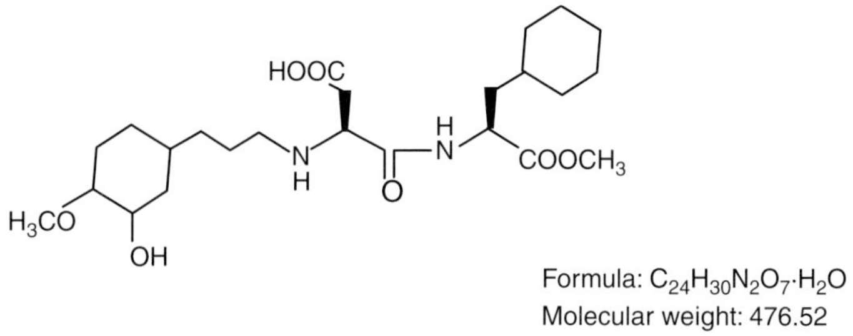

Formula: $C_{24}H_{30}N_2O_7 \cdot H_2O$
Molecular weight: 476.52

Fig. 6.14 Advantame structure.
Source: Ajinomoto.

Advantame is N-[N-[3-(3-hydroxy-4-methoxyphenyl) propyl-α-aspartyl]-L-phenylalanine 1-methyl ester monohydrate (Figure 6.14).

Advantame shares certain structural characteristics with some natural sweeteners, for example Phyllodulcin.[57]

Advantame has the empirical formula $C_{24}H_{30}N_2O_7 \cdot H_2O$ and a molecular weight of 476.52 Its melting point is 101.5°C and it is a white/yellow powder.

6.3.1 Synthesis

Advantame is synthesised from aspartame and (3-hydroxy-4-methoxy-phenyl) propylaldehyde in a one-step reductive N-alkylation process in which aspartame and the aldehyde react with hydrogen in methanolic solution in the presence of a platinum catalyst. Vanillin is used to derive the important intermediate (3-hydroxy-4-methoxyphenyl) propylaldehyde.[56]

6.3.2 Sensory properties

Advantame is reported to have a RS of 7000–47,000 times that of sucrose over the range 3–14% SE and 70–120 times the sweetness of aspartame over the same range.[57]

Sweetness quality is described as clean, sweet and similar to aspartame; weak bitter and sour notes are present.[57]

Flavour enhancement has been found with a number of fruit flavours and vanilla at very low use levels (about 0.005 mg/100 mL).[57]

6.3.3 Stability

Advantame is stable in dry form while in liquid systems (e.g. soft drinks), degradation does occur over time. In conditions of 25°C/60% relative humidity, 60% of the Advantame remained after 20 weeks. The perceived decrease in sweetness was less than the actual decrease, which is to be expected since RS increases in high-intensity sweeteners as concentration of sweetener decreases.[57]

6.3.4 Solubility

Solubility is quoted as 0.009 g/dL at 25°C in 30 minutes and because of the very high RS, this is sufficient for most applications. A co-crystalline product with aspartame

(Advantame:aspartame = 0.022:1), has been shown to have an increased rate of dissolution.[57] It is suggested by Ajinomoto that applications for Advantame would be as a sole or blended sweetener in a wide range of food products.

6.3.5 Safety

Safety studies have indicated a NOEL = 667 mg/kg per day. Maximum human exposure for general use is estimated at 0.05 mg/kg per day giving a safety factor in excess of 13,000 times the NOEL.[57]

6.3.6 Regulatory

Ajinomoto applied for FDA and FSANZ approval of Advantame in 2009 and expect approval within 2 years.[58] It is expected that they will apply for approval in other major markets in due course.[58]

REFERENCES

1. Ajinomoto Co Inc British Patent, 1,243,169; 1970.
2. Ajinomoto Co Inc British Patent, 1,309,605; 1971.
3. J.D. Higginbotham. Recent developments in non-nutritive sweeteners. In: T.H. Grenby, *et al.* (ed). *Developments in Sweeteners 2*, Applied Science Publishers, London; 1983, pp. 119–156.
4. W. Vetsch. Aspartame: technical considerations and predicated use *Food Chemistry* 1985, 16(3/4), 245–258.
5. Sweetener in Food Regulations. *Statutory Instrument No. 1211, Food Composition and Labelling*, HMSO, London; 1983.
6. B.E. Homler. The properties and stability of aspartame. *Food Technology* 1984, 38(7) 50–55.
7. Ajinomoto. *Aspartame Technical Bulletin*, Ajinomoto AG, Zug, Switzerland; 2001/2002.
8. H.R. Schade. Taste modifier for artificial sweeteners, US Patent, 3934047; 1976.
9. D. Kilcast. *Food Industry Updates*, Leatherhead Food Research Association, Leatherhead, England; 2002.
10. P. Van Tomout, *et al.* Sweetness evaluation of mixtures of fructose with saccharin, aspartame and acesulfame K. *Journal of Food Science* 1985, 50, 469–472.
11. R. Franta, *et al.* Sweetness, 3 alternatives to can and beet sugar. *Food Technology* 1986, 40(1) 116–128.
12. B.T. Carr, *et al.* Sensory methods for sweetener evaluation. In: C.T. Ho and C.H. Manley (eds). *Flavour Measurement*, Marcel Dekker, New York; 1993, pp. 226–227.
13. Sweeteners – the taste of things to come. Soft Drinks World Supplement. www.zipublishing.com; 2004.
14. B. Homler. Aspartame: Implications for the food scientist. In: L.D. Stegink and L.J. Filer (eds). *Aspartame – Physiology & Biochemistry*, Marcel Dekker, New York; 1984, pp. 247–262.
15. TwinSweet. www.HSC.com; 2005.
16. W. Wafwoyo, *et al.* Interaction of aspartame with selected hydrocolloids: solubility of aspartame. *Food Hydrocolloids* 1999, 13, 299–302.
17. US Patent Application 20070160731A1; 12 July 2007.
18. L.D. Stegink. Aspartate and glutamate metabolism. In: L.D. Stegink and L.J. Filer (eds). *Aspartame: Physiology and Biochemistry*, Marcel Dekker, New York; 1984, pp. 47–109.
19. L.D. Stegink, *et al.* Aspartame metabolism in human subjects. In: *Health and Sugar Substitutes (Proc ERGOB CoP nf)*, Karger AG, Basel, Switzerland; 1978.
20. US Patent application #20100297305; 11 September 2008.
21. A.E. Harper. Phenylalanine metabolism. In: L.D. Stegink and L.J. Filer (eds). *Aspartame: Physiology and Biochemistry*, Marcel Dekker, New York; 1984, pp. 77–110.

22. R.E. Ranney and J.A. Opperman. A review of the moiety of aspartame in experimental animals and man. *Journal of Environmental Pathology* 1979, 2, 797–985.
23. L.D. Stegink, *et al.* Blood methanol concentration in normal adult subjects – administered abuse doses of aspartame. *Journal of Toxicology and Environmental Health* 1981, 7, 281–290.
24. L.D. Stegink, *et al.* Blood methanol concentration in one year old infants – administered graded doses of aspartame. *Journal of Nutrition* 1983, 113, 1600–1606.
25. W.H. Bowen. Role of sugar and other sweeteners in dental caries. In: *Aspartame – Physiology and Biochemistry*, Marcel Dekker, New York and Basel; 1984, pp. 263–267.
26. M.G. Tordoff and A.M. Alleva. Effect of drinking soda sweetened with aspartame or high-fructose corn syrup on food intake and body weight. *The American Journal of Clinical Nutrition* 1990, 5, 963–969.
27. A. Drewnowska. Review: Intense sweeteners and energy density of foods, implications for weight control. *European Journal of Clinical Nutrition* 1999, 53, 757–763.
28. D.S. Ludwig, K.E. Peterson and S.L. Gortmaker. Relation between consumption of sugar sweetened drinks and childhood obesity: a perspective and observational analysis. *Lancet* 2001, 357, 505–508.
29. Diabetes and Nutrition Study Group (DNSG) of the European Association for the Study of Diabetes. Recommendations for the nutritional management of patients with diabetes mellitus. *European Journal of Clinical Nutrition* 2000, 54, 353–355.
30. A. de la Hunty, *et al.* A review of the effectiveness of aspartame in helping with weight control' British Nutrition Foundation. *Nutrition Bulletin* 2006, 31 115–128.
31. H.H. Butchko and F.N. Kotsonis. Aspartame: review of recent research. *Comments on Toxicology* 1989, 3(4), 253–278.
32. Euromonitor International. Passport, making sense of global markets. *Ingredients 2011.* Published 2010.
33. K. O'Donnell. Carbohydrate and intense sweeteners. In: P.R. Ashurst (ed). *Chemistry and Technology of Soft Drinks and Fruit Juices*, 2nd edn, Blackie Publishing, Oxford, England; 2005, pp. 68–89.
34. P. Graf-Spar. *Dynamik des Aspartamabbu bei der Joghurtherstellung*, M.S. Hochschule, 1999 Wädenswil, Switzerland
35. P. Langguth, *et al.* Studies on the stability of aspartame (1) specific and reproducible HPLC assay for aspartame and its potential degradation products and applications to aid hydrolysis of aspartame. *Pharmazie* 1990, 46, 188–192.
36. R. MacArthur, *et al. Development and Validation of an HPLC Method for Simultaneous Determination of Intense Sweeteners in Food Stuffs (A01012).* DEFRA Central Science Laboratory, New York; 2002.
37. Technical Information. *Alpha-apm Determination of Perchloric Acid Titration*, The NutraSweet Company, Chicago, IL; 1985.
38. Reference Guide to Aspartame Scientific Research. The NutraSweet Company; 1996.
39. SCF Opinion of the Scientific Committee for Food - Update on the Safety of Aspartame (expressed on 4 December 2002), SCF/CS/ADD/EDUC/222. Final 10 December 2002.
40. B.A. Magnusson, *et al.* Aspartame: a safety evaluation based on current use levels, regulations, and toxicological and epidemiological studies. *Critical Reviews in Toxicology* 2007, 37, 629–727.
41. EFSA. Statement of EFSA on the scientific evaluation of two studies related to the safety of artificial sweeteners *EFSA Journal* 2011, 9(2), 2089.
42. EFSA. www.efsa.europa.eu/en/topics/aspartame; 2011.
43. US Patent, 4,935,517; 1990.
44. European Patent Application, 894201359; 14 April 1989.
45. US Patent Application, 601,623; 23 October 1990.
46. J.M. Tinti and C. Nofre. Presentation at FIE Show Paris; 8 October 1991.
47. C. Nofre and J.M. Tinti. Neotame: discovery, properties, utility, *Food Chemistry* 2000, 69, 245–297.
48. W.W. Stargel, *et al.* Neotame. In: L.O. Nabors (ed). *Alternative Sweeteners*, 3rd edn, Marcel Dekker, New York; 2001, pp. 129–145.
49. J. Bergman, The NutraSweet Company – Personal Communication; 2005.
50. Neotame. Ingredient Overview, www.Neotame.com.
51. I. Prakash, *et al.* Neotame: the next-generation sweetener. *Food Technology* 2002, 56(7) 36–40.
52. I. Prakash, I.E. Bishay, N. Desai and D.E. Walters. Modifying the temporal profile of the high-potency sweetener neotame. *Journal of Agricultural and Food Chemistry* 2001, 49(2), 786–789.
53. J.R. Garbow, *et al.* Structure, dynamics, and stability of cyclodextrin inclusion complexes of aspartame and neotame. *Journal of Agricultural and Food Chemistry* 2001 49(4), 2053–2060.
54. US Patent Application, 089319 A2; August 2007.

55. I. Bishay, J. Fotos, N. Desai, The Use of Cyclodextrin to stabilise N-[N-(3,3-dimethylbutyl)-L-α-aspartyl]-L-phenylalanine L-methyl ester *PCT International Application* 2001; WO 00/15045.
56. Codex Alimentarius Commission. CL2004/44- FAC; September 2004.
57. Y. Amino, K. Mori In: D.K. Weerasinghe and G.E. DuBois (eds). *Sweetness and Sweeteners: Biology, Chemistry and Psycophysics (ACS Symposium Series)*, Oxford University Press, Washington DC; 2006, pp. 463–480.
58. J. Halliady. Ajinomoto seeks approval for new sweetener. *Food Navigator*; 8 April 2009.

7 Saccharin and Cyclamate

Grant E. DuBois

Sweetness Technologies, LLC, Roswell, GA, USA

7.1 INTRODUCTION

The commercial development of non-caloric and reduced-calorie foods and beverages began to accelerate dramatically in the late 1950s and early 1960s. The bioavailable energy contents of foods and beverages are, of course, due to the combined calories contributed by carbohydrates, proteins and fats. In some food categories however, such as many types of beverages, the caloric levels are due principally or entirely to carbohydrate sweeteners and, particularly for such products, the technology was in place by the late 1950s to formulate products with good-quality taste. The ingredients that made this possible were non-caloric sweeteners, specifically saccharin and cyclamate. The discovery of saccharin was reported in 1879 and the discovery of cyclamate in 1944, but until the late 1950s, saccharin and cyclamate were primarily sugar substitutes for diabetics. Neither was a very good match for the taste of sugar and, for this reason, they were not widely utilised. Then, in 1957, Helgren of Abbott Laboratories made a key discovery facilitating a burst of growth in low-calorie foods and beverages.[1] He found that blending saccharin with cyclamate, in a ratio such that each sweetener contributes equal sweetness, results in a product with dramatically improved taste quality. This blend is one of approximately 10 parts cyclamate to 1 part of saccharin, since saccharin is roughly 10 times more potent than cyclamate. The taste quality of this cyclamate/saccharin blend was so good that, for the first time, zero- and low-calorie food and beverage products were possible without major compromise in taste. In this chapter, after a brief review of the present understanding of how sweeteners like saccharin and cyclamate activate our taste buds, the current state of knowledge on these two commercially important sweeteners is summarised.

7.2 CURRENT UNDERSTANDING OF SWEETNESS

It is not obvious why a blend of saccharin and cyclamate should taste more sugar-like than either sweetener by itself and it is logical to question why that should be the case. The answer to this question is likely to follow from an understanding of how sweeteners like sugar, saccharin and cyclamate activate taste bud cells, and as a result of a series of

---

Sweeteners and Sugar Alternatives in Food Technology, Second Edition.
Edited by Dr Kay O'Donnell and Dr Malcolm W. Kearsley.
© 2012 John Wiley & Sons, Ltd. Published 2012 by John Wiley & Sons, Ltd.

breakthroughs, just within the last decade, we are beginning to understand sweet taste. First, in 2001, a collaborative team of the research groups of Zuker (University of California at San Diego) and Ryba (National Institutes of Health) reported the discovery of the rat sweetener receptor;[2] this landmark discovery was quickly followed in 2002 by the identification of the human sweetener receptor by Li and co-workers (Senomyx).[3] In both the rodent and human systems, strong support was obtained for the idea that the sweet taste of all sweeteners follows activation of sweet-sensitive taste bud cells through *a single receptor*. The findings of the Zuker/Ryba team and Li and co-workers provided strong support for the concept that the sweetener receptor is a heterodimer of two proteins, which they named T_1R_2 and T_1R_3, both of which are 7-transmembrane domain (TMD) proteins and both of which are of the receptor class commonly referred to as G protein-coupled receptors (GPCRs). This heterodimeric receptor is written as T_1R_2/T_1R_3. Both T_1R_2 and T_1R_3 are members of the small family of class C GPCRs. Class C GPCRs are unique among the super-family of GPCRs in that they possess very large N-terminal (i.e. extracellular) domains that appear to function like Venus flytraps (VFDs). Supporting this Venus flytrap idea, the VFD of one class C GPCR, a metabotrophic glutamate receptor, has been demonstrated to close on glutamate binding, just like a Venus flytrap closes on its prey.[4] This precedent and parallel work by the Zuker/Ryba team on the umami (i.e. savoury) taste receptor leads to the expectation that sweeteners likely bind in the VFD of T_1R_2. Subsequent work by Li and co-workers in 2004 probed the fundamental question of sweetener binding locus with the finding that, while the two sweeteners, aspartame and neotame, do bind in the VFD of T_1R_2, cyclamate does not.[5] Cyclamate was shown to bind within the TMD of T_1R_3. At about the same time, Margolskee, Max, Osman and their co-workers (Mount Sinai School of Medicine) obtained similar results to those of Li and co-workers, but went further and determined the specific locus of cyclamate binding in the TMD of T_1R_3.[6] Interestingly, their work showed that the cyclamate binding site partially overlaps the site at which lactisole, a sweetness inhibitor, binds to the sweetener receptor. In 2006, in a continuation of work by the Margolskee, Max and Osman groups to understand loci of sweetener/receptor binding, they reported that there are at least three such loci, one in the T_1R_2 VFD region (aspartame and neotame), one in the T_1R_3 TMD (cyclamate) and one in the cysteine-rich domain of T_1R_3 for brazzein, a protein sweetener.[7] The finding that the cyclamate locus of binding is at a site distinct from most other sweeteners is also supported by the finding by Breslin and co-workers (Monell Chemical Senses Center) that, while $ZnSO_4$ strongly inhibits the sweetness activities of sucrose, aspartame, saccharin and other sweeteners evaluated, it has no effect on cyclamate sweetness.[8]

At the present time, the loci of T_1R_2/T_1R_3 binding by saccharin, sucrose and other carbohydrate sweeteners are not known with certainty. It seems most likely, however, that the carbohydrate sweeteners bind within the T_1R_2/T_1R_3 VFDs, and in 2005 and 2006, support was provided by Munger and co-workers (University of Maryland) that this is the case, based on binding studies to the individual T_1R_2 and T_1R_3 VFDs.[9,10]

It is now clear that sweeteners act at a plurality of sites on the sweetener receptor and it seems most likely that saccharin and cyclamate act at two different sites. If this is the case, it seemed plausible that the binding of either sweetener may enhance the binding of the other (i.e. a positive cooperativity effect), thus explaining the synergism observed in saccharin/cyclamate blends.[11] This rationale for sweetness synergy was also proposed by the collaborative team of Morini and Bassoli (University of Milan) and Temussi (University of Naples).[12] It is important to note that, by this cooperative binding mechanism, both sweeteners must bind to the receptor concurrently and therefore sweetness synergy should only be possible if the sweeteners bind to the receptor at different sites.

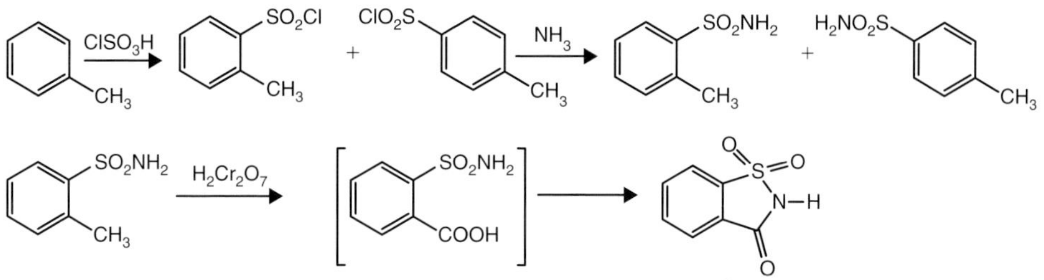

Fig. 7.1 The Remsen–Fahlberg process for saccharin manufacture.

Saccharin/cyclamate sweetness synergy allows both sweeteners to be used at less than half of their additive sweetness levels, which minimises the 'off' tastes that these sweeteners exhibit at higher concentrations and thus enables a much improved quality of taste.

7.3 SACCHARIN

7.3.1 History, manufacture and chemical composition

Saccharin (see Figure 7.1), sometimes referred to as o-benzoic sulfimide, was the first commercially developed, sweet-tasting organic compound, which was significantly more potent than sucrose. The story of its commercial development and use as a non-caloric sweetener was reviewed in 2001 by Pearson (PMC Specialties Group, Inc.), a company long involved in saccharin manufacture,[13] and more recently by Hicks (Pennsylvania State University).[14] Today, major manufacturers include Kaifeng and Shanghai Fortune of China and JMC of South Korea.

The discovery of saccharin was reported in 1879 by Remsen and Fahlberg.[15] Fahlberg was a postdoctoral associate working in the laboratory of Professor Remsen at Johns Hopkins University. Saccharin was originally obtained by Fahlberg through the oxidation of o-methyl-benzenesulfonamide, a starting material obtained by chlorosulfonation of toluene with chlorosulfonic acid. This oxidation led directly to the heterocyclic compound now known as saccharin, rather than the expected o-carboxy-benzenesulfonamide. This chemistry, illustrated in Figure 7.1, formed the basis for the original saccharin manufacturing process first commercialised in the United States by Queeny. In 1901, Queeny took his personal savings of $1500 and a loan of $3500 to found the Monsanto Chemical Company with saccharin as its first product (Monsanto was his wife's maiden name). The original Remsen–Fahlberg process has the disadvantage that the chlorosulfonation reaction gives a mixture of ortho and para substitution products. In the early 1950s, a new process was discovered that is not limited in this way. This process, based on the common grape flavourant methyl anthranilate as starting material, was invented by Senn[16] and Schlaudecker[17] and was commercialised in the mid-1950s by the Maumee Chemical Company. The Maumee process is illustrated in Figure 7.2. In 1966, the Sherwin Williams Company acquired Maumee and subsequently, in 1985, sold the saccharin business to PMC Specialities Group, Inc.

Saccharin is commercially available in sodium and calcium salt forms, both of which dissolve readily in water. Saccharin is exemplary of the *N-sulfonyl amide* structural class of

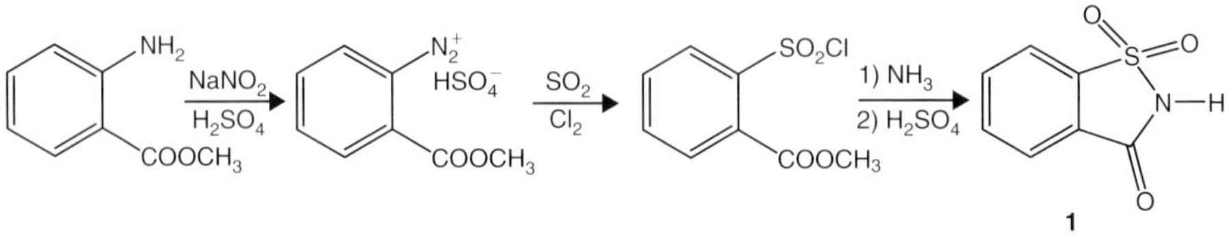

Fig. 7.2 The Maumee process for saccharin manufacture.

sweeteners, which is distinguished by the common -CONHSO$_2$- substructure. Acesulfame K is another *N-sulfonyl amide* non-caloric sweetener that has been commercially developed.

7.3.2 Organoleptic properties

Non-caloric sweeteners differ in taste from carbohydrate sweeteners in a number of ways. Generally, they are more potent, enabling them to be used at significantly lower concentrations. In addition, many of them, including saccharin, exhibit 'off' tastes such as bitter, metallic, cooling or liquorice-like and finally, nearly all of them exhibit sweetness that is slower in onset and that lingers, relative to that of carbohydrate sweeteners. Saccharin's performance on these organoleptic metrics is as follows:

- *Sweetness potency*: The use of non-caloric sweeteners in foods and beverages generally enables significant cost reductions over the costs of sugar or other carbohydrate sweeteners. These cost improvements are due to the higher potencies of non-caloric sweeteners. It is common practice to express sweetness potency (P) for a non-caloric sweetener as a multiple of the P of sucrose, which is defined as unity. And, although sometimes, P is expressed on a molar basis, most commonly P is expressed on a weight basis (P_w). It is important to recognise that P_w is not constant for high-potency sweeteners like saccharin as it is for carbohydrate sweeteners.[18] Thus, the P_w for saccharin varies according to the sucrose reference concentration. A plot of saccharin's concentration/response (C/R) function illustrating the hyperbolic relationship between saccharin's perceived sweetness intensity or response (R) and saccharin concentration (C) is given in Figure 7.3. It has been

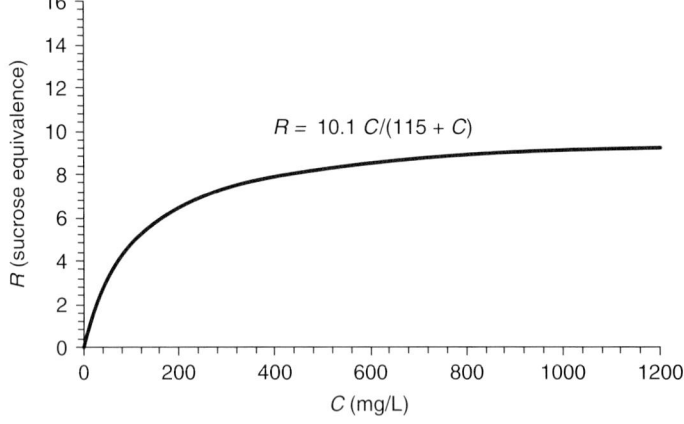

Fig. 7.3 Saccharin-Na C/R function.

demonstrated by The NutraSweet Company that the Law of Mass Action function $R = R_m C/(k_d + C)$ provides a good fit for saccharin as well as other non-caloric sweetener C/R function data.[19] When saccharin's C/R data is fit to the Law of Mass Action equation, the constant terms provide very useful information. R_m is the maximal response, in sucrose equivalents, that is possible for saccharin. Thus, on inspection of the C/R function for saccharin given in Figure 7.3, it is clear that the maximal sweetness possible for saccharin is equivalent to that of 10.1% sucrose. This R_m for saccharin is rather low and indicates that it would be difficult to formulate a food or beverage product based on saccharin as the sole sweetener. The constant k_d also has physical meaning. It is the saccharin concentration in mg/L (ppm), which gives 50% of its R_m. Thus, for saccharin, since $k_d = 115$, a concentration of 115 mg/L is equivalent in sweetness to 5.0% sucrose. In addition, k_d is the apparent saccharin/receptor dissociation constant and thus at a concentration of 115 mg/L, to a first approximation, it can be said that half of the sweetener receptors are occupied by saccharin. Sweetener C/R functions are useful in assessment of the potential of a sweetener as a sole sweetener or as a component of a blend system. Inspection of saccharin's C/R function clearly shows a strong dependence of P_w on sucrose reference concentration. Thus, saccharin P_w values of 710, 180 and 9 may be calculated relative to sucrose references of 2%, 8% and 10%, respectively. The abrupt drop in P_w shown here for saccharin is due to its low R_m of 10.1. And thus, it is obvious that the major utility of saccharin will be in blends with other non-caloric sweeteners since saccharin will only provide common levels of sweetness at very high concentrations.

- *Flavour profile*: The taste quality of a sweetener is really only meaningful in the context of a food or beverage product and the taste qualities of such products are best assessed by consumer studies. However, less resource-intensive techniques are often used to predict taste quality. Flavour Profile Analysis (FPA) is such a technique.[20] In FPA, a methodology pioneered at the Arthur D. Little Company in the 1940s, expert sensory panels were trained to break down complex and multiple-flavour attribute systems and to rate component intensities. The FPA technique has also been used to assess the taste attributes of sweeteners and the flavour profile of saccharin is illustrated in Figure 7.4. Methodology employed for the generation of the data has been described by Carr and co-workers (The NutraSweet Company).[21]

As is obvious on inspection of Figure 7.4, the flavour profile of saccharin is substantially inferior to that of sucrose, which shows only sweetness at any concentration we have

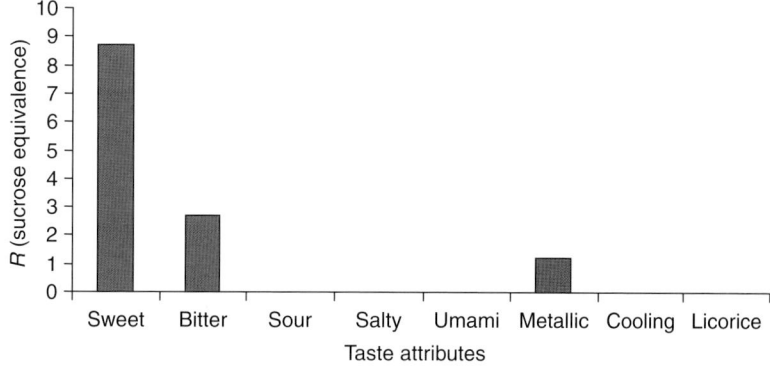

Fig. 7.4 Flavour profile of saccharin-Na.

evaluated by this methodology. Its sweet taste is accompanied by significant bitter and metallic taste attributes. Interestingly, it has been found that the population is heterogeneous in this respect with approximately one-third being hypersensitive to these 'off'-tastes, while the remaining two-thirds of people vary from moderately sensitive to insensitive.[22–24] From saccharin's C/R function, we see its maximal sweetness is equivalent to that of 10.1% sucrose. People hypersensitive to saccharin's bitter and metallic tastes do not experience an R_m which even approaches 10, while individuals insensitive to these off-tastes likely experience a higher R_m. However, the average result, for the expert sensory panel used to generate the data in this study is 10.1. While the reason for a low R_m for some panellists is not known, it seems likely that it may be due to mixture suppression effects. This phenomenon of suppression of one taste attribute (e.g. sweetness) by another (e.g. bitter or sour) is well known.[25]

- *Temporal profile*: Some non-nutritive sweeteners are quite similar to sucrose in flavour profile. One might expect that such sweeteners would make zero- and reduced-calorie products possible, which are equivalent in taste to sucrose-sweetened products. However, this is not the case. The major factor contributing to the difference in taste between the non-nutritive sweetener and sucrose-sweetened products is temporal profile. A method was developed to compare the temporal profiles of non-caloric sweeteners.[26] Here, the time to reach a sweetness response equivalent to 10% sucrose was defined as the sweet-taste appearance time (AT) and 10% sucrose was found to exhibit an AT of 4 seconds. Another parameter determined by this method was the time required for the perceived sweetness intensity to decline from a 10% sucrose equivalent to the low, but greater than threshold, 2% sucrose equivalent. This time was defined as the extinction time (ET) and 10% sucrose was determined to exhibit an ET of 14 seconds. Temporal profile plots of sucrose, saccharin and three other non-caloric sweeteners are illustrated in Figure 7.5. The AT and ET values for saccharin were found to be the same as those of sucrose and thus, in temporal profile, saccharin was quite sucrose-like. Interestingly, the AT value for aspartame was found to be slightly greater than that of sucrose and its ET value to be substantially greater, consistent with the observation that aspartame has a slight delay in sweetness onset and with noticeable lingering sweetness. The temporal profile of monoammonium glycyrrhizinate, the natural liquorice-derived sweetener was also

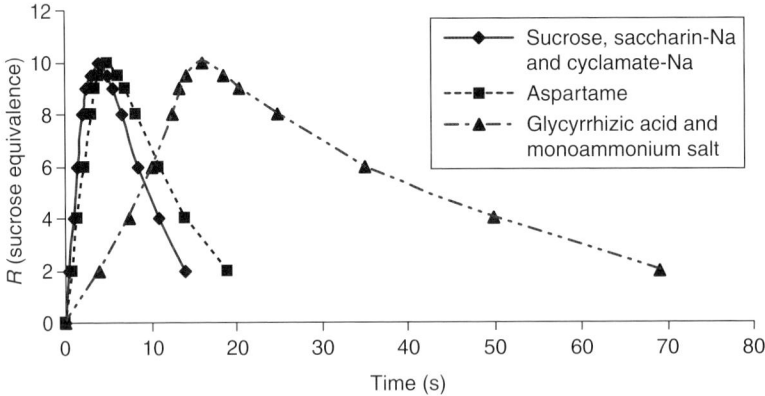

Fig. 7.5 Temporal profiles of sucrose, saccharin-Na, cyclamate-Na, aspartame and glycyrrhizic acid, mono-ammonium salt.

evaluated. It has a strongly delayed sweetness onset and exhibits a very much prolonged sweetness, thus explaining why it has never experienced significant usage in foods and beverages.

- *Synergy and compatibility with other sweeteners*: As a consequence of its bitter and metallic off-taste attributes, saccharin is most commonly employed in blends with other sweeteners, both caloric and non-caloric. Such blends typically derive about 50% of their sweetness from saccharin and in many cases are advantaged by synergistic interactions. In efforts to better understand the phenomenon of sweetness synergy, Schiffman and co-workers (Duke University) conducted a study to determine the interaction effects of 14 sweeteners; 5, caloric and 9, non-caloric.[27] In this study, interaction effects were quantified for all possible binary blends of the 14 sweeteners, at 6%, 10% and 14% levels of additive sweetness intensity in sucrose-equivalent units, in an effort to cover the range of sweetness commonly encountered in food and beverage systems. The findings on saccharin are summarised in Table 7.1. The following observations may be made from these results:
  ○ Sweetness synergy of saccharin with other sweeteners is not universal or predictable. It occurs with some (principally) other non-caloric sweeteners but not with others. As noted previously, since the activities of all sweeteners appear to be mediated by a single receptor, and since sweetness synergy is most easily rationalised as a result of cooperative binding effects at multiple sites on the sweetener receptor, saccharin synergy is evidence for complementary binding loci on the receptor.
  ○ Percent sweetness synergy is inversely related to additive sweetness intensity. In other words, sweetness synergy is highest at low levels of sweetness intensity and lowest at high levels. This is an expected consequence of the hyperbolic C/R functions for non-caloric sweeteners.
  ○ If sweetness synergy is a consequence of cooperative binding of two sweeteners at two different sites on the receptor, some of the data in Table 7.1 makes sense while

Table 7.1 Interaction effects in blends of saccharin-Na with other sweeteners.

Sweetener blended with saccharin-Na	Sweetness synergy (%) as a function of additive sweetness intensity (% Sucrose Equivalent)[a]		
	6	10	14
Sucrose	**0** (0)	**0** (−5)	**−19**
Glucose	**0** (14)	**0** (8)	**0** (−4)
Fructose	**0** (8)	**0** (9)	**−10**
Sorbitol	**+26**	**0** (1)	**−11**
Mannitol	**+18**	**0** (−2)	**−10**
Cyclamate	**+24**	**0** (10)	**−10**
Aspartame	**+28**	**+17**	**−11**
Alitame	**0** (−14)	**−20**	**−32**
Acesulfame K	**−31**	**−33**	**−52**
Neohesperidin Dihydrochalcone	**+41**	**0** (5)	**−12**
Stevioside	**+43**	**0** (7)	**−11**
Rebaudioside A	**+31**	**+12**	**−18**
Thaumatin	**0** (0)	**−14**	**−28**

[a]Sweetness synergy levels given here are calculated from data in the Schiffman *et al.* reference and are given in bold ($p < 0.05$); when not significantly different from zero, sweetness synergy is listed as 0% with the empirically-determined level given in parentheses.

Table 7.2 Physical properties of saccharin in acid, sodium salt and calcium salt forms.

Property	Saccharin form		
	Acid form	Sodium salt	Calcium Salt
Chemical Structure	(structure)	(structure) ·2H$_2$O	(structure) ·2H$_2$O
Molecular Formula	C$_7$H$_5$NO$_3$S	C$_7$H$_4$NO$_3$SNa·2H$_2$O	C$_{14}$H$_8$N$_2$O$_6$S$_2$Ca·2H$_2$O
Molecular Weight	183.18 (anhydrous)	241.20 (dihydrate)	440.48 (dihydrate)

some does not. The absence of sweetness synergy between saccharin and acesulfame K is expected in view of their similar chemical structures since they likely compete for the same receptor binding site. On the other hand, the finding of strong synergy with aspartame and an absence of synergy with alitame, an aspartame structural analogue, is not expected. Further, synergy between saccharin and the two carbohydrate sweeteners sorbitol and mannitol, but no synergy with the three other carbohydrate sweeteners sucrose, glucose or fructose is also unanticipated. Further work is needed to explain these apparent anomalies.

7.3.3 Physical and chemical properties

Saccharin is commercially available in acid form as well as in sodium and calcium salt forms. All of these are white odourless solids. Saccharin in acid form is a strong acid with pK_a of 2.32.[28] Chemical compositions of the three commercially available forms of saccharin are summarised in Table 7.2.

To be used in foods and beverages, a non-caloric sweetener must be sufficiently soluble and many non-caloric sweeteners do not meet this requirement. Commonly, sweetness intensity levels at least equivalent to 10% sucrose are required and in some systems (e.g. frozen desserts), sweetener levels matching the sweetness of 15–20% sucrose are needed. In addition, for many food systems, rapid dissolution is critical to comply with manufacturing requirements. For example, in carbonated soft drinks, concentrates of the sweetener-flavour system complex are prepared and it is important that all components rapidly dissolve. Thus, high solubilities and rapid dissolution rates are very desirable properties for non-nutritive sweeteners. In addition, a commercially viable non-caloric sweetener must be sufficiently stable to hydrolysis as well as to thermal and photochemical breakdown to be used in beverages, baked goods and confectionery. The performance of sodium saccharin relative to these metrics is as follows:

- *Solubility*: Saccharin in its acid form is poorly soluble in water while its salt forms are highly soluble. Data on the solubility of the acid and salt forms of saccharin are summarised in Table 7.3. The most commonly used form of saccharin is the sodium salt and in Table 7.3 it is shown that 1200 g is soluble in 1 L of water at ambient temperature. As an example to illustrate the ample solubility of saccharin for common food or beverage

Table 7.3 Saccharin water solubility as a function of temperature in acid, sodium salt and calcium salt forms.

Temperature	Solubility (g/L)[29]		
	Acid form	Sodium salt	Calcium salt
Ambient	3.4	1200	–

Temperature (°C)	Solubility (g/100 g)[30]		
	Acid form	Sodium salt	Calcium salt
20	0.2	100	37
35	0.4	143	82
50	0.7	187	127
75	1.3	254	202
90	–	297	247

products, if an application requires a sweetness level equivalent to 4% sucrose (a common situation in binary blends), from saccharin's *C/R* function, it can be calculated that a saccharin concentration of 75 mg/L is necessary. However, with a solubility of 1200 g/L, saccharin as its sodium salt is 16,000 times more soluble than necessary to meet this need. It is noteworthy that saccharin in its acid form is much less soluble than its salt forms and it is logical to be concerned about precipitation in acidic products such as soft drinks and reduced-calorie juices. However, this does not appear to be a problem. The only solubility problem observed is with the calcium salt in phosphoric acid-based beverages where $Ca_3(PO_4)_2$ may precipitate.

- *Stability*: To be commercially viable, a non-caloric sweetener must be stable to degradation from hydrolytic, pyrolytic or photochemical processes that may be encountered in food or beverage applications. Stability is critical for three reasons. First, the rate of degradation must not be such that product shelf life is affected. Second, degradation must not cause any 'off'-taste or odour. And third, since non-caloric sweeteners are food additives, any degradation products formed must also be safe. In the United States, for any food or beverage application, if exposure to the degradation product may reach or exceed 12.5 μg/kg, then safety assessment studies equivalent to those required for the sweetener itself must be conducted before regulatory approval is granted.[31]

Saccharin is very stable to all the conditions to which it may be exposed in food and beverage applications. Accelerated stability studies on saccharin as a function of pH and temperature (100°C, 125°C and 150°C) were first reported in 1952 by DeGarmo and coworkers (Monsanto Chemical Company).[32] Later, accelerated studies at a single temperature (120°C) were carried out at the Sherwin Williams Company.[33] The results of all of these studies are summarised in Table 7.4. Interestingly, the degradation pathway was found to be pH dependent. At acidic pH, the exclusive hydrolysis product is 2-sulfobenzoic acid, while under alkaline conditions, the sole degradation product is 2-sulfonamidobenzoic acid. Both of these compounds are sometimes found as trace contaminants in commercial samples of saccharin. As a consequence of saccharin's high stability, neither loss of sweetness during food or beverage product lifetime nor degradation product safety is a significant concern.

Table 7.4 Hydrolytic stability of saccharin as a function of temperature, pH and time.

Temperature (°C)	pH	Time (h)	Saccharin loss (%)
100[32]	2.0	1	2.9
	3.3		0
	7.0		0.3
	8.0		0
125[32]	2.0		8.5
	3.3		1.9
	7.0		1.6
	8.0		0
150[32]	2.0		18.6
	3.3		1.9
	7.0		1.6
	8.0		0
120[33]	3.3	27	18
		219	69
	7.0	27	0
		219	6
	9.0	27	2
		219	12

7.3.4 Physiological properties

Renwick (University of Southampton) has reviewed the state of knowledge of the absorption, distribution, metabolism and excretion of saccharin in humans, as well as in a number of animal models.[34–37] Findings with high relevance to humans are as follows:

- *Absorption*: Saccharin is largely absorbed from the small intestine. In a human study with radio-labelled saccharin, only 5% of the dose was recovered in the faeces, thus indicating that 95% of the dose is absorbed into the circulation. In rat studies, the level of absorption was demonstrated to be dose dependent.
- *Distribution*: In rat studies, at very high doses, saccharin was found to concentrate slightly in the organs of excretion, the kidney and bladder, while most was concentrated in the plasma and with only low levels in other tissues. These findings are consistent with expectation for a highly polar molecule that is cleared rapidly in the urine.
- *Metabolism*: Saccharin is not metabolised in humans or rats. In addition, saccharin does not covalently bind to DNA in the bladder, as would be expected if saccharin is a carcinogen of the classical type (suggested by rodent studies that are discussed later in this review). For saccharin to bind with DNA, metabolism to an electrophilic species would be necessary and the failure of this to occur is consistent with an absence of carcinogenicity.
- *Excretion*: In humans, oral doses are excreted almost completely by the kidneys with the balance recovered in the faeces.

As is the case for nearly all biologically active compounds that initiate their biological activities through interactions with the proteome, saccharin has also been observed to have activities with proteins in addition to the sweetener receptor. For example, Naim and co-workers have demonstrated that saccharin inhibits the activities of G protein receptor kinases, which are believed to desensitise the sweetener receptor.[38] And as a second example, saccharin has been observed to inhibit the activity of adenylate cyclase in a number of cell types including adipocytes, astrocytes and thyrocytes.[39] The significance of these biological activities is not clear at this time.

7.3.5 Applications

The common caloric sweetener sucrose is used in many applications including baked good, frozen dessert, confectionery, processed food, beverage and table-top applications. Saccharin, as well as other high-potency non-caloric sweeteners, however, only find significant usage in the latter two categories, beverages and table-top applications. In these applications, saccharin provides only sweetness. Other foods require additional physical properties that are provided by sucrose but not saccharin. In baked goods, for example, sucrose elevates the gelatinisation point of starch, provides structure and, following hydrolysis, undergoes the Maillard browning reaction, thus leading to finished products with the expected structure and appearance. As a second example, in frozen desserts, sucrose depresses the freezing point of the product as well as providing structure such that a product with the common malleable composition is obtained. To formulate acceptable-quality baked goods, frozen desserts and confectionary with a reduction in sugar-derived calories, it is necessary to blend sweeteners like saccharin with a zero- or reduced-calorie sugar substitute, which can provide the physical properties of sugar that have already been noted. Today, there are several such ingredients as reviewed in 2000.[40] All of these substances are carbohydrates or carbohydrate-like ingredients and included among them are maltitol, lactitol, isomalt, fructooligosaccharide sweetener and tagatose. However, none of these substances are non-caloric; they are only reduced in caloric content. They are also poorly absorbed from the small intestine leading to passage into the large intestine where they are fermented to short-chain fatty acids and gases, principally hydrogen and methane. The short-chain fatty acids are substantially absorbed from the large intestine, thus accounting for their significant bioavailable energy contents. The short-chain fatty acids also have high osmotic activity, thus leading to laxative effects and, of course, the fermentation gases lead to flatulence. As a consequence, only small amounts of any of these ingredients can be consumed and therefore, they may only be used in combination with saccharin in low-usage applications such as confectionery.

Erythritol is a carbohydrate-like ingredient, and the only one that is non-caloric as well as natural and that shows exceptional promise for blending with saccharin in nearly all sucrose-substitution applications. It has been reviewed by Embuscado and Patil[41] as well as by de Cock and Bechert (Cerestar).[42] Interestingly, recent work by den Hartog and co-workers (Maastricht University) has demonstrated anti-oxidant and free-radical quenching activities for erythritol suggesting it may offer substantial health benefits when used in foods and beverages.[43] Erythritol is unique among carbohydrate-like ingredients owing to its non-caloric nature and because it is well absorbed from the small intestine and excreted unchanged in the urine. As a result, erythritol is very well tolerated. At the present time, erythritol is accepted in the United States to have a bioavailable calorie content of 0.2 cal/g. However, available evidence supports a zero-calorie bioavailable energy content and a change to regulate it as a zero-calorie sweetener is likely.

The major applications of saccharin are in beverages, either in finished products such as carbonated soft drinks, or in beverages sweetened with saccharin as a table-top sweetener. In these applications, the only requirement is sweet taste and so a second ingredient providing additional sucrose functionality is unnecessary. Saccharin is limited, however, by bitter and metallic off-tastes when employed as a sole sweetener and, as a result, it is most commonly employed in blends with other high-potency sweeteners such as aspartame or cyclamate. Recently, advantaged by the understanding that human bitter taste is initiated by a family of 25 bitterant receptors known as T2Rs expressed in a subset of taste bud cells,[44,45] the specific receptors for saccharin bitterness were identified.[46] In 2010, a collaborative team of Slack and co-workers (Givuadan) and Meyerhof and co-workers (German Institute of Human Nutrition) reported the results of a cell-based HTS program targeted at saccharin bitterness antagonist discovery that led to GIV3727, an antagonist for saccharin bitterness.[47] Interestingly, however, while GIV3727 is a highly effective antagonist *in vitro*, it is not fully effective in sensory panel testing. The authors suggest that genetic diversity among human subjects, as has been reported for T2Rs by several groups,[48–50] may be responsible for the limited sensory panel effectiveness of GIV3727. Interestingly, the potential for a very high level of complexity of the receptors that mediate human bitter taste was demonstrated in 2010 by Meyerhof and colleagues.[51] They showed that the 25 human bitterant receptors may hetero- as well as homo-dimerise and thus, that if this occurs *in vivo*, the real-life situation could be that bitter taste is mediated by a combination of 25 homo- and 300 hetero-dimers for a total of 325 bitterant receptors. It seems possible that an antagonist that is highly effective against the homo-dimeric species present in a cell-based assay may have different efficacies against a highly diverse population of bitterant receptors that may be present in human taste bud cells. This difference could be the explanation for the limited efficacy of GIV3727 in sensory testing. In summary, while the discovery of potent antagonists against individual T2Rs by HTS of large chemical libraries may be routine, it seems likely that the challenge will remain to discovery antagonists, which are highly effective in human sensory testing.

GIV3727

Substantial work over the last century has been carried out to find antagonists for saccharin bitterness. However, for research scientists considering more work of this type, it may be important to consider the likely value of a saccharin bitterness antagonist. As noted previously, the C/R function of saccharin is well modelled by the equation $R = 10.1\ C/(115 + C)$, thus showing saccharin to have an R_m of 10.1% sucrose equivalency. The practical ramification of this low R_m is that it is not sensible to try to reach a 10% sucrose level of sweetness intensity with saccharin alone, as is required in many foods and beverages. To reach a sweetness intensity equivalent to 10% sucrose, this equation predicts that a saccharin concentration of 11,500 mg/L would be needed, a level that is totally impractical. As a bottom line, for food or beverage systems requiring a high level of sweetness, the bitterness antagonist approach may be of little value and the only practical approach may be to employ saccharin in blends with other non-caloric sweeteners where the bitter off-taste is imperceptible.

In addition to bitter off-taste, saccharin has also been reported to exhibit a metallic off-taste, a sensory perception that is not really understood. Recently, however, le Coutre and co-workers (Nestle) provided evidence that metallic taste may be initiated by saccharin activity at the capsaicin receptor, also known as TRPV1.[52,53] In other work by Klebe and co-workers (Philipps-Universität Marburg), evidence for saccharin metallic taste initiation by modulation of carbonic anhydrase enzymes was reported.[54] While these results are intriguing, more work is needed to clearly understand the mechanistic pathways that mediate saccharin metallic off-taste.

Interestingly, saccharin is synergistic with many sweeteners including aspartame or cyclamate and it has been found that the optimal blends derive equal sweetness from each blend component. In such blends, the sweetness contribution requirement from saccharin is reduced with the result that little to no bitter/metallic 'off'-taste is perceptible. Binary blends such as saccharin/aspartame and saccharin/cyclamate and the ternary blend saccharin/aspartame/cyclamate are particularly good-quality sweetener systems. The compositions of saccharin sweetener blend systems with optimal taste qualities can be predicted based on the component sweetener C/R functions and expected synergy levels.

7.3.6 Safety

In the United States, the use of sweeteners is regulated by the 1958 Food Additives Amendment to the Food, Drug and Cosmetic Act of 1938. This legislation and its effects on the regulation of sweeteners and other food additives have been reviewed.[55,56] In the original 1958 legislation, both saccharin and cyclamate were exempted as generally recognised as safe (GRAS) food ingredients, although their status as GRAS were later reversed by the FDA. For the cases of substances not included on the original GRAS list, two tracks towards regulatory approval for use in foods and beverages are defined, both of which are to be of equal rigour. First is the GRAS process. To qualify for this track, one of the following two criteria must be met:

1. It must be demonstrated that the substance was in common use in food in the United States prior to 1958.
2. In the judgment of a qualified panel of experts on chemical safety, following review of published safety assessment studies, a consensus is reached that the ingredient is safe for consumption.

The second track towards approval of a sweetener for use in foods and beverages is the Food Additive Petition (FAP) process. And both the GRAS and FAP processes are based on equivalent rigorous and extensive safety studies in animals and humans to ensure safety. One objective of the safety assessment studies is determination of the highest dose that may be given without adverse effects. This dose, in the most sensitive species evaluated, is termed the No Observed adverse Effect Level (NOEL). The NOEL is then used to regulate human exposure by defining an acceptable daily intake (ADI) as 1/100th of the NOEL. NOEL and ADI levels are given in mg/kg body weight (BW). Thus, as an example, if the NOEL of a sweetener is found to be 500 mg/kg, the ADI would be 5 mg/kg. The ADI is used to determine the food categories, and levels in those categories, in which the sweetener may be used. The objective is to ensure that the ADI level is not exceeded on a chronic basis. In order to do this, 90th percentile, 14-day average food category consumption data are employed. The sweetener is then allowed for use in food categories where aggregate consumption does not exceed the ADI. Marshall and Pollard have comprehensively reviewed the category and level approvals granted in both the United States and abroad for saccharin and other non-caloric sweeteners.[57] Regulatory issues related to new sweetener development in the United States have been reviewed by Broulik[58] in 1996 and DuBois[59] in 2008.

Although individual countries assume responsibility for the regulation of food additives within their boundaries, there has been an attempt at international standardisation.[60] In 1956, the Food and Agriculture Organization of the United Nations (FAO) and the World Health Organization (WHO) established the Joint FAO/WHO Expert Committee on Food Additives (JECFA). The fundamental objective of JECFA is the establishment of ADIs for food additives following the assembly and interpretation of all relevant biological and toxicological data. It is important to recognise that ADIs provide a wide margin of safety. According to JECFA, 'an ADI provides a sufficiently large safety margin to ensure that there need be no undue concern about occasionally exceeding it provided the average intake over longer periods of time does not exceed it'.[61]

Questions concerning the safety of saccharin date back to the early 1900s. A staunch defender of saccharin safety was President Theodore Roosevelt. In response to those who questioned its safety, Roosevelt stated, 'Anybody who says saccharin is injurious to health is an idiot'.[62] Despite such high-level support, however, saccharin has continued to come under fire. The chronologies of safety studies and regulatory agency actions concerning saccharin have been comprehensively reviewed.[63,64] The principal events are summarised here. In 1958, saccharin was listed as one of the 675 substances on the original GRAS list. In 1972, however, the FDA retracted the GRAS status of saccharin based on concern over results in a long-term rat feeding study. Then, in 1977, as a result of a multi-generation rat study in which bladder tumours were found in the second-generation animals, the FDA announced its intention to ban saccharin. The FDA held that it had no choice other than to ban saccharin because of the Delaney Clause of the Food Additive Amendment to the Food, Drug and Cosmetic Act. The Delaney Clause requires that if any food additive, at any dose, in any animal species, is found to cause cancer, its further use is to be outlawed. With respect to saccharin, acting FDA Commissioner Gardner stated, 'We have no evidence that saccharin has ever caused cancer in human beings. But we do now have clear evidence that the safety of saccharin does not meet the standards for food additives established by Congress'.

Acting in response to strong consumer and industry pressure to ensure the continued availability of saccharin, in 1977, Congress placed a moratorium on the FDA ban pending additional research, a moratorium that was extended seven times. However, during the moratorium period, all saccharin-containing foods and beverages in the United States were required to carry a warning label advising consumers that saccharin had been determined to cause cancer in laboratory animals. Since the initial ban on saccharin, the US regulatory climate has evolved considerably such that, today:

- a policy that considers both risks and benefits is considered more appropriate than the inflexible Delaney Clause;
- risk is recognised to be more of a function of dose than an all-or-none phenomenon;
- the mechanism of action is recognised as an important factor to consider.

Since the 1970s, substantial research has been conducted to elucidate the mechanism whereby saccharin initiated bladder cancer in the strain of rats tested. Interestingly, in one important study, evidence was obtained that the effect, which could only be demonstrated in rats, and mainly in male rats, is an effect common to many sodium salts.[65] In this and related studies, sodium ascorbate, sodium chloride and several other sodium salts were found to promote similar rat bladder neoplasia. The bladder cancer effect appears to be unique to rats and is thought to be related to the high osmolarity of rat urine and crystallisation of $Ca_3(PO_4)_2$ that causes a cytotoxic effect on the superficial layer of the bladder epithelium thus leading to regenerative hyperplasia and tumours. As a result of all the work carried out showing that the bladder carcinogenicity effect observed at very high doses in rats is not relevant to humans, in 2000, the National Toxicology Program of the US Department of Health and Human Services removed saccharin from its list of potential carcinogens. On 15 December 2000, President Clinton signed legislation removing the requirement for a warning label on saccharin-containing foods and beverages. At the present time, the ADI for saccharin in the United States is 0–5 mg/kg BW, which is sufficient for use in all food and beverage applications at levels that reproduce the sweetness of common caloric sweeteners.

The content of saccharin in foods and beverages can be determined in many ways. A validated Association of Official Analytical Chemists (AOAC) method has been reported employing reverse phase HPLC with UV detection at 254 nm.[66] Other methods reported include ion-exclusion HPLC[67] and ion chromatography,[68] both employing UV detection.

7.3.7 Regulatory Status

Comprehensive information on regulatory approvals of food additives such as sweeteners, on a worldwide basis, is difficult to obtain and verify, and is also often subject to differences in interpretation. Contributing to this challenge are the ever-changing regulatory processes and the ambiguous nature of many regulations. Nonetheless, the Calorie Control Council has undertaken a major effort to maintain a database of such information on saccharin and other commercially important non-caloric sweeteners and this information on saccharin regulatory approvals is provided in Table 7.5.[69] The Calorie Control Council does not guarantee the accuracy of this database but states that its information has been obtained from sources that are generally considered reliable and that all of its information is consistent with the EU Sweetener Directive.

7.4 CYCLAMATE

7.4.1 History, manufacture and chemical composition

Although many sweet-tasting organic compounds were found following the discovery of saccharin in the late nineteenth century and early twentieth century, none achieved significant usage in foods. Then in 1944, the discovery that salts of cyclohexylsulfamic acid are sweet was reported by Sveda and Audrieth (University of Illinois).[70] These cyclohexylsulfamic acid salts were prepared as illustrated in Figure 7.6 using chlorosulfonic acid as a sulfamating agent. Since this initial work, many alternative SO_3 sources (i.e. sulfamating agents) have been used including fuming sulfuric acid. These cyclohexylsulfamates are commonly referred to as cyclamic acid salts or, more simply, just as cyclamates. In the acid form, cyclamate

Table 7.5 Worldwide regulatory status of saccharin.

Country	Food[a,b]	Beverage[b]	Table-top[b]	Pharmaceutical[b]
Afghanistan		+		
Algeria	−	−	−	−
Antigua		+		
Argentina	+	+	+	+
Australia	+	+	+	+
Austria	+	+	+	+
Bahamas	+	+		
Barbados		+		
Belgium	+	+	+	
Bermuda		+		
Bolivia	−	−	+	−
Brazil	+	+	+	
Bulgaria	+	+		
Burundi		+		
Canada	−	−	+	+
Caribbean (Ind.)	+	+	+	
Chile		+	+	+
China	+	+		
Columbia	+	+	+	
Costa Rica	+	+	+	
Cyprus	+	+	+	+
Czech Republic	+	−	+	
Denmark	+	+	+	+
Dominica	+	+	+	
Ecuador	+	+	+	+
Egypt		+		
El Salvador	+	+	+	
Ethiopia		+	+	
Fiji	+	−	−	−
Finland	+	+	+	+
France	+	+	+	+
French Guiana	−	−	−	−
Germany	+	+	+	+
Greece	+	+	+	
Guadeloupe	+	+	+	+
Guam	+	+		
Guatemala	+	+	+	
Guiana	+	+		
Haiti		+	+	
Honduras	+	+		
Hong Kong	+	+	+	
Hungary	+	+	+	
Iceland	+	+	+	−
India		+	+	+
Indonesia	+	+	+	
Iran	+	−	−	−
Ireland	+	+	+	+
Israel	+	+	+	+
Italy	+	+	+	+
Jamaica				
Japan	+	+	+	+
Korea	+	+	+	+
Kenya	+	+	+	
Kuwait	+	+	+	
Lebanon	+	+	+	

(Continued)

Table 7.5 (Continued)

Country	Food[a,b]	Beverage[b]	Table-top[b]	Pharmaceutical[b]
Luxembourg	+	+	+	−
Malaysia	+	+	+	+
Malta	+	+	+	
Martinique	+	+	+	+
Mexico	+	+	+	
Monserrat		+		
Morocco	+	+	+	−
Nassau		+		
Netherlands	+	+	+	+
New Zealand	+	+	+	+
Nicaragua	+	+	+	
Nigeria	+	+	+	
Norway	+	+	+	+
Oman	−	−	+	+
Pakistan	+	+	+	
Panama	+	+	+	
Papua New Guinea	+	+	+	
Paraguay	+	−	+	
Peru	+	+	+	−
Philippines	+	+	+	+
Poland	+		+	
Portugal	+	+	+	+
Puerto Rico	+	+		
Qatar			+	
Romania				
Russia	+	−	+	+
Rwanda		+		
Samoa	+	+		
Saudi Arabia	+	−	+	
Sierra Leone	+		+	
Singapore	+	+	+	+
Slovakia	+	−	+	
South Africa	+	+	+	
Spain	+	+	+	+
Sri Lanka		+	+	
Sudan				
Surinam	+	+	+	
Sweden	+	+	+	
Switzerland	+	+	+	
Taiwan	+	+	+	−
Thailand	+	+	+	
Trinidad	+	+		
Tunisia	−	−	−	−
Turkey	+	+	+	+
Uganda				
United Kingdom	+	+	+	+
United Arab Emirates			+	
United States	+	+	+	+
US Virgin Islands	+	+		
Uruguay	+	+	+	
Venezuela	+	+	+	
Yugoslavia	+	+		
Zaire		+		
Zambia		+		

[a] May not apply to all food categories.
[b] + = permitted; − = prohibited.

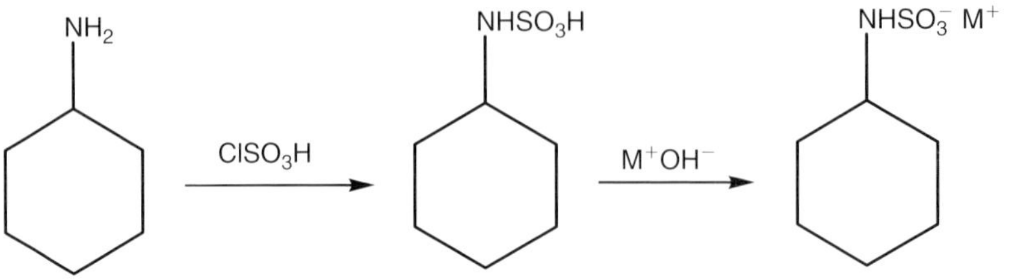

Fig. 7.6 Audrieth-Sveda process for cyclamate manufacture.

has the chemical structure shown in Figure 7.6 and was the first of the *sulfamate* structural class of sweeteners to be discovered. Sweeteners of this class possess the -NH-SO$_3^-$ moiety as the common structural subunit. Cyclamate is generally used in foods as either the sodium or calcium salt. Since 1944, many other sulfamates have been synthesised, although none have been developed for use in foods. Interestingly, however, in 2007, work was reported by Cavicchioli and co-workers on the Ag$^+$ salt of cyclamate exploring its therapeutic potential as an antibacterial.[71] It was found to have significant inhibitory activity against *Mycobacterium tuberculosis*, and, of course, in this material the antibacterial activity is due to the Ag$^+$ and not cyclamate, as the investigators also showed antibacterial activities for Ag$^+$ salts of saccharin and aspartame.

7.4.2 Organoleptic properties

As discussed previously for saccharin, the organoleptic characteristics of a non-caloric sweetener are best appreciated in terms of their *C/R* functions, flavour profiles and temporal profiles. Cyclamate's performance on these organoleptic metrics is as follows:

- *Sweetness potency*: The *C/R* function for sodium cyclamate is illustrated in Figure 7.7.[19] In common with all high-potency sweeteners, the potency of sodium cyclamate is dependent on sucrose reference concentration. Thus, $P_w(2) = 42$, $P_w(8) = 23$ and $P_w(10) = 17$ may be calculated from the *C/R* function given in Figure 7.7.
- *Flavour profile*: Sensory panel studies on cyclamate salts demonstrate them to be better in reproducing the taste quality of sucrose than is saccharin. Nonetheless, significant bitter and salty taste attributes are noted for the most commonly employed form of cyclamate that is the sodium salt. The flavour profile of sodium cyclamate is illustrated in Figure 7.8.[21] The weak salty taste attribute in sodium cyclamate is due to the high sodium ion concentration present at the concentrations providing high sweetness. Thus, although sodium cyclamate exhibits a flavour profile with some weak off taste notes at high levels of sweetness, since it is generally used only in blends where these weak 'off'-taste notes are not perceptible, it is a very good non-caloric sweetener for use in food and beverage applications.
- *Temporal profile*: The temporal profile of sodium cyclamate is not distinguishable from that of sucrose with ATs of 4 seconds and ETs of 14 seconds for both substances as is illustrated in Figure 7.5.[21] As already discussed, the major drawback of most non-caloric sweeteners relates to their failure to reproduce the temporal behaviour of sucrose. The

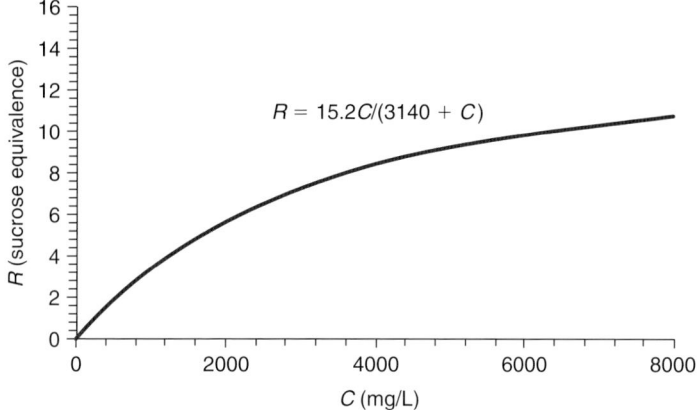

Fig. 7.7 Cyclamate-Na C/R function.

fact that cyclamate salts do accurately reproduce sucrose taste on this metric explains why foods and beverages formulated with cyclamates are so well accepted by consumers.

- *Synergy and compatibility with other sweeteners*: As a consequence of its bitter and salty 'off'-taste attributes, as well as its low sweetness potency, cyclamate salts are most commonly employed in blends with other non-caloric sweeteners, principally saccharin, aspartame and acesulfame K. Optimal binary blends typically derive 50% of their sweetness from cyclamate and similarly, optimal ternary blends derive 33% of their sweetness from cyclamate. As discussed for saccharin, Schiffman and co-workers also determined the interaction effects of sodium cyclamate with 13 other sweeteners at 6%, 10% and 14% sucrose levels of total sweetness intensity.[27] On inspection of the results, which are summarised in Table 7.6, the following observations may be made:
  - As for saccharin, percent sweetness synergy for cyclamate is inversely related to additive sweetness intensity. In other words, sweetness synergy is highest at low levels of sweetness intensity and lowest at high levels. This is an expected consequence of the sweetener hyperbolic *C/R* functions.

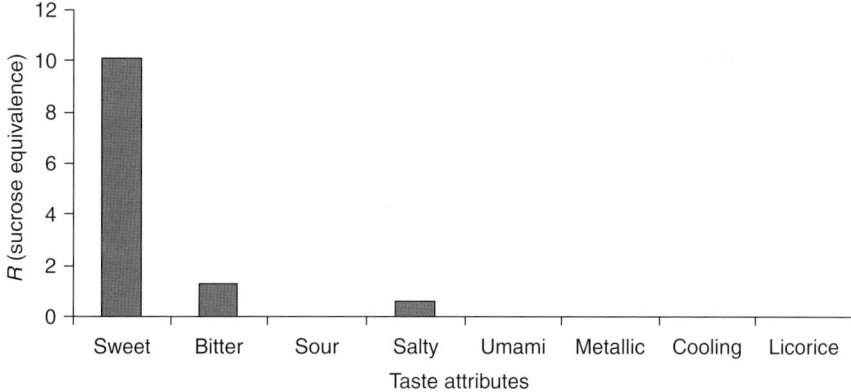

Fig. 7.8 Flavour profile of cyclamate-Na.

Table 7.6 Sweetener interaction effects in sodium cyclamate blends with other sweeteners.

Sweetener blended with cyclamate	Percent sweetness synergy as a function of additive sweetness intensity (% Sucrose Equivalent)[a]		
	6	10	14
Sucrose	**+17**	0 (+5)	**−9**
Glucose	**+24**	0 (−1)	0 (−4)
Fructose	**+41**	0 (+8)	0 (−6)
Sorbitol	**+23**	0 (2)	**−9**
Mannitol	0 (+1)	0 (−2)	**−16**
Saccharin-Na	**+24**	0 (10)	**−10**
Aspartame	**+26**	0 (+3)	0 (−6)
Alitame	0 (+12)	0(−2)	**−9**
Acesulfame K	**+36**	**+11**	**−9**
Neohesperidin Dihydrochalcone	**+26**	0 (−3)	**−22**
Stevioside	**+32**	**+14**	**−22**
Rebaudioside A	**+22**	0 (−3)	**−22**
Thaumatin	**+19**	0 (−5)	**−24**

[a]Sweetness synergy levels given here are calculated from data in the Schiffman et al. reference and are given in bold ($p < 0.05$); when not significantly different from zero, sweetness synergy is listed as 0% with the empirically-determined level given in parentheses.

○ Sweetness synergy of cyclamate with other sweeteners is universal, or at least nearly so. Eleven of the thirteen sweeteners evaluated were observed to be synergistic with cyclamate. According to the Schiffman group findings, only one carbohydrate-type sweetener (mannitol) and one high-potency sweetener (alitame) are not synergistic. However, on reflection on the mechanistic model for sweetness synergy discussed previously (i.e. cooperative binding effects between two different binding sites on the receptor), these findings are surprising. As has already been discussed, we know that cyclamate binds within the TMD of the T_1R_3 component of the receptor and that aspartame binds within the VFD of T_1R_2. Postulation of a cooperative binding interaction between these two sites provides a rational explanation for cyclamate/aspartame synergy. However, since alitame is a structural analogue of aspartame, it is most likely that it binds in the aspartame binding site, thus leading to the expectation of cyclamate/alitame synergy. Also, it seems most likely that mannitol binds to the same site on the receptor as other carbohydrate-type sweeteners and, since cyclamate was found synergistic with all of the other four carbohydrate-type sweeteners evaluated, it is surprising that mannitol was not also found to be synergistic. Further work is needed to resolve these apparent anomalies.

7.4.3 Physical and chemical properties

Cyclamate is commercially available in the sodium and calcium salt forms. Both of these are colourless and odourless solids. Cyclamate in its acid form is a strong acid with pK_a of 1.71.[72] Interestingly, the acid form of cyclamate has been demonstrated by X-ray crystallography

Table 7.7 Physical properties of cyclamate in sodium salt and calcium salt forms.

Property	Cyclamate form	
	Sodium salt	**Calcium salt**
Chemical structure	cyclohexyl-NHSO$_3^-$ Na$^+$	[cyclohexyl-NHSO$_3^-$]$_2$ Ca^{2+}
Molecular formula	C$_6$H$_{12}$NO$_3$SNa	C$_{12}$H$_{24}$N$_2$O$_6$S$_2$Ca
Molecular weight	201.20	396.50

to exist in the zwitterionic state.[74] However, given its very low pK$_a$, cyclamate must exist in the anionic form in all food and beverage systems. The chemical compositions of the two commercially available forms of cyclamate are summarised in Table 7.7.

To have utility for use in foods and beverages, a non-caloric sweetener must (1) be sufficiently soluble, (2) have a rapid rate of dissolution, (3) be hydrolytically stable, (4) be thermally stable and (5) be stable to sunlight. The performance of sodium cyclamate on these metrics is as follows:

- *Solubility*: Cyclamates exhibit excellent solubility characteristics for use in essentially all imaginable applications. Although the acid form is sufficiently water-soluble (133 g/L), its high acidity results in preference for the very soluble sodium (200 g/L) or calcium (250 g/L) salts.[75] To illustrate the more than adequate solubility of sodium cyclamate, consider an application in which cyclamate is used in a binary blend with a sweetener such as saccharin. In such a situation, it is generally desired that cyclamate should provide half of the total sweetness desired that would typically be, allowing for sweetness synergy, sweetness equivalent to approximately 4% sucrose. Thus, since, from the C/R function given in Figure 7.7, the concentration of sodium cyclamate equi-sweet with 4% sucrose is 1120 mg/L, sodium cyclamate is about 180 times more soluble than necessary for this level of sweetness. Cyclamate salts therefore have ample solubility for all food and beverage applications.
- *Stability*: Hydrolytic degradation of cyclamate salts yields cyclohexylamine and inorganic sulfate. As a consequence of the adverse biological activity of cyclohexylamine, FDA scientists conducted a comprehensive evaluation of cyclohexylamine levels in a range of food products.[76] Cyclohexylamine was found in the majority of these products, albeit at low levels. Interestingly, even in the most acidic samples (cola soft drinks), cyclohexylamine levels did not significantly change over 4 months of ambient-temperature storage. The trace levels of cyclohexylamine in food products appear to be substantially derived from trace levels present in the cyclamate sweeteners employed. Data have been reported on the hydrolysis of cyclamate under extreme conditions and are reported in Table 7.8.[77] In summary, cyclamate sweeteners are quite stable. No significant loss of sweetness or generation of unsafe degradation products is expected in any common applications.

Table 7.8 Cyclamate hydrolysis as a function of pH in 1 hour at 100°C.

PH	Cyclamate loss (%)
0.9	13.7
1.6	8.1
2.5	0.98
4.5	0.1
5.3	0.52
6.5	0.58

7.4.4 Physiological properties

A comprehensive review of the state of knowledge of the absorption, distribution, metabolism and excretion of cyclamate in humans, as well as a number of animal models, has been provided by Bopp and Price.[78] Findings in humans, and in animal models with high relevance to humans, are as follows:

- *Absorption*: Cyclamate is slowly and incompletely absorbed from the small intestine. In a study of nearly 200 subjects, an average of 37% of the dose of cyclamate was absorbed. In contrast, cyclohexylamine, which is present as a trace contaminant in cyclamate and is formed on interaction with the microflora of the large intestine in a minor fraction of the population, is completely absorbed.
- *Distribution*: In rat studies, cyclamate, following absorption, has a volume of distribution equivalent to the total body water content and thus, cyclamate does not concentrate in tissues. In contrast, however, cyclohexylamine has tissue concentrations that exceed those of plasma.
- *Metabolism*: Cyclamate is not metabolised by mammalian tissue but is, however, metabolised to cyclohexylamine by the microflora present in the large intestine of some individuals. Approximately 25% of the population have the capacity to metabolise cyclamate to cyclohexylamine and it appears that this ability is somewhat lower (about 20%) in Europeans and North Americans and higher (about 80%) among Japanese people. This metabolic capability of the microflora of the large intestine varies considerably between subjects (<0.1% to >60%) and in the same subjects, varies significantly from day to day. Following absorption, cyclohexylamine is metabolised to an extent of 1–2% to cyclohexanol and cyclohexane-1,2-diol. The NOEL and ADI values that have been established for cyclamate are based on cyclohexylamine levels of high cyclamate converters.
- *Excretion*: All cyclamate that is absorbed is excreted unchanged in the urine. And likewise, the cyclohexylamine formed by some individuals is almost completely excreted unchanged in the urine.

7.4.5 Applications

The major applications of cyclamate are in beverages, either in finished products such as carbonated soft drinks or in beverages sweetened with cyclamate as a table-top sweetener. In such applications, the only functionality required is sweet taste and so a second ingredient to provide additional functionality (e.g. freezing point depression, etc.) is unnecessary. As described previously in some detail for saccharin, cyclamate may also be used as a sugar substitute in all other applications but with the requirement that it be used in combination with a carbohydrate-type, zero- or reduced-calorie sweetener that provides the other physical characteristics of sugar (i.e. freezing point depression in frozen desserts, starch gelatinisation point elevation in baked goods, etc.). In all applications, however, cyclamate is limited, by a weak bitter off-taste as well as limited regulatory approvals and so is never used as a sole sweetener. Thus, cyclamate is nearly always used in blends with other high-potency sweeteners such as saccharin, aspartame and acesulfame K. As already noted for saccharin, cyclamate is synergistic with other sweeteners and it has been found that the optimal blends derive equal sweetness from each blend component. In such blends, due to sweetness synergy, the sweetness contribution requirement from cyclamate is reduced with the result that no bitter or salty 'off'-tastes are perceptible. Thus, binary blends such as saccharin/cyclamate and the ternary blends aspartame/saccharin/cyclamate and aspartame/acesulfame K/cyclamate are particularly good-quality sweetener systems. The compositions of cyclamate sweetener blends, in the ideal situation, derive equivalent sweetness contributions from all blend contributors (i.e. 50% contribution in binary blends, 33% contribution in ternary blends, etc.). However, in countries where regulatory approvals limit cyclamate usage to less than ideal contribution levels, cyclamate should be used at its maximum allowed level and the residual sweetness provided by the other sweetener(s). The sweetener concentrations to be used can be predicted through use of the *C/R* functions and expected synergy levels.

7.4.6 Safety

The history of the evaluation of cyclamate safety was reviewed comprehensively in 1986 by Barbara Bopp and co-workers[79] and again by Barbara Bopp and Paul Price in 2001.[78] Key elements of cyclamate safety studies are summarised here.

Cyclamates were first approved in 1951 for use as drugs by people with diabetes and others who had to restrict their sugar intake. In 1958, they were listed by the FDA on the original GRAS list. Cyclamate usage experienced explosive growth during the 1960s following the finding that a 10/1 mixture of cyclamate and saccharin exhibited taste quality superior to either sweetener individually and approached that of sucrose.

In 1969, the FDA became concerned about cyclamate safety based on findings in a rat study in which the commonly used 10/1 cyclamate/saccharin mixture was shown to induce bladder tumours. These results were interpreted by the FDA to indicate that cyclamate salts are bladder carcinogens in rats. As a consequence, cyclamates were immediately removed from the GRAS list and, in 1970, banned from use in the United States. At the same time, however, cyclamates continued to be allowed in over 50 other countries. Since the 1970 FDA action, a great deal of work has been done to have cyclamates reapproved in the United States. Many well-controlled carcinogenicity studies have now been completed in several animal models (i.e. mice, rats, dogs and monkeys) on both sodium and calcium salt forms of cyclamate as well as on cyclohexylamine. These studies have failed to demonstrate any carcinogenic effects. Similarly, epidemiological studies in humans have failed to show any evidence of carcinogenicity. In addition, many *in vitro* studies have been carried out to determine if either cyclamate salts or cyclohexylamine can cause heritable genetic damage (i.e. mutagenicity) and all have failed to produce any evidence of mutagenic effects. As a result of all of this very large body of work, it is now uniformly accepted that cyclamates are not carcinogens.

It has been already noted that some individuals have the ability to metabolise cyclamate to cyclohexylamine and because cyclohexylamine has a higher level of biological activity than cyclamate, the safety of cyclohexylamine has also been studied extensively. Studies in the rat have shown the testes to be the organ most sensitive to cyclohexylamine. To quantify the level of this toxicity, extensive studies have been carried out in both mice and rats with the finding that, in the more sensitive rat model, a clearly defined NOEL of 100 mg/kg per day is observed.

The other major area of cyclohexylamine biological activity relates to its cardiovascular effects. Cyclohexylamine exhibits sympathomimetic activity similar to that of tyramine, albeit with 100-fold lower potency. Intravenous administration of cyclohexylamine in test animals causes vasoconstriction and increases in both blood pressure and heart rate. In humans, it has been found that a single bolus dose of cyclohexylamine can induce an increase of blood pressure. At the same time, however, exposure to cyclohexylamine as a metabolite of cyclamate has failed to demonstrate any effect on blood pressure or heart rate even among high converters of cyclamate and at high cyclamate doses. In summary, cardiovascular effects are not observed in humans from cyclamate exposure.

The ADI established by JECFA for cyclamate is 11.0 mg/kg BW. In the European Union, cyclamate levels in cola soft drinks are limited to 250 ppm.[80] From the equation given in Figure 7.7, it can be calculated that 250 mg/L cyclamate is equivalent to 1.1% sucrose in sweetness intensity. Clearly, 1.1% sucrose equivalent level of sweetness is far from the optimal sweetness contribution level of 5% for binary blends and 3.3% for ternary blends. However, given the significant sweetness synergy that cyclamate exhibits with saccharin, aspartame and acesulfame K, cyclamate is still helpful, even at this less than optimal level, in improving the taste quality of ternary non-caloric sweetener blends. In beverage applications, cyclamate/saccharin/aspartame and cyclamate/acesulfame K/aspartame blends are surprisingly sugar-like in taste quality and these sweetener systems are used in countries where regulations permit. In the United States, cyclamate is still not permitted even following the extensive studies demonstrating that cyclamate is not carcinogenic or unsafe in other ways if used at the ADI established by JECFA. A petition for cyclamate reapproval in the United States was filed in 1982 and FDA action on this petition remains pending.

The content of cyclamate in foods and beverages can be determined in many ways. A validated AOAC method has been reported involving hydrolytic digestion and gravimetric determination as $BaSO_4$.[81] A method based on chromatography with conductimetric detection has been reported very recently.[68] This method has the advantage that it allows for the simultaneous analysis of cyclamate, aspartame, acesulfame K and saccharin.

7.4.7 Regulatory status

As is the case for saccharin, the Calorie Control Council maintains a database of worldwide regulatory approval information on cyclamate and this information on cyclamate regulatory approvals is provided in Table 7.9.[82]

Table 7.9 Worldwide regulatory status of cyclamate.

Country	Food[a,b]	Beverage[b]	Table-top[b]	Pharmaceutical[b]
Albania		+		
Angola		+	+	
Antigua	+	+		
Argentina	+	+	+	+
Armenia		+		
Australia	+	+	+	(Assessed on individual basis)
Austria[c]	+	+	+	
Azerbaijan		+		
Bahamas	+	+		
Barbados		+		
Belarus		+		
Belgium[c]	+	+	+	+
Bolivia		+		
Bosnia and Herzogovina		+		
Botswana		+		
Brazil	+	+	+	
Bulgaria	+	+	+	+
Canada			+	+
Caribbean (Ind.)	+	+	+	
Chile	+	+	+	
China	+	+		
Comoros		+		
Costa Rica		+		
Croatia		+		
Cyprus	+	+	+	+
Czech Republic	+	+	+	
Denmark[c]	+	+	+	+
Djibouti		+		
Dominica	+	+	+	+
Dominican Republic		+		
Ecuador			+	+
El Salvador		+		
Estonia	+	+	+	
Finland[c]	+	+	+	+
France[c]	+	+	+	+
French Guiana		+		
Georgia		+		
Germany[c]	+	+	+	+
Great Britain		+		
Greece[c]	+	+	+	+
Grenada		+		
Guadeloupe		+		
Guatemala		+	+	
Guyana		+		
Haiti	+	+		
Honduras		+		
Hong Kong	+	+	+	+
Hungary	+	+	+	
Iceland	+	+	+	
India	−	−	−	
Indonesia	+	+	+	
Iran		−		

(Continued)

Table 7.9 Worldwide regulatory status of cyclamate.

Country	Food[a,b]	Beverage[b]	Table-top[b]	Pharmaceutical[b]
Iraq		+		
Ireland[c]	+	+	+	+
Israel	+	+	+	+
Italy[c]	+	+	+	+
Jamaica		+		
Japan	−	−	−	−
Jordan				+
Kazakhstan		+		
Kenya		+		
Korea	−	−	−	
Kuwait			+	
Kyrgyzstan		+		
Latvia	+	+	+	
Lesotho		+		
Lithuania	+	+	+	
Luxembourg[c]	+	+	+	
Macau (Macao)		+		
Macedonia		+		
Madagascar		+		
Malawi		+		
Malaysia		−		
Malta[c]	+	+	+	
Martinique	+	+	+	
Mauritius		+		
Mayotte		+		
Mexico	+	+	+	
Moldova		+		
Montserrat		+		
Montenegro		+		
Morocco		+		
Mozambique		+		
Namibia		+		
Netherlands[c]	+	+	+	
New Zealand	+	+	+	+
Nicaragua	+	+	+	+
Norway	+	+	+	+
Oman			+	+
Pakistan	+		+	
Panama		+		
Papua New Guinea	+	+	+	
Paraguay	+	+	+	
Peru	+	+		
Poland	+	+	+	
Portugal[c]	+	+	+	
Romania	+	+	+	
Russian Federation	+	+	+	
Rwanda		+		
Saint Helena		+		
Saint Kitts and Nevis		+		
Saint Lucia		+		
Saint Vincent		+		
Saudi Arabia	+		+	

(*Continued*)

Table 7.9 Worldwide regulatory status of cyclamate.

Country	Food[a,b]	Beverage[b]	Table-top[b]	Pharmaceutical[b]
Serbia		+		
Seychelles		+		
Sierra Leone	+		+	
Slovakia	+	+	+	
Slovenia	+	+	+	
South Africa	+	+	+	+
Spain[c]	+	+	+	
Sri Lanka			+	
Suriname		+		
Swaziland		+		
Sweden[c]	+	+	+	
Switzerland	+	+	+	
Taiwan	+	+	+	+
Tajikistan		+		
Tanzania		+		
Thailand			+	
Trinidad and Tobago	+	+		
Tunisia		+		
Turkey	+	+	+	+
Turkmenistan		+		
Turks and Caicos Islands		+		
Ukraine		+		
United Arab Emirates			+	
United Kingdom[c]	+	+	+	
United States	−	−	−	−
Uruguay	+	+	+	+
Uzbekistan		+		
Venezuela		+	+	+
Vietnam[d]	+			
West Bank – Gaza		+		
Yugoslavia		+	+	
Zambia		+		
Zimbabwe	+		+	

[a]May not apply to all food categories.
[b]+ = permitted; − = prohibited.
[c]EU regulations apply.
[d]Products must comply with regulations of the exporting country.

REFERENCES

1. F.J. Helgren. Sweetening composition and method of producing the same. US Patent, 2,803,551; 1957.
2. G. Nelson, M.A. Hoon, J. Chandrashekar, Y. Zhang, N.J.P. Ryba and C.S. Zuker. Mammalian sweet taste receptors. *Cell* 2001, 106, 381–390.
3. X. Li, L. Staszewski, H. Xu, K. Durick, M. Zoller and E. Adler. Human receptors for sweet and umami taste. *Proceedings of the National Academy of Sciences of the United States of America* 2002, 99, 4692–4696.
4. N. Kunishima, Y. Shimada, Y. Tsuji, T. Sato, M. Yamamoto, T. Kumasaka, S. Nakanishi, H. Jingami and K. Morikawa, Structural basis of glutamate recognition by a dimeric metabotrophic glutamate receptor. *Nature* 2000, 407, 971–977.
5. H. Xu, L. Staszewski, H. Tang, E. Adler, M. Zoller and X. Li. Different functional roles of T1R subunits in the heteromeric taste receptors. *Proceedings of the National Academy of Sciences of the United States of America* 2004, 101, 14258–14263.

6. P. Jiang, M. Cui, B. Zhao, L.A. Snyder, L.M.J. Bernard, R. Osman, M. Max and R.F. Margolskee. Identification of the cyclamate interaction site within the transmembrane domain of the human sweet taste receptor subunit T1R3. *Journal of Biological Chemistry* 2005, 280(40), 34296–34305.
7. M. Cui, P. Jiang, E. Maillet, M. Max, R.F. Margolskee and R. Osman. The heterodimeric sweet taste receptor has multiple potential ligand binding sites. *Current Pharmaceutical Design* 2006, 12, 4591–4600.
8. R.S.J. Keast, T. Canty and P.A.S. Breslin. Oral zinc sulfate solutions inhibit sweet taste perception. *Chemical Senses* 2004, 29, 513–521.
9. Y. Nie, S. Vigues, J.R Hobbs, G.L Conn and S.D. Munger. Distinct contributions of T1R2 and T1R3 taste receptor subunits to the detection of sweet stimuli. *Current Biology* 2005, 15, 1948–1952.
10. Y. Nie, J.R. Hobbs, S. Vigues, W.J. Olson, G.L. Conn and S.D. Munger. Expression and purification of functional ligand–binding domains of T1R3 taste receptors. *Chemical Senses* 2006, 31, 505–513.
11. G.E. DuBois. Unraveling the biochemistry of sweet and umami tastes. *Proceedings of the National Academy of Sciences of the United States of America* 2004, 101, 13972–13973.
12. G. Morini, A. Bassoli and P.A. Temussi. Multiple receptors or multiple binding sites? Modeling the Human T1R2-T1R3 sweet taste receptor. In: D.K. Weerasinghe, G.E. DuBois (eds). *Sweetness and Sweeteners: Biology, Chemistry and Psychophysics, ACS Symposium Series 979*, American Chemical Society, Washington, DC; 2008, pp. 147–161.
13. R.L. Pearson. Saccharin. In: L. O'Brien Nabors (ed). *Alternative Sweeteners*, 3rd edn, Marcel Dekker, New York; 2001, pp. 147–165.
14. J. Hicks. The pursuit of sweet: A History of Saccharin. *Chemical Heritage* 2010, 28, 26–31.
15. C. Fahlberg and I. Remsen. Uber die oxydation des orthotoluolsulfamids. *Chemische Berichte* 1879, 12, 469–473.
16. O.F. Senn. Method of preparing ortho sulfonyl chloride benzoic acid esters. US Patent, 2,667,503; 1954.
17. G.F. Schlaudecker. Esterification of dithiosalicylic acid. US Patent, 2,705,242; 1955.
18. H.R. Moskowitz. *Product Testing and Sensory Evaluation of Foods*, Food and Nutrition Press, Westport, CT; 1983, pp. 110–120.
19. G.E. DuBois, D.E. Walters, S.S. Schiffman, Z.S. Warwick, B.J. Booth, S.D. Pecore, K.M. Gibes, B.T. Carr and L.M. Brands. A systematic study of concentration-response relationships of sweeteners. In: D.E. Walters, F.T. Orthoefer and G.E. DuBois (eds). *Sweeteners: Discovery, Molecular Design, and Chemoreception*, American Chemical Society, Washington, DC; 1991, pp. 261–276.
20. M. Meilgaard, G. Vance Civille, and B.T. Carr. *Sensory Evaluation Techniques*, Vol. II, CRC Press, Boca Raton, FL; 1987, pp. 5–6.
21. B.T. Carr, S.D. Pecore, K.M. Gibes and G.E. DuBois. Sensory methods for sweetener evaluation. In: C.-T. Ho and C.H. Manley (eds) *Flavor Measurement*, Marcel Dekker, New York; 1993, pp. 219–237.
22. F.J. Helgren, M.F. Lynch, and F.J. Kirchmeyer. A taste panel study of the saccharin "off- taste". *Journal of the American Pharmaceutical Association (Science Edition)* 1955, 44, 353–355.
23. L.M. Bartoshuk. Bitter taste of saccharin related to the genetic ability to taste the bitter substance 6-*n*-propylthiouracil. *Science*, 1979, 205, 934–935.
24. J.F. Gent and L.M. Bartoshuk. Sweetness of sucrose, neohesperidin dihydrochalcone, and saccharin is related to genetic ability to taste the bitter substance 6-*n*-propylthiouracil. *Chemical Senses* 1983, 7, 265–272.
25. H.R. Moskowitz. *Product Testing and Sensory Evaluation of Foods*, Food and Nutrition Press, Westport, CT; 1983, p. 133–134.
26. G.E. DuBois and J.F. Lee. A simple technique for the evaluation of temporal taste properties. *Chemical Senses* 1983, 7, 237–246.
27. S.S. Schiffman, B.J. Booth, B.T. Carr, M.L. Losee, E.A. Sattely-Miller and B.G. Graham. Investigation of synergism in binary blends of sweeteners. *Brain Research Bulletin* 1995, 38(2), 105–120.
28. J.A. Dean (ed.). *Lange's Handbook of Chemistry*, 13th edn, McGraw-Hill Book Company, New York; 1985, Section 5, p. 56.
29. M. Windholz (ed.). The Merck Index, 10th edn, Merck & Co., Rahway, NJ; 1983 p. 1197.
30. A. Salant (ed). *Handbook of Food Additives*, The Chemical Rubber Co., Cleveland, OH; 1968, pp. 503–512.
31. *Toxicological Principles for the Safety Assessment of Direct Food Additives and Color Additives Used in Food*, US Food and Drug Administration, Bureau of Foods; 1982, pp. 1–19.
32. O. DeGarmo, G.W. Ashworth, C.M. Eaker and R.H. Munch. Hydrolytic stability of saccharin. *Journal of American Pharmacists Association* 1952, 41, 17–18.

33. M.L. Mitchell. Unpublished Results. PMC Specialties Group, Cincinnati, OH.
34. A.G. Renwick. The metabolism of intense sweeteners. *Xenobiotica* 1986, 16(10/11), 1057–1071.
35. A.G. Renwick. The disposition of saccharin in animals and man – A review. *Food and Chemical Toxicology* 1985, 23(4/5), 429–435.
36. A.G. Renwick. The fate of intense sweeteners in the body. *Food Chemistry* 1985, 16, 281–301.
37. A.G. Renwick. The intake of intense sweeteners – an update review. *Food Additives and Contaminants* 2006, 23(4), 327–338.
38. M. Zubare-Samuelov, M.E. Shaul, I. Peri, A. Aliluiko, O. Tirosh and M. Naim. Inhibition of signal termination-related kinases by membrane-permeant bitter and sweet tastants: potential role in taste signal termination. *American Journal of Physiology, Cell Physiology Section* 2005, 289, C483–C492.
39. K. Dib, F. Wrisez, A. El Jamali, B. Lambert and C. Correze. Sodium saccharin inhibits adenylyl cyclase in non-taste cells. *Cellular Signaling* 1997, 9(6), 431–438.
40. G.E. DuBois. Non-nutritive sweeteners. In: *Encyclopedia of Food Science & Technology*, John Wiley & Sons, New York; 2000, 2245–2265.
41. M.E. Embuscado and S.K. Patil. Erythritol. In: L. O'Brien Nabors (ed.). *Alternative Sweeteners*, 3rd Edition, Marcel Dekker, New York; 2001, 235–254.
42. P. de Cock and C.-L. Bechert. Erythritol. Functionality in noncaloric functional beverages. *Pure and Applied Chemistry* 2002, 74(7), 1281–1289.
43. G.J.M. den Hartog, A.W. Boots, A. Adam-Perrot, F. Brouns, I.W.C.M. Verkooijen, A.R. Weseler, G.R.M. M. Haenen, and A. Bast. Erythritol is a sweet antioxidant. *Nutrition* 2010, 26(4), 449–458.
44. E. Adler, M.A. Hoon, K.L. Mueller, J. Chandrashekar, N.J.P. Ryba and C.S. Zuker. A novel family of mammalian taste receptors. *Cell* 2000, 100, 693–702.
45. H. Matsunami, J.P. Montmayeur and L.B. Buck. A family of candidate taste receptors in human and mouse. *Nature* 2000, 404, 601–604.
46. C. Kuhn, B. Bufe, M. Winnig, T. Hofmann, O. Frank, M. Behrens, T. Lewtschenko, J.P. Slack, C.D. Ward and W. Meyerhof. Bitter taste receptors for saccharin and acesulfame K. *Journal of Neuroscience* 2004, 24, 10260–10265.
47. J.P. Slack, A. Brockhoff, C. Batram, S. Menzel, C. Sonnabend, S. Born, M. Mercedes Galindo, S. Kohl, S. Thalmann, L. Ostopovici-Halip, C.T. Simons, I. Ungureanu, K. Duineveld, C.G. Bologna, M. Behrens, S. Furrer, T.I. Oprea and W. Meyerhof. Modulation of bitter taste perception by a small molecule hTAS2R antagonist. *Current Biology* 2010, 7, 1104–1109.
48. B. Bufe, P.A.S. Breslin, C. Kuhn, D.R. Reed, C.D. Tharp, J.P. Slack, U.K. Kim, D. Drayna and W. Meyerhof. The molecular basis of individual differences in phenylthiocarbamide and propylthiouracil bitterness perception. *Current Biology* 2005, 15, 322–327.
49. U. Kim, S. Wooding, D. Ricci, L.B Jorde and D. Drayna. Worldwide haplotype diversity and coding sequence variation at human bitter taste receptor loci. *Hum. Mutat.* 2005, 26, 199–204.
50. A.N. Pronin, H. Xu, H. Tang, L. Zhang, Q. Li and X. Li. Specific alleles of bitter receptor genes influence human sensitivity to the bitterness of aloin and saccharin. *Current Biology* 2007, 17, 1403–1408.
51. C. Kuhn, B. Bufe, C. Batram and W. Meyerhof. Oligomerization of TAS2R bitter taste receptors. *Chemical Senses* 2010, 35, 395–406.
52. C.E. Riera, H. Vogel, S.A. Simon and J. le Coutre. Artificial sweeteners and salts producing a metallic taste sensation activate TRPV1 receptors. *American Journal of Physiology. Regulatory, Integrative and Comparative Physiology* 2007, 293, R626–R634.
53. C.E. Riera, H. Vogel, S.A. Simon, S. Damak and J. le Coutre. The capsaicin receptor participates in artificial sweetener aversion. *Biochemical and Biophysical Research Communications* 2008, 376, 653–657.
54. K. Kőhler, A. Hillebrecht, J. Schulze Wischeler, A. Innocenti, A. Heine, C.T. Supuran and G. Klebe. Saccharin inhibits carbonic anhydrases: possible explanation for its unpleasant metallic aftertaste. *Angewandte Chemie International Edition* 2007, 46(40), 7697–7699.
55. R.J. Ronk. Regulatory constraints on sweetener use. In: J.H. Shaw and G.G. Roussos (eds). *Sweeteners and Dental Caries*. Information Retrieval, Washington, DC.; 1978, pp. 131–144.
56. H.W. Schultz, *Food Law Handbook*, Avi Publishing Company, Westport, CT.; 1981.
57. J.P. Marshall and J.A. Pollard. *Sweeteners: International Legislation, Issue 1: Food Legislation Surveys*, 4th ed., British Food Manufacturing Industries Research Association, Leatherhead Food RA, UK; 1987.
58. F.J. Broulik. Regulatory processes for new sweeteners in the United States of America. In: T.H. Grenby (ed.) *Advances in Sweeteners*, Blackie Academic & Professional, London; 1996, pp. 35–55.

59. G.E. DuBois. Sweeteners and sweetness modulators: requirements for commercial viability. In: G.E. DuBois and D. Weerasinghe (eds). *Sweetness and Sweeteners, ACS Symposium Series 979*, Chapter 29, ACS Books, Washington, DC; 2008, pp. 444–462.
60. G. Vettorazzi. Role in international scientific bodies. In: R.D. Middlekauf and P. Shubik (eds). *International Food Regulation Handbook*, Marcel Dekker, New York; 1989, pp. 481–505.
61. *World Health Organization Technical Report Series*, No. 539, Geneva, Switzerland; 1974, p. 11.
62. E.M. Whelan. *The Conference Board Magazine* 1977, 16, 54.
63. G.J. Walter and M.L. Mitchell. Saccharin. In: L. O'Brien Nabors and R.C. Gelardi (eds). *Alternative Sweeteners*, Marcel Dekker, New York; 1986, pp. 15–41.
64. A.G. Renwick. Saccharin: a toxicological evaluation. *Comments on Toxicology* 1989, 3(4), 289–305.
65. S.M. Cohen, T.A. Anderson, L.M. de Oliveira and L.L. Arnold. Tumorigenicity of sodium ascorbate in male rats. *Cancer Research* 1998, 58, 2557–2561.
66. *Official Methods of Analysis of AOAC International*, 17th edn, AOAC International, Gaithersburg, MD; 2000; AOAC Official Method 979.08.
67. M. Lourdes Morales, R. Ferreira, A.G. Gonzalez and A.M. Troncoso. Simultaneous determination of organic acids and sweeteners in soft drinks by ion-exclusion HPLC. *Journal of Separation Science* 2001, 24, 879–884.
68. Y. Zhu, Y. Guo, M. Ye and F.S. James. Separation and simultaneous determination of four artificial sweeteners in food and beverages by ion chromatography. *Journal of Chromatography A* 2005, 1085, 143–146.
69. Lyn O'Brien Nabors. Worldwide status of saccharin. Personal Communication, Calorie Control Council, Suite 500, Building G, 5775 Peachtree-Dunwoody Road, Atlanta, GA 30342; www.caloriecontrol.org; Update April 2009.
70. L.F. Audrieth and M. Sveda. Preparation and properties of some N-substituted sulfamic acids. *Journal of Organic Chemistry* 1944, 9, 89–101.
71. M. Cavicchioli, C.Q.F. Leite, D.N. Sato and A. Massabni. Synthesis, characterization and antimycobacterial activity of Ag(I)-aspartame, Ag(I)-saccharin and Ag(I)-cyclamate complexes. *Archiv Der Pharmazie – Chemistry in Life Science* 2007, 340, 538–542.
72. W.J. Spillane and J.B. Thomson. Studies of the protonation equilibria of sulphamates using ^{13}C and ^{1}H nuclear magnetic resonance spectroscopic, potentiometric, and conductimetric methods. *Journal of Chemical Society, Perkin Transactions 2* 1977, 580–584.
73. G.A. Benson and W.J. Spillane. Sulfamic acid and its N-substituted derivatives. *Chemical Reviews* 1980, 80, 151–186.
74. I. Leban, D. Rudan-Tasič, N. Lah and C. Klofutar. Structures of artificial sweeteners – cyclamic acid and sodium cyclamate with other cyclamates. *Acta Crystallographica* 2007, B63, 418–425.
75. K.M. Beck. Nonnutritive sweeteners: saccharin and cyclamate. In: T.E. Furia (ed). *CRC Handbook of Food Additives*, Vol. 11, 2nd edn, CRC Press, Boca Raton, FL; 1980, p. 125.
76. T. Fazio, J.W. Howard and E.O. Haenni. Survey of cyclohexylamine content of food products containing cyclamates. *Journal of the Association of Official Analytical Chemists* 1970, 3, 1120–1128.
77. United States Congress, House of Representatives, Committee on the Judiciary Subcommittee No. 2. Cyclamates. Hearings before subcommittee no. 2 of the committee on the Judiciary House of Representatives, Ninety-Second Congress, 29, 30 September and 6 October 1971, Serial No. 22. US Government Printing Office, Washington, DC; 1972.
78. B.A. Bopp and P. Price. Cyclamate. In: L. O'Brien Nabors (ed). *Alternative Sweeteners*, 3rd edn, Marcel Dekker, New York; 2001, pp. 63–85.
79. B.A. Bopp, R.C. Sonders and J.W. Kesterson. Toxicological aspects of cyclamate and cyclohexylamine. *CRC Critical Reviews in Toxicology* 1986, 16(3), 213–306.
80. *Food Chemical News*, 13 November; 1989, p. 19.
81. *Official Methods of Analysis of AOAC International*, 17th edn, AOAC International, Gaithersburg, MD; 2000; AOAC Official Method 957.10.
82. Lyn O'Brien Nabors. Worldwide status of cyclamate. Personal Communication, Calorie Control Council, Suite 500, Building G, 5775 Peachtree-Dunwoody Road, Atlanta, GA 30342, www.caloriecontrol.org; Update 29 September 2009.

8 Sucralose

Samuel V. Molinary[1] and Mary E. Quinlan[2]

[1] Scientific & Regulatory Affairs, Beaufort, SC, USA
[2] Tate & Lyle, London, UK

8.1 INTRODUCTION

Sucralose is one of the more recent high-potency sweeteners made available to the food industry. Sucralose is uniquely made from sucrose by a process of chemical modification that results in the enhancement of the sweetness intensity, retention of a pleasant sugar-like taste and creation of a very stable molecule. This latter property makes sucralose suitable for use in low pH and neutral products as well as heat-processed foods. It is a very versatile sweetener that can be used in a wide variety of foods and beverages, and it allows the development of an increasing range of good-tasting, low-calorie foods, including baked goods. Sucralose is approved for use in food products in most countries around the world and has become very popular with both consumers and the food and beverage industry.

8.2 HISTORY OF DEVELOPMENT

Sucralose is the direct result of an intensive research programme carried out in the 1970s by Tate & Lyle, PLC. The primary objective of this research effort was to discover new non-food uses for sugar, a relatively inexpensive and widely available agricultural commodity. Many potential product areas were explored including detergents, plastics and fine chemicals. These areas met with only modest success.

However, there was a separate, small, basic research programme that sought to examine the nature of sucrose's sweet taste and the factors responsible for its pleasant taste sensation. This research was carried out by Tate & Lyle scientists in Reading, England, in association with Prof. Les Hough at the University of London's Queen Elizabeth College. The key observation from this research, after making and evaluating many hundreds of derivatives of sucrose, was that halogenating sucrose at selected sites dramatically increased the perceived sweetness of the resulting molecule.[1,2] Prior to this, chemical modification of sucrose led to compounds that were either tasteless or bitter.[3] This unexpected sweetness enhancement led to the laboratory synthesis of many halogenated derivatives of sucrose and an intense programme of structure/activity studies of the resulting compounds. The structure of each derivative was verified by nuclear magnetic resonance spectroscopy and mass spectroscopy

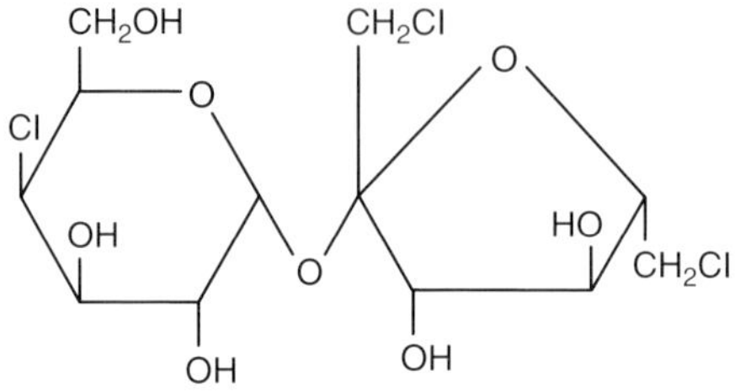

Fig. 8.1 Sucralose.

before being tasted by volunteers – the only certain method to determine if a molecule is sweet or not. It was found that sweetness enhancement was limited to a relatively small class of halogenated compounds.

The key factors found to be required for sweetness enhancement of sucrose were the retention of the hydroxyl groups at the 2, 3, 6 and 3′ positions with large electronegative (halogen) groups at the remaining positions of the molecule. Tetrachloro- and tetrabromo-derivatives produced the greatest increase in sweetness intensity, 2200× and 7500× sucrose, respectively, but the quality of the sweet taste of these compounds did not meet that required for a commercial sweetener.[4] The enhancement of sweetness coupled with a loss of sweet taste quality in these highly substituted compounds is thought to be due to the increased lipophilicity, which causes a lingering taste. The increased lipophilicity presumably is due to increased binding of these highly halogenated molecules to the taste bud receptor.[5] After evaluating many of these halogenated derivatives of sucrose, Tate & Lyle selected sucralose for further development as a commercial product (Figure 8.1). The selection process had identified sucralose as retaining the high sugar-like sweetness quality, in addition to its having a high heat and chemical stability, high water solubility and low toxicity.[6]

8.3 PRODUCTION

Sucralose is manufactured by the selective replacement of three hydroxyl groups on the sucrose molecule by three chlorine atoms to produce 1,6-dichloro-1,6-dideoxy-beta-D-fructofuranosyl-4-chloro-4-deoxy-alpha-D-galactopyranoside. The basic process is the selective protection of the essential hydroxyl groups, followed by chlorination, deblocking and purification (Figure 8.2). Sucralose can be crystallised from an aqueous solution, and can be produced to a high level of purity and consistency.

8.4 ORGANOLEPTIC PROPERTIES

The sweetness intensity of sucralose was measured against a range of sucrose reference solutions with the result that the potency of sucralose was shown to decrease slightly from the lowest to the highest sucrose concentration. Sucralose is approximately 750× sweeter than sucrose at a concentration equi-sweet to a 2% sucrose solution, while at 9% sucrose, sucralose is about 500× sweeter than sucrose. The general rule is that on average, sucralose is around 600× sweeter than sucrose.[7]

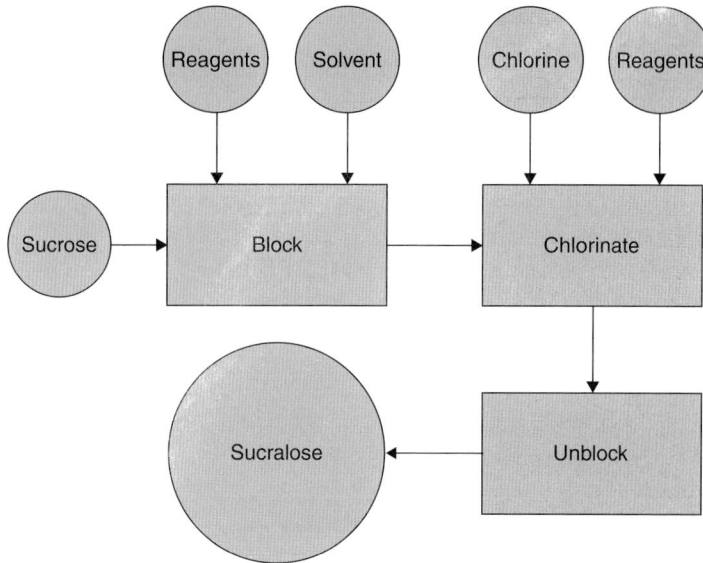

Fig. 8.2 Sucralose manufacturing process overview.

The potency and consequently the optimum use level of sucralose will vary with the food matrix in which it is employed, as, along with all intense sweeteners, potency depends on factors such as the sweetness level, pH, temperature and the presence of other ingredients. However, as sucralose is intensely sweet, only a very small quantity is needed to reach an isosweet level with sucrose. For example, approximately 200 mg/L sucralose is needed to sweeten a beverage to the same sweetness as 9–10% sugar. Table 8.1 shows examples of sucralose sweetness factors in some typical food products.

An extensive sensory research programme was conducted to determine the taste characteristics and quality of sucralose. Sweeteners have flavour attributes other than sweetness, and the profile for sucralose was studied by having panellists evaluate isosweet aqueous solutions of sucrose and sucralose, and to assign scores for the key taste attributes. Figure 8.3 compares the flavour profiles of sucrose and sucralose solutions equi-sweet to 9% sucrose. It can be seen that sucralose compares favourably with 9% sucrose in having high-quality sweetness with no bitter aftertaste or metallic notes. There is a minimal difference in that sucralose sweetness has a slightly longer duration than sucrose,[8] but this can be reduced when correctly formulated in food and beverage systems.

Table 8.1 Relative sweetness factors versus sucrose (=1) for sucralose in food.

Cola	475
Jam	540
Strawberry milk	680
Yoghurt	450
Canned peaches	530
Beans in tomato sauce	680

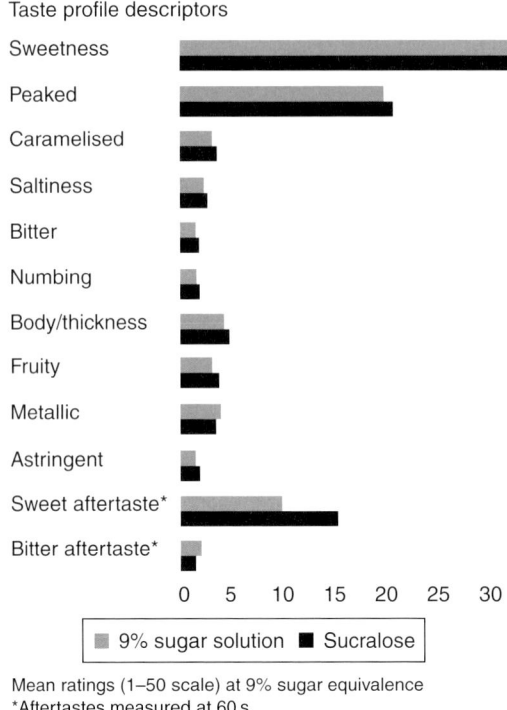

Fig. 8.3 Flavour profiles of sucrose and sucralose (9% sucrose equivalence).

While sucralose functions well as a sole sweetener in food systems, it can be blended with other non-nutritive and nutritive sweeteners. When it is part of a non-nutritive sweetener blend, there is synergy with acesulfame K, cyclamate and saccharin, but little or no synergy with aspartame. In nutritive sweetener blends, sucralose is synergistic with fructose and other carbohydrate sweeteners, although not with sucrose. Despite this, sucrose and sucralose blends perform well and there are no incompatibilities between sucralose and any sweetener, nutritive or non-nutritive.

8.5 PHYSICO-CHEMICAL PROPERTIES

The usefulness of an ingredient, such as sucralose, to the food-manufacturing industry depends not only on its sensory qualities but also on its physico-chemical properties. These properties will determine the types of food-manufacturing processes in which the ingredient can be used as well as the types of products in which it may function. These properties will also be important to the shelf life of the products in which the ingredient is used. Before its regulatory approvals were gained, the physico-chemical properties of sucralose were predicted to make it a very useful sweetener for the food industry,[9] and the intervening years since its approval have fully borne out this prediction as sucralose is an ingredient in well over 4000 products worldwide.

Pure sucralose is a white, free-flowing powder. It is intensely sweet, practically odourless and freely soluble in water. Sucralose powder is non-hygroscopic at humidities below 80%

Table 8.2 Physico-chemical properties of sucralose.

Physical form	White crystalline powder
Odour	Practically odourless
Taste	Intense sugar-like sweetness
Taste intensity	400–800 times sugar
Caloric content	Zero
Solubility	Freely soluble in water (28.2 g/100 mL at 20°C)
	Insoluble in corn oil (0.1 g/100 g at 20°C)
Specific optical rotation	$[\alpha]^D_{20}$ +85.8° (C_{10}, H_{20})
Octanol/water partition coefficient	0.32 (20°C)
Surface tension of aqueous solutions	71.8 mN/m (20°C, 0.1 g/100 mL)
Melting (decomposition) point	125°C (when heated from 115°C at 5°C/min)
Specific gravity (10% aqueous solution)	1.04 (20°C)
Molecular weight	397.64

and does not require special storage conditions to prevent moisture pick up. The principal properties of sucralose are shown in Table 8.2.

Being a modified carbohydrate, sucralose retains the high aqueous solubility of this class of compounds, which, along with its stability, make it highly functional in food systems. This was shown experimentally by measuring its solubility over a temperature range from 20°C to 60°C with the result that its solubility increased from 28.2 g/100 mL at 20°C to 66 g/100 mL at 60°C. Such a high degree of solubility means that sucralose can be easily incorporated into any aqueous food system. It was also shown to be highly soluble in methanol, ethanol and propylene glycol, other major solvents used in food manufacturing. However, sucralose is insoluble in corn oil; a result consistent with its hydrophilic nature. Therefore, in any food system with aqueous and lipid phases, sucralose will always partition with the aqueous phase.[9]

Viscosity is another important property affecting the usefulness of a food ingredient. Experimentally, sucralose solutions were compared with distilled water at different shear rates in a Rheomat 30 viscometer. The dynamic viscosity of these aqueous solutions was measured over a temperature range from 20°C to 60°C, with concentrations of sucralose ranging from 10% to 50%. The results indicated that the viscosity remained constant at all shear rates and depends only on the temperature and pressure of the test system. This is described as Newtonian behaviour. Sucrose studies yield very similar results, again, in agreement with the chemical nature of both sucrose and sucralose. The low viscosity of aqueous sucralose indicates that it will not negatively affect food processes that involve mixing and dispersion of food components, and this has proved to be the case in practice.[7]

The effect of aqueous sucralose solutions on surface tension was compared to water at 20°C. These dilute solutions (0.1 g and 1 g/100 mL) caused a negligible lowering of the surface tension. The conclusion is that sucralose has no surfactant activity and will not cause excessive foaming in products, such as soft drinks, that require high-speed filling.

Sucralose has very little effect on the pH of aqueous solutions. This is consistent with the chemical nature of sucralose, which has no reactive functional groups.

The refractive index (RI) of aqueous sucralose solutions is linear with respect to concentration, as shown in Figure 8.4. This finding is particularly important in that it provides a simple, rapid means for determining sucralose concentrations in solution. The presence of other solids will obviously affect the reading, and corrections have to be made. However, when an RI detector is coupled to a high-performance liquid chromatography (HPLC) system, the concentration of sucralose can be very accurately measured.[10]

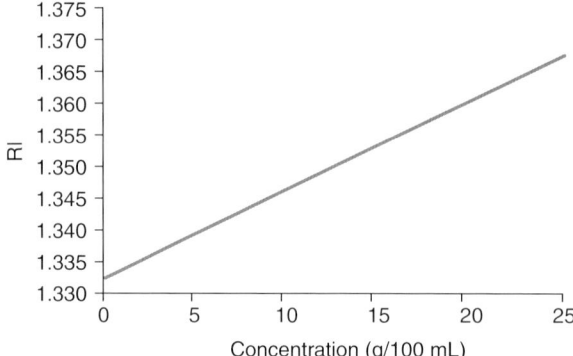

Fig. 8.4 Refractive Index of aqueous sucralose solutions.

The stability of sucralose is of particular interest to the food scientist. Principally, whether there is any potential for sucralose to undergo chemical interaction with other components in the food matrix in which it is used, and whether it has any potential to decompose or hydrolyse in dilute aqueous systems. As noted previously, the selective chlorination of sucrose to synthesise sucralose alters the resultant molecule by both intensifying the sweetness and making it more chemically stable. The added stability of sucralose manifests itself both in its non-reactivity with other food components and in its resistance to hydrolysis under extremes of acid and heat.

Sucralose was evaluated in food model systems to determine its potential to react with other classes of food chemicals. Four categories of food chemicals were tested: (1) bases, (2) oxidising and reducing agents, (3) aldehydes and ketones and (4) metal salts. The specific compounds evaluated were: niacinamide and monosodium glutamate (bases), hydrogen peroxide and sodium bisulfite (redox agents), acetaldehyde and ethyl acetoacetate and ferric chloride. These compounds were prepared as 0.1% each in a 10% sucralose solution. The solutions were stored for 7 days at 40°C, at pH 3, 4, 5 and 7. At the end of 7 days, the sucralose concentration was determined by HPLC. Sucralose was recovered at 98% from all solutions, except the ferric chloride solution at pH 3 (95.9%). However, in the ferric chloride/sucralose solution at pH 5, sucralose was recovered at 98%. In re-examining these ferric chloride results, it was seen that the concentration of ferric chloride was significantly higher than that normally present in food products. Consequently, an additional study was undertaken to assess the effect of lower levels of ferric ions (5 ppm). In this study, there was no noticeable effect on the stability of sucralose. The data strongly suggest that sucralose would not interact with other compounds typically used in food systems.[7] Subsequent experience with actual products has amply borne this out.

Another attractive feature of sucralose for the food manufacturer is its stability in aqueous solutions. Under extreme conditions of acid and heat, sucralose in a dilute aqueous medium will slowly hydrolyse to its component monosaccharides, 1,6-dichlorofructose (1,6-DCF) and 4-chlorogalactose (4-CG) (Figure 8.5). The acid hydrolysis mechanism for sucralose is the same as that for sucrose when it hydrolyses to its monosaccharides, glucose and fructose. However, the chlorine atoms of sucralose create a major difference in that they stabilise the glycosidic linkage towards protonation and thereby reduce the reaction rate by approximately two orders of magnitude.[5] For all practical purposes, the loss of sucralose in foods will be

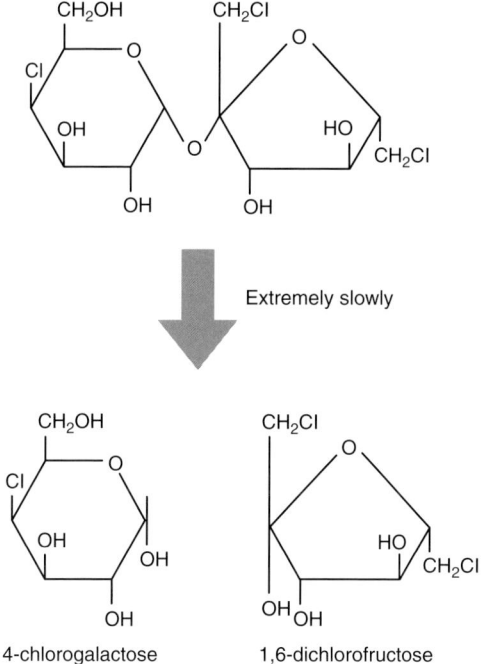

Fig. 8.5 Sucralose hydrolysis in acidic conditions.

negligible, which will assure that the original sweetness imparted by the sucralose will not be lost over time (Figure 8.6).

To confirm that the only breakdown products of sucralose were 1,6-DCF and 4-CG, a series of studies were carried out using sucralose labelled with radioactive ^{36}Cl. The resulting radio-labelled sucralose was then dissolved in glycine/HCl buffered solutions at pH 2.5, 3.0 and 3.5. To simulate a true food system, two formulations of a cola carbonated beverage were prepared at pH 3.0 and 3.5. The total sucralose concentration in all solutions and beverages

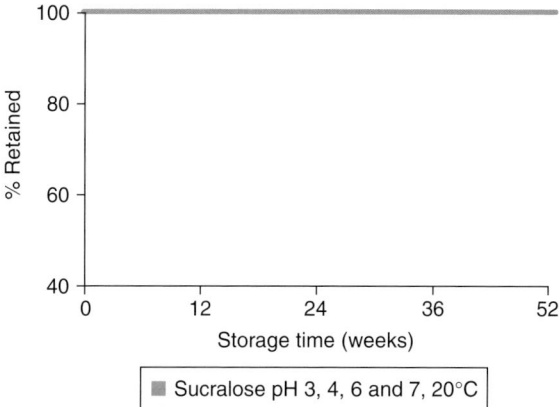

Fig. 8.6 Stability of sucralose.

was 200 ppm. All of these preparations were stored at 30°C and 40°C for 1 year, and samples were taken and analysed at 0, 8, 16, 26 and 52 weeks. The analysis of these samples consisted of thin layer chromatography, followed by scintillation counting. After correcting for any 1,6-DCF and 4-CG present in the day 0 samples, the amounts of sucralose, 1,6-DCF and 4-CG were determined.[7]

From the results of these analyses, it was clear that no new compounds formed during the 1-year storage in any sample nor was any inorganic chloride present. These data were fully consistent with the concept that sucralose yields only two compounds on acid hydrolysis, 1,6-DCF and 4-CG. Furthermore, the results from the two cola beverages verified that sucralose does not interact with other food components. Additionally, mass balance calculations based on these data confirmed that sucralose and its two hydrolysis products, 1,6-DCF and 4-CG, accounted for all the sucralose present in the original day 0 samples. These observations also substantiate that 1,6-DCF and 4-CG do not further break down.[2,7]

8.6 PHYSIOLOGICAL PROPERTIES

As part of the very large and comprehensive database compiled on sucralose during the food additive petition preparation period, extensive biological data on sucralose were generated. While most of these studies were concerned with the safety in use of sucralose, there were other questions to be answered, such as the potential for sucralose to be metabolised as a carbohydrate, whether it had cariogenic potential and if its presence would have an impact on the environment. The most obvious physiological property of sucralose has already been discussed in Section 8.2, that is, its sweet taste when ingested. Given the basic carbohydrate structure of sucralose, it was essential to establish whether or not the body metabolism treated ingested sucralose as it would other disaccharides and break it down to derive energy and carbon dioxide from it.

A series of metabolism-related studies were undertaken using the common test species used in toxicology studies, rats, mice, dogs and rabbits.[11–14] These studies explored the absorption of ingested sucralose from the gastrointestinal tract, the distribution of the fraction that was absorbed within the body organs and tissues, any metabolic transformations of this circulating fraction and, finally, its elimination from the body. Eventually, similar studies were carried out in human volunteers to assure that all these species metabolised sucralose in a comparable manner. The results of these studies were unequivocal; sucralose passes through the body without being broken down or losing any of the attached chlorine atoms. Data from the human volunteer study showed that ∼85% of the ingested sucralose dose was eliminated in the faeces largely unchanged, while ∼15% of the dose is absorbed from the gastrointestinal tract and enters into the general circulation and is eliminated intact in the urine.[15] These studies all indicate that a small fraction (∼1–5%) of the ingested dose appears in the urine as highly polar glucuronide conjugates. These compounds are chemically more polar than sucralose and represent biotransformation to glucuronide adducts at different available hydroxyl sites on the sucralose molecule. While these are referred to as metabolites, the sucralose itself remains intact. All the test species showed comparable results.[11–15]

Additionally, similar studies were carried out in rats using sucralose labelled with ^{14}C and ^{36}Cl radioisotopes. These studies confirmed that ingested sucralose does not break down, and that the chlorine atoms stay attached to the molecule. They also confirmed that the fraction of ingested sucralose that is absorbed moves freely through all body compartments by simple diffusion and that it does not get accumulated in any tissue or organ.[11] The collective result of these studies is that sucralose does not behave metabolically like a carbohydrate and that it is calorie free.

Another aspect of this rat study compared rats that had been given high concentrations of sucralose (3%) in the diet for 18 months with rats treated acutely. The chronically treated rats had a similar metabolic profile to the other study groups. It can be concluded from these observations that neither metabolic adaptation of the gut flora nor mammalian enzyme induction or adaptation had occurred during the 18-month treatment period.[11]

Another characteristic of some carbohydrates is that they can support the growth of *Streptococcus mutans* in the oral cavity. This is an important step in dental caries formation. Sucralose was studied for its ability to support growth of this microorganism using a rat model system.[16] *Streptococcus* was unable to grow in the presence of sucralose, and it was concluded that sucralose is non-cariogenic.

Sucralose is derived from sucrose and tastes sweet, but the chlorine atoms clearly modified the molecule such that it is not perceived biochemically as a sugar. This suggests that the chlorine atoms have also conferred a biological stability on the molecule as well as chemical stability. Since sucralose does not contribute calories to the diet, it can be useful in the production of calorie-reduced and calorie-free foods and beverages with the expectation that these products will retain their initial sweetness through their useful shelf life.

As required by the US FDA and the Organisation for Economic Co-operation and Development (OECD), a complete battery of environmental studies was carried out to determine if sucralose had any environmental impact. A series of studies of the physico-chemical events surrounding introduction of sucralose into the environment through the sewage system was carried out. In addition, a series of biological studies were conducted in environmentally relevant aquatic species to determine whether sucralose was bioaccumulative or toxic to these species. The resulting environmental assessment was included in all sucralose food additive petitions submitted to regulatory authorities worldwide.[17, 18] The resulting sucralose environmental database demonstrated that the trace amounts present in receiving waters did not bioaccumulate in any species and was not toxic. It was concluded that sucralose has no negative environmental impact either physico-chemically or biologically. These conclusions have subsequently been confirmed by additional studies conducted at several government and academic laboratories.

8.7 APPLICATIONS

The sweetness quality and excellent stability of sucralose means that it is an extremely versatile sweetener, suitable for use in a broad range of food applications. As well as being stable across a wide pH range, sucralose can withstand the high-temperature food manufacturing processes as shown in Table 8.3.

These functional characteristics mean that sucralose is ideal for use in food products from beverages to biscuits. Some general usage levels are given in Table 8.4.

8.7.1 Beverages

One of the largest areas of use is in soft drinks, where sucralose already sweetens many beverages around the world. As previously noted, the potency of sucralose, as with other

Table 8.3 Sucralose stability during food manufacturing processes.

	pH	Process conditions	Sucralose remaining post-processing (%)
Pasteurisation			
Tropical beverage	2.8	93°C for 24 s	100
Tomato ketchup	3.8	93°C for 51 min	100
Canned fruits – pears	3.3	100°C for 12 min	100
Sterilisation			
Beans in sauce	5.6	121°C for 80 min	100
UHT			
Dairy dessert	6.7	140°C for 15 s	100
Vanilla milk	6.5	141°C for 3.5 s	100

UHT, ultra-high temperature.

intense sweeteners, is influenced by many factors and where sucralose is used as the only sweetener in a beverage it will generally exhibit a potency of about 450–550 times that of sugar. Some typical usage levels for sucralose in different types of beverages are given in Table 8.5.

The fact that sucralose is both very stable and can provide a good sweetness quality means that it performs well as a sole sweetener and does not need to be used in a blend with other sweeteners. However, where beverage manufacturers may wish to use sweetener blends to further customise the flavour profile or to meet a particular ingredient budget, sucralose will work well with the other intense sweeteners. Sucralose can also be used in combination with nutritive carbohydrate sweeteners to provide excellent tasting reduced-sugar beverages. Sensory studies have confirmed that 30–40% of sugar can be replaced with sucralose, without significantly altering the taste or sweetness of the beverage.

The good water solubility of sucralose, even at ambient temperature, means that it is quick and easy to prepare concentrated stock solutions without the need to acidify the solution or stir for prolonged periods. Sucralose solutions can also be pumped or mixed without excessive foaming.

The stability of sucralose allows it to be used in low-pH beverages and those that require heat treatment such as pasteurisation or ultra-high temperature (UHT). Studies have confirmed that there is no measurable loss of sucralose from a diet cola drink during its 6-month shelf life, which in turn means that the beverage will maintain its flavour and sweetness quality throughout this period, as shown in Figure 8.7. Sucralose is one of the most stable high-potency sweeteners available for use in beverages, as demonstrated in Figure 8.8. This compares the stability of sucralose, acesulfame K and aspartame in a cola drink (pH 3.2) when stored at 35°C for 6 months.

Table 8.4 Typical dosage levels for sucralose in food products.

Category	Use level (%)
Yoghurts	0.010–0.015
Ice cream	0.010–0.014
Canned fruit	0.012–0.025
Boiled sweets	0.020–0.030
Chewing gum	0.100–0.300
Biscuits	0.010–0.035

Table 8.5 Typical usage levels for beverages.

Category	Use level (%)
Carbonated soft drinks	0.015–0.022
Flavoured waters	0.008–0.015
Isotonic drinks	0.010–0.020
Still beverages	0.012–0.018
Flavoured drink mixes[a]	1.500–3.500

[a]Dry blend before mixing.

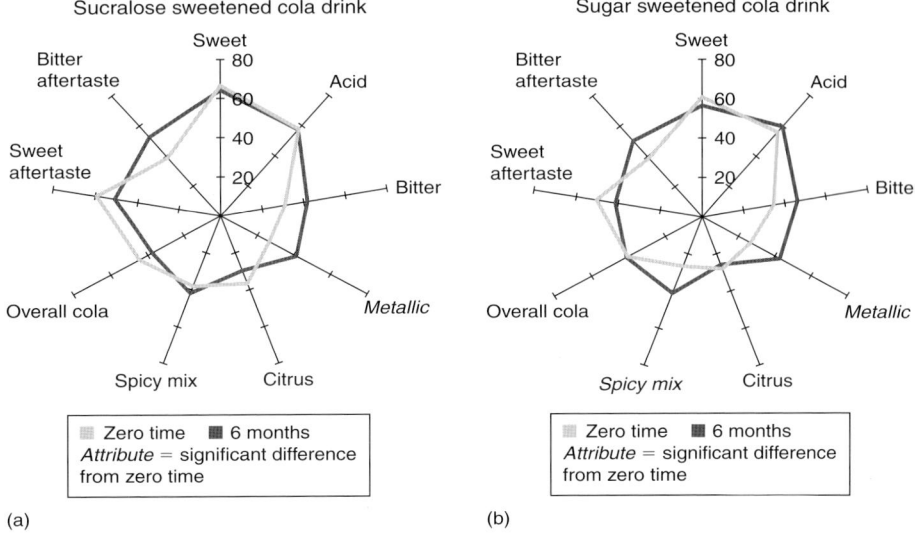

Fig. 8.7 Flavour profile of cola drinks when freshly prepared and after 6-month shelf life at 35 degrees C.

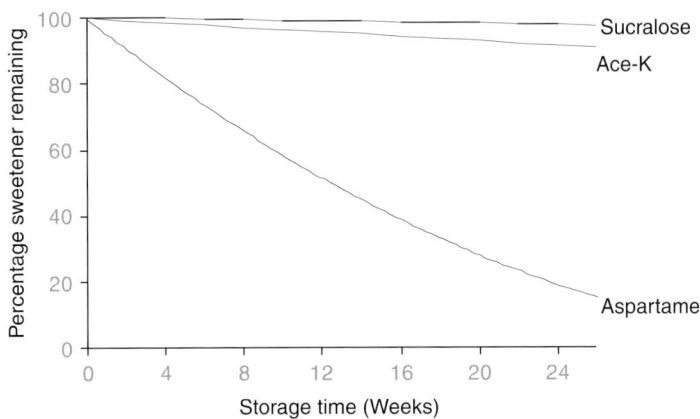

Fig. 8.8 Sweetener stability in cola.

8.7.2 Dairy products

Sucralose is suitable for use in all types of dairy products from milk drinks to ice cream. In the case of yoghurts, sucralose can be added either in the fruit preparation or to the yoghurt culture. This is because sucralose is not metabolised by, nor interferes with, the action of food-related microorganisms, enabling it to be present during the fermentation of the yoghurt without being broken down.

For other dairy applications such as ice cream, where sugar provides bulk and textural characteristics as well as sweetness, sucralose must be used in combination with other ingredients to successfully replace the sugar. The high potency of sucralose means that it can only provide sweetness in these types of products and so it is used alongside bulking agents such as polydextrose or the sugar alcohols. In ice cream, polydextrose acts in a similar manner to glucose syrup and will give a harder texture; sorbitol behaves in a similar way to dextrose and so gives softer textures, while maltitol provides a texture similar to sucrose. Combinations of polydextrose with either sorbitol or maltitol have been found to work well with sucralose to give a good overall sweetness and consistency. In these types of products, the usage level of sucralose will vary depending on the actual bulking agent(s) used, as the different bulking agents provide different levels of sweetness.

8.7.3 Confectionery

Sucralose is used in many commercial confectionery products, but once again it is only suitable for replacing the sweetness of the sugar and glucose syrups commonly used in standard products. Due to the very low levels of sucralose addition needed for reduced-sugar or no-added sugar confectionery, the physical characteristics of the nutritive sweeteners must be replaced by combinations of ingredients like polydextrose and the sugar alcohols. Sucralose has been found to work well with all these types of ingredients, giving a good flavour and sweetness quality. In addition, as sucralose does not support the growth of oral bacteria that cause tooth decay, it is suitable for use in 'tooth-friendly' products.

8.7.4 Baked products

The potential to use sucralose in baked products was originally assessed by confirming its stability to three different types of products that reflected the diverse baking conditions that could be employed in industrial production. These were:

1. Sponge cake cooked at 180°C for 25 minutes.
2. Biscuits baked at 210°C for 8 minutes.
3. Crackers baked at 230°C for 4 minutes.

Subsequent analysis of these products confirmed that there had been no loss of sucralose during any of the cooking processes.[19]

As discussed previously, sucralose will only provide sweetness and must be used with bulking agents in these applications. Again, the level of sucralose needed will be dependent on which bulking agents are used and how much sweetness is provided by them.

Sucralose can also be used to sweeten jams, fruit fillings and icings that are often used with baked products.

8.7.5 Pharmaceuticals

Sucralose is included in both the US Pharmacopoeia and the European Pharmacopoeia and can be used as an excipient in pharmaceutical applications. However, local regulations should be checked to ensure that additional regulations are not required.

To date, sucralose has been successfully used in products such as vitamin tablets, throat lozenges and cold cures.

8.8 ANALYTICAL METHODS

The determination of sucralose in food products is typically conducted by HPLC analysis with a RI detector using the following conditions:

Column	C_{18} reverse phase
Mobile phase	30% methanol/70% water
Flow rate	1–1.5 mL/min
Detection	RI
Injection volume	50–100 µL
Retention time	About 10 min

Water-based products such as soft drinks can be analysed directly, although carbonated beverages should be de-carbonated prior to analysis. For more complex products, sucralose can be extracted with a polar solvent such as water or methanol.

Generally, some form of sample clean-up is required prior to analysis, to remove potentially interfering components and this is often achieved via solid phase extraction cartridges. The cartridge packings that have proved to be the most effective are C_{18} reverse phase and alumina. By using extraction and clean-up procedures based on the aforementioned, recoveries of about 100% are normally achieved for sucralose.

8.9 SAFETY

Sucralose is legally considered a food additive, as are all intense sweeteners. As such, it comes under food additive legislation. Food additive legislation requires that an extensive data package be developed and submitted for regulatory review and approval prior to placing the substance on the market. While this data package (food additive petition) covers all aspects of the proposed food additive from manufacturing, proposed uses, analytical methods, environmental assessment, etc., the largest section of the petition is that which presents studies establishing the safety in use of the additive. Positive evaluation by the regulatory authorities of this section is crucial to gaining approval to market a food additive.

The full range of toxicology studies were conducted on sucralose and, separately, its two hydrolysis products, 1,6-DCF and 4-CG. These studies eventually numbered over 150 animal and clinical studies, the major categories of which are shown in Table 8.6.

Clearly, a database this large must be built up over many years and develops incrementally from acute to chronic animal studies, eventually leading to clinical studies. These clinical studies are not required by any regulatory authority for food additive approval, but have

Table 8.6 Major categories of studies conducted in support of safety.

Safety database
- Acute, subchronic, chronic and oncogenicity studies
- Mutagenicity studies
- Reproduction and teratology studies
- Absorption, distribution, metabolism and elimination studies, in animals and man
- Special studies usually include the following:
  - Neurotoxicity
  - Immunotoxicity
  - Unique studies/related to specific molecule
- Clinical studies in normal and diabetic subjects, not required but usual in sweetener FAP

FAP, food additive petition.

become standard procedure for sweetener applications. It is also evident that building such an extensive database requires a major commitment in time and resources.

It is not appropriate here to discuss in detail the extensive data produced in support of the safety of sucralose for human consumption as these safety studies have been fully reviewed in other publications.[20,21] A wide-ranging summary follows, which will illustrate the major results.

The data derived from these studies show no evidence that sucralose, or its two hydrolysis products, have any adverse effect. As shown in Figure 8.5, sucralose will very slowly hydrolyse in acidic media into its monosaccharide components, 1,6-DCF and 4-CG. Even though these compounds would only be present in foods in microgram quantities, given the stability of sucralose, these two compounds were subjected to almost the same toxicology-testing programme as was sucralose itself.

As discussed in Section 8.6, extensive metabolic studies were undertaken with sucralose in a variety of animal species and human volunteers to determine the disposition of ingested sucralose in the body. From these studies, it was shown that sucralose passes through the body intact, and is not broken down. The bulk of ingested sucralose is eliminated in the faeces, and the small fraction that is absorbed into the body passes through the tissues and organs by passive diffusion, eventually being excreted in the urine. It was established that the mouse, rat, dog and rabbit dispose of sucralose in a comparable fashion to man. These observations provide assurance that data derived from animal safety studies can be extrapolated to man.

Sucralose is extremely non-toxic when administered acutely to mice and rats. The LD-50 is >16 g/kg body weight (BW) in mice and >10 g/kg BW in rats.[22] Other studies in rodents lasting up to 26 weeks were conducted in preparation for long-term/life-time toxicology and carcinogenicity tests in rats and mice. No evidence of carcinogenicity was observed in the rat and mouse studies even at the highest dietary level of sucralose, 3% of the diet (equivalent to 1500 mg/kg BW per day), and there were no toxicologically significant findings in the long-term rat study[23] at this dosage level. A 1-year toxicity study in the dog was also conducted, and no remarkable results were observed in this study.[24]

When studied in rats and rabbits, neither sucralose nor its hydrolysis products had an effect on fertility or reproduction. There was also no evidence of developmental toxicity in rats or rabbits.[25]

Special tests were conducted to determine whether sucralose or its hydrolysis products had any neurotoxic potential. It was found that neither sucralose nor its two breakdown products represented any neurotoxic risk.[26] A test for immunotoxicity potential in developing rats demonstrated that sucralose had no effect on the developing immune system even at 1500 mg/kg BW per day.[27] Additionally, an extensive battery of genotoxicity tests was carried out on both sucralose and its two hydrolysis products to determine whether these compounds had any mutagenic potential. The results of these studies fully supported the overall conclusion that sucralose was safe for its intended use.[28]

A number of clinical studies were performed in human volunteers over a number of years. These studies gained in complexity over time, varying from tolerance studies lasting up to 3-month duration with doses approaching the acceptable daily intake (ADI) of 15 mg/kg BW per day (13 mg/kg BW per day) in normal volunteers[29,30] to studies lasting up to 3 months in diabetic subjects[31,32]. No adverse experiences or clinically detectable effects were attributed to sucralose ingestion in any study. These studies, plus the animal studies described previously, firmly established that sucralose has no effect on glucose homeostasis. It did not interfere with normal carbohydrate metabolism or alter insulin secretion or action in animals or man.

8.10 REGULATORY STATUS

To be approved as a food additive, the sucralose safety database had to be evaluated by international and governmental regulatory agencies to determine the safety of sucralose for its intended use as a non-caloric sweetener. The first agency to complete an evaluation of the extensive sucralose safety database was the Joint Expert Commission for Food Additives (JECFA), a joint committee between the World Health Organization and the Food and Agriculture Organization. JECFA reviews the safety data of food additives, sets an ADI and publishes specifications for each additive for which an ADI has been assigned.

JECFA allocated an ADI for sucralose of 0–15 mg/kg BW per day. This ADI was based primarily on the pivotal long-term rat studies in which 1500 mg/kg BW per day was the no-effect level.[33] Specifications were also established for sucralose and were subsequently published.[34] The JECFA is not a regulatory body, but its recommendations are usually adopted by countries around the world, and this has been the case for sucralose as it is approved in over 100 countries, including the United States[17,18] and the European Union.[35] The Codex Alimentarius has also adopted sucralose. The wide availability of sucralose to the food, beverage, and pharmaceutical industry allows them the opportunity to improve existing products and to develop new ones that will increase consumer choice when seeking high-quality, good-tasting, low-sugar food and beverages.

REFERENCES

1. L. Hough and S.P. Phadnis. Enhancement in the sweetness of sucrose. *Nature* 1976, 263, 800.
2. L. Hough and R. Khan. Intensification of sweetness. *Trends Biochemical Sciences* 1978, 3, 61.
3. M.G. Lindley, G.G. Birch and R. Khan. Sweetness of sugar and xylitol. Structural considerations. *Journal of Science and Food Agriculture* 1976, 27 140.
4. M.R. Jenner. The uses and commercial development of sucralose. In: T.H. Grenby (ed). *Advances in Sweeteners*, Blackie Academic and Professional, Glasgow; 1996, p. 253.
5. S.V. Molinary and M. Jenner. History and development of sucralose. *Food and Food Ingredients Journal of Japan* 1999, 182, 6–12.

6. M.R. Jenner. Sucralose: unveiling its properties and applications In: T.H. Grenby (ed.). *Progress in Sweeteners*, Elsevier, London; 1989, p. 121.
7. G.A. Miller. Sucralose In: L. O'Brien Nabors and R.C. Gelardi (eds). *Alternative Sweeteners*, 2nd edn. Marcel Dekker, New York; 1991; p. 173.
8. S.G. Wiet and P.K. Beytes. Sensory characteristics of sucralose and other high intensity sweeteners. *Journal of Food Science* 1992, 57(4), 1014.
9. M.R. Jenner and A. Smithson. Physiochemical properties of the sweetener sucralose. *Journal of Food Science* 1989, 54(6), 1646–1649.
10. M.E. Quinlan and M.R. Jenner. Analysis and stability of the sweetener sucralose in beverages. *Journal of Food Science* 1990, 55(1), 244.
11. J. Sims, A. Roberts, J.W. Daniels and A.G. Renwick. The metabolic fate of sucralose in rats. *Food and Chemical Toxicology* 2000, 38(Suppl. 2), S115–S121.
12. B.A. Johns, S.G. Wood and D.R. Hawkins. The pharmacokinetics and metabolism of sucralose in the mouse. *Food and Chemical Toxicology* 2000, 38(Suppl. 2), S107–S110.
13. S.G. Wood, B.A. John and D.R. Hawkins. The pharmacokinetics and metabolism of sucralose in the dog. *Food and Chemical Toxicology* 2000, 38(Suppl. 2), S99–S106.
14. B.A. John, S.G. Wood and D.R. Hawkins. The pharmacokinetics and metabolism of sucralose in the rabbit. *Food and Chemical Toxicology* 2000, 38(Suppl. 2), S111–S113.
15. A. Roberts, A.G. Renwick, J. Sims and D.J. Snodin. Sucralose metabolism and pharmacokinetics in man. *Food and Chemical Toxicology* 2000, 38(Suppl. 2), S31–S41.
16. W.H. Bowen, D.A. Young and S.K. Pearson. The effect of sucralose on coronal and root-surface caries. *Journal of Dental Research* 1990, 69(8), 1485–1487.
17. Food additives permitted for direct addition to food for human consumption: sucralose. *Federal Register* 1998, 63(64), p. 16417–16433.
18. Food additives permitted for direct addition to food for human consumption: sucralose. *Federal Register* 1999, 64(155), p. 43908–439009.
19. R.L. Barndt and G. Jackson. Stability of sucralose in baked goods. *Food Technology* 1990, 44(1), 62–66.
20. J.F. Borzelleca and H. Verhagen (eds.) *Food Chemical Toxicology* 2000, 38(Suppl. 2), S1–S129
21. V. L. Grotz and I.C. Munro. An overview of the safety of sucralose. *Regulatory Toxicology and Pharmacology* 2009, 55, 1–5
22. WHO Food Additive Series: 24. Toxicological evaluation of certain food additives and contaminants. The 33rd Meeting of the Joint FAO/WHO Expert Committee on Food Additives, Geneva, Switzerland; 1989, p. 45.
23. S.W. Mann, M.M. Yuschak, S.J.G. Amyes, P. Aughton and J.P. Finn. A combined chronic toxicity/carcinogenicity study of sucralose in Sprague-Dawley rats. *Food and Chemical Toxicology* 2000, 38(Suppl. 2), S71–S89.
24. L.A. Goldsmith. Acute and subchronic toxicity of sucralose. *Food and Chemical Toxicology* 2000, 38(Suppl. 2), S53–S69.
25. J.W. Kille, J.M. Tesh, P.A. McAnulty, F.W. Ross, C.R. Willoughby, G.P. Bailey, O.K. Wilby and S.A. Tesh. Sucralose: assessment of teratogenic potential in rat and rabbit. *Food and Chemical Toxicology* 2000, 38(Suppl. 2), S43–S52.
26. J.P. Finn and G.H. Lord. Neurotoxicity studies on sucralose and its hydrolysis products with special reference to histopathologic and ultrastructure changes. *Food and Chemical Toxicology* 2000, 38(Suppl. 2), S7–S17.
27. Opinion of the Scientific Committee on Food, SCF/CS/ADDS/EDUL/190Final, European Commission, Health and Consumer Protection Directorate-General; 2000.
28. D. Brusick, V. L. Grotz, R. Slesinski, C.L. Kruger and A.W. Hayes. The absence of genotoxicity in sucralose. *Food and Chemical Toxicology* 2010, 48(11), 3067–3072
29. I.M. Baird, N.W. Shephard, R.J. Merritt and P. Hildick-Smith. Repeated dose study of sucralose tolerance in human subjects. *Food and Chemical Toxicology* 2000, 38(Suppl. 2), S123–S130.
30. A. Roberts. Sucralose and diabetes. *Food and Food Ingredients Journal of Japan* 1999, 182, 49–53.
31. H.E. Mezitis, C.A. Maggio, P. Koch, A. Quoddoos, D.B. Alison and F.X. Pi-Sunyer. Glycemic effect of a single high dose of the novel sweetener sucralose in patients with diabetes. *Diabetes Care* 1996, 19, 10004–10005.

32. V.L. Grotz, R.R. Henry, J.B. McGill, M.J. Prince, H. Schmoon, J.R. Trout and F.X. Pi-Sunyer. Lack of effect of sucralose on glucose homeostasis in subjects with type 2 diabetes (T2DM). *Journal of the American Dietetic Association* 2003, 103, 1607–1612.
33. WHO Food Additive Series: 28. Toxicological evaluation of certain food additives and contaminants. The 37th Meeting of the Joint FAO/WHO Expert Committee on Food Additives, Geneva, Switzerland; 1991, p. 219.
34. Compendium of Food Additive Specifications, Addendum 2, Joint FAO/WHO Expert Committee on Food Additives, 41st Session, Geneva, Switzerland; 1993, p. 119.
35. Official Journal of the European Union, Directive 2003/115/EC; 2004, p. L24/65.

9 Natural High-Potency Sweeteners

Michael G. Lindley
Lindley Consulting, Crowthorne, UK

9.1 INTRODUCTION

Many high-potency sweeteners of diverse chemical structures are known to occur naturally. The presence of potent sweeteners in plants, presumably formed as secondary plant metabolites, may, in some instances, induce benefits for the plant with respect to its propagation by assisting seed dispersal, but in most cases the presence of a sweetener within some part of the plant seems to have no obvious function. There also appears to be no obvious pattern to the plant species in which potent sweeteners have been found; thus, the synthesis of a potently sweet compound within plants generally appears to be little more than happenstance.

Natural sweeteners show many similarities to their synthetic counterparts in terms of their overall taste characteristics. They range in potencies from approximately 10 times that of sucrose to 1000s of times the sweetening power of sucrose. Overall sweet taste quality, measured by traditional methods that assess temporal characteristics and others that quantify the intensity of side- and aftertastes, also range across the spectrum in much the same way as has been found with synthetic sweeteners. As a consequence, many of the challenges inherent in formulating sugar-reduced foods and beverages so that they will deliver taste experiences fully acceptable to consumers are similar, irrespective of the origin of the potent sweetener.

Consumer interest in natural high-potency sweeteners has grown dramatically in recent years, fuelled by concerns about the use of artificial additives in foods. How far the use of natural potent sweeteners will go remains to be seen, but within the sweetener industry, the level of development activity is high, as is the activity involved in a continuing search for other natural sweeteners through the use of various screening techniques.

Knowledge of the existence of potent sweeteners in plants is not new; there have been literature references to examples dating back to the mid-nineteenth century. However, the majority of references to potent natural sweeteners also note overall taste characteristics that are basically incompatible with successful commercialisation. Thus, although of intrinsic interest and also perhaps providing leads for the identification of other potent sweeteners within plants of the same family, interest in natural (and synthetic) sweeteners is essentially proportional to their commercial viability, and so most natural sweeteners listed here (Table 9.1, references 1–28) have not been considered seriously for inclusion in commercial

Sweeteners and Sugar Alternatives in Food Technology, Second Edition.
Edited by Dr Kay O'Donnell and Dr Malcolm W. Kearsley.
© 2012 John Wiley & Sons, Ltd. Published 2012 by John Wiley & Sons, Ltd.

Table 9.1 Examples of high-potency sweeteners of plant origin.

Sweetener	Structural class	Plant source	Country of origin	Reference
Brazzein	Protein	*Pentadiplandra brazzeana*	West Africa	1
Curculin	Protein	*Curculigo latifolia*	Malaysia	2
Mabinlin	Protein	*Capparis masakai*	China	3
Monellin	Protein	*Discoreophyllum cumminsii*	West Africa	4
Pentadin	Protein	*Pentadiplandra brazzeana*	West Africa	5
Thaumatin	Protein	*Thaumatococcus daniellii*	West Africa	6
Monatin	Amino acid	*Schlerochiton ilicifolius*	South Africa	7
Abrusoside	Glycoside	*Abrus precatorius*	Thailand	8
Albiziasaponins	Glycoside	*Albizia myriophylla*	Thailand	9
Baiyunoside	Glycoside	*Phlomis betonicoides*	China	10
Bryoside	Glycoside	*Bryonia dioica*	Italy	11
Cussoracosides	Glycoside	*Cussonia racemosa*	Madagascar	12
Cyclocarioside	Glycoside	*Cyclocarya paliurus*	China	13
Glycyrrhizin	Glycoside	*Glycyrrhiza glabra*	China	14
Lo han guo	Glycoside	*Siratia grosvenorii*	China	15
Mukuroziside	Glycoside	*Sapindus mukurossi*	China	16
Osladin	Glycoside	*Polypodium vulgare*	Widespread, Inc., USA, Eur	17
Periandrin	Glycoside	*Periandra dulcis*	Brazil	18
Phlomisoside	Glycoside	*Phlomis younghusbandii*	China	19
Polypodoside	Glycoside	*Polypodium glycyrrhiza*	USA	20
Pterocaryoside	Glycoside	*Pterocarya paliurus*	China	21
Rubusoside	Glycoside	*Rubus suavissimus*	China	22
Steviol glycosides	Glycoside	*Stevia rebaudiana*	Paraguay	23
Telosmosides	Glycoside	*Telosma procumbens*	Philippines	24
Selligueain A	Proanthocyanidin	*Selliguea feei*	Indonesia	25
Hernandulcin	Bisabolane sesquiterpene	*Lippia dulcis*	Mexico	26
Phlorizin	Dihydrochalcone	*Lithocarpus litseifolius*	China	27
Trilobatin	Dihydrochalcone	*Lithocarpus litseifolius*	China	27
Phyllodulcin	Flavonoid	*Hydrangea macrophylla*	Japan	28

development programmes. Therefore, this chapter will review the characteristics of only those natural potent sweeteners that are commercial or whose commercial development has been reported to be in progress. It is also important to acknowledge that, frustratingly, it is a feature of the natural high-potency sweeteners that their sensory, physical and functionality properties have rarely been studied academically and as a consequence there are few peer review publications in the literature that provide detailed objective data.

Although a commercial product, specifically excluded from this discussion is the triterpenoid sweetener, glycyrrhizin. Salts of glycyrrhizic acid occur to a total level of 6–14% in the roots of a shrub, *Glycyrrhiza glabra* L. and extracts of the roots are well known as liquorice. Although reportedly about 30 times as sweet as sucrose at 10% sucrose equivalence,[29] glycyrrhizin is excluded from further discussion here because it has regulatory approval in the United States through FEMA for use only as a flavouring agent and flavour enhancer, not as a sweetener. In addition, it is not approved for use as a sweetener in Europe.

9.2 THE SWEETENERS

9.2.1 Thaumatin

9.2.1.1 Introduction

Thaumatin is the common name for a mixture of potently sweet proteins that can be extracted from the West African plant *Thaumatococcus daniellii* (Bennett) Benth, known indigenously as the katemfe berry. This plant produces fruit, the arils of which contain the sweet proteins. The plant grows naturally in most West African countries and has also been shown to thrive and produce protein-containing fruit in Malaysia where it grows particularly well as an intercrop within rubber plantations.

The first references to thaumatin in the scientific literature are from the nineteenth century,[6] but it was not until the latter part of the twentieth century that research was undertaken to determine the structure of the protein and to understand the fundamentals of its physical and sensory characteristics.[30] With subsequent development of horticultural procedures, progress was made in the commercialisation of the sweetener.

The plant produces a mixture of proteins, thaumatins I and II being the major constituents, each with almost identical molecular weight of 22,000 Daltons. Thaumatins I and II have very similar amino acid sequences, differing only in five residues.[31] The amino acid composition of thaumatin I is presented in Table 9.2. The protein is stabilised by eight disulphide bridges, thus conferring a greater stability to heat and pH denaturation to the molecule than might be expected for a protein. These disulphide bridges are also believed to be responsible for holding the protein chain in the correct conformation to elicit sweetness, a conclusion supported by the observation that cleavage of even a single disulphide bridge results in a loss of sweetness.[32]

These proteins have been the subject of much academic study and a number of detailed reviews have been published.[33] In addition, thaumatin was used as a tool in early studies that sought to understand the structure of the mammalian receptor for sweetness.[34] In this study, antibodies raised against thaumatin were shown to compete for another sweet protein, monellin, and for many other sweet compounds, suggesting that all sweeteners are all recognised by the same receptor.

Table 9.2 Amino acid composition of thaumatin I.[30]

Amino acid	Number of residues
Glycine	24
Threonine	20
Alanine	16
Half-cystine	16
Serine	14
Arginine	12
Aspartic acid	12
Proline	12
Lysine	11
Phenylalanine	11
Asparagine	10
Valine	10
Leucine	9
Isoleucine	8
Tyrosine	8
Glutamic acid	6
Glutamine	4
Tryptophan	3
Methionine	1

Thaumatin is not, of course, a synthetic sweetener and is extracted from source using only aqueous extraction processes followed by physical separation processes to remove unwanted material. The resulting product is a light tan coloured powder.

9.2.1.2 Sensory properties

The sweetness potency of thaumatin is normally described as being approximately 2000 times the sweetness of sucrose, although, given the temporal profile of thaumatin, precise determination of potency is difficult to determine. DuBois *et al.* have generated dose–response data under controlled conditions,[35] leading to estimations of sweetness potency for thaumatin of about 8600 times that of sucrose at 2% sugar equivalent sweetness, about 6000 times sucrose at 6% sugar equivalence and about 2000 times that of sucrose at 8% sucrose equivalence, indicating clearly the rapid decline in relative sweetness factor as concentration increases (Figure 9.1).

The temporal taste profile of thaumatin is characterised by a delay in perceiving sweetness, a lengthy sweetness growth phase until maximum sweetness is perceived, followed by a lingering sweet/liquorice aftertaste.[36] These temporal properties have resulted in only limited generation of detailed sensory data describing its use at sweetening concentrations, largely because those temporal properties are incompatible with mainstream food and drink applications. Similarly, there are no detailed reports of synergy between thaumatin and other bulk or potent sweeteners. Many attempts to move the taste of thaumatin to be more 'sucrose-like' have been described, largely in Japanese literature, but any improvements achieved were acknowledged as being subtle. Combinations of thaumatin with glycyrrhizin, amino acids, citric and succinic acids, monosodium glutamate and lactose were marketed as a table-top sweetener in Japan during the 1980s. In addition, thaumatin does exhibit some interesting and commercially relevant flavour-modifying characteristics.[37]

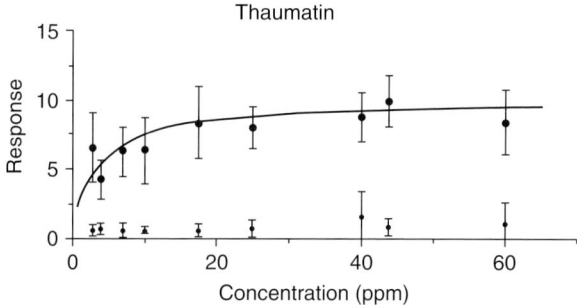

Fig. 9.1 Dose–response relationships of thaumatin. Closed circles show sweetness ratings and open circles show bitterness ratings.
Source: Reproduced with permission from ref. 35. Copyright 1991, American Chemical Society.

9.2.1.3 Physical and chemical properties

As has been described, thaumatin is stabilised by the eight disulphide bridges that result in a cross-linked network of amino acid chains. This confers a measure of stability to heat and extremes of pH, although there are no published data to this effect. These disulphide bridges are also responsible for holding the protein chain in the correct conformation to elicit sweetness, as has been confirmed[38] by demonstrating that cleavage of a single disulphide bridge results in loss of sweetness. The resulting modified thaumatin then possessed protease, amidase and esterase enzymic activities.

Again, specific data on the solubility of thaumatin are not published, but the sweetener is prepared as a freeze-dried powder that is readily soluble in water.

9.2.1.4 Physiological properties

Thaumatin is a natural plant protein of known structure containing normal amino acids. Therefore, it would be anticipated that thaumatin should be metabolised in the same way as other plant proteins, and this has been demonstrated to be the case.[39]

9.2.1.5 Applications

During the development of thaumatin as a novel food ingredient, it was anticipated that it would be used in applications on the basis of its potent sweetening attribute. In the event, this has not proved to be the case, largely because, as has been explained, the temporal profile of the sweetener is incompatible with the majority of mainstream food and drink product applications. Although it is sometimes possible to use thaumatin to contribute a small portion of the total sweetness in products, the marketing cachet of 'naturally sweetened' may then be lost, thus undermining the primary justification for selecting thaumatin in the first place.

Consequently, the main commercial applications for thaumatin have capitalised on its claimed flavour modifying and enhancing functionalities. Beneficial effects on flavour intensity and quality in a wide array of foods and drinks have been summarised by Higginbotham.[37] Japanese food manufacturers pioneered the commercial introduction of thaumatin as a flavour

Table 9.3 Bitterness reduction by thaumatin.[40]

Bitterant	Relative bitterness rating at the following concentrations of thaumatin (ppm)				
	0	0.5	1.0	2.0	5.0
Vitamin B complex (0.44%)	100	92	75	60	45
Caffeine (0.1%)	100	80	60	50	40
Soy peptides (0.1%)	100	85	65	60	55

enhancer across a wide spectrum of food and beverage applications and it has been used successfully in Western markets in products such as chewing gum, where it is claimed to boost sweetness and flavours such as spearmint and peppermint, and in selected soft drinks. Current uses are mainly as a flavour ingredient for its enhancing properties.

In addition to its specialist uses in foods, thaumatin has found application in liquid medicines, oral care products and in the nutraceutical/fortified foods industries, mainly for its ability to help mask bitter and astringent tastes (Table 9.3). As can be seen, increasing concentrations of thaumatin appear to have increasing effects on the bitterness associated with vitamin B complex preparations, caffeine and soybean peptides. In these applications, thaumatin's long-lasting sweetness presumably helps to counteract some of the bitterness and astringency frequently associated with such products.[40]

9.2.1.6 Safety

A comprehensive safety evaluation programme was conducted on thaumatin prior to its approval for use, the details of which have been published.[39] During this programme, thaumatin was studied for its subacute toxicity in rats and dogs and its ability to produce anaphylactic antibodies following oral administration to rats and normal human subjects. It was found to be readily digested prior to absorption in rats and no adverse effects resulted from its continuous administration to rats and dogs at dietary concentrations of 0%, 0.3%, 1.0% and 3.0% for 13 weeks. It was shown to be non-teratogenic when administered orally to rats at 0, 200, 600 and 2000 mg/kg body weight/day from day 6 to 15 of gestation and was without effect on the incidence of dominant lethal mutations when administered on five consecutive days to male mice at 200 and 2000 mg/kg per day.

The lack of mutagenic potential was confirmed in bacterial mutagenic assays with *Salmonella typhimurium* (strains TA1535, TA1537, TA1538, TA98 and TA100) and *Escherichia coli* WP2, at levels of addition of 0.05–50 mg per plate.

In rats, thaumatin was found to be a weak sensitiser, comparable with egg albumen, when administered systemically but to be inactive when administered orally. Prick testing of laboratory personnel who had been intermittently exposed by inhalation to thaumatin for periods up to 7 years showed that 9.3% (13/140) responded positively to commercial thaumatin, while 30.7% were positive to *Dermatophagoides pteronyssinus* (house dust mite). None of the subjects who gave a positive skin reaction to commercial thaumatin responded to the plant components remaining after removal of the specific sweet thaumatin proteins. Challenge tests in man did not demonstrate any oral sensitisation.

Overall, it was concluded that the results indicate that thaumatin when used as a flavour modifier and extender, and partial sweetener, is unlikely to be hazardous at the anticipated level of consumption.

Table 9.4 Regulatory approval of thaumatin in the EU.

Sweetener	Product category	Maximum usable dose
Thaumatin (E957)	Confectionery	50 mg/kg
	• Confectionery with no added sugar	50 mg/kg
	• Cocoa or dried fruit based	50 mg/kg
	• Confectionery; energy reduced or with no added sugar	
	• Chewing gum with no added sugar	
	Food supplements	400 mg/kg
	Edible ices, energy reduced or with no added sugar	50 mg/kg

9.2.1.7 Regulatory status

Thaumatin was originally permitted as a natural food additive in Japan in 1979. It was then approved as a sweetener in the United Kingdom and Australia, and was incorporated into the EU Sweeteners Directive where it has approval for use in a small number of food categories (Table 9.4). In the United States, thaumatin was accorded GRAS status as a flavour adjunct for chewing gum in 1984 and this has since been extended by FEMA to general use across all food categories.

A JECFA review of the safety of thaumatin resulted in JECFA according an acceptable daily intake (ADI) 'not specified' status to the sweetener.

9.2.2 Steviol glycosides

9.2.2.1 Introduction

Stevia rebaudiana Bertoni, a plant indigenous to the northern part of Paraguay in South America, is the source of the potently sweet *ent*-kaurenoid diterpene glycosides stevioside, rebaudioside A and several other steviol glycosides, some of which are also being evaluated for potential commercialisation. Although stevioside, the major sweet glycoside of the species, has been known to be sweet for more than 100 years, it was only in the 1970s that it was developed for use as a commercial sweetener in Japan.[41] Several other sweet analogues are known (there are at least eight steviol glycosides), although until recently, none has generated the levels of academic and commercial interest as have stevioside and rebaudioside A. Recent reports that rebaudioside C is a sweet taste potentiator have led to a flurry of media stories of its development for use in this sweetness-enhancing capacity.

The steviol glycoside sweeteners share a common aglycone, steviol. Linked to steviol are carbohydrate moieties and it is the number and linkages of these that differentiate the steviol glycoside sweeteners, as well as influencing their individual sensory properties. Stevioside itself comprises steviol to which are linked a glucose (β-linked) and one glucose-glucose disaccharide (comprising β-D-glucose linked 2-1 with β-D-glucose, or sophorose) moieties (Figure 9.2); rebaudioside A comprises steviol to which are linked a glucose (β-linked) and one glucose-glucose-glucose trisaccharide (comprising β-D-glucose linked 2-1 and 3-1 with β-D-glucose) moieties (Figure 9.3). Thus, rebaudioside A, containing more polar groups than stevioside, is more soluble and is claimed to deliver a cleaner, more sucrose-like taste than stevioside. Stevioside itself is generally agreed to deliver a clear bitter taste component in addition to being sweet. The claimed sweetness potentiator, rebaudioside C,

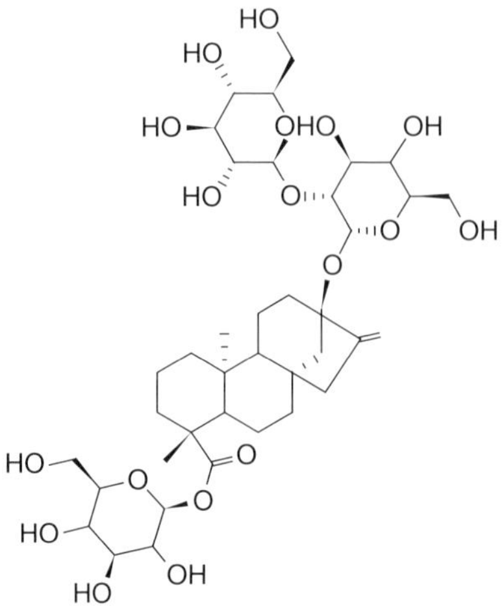

Fig. 9.2 Stevioside.

also comprises steviol to which are linked a glucose (β-linked) and one glucose-glucose-rhamnose trisaccharide (comprising β-D-glucose linked 3-1 with β-D-glucose linked 2-1 with an α-L-rhamnose) moieties.

Although early work attempting to propagate *Stevia rebaudiana* was largely unsuccessful, substantial efforts in Japan ultimately proved successful and now the plant is cultivated from root stock and grown commercially in many countries, including Japan, Taiwan, South Korea, China and Brazil. The steviol glycoside composition depends on the plant source, but in all cases, stevioside is the predominant sweet component. With stevioside contributing a clear bitter taste component, it is not surprising therefore that the sweetener remained a relatively niche product for many years. Now, plant-breeding developments are helping to make available steviol glycoside preparations that are rich in rebaudioside A. As a result,

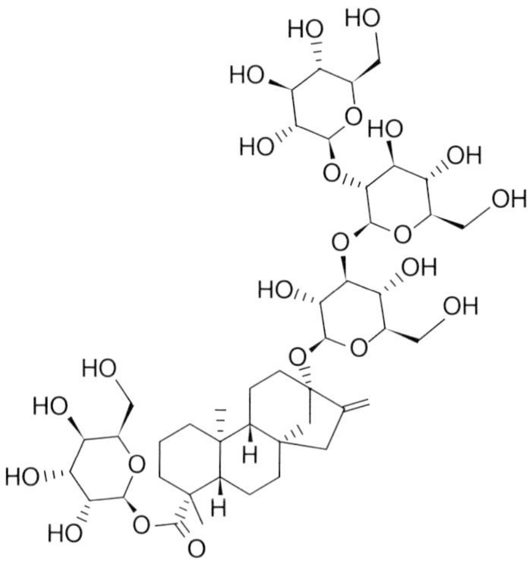

Fig. 9.3 Rebaudioside A.

application of appropriate purification processes has ensured that at least one commercial product is available that is claimed to consist of >99% rebaudioside A.

High concentrations (10%; w/w) of steviol glycosides occur in the dried leaves of *Stevia rebaudiana*. Although there will be many variations in the details of actual purification processes adopted by producers, preparation of sweetener from the leaves typically involves some or all of the following unit operations: aqueous extraction followed by selective extraction into a polar organic solvent, decolourisation, removal of impurities through flocculation and filtration, ion exchange and finally crystallisation.[42]

There is an extensive body of literature on the properties of stevioside, in particular. However, it has been accepted that a limitation of much of this information is that only rarely is the exact composition known of the material actually tested. Since the early steviol glycoside extraction and purification procedures did not selectively isolate specific glycosides, the great majority of work carried out has used samples that contained mixtures of steviol glycosides, rather than pure stevioside or other steviol glycoside. This has particular relevance for the conduct of toxicology testing since an incomplete and imprecise definition of the composition of the material being tested makes interpretation of results generated difficult if not impossible. This issue surrounding stevioside in particular and the impact it had on its regulatory approvals are discussed later in this chapter.

9.2.2.2 Sensory properties

Few studies have evaluated the sensory properties of all of the sweet steviol glycosides, but literature data have been collated and summarised by Kim and DuBois.[43] Information on sweetness potencies is presented in Table 9.5.

Most other literature reports of the sensory properties of stevioside and rebaudioside A are, as has been noted, complicated due to uncertainties about the purity of samples being evaluated. However, there is a consistent theme running through all these literature reports that the overall quality of taste elicited by rebaudioside A is superior to that elicited by stevioside. This is generally ascribed to a significantly lower bitter taste component of rebaudioside A and a reduced level of cooling in its aftertaste. Quantitative confirmation of the reduced level of bitterness can be seen in data generated by DuBois and co-workers who examined the dose–response sweetness and bitterness characteristics of most commercial high-potency sweeteners. Their data clearly demonstrate that rebaudioside A is not only more potent than stevioside, but it also delivers a substantially lower level of bitterness at iso-sweet concentrations (Figure 9.4).[35]

Table 9.5 Sweetness potencies of steviol glycosides.

Compound	Relative sweetness[a]
Stevioside	300
Rebaudioside A	250–450
Rebaudioside B	300–350
Rebaudioside C	50–120
Rebaudioside D	250–450
Rebaudioside E	150–300
Dulcoside A	50–120
Steviolbioside	100–125

[a]Sweetness potency measured relative to 0.4% (w/v) sucrose.

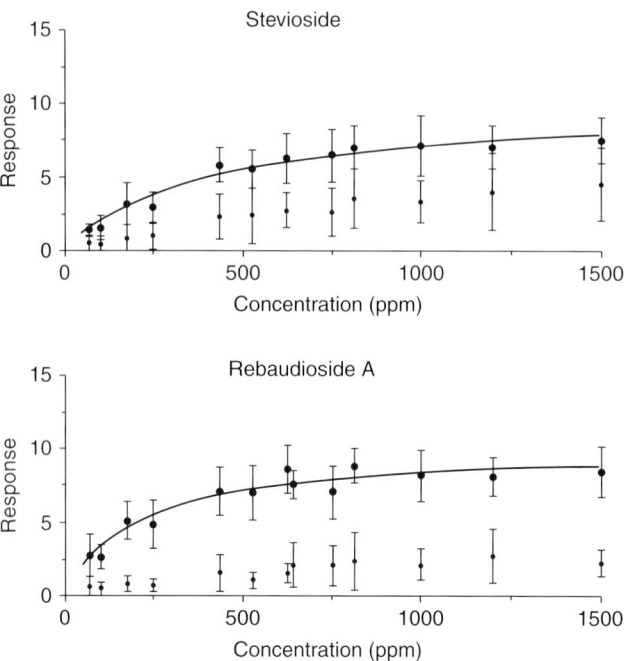

Fig. 9.4 Dose–response relationships of stevioside and rebaudioside A. MG Closed circles show sweetness ratings and open circles show bitterness ratings.
Source: Reproduced with permission from ref. 35. Copyright 1991, American Chemical Society.

The qualitative flavour profiles of stevioside and rebaudioside A have been studied and the results reported in Table 9.6.[44] These demonstrate the qualitative improvement of rebaudioside A relative to stevioside, and also confirm that the overall taste qualities of these sweeteners do not mimic that of sucrose. Interestingly, and in contrast to most literature, this particular report suggests rebaudioside A is less sweet than stevioside that may be a consequence of the methodologies employed or due to the actual materials evaluated being mixtures of steviol glycosides rather than the pure glycoside.

Not surprisingly, there are few reports of the sensory performance of stevioside and rebaudioside A blended with other sweeteners, particularly other high-potency sweeteners. This is probably because the primary commercial driver for using these sweeteners is their natural status that would be compromised if either was to be blended with any of the so-called artificial sweeteners. However, one study did examine binary blends of stevioside with aspartame, acesulfame K and cyclamate, finding sweetness synergy between stevioside

Table 9.6 Potencies and taste profiles of stevioside and rebaudioside A.[36]

Compound	Sweetness potency (x sucrose, at 10% sucrose equivalence)	Taste quality (sweet/bitter/other)[a]
Stevioside	190	62/30/8
Rebaudioside A	170	85/12/3

[a]Percentage of the total taste sensation.

and aspartame (17%), stevioside and acesulfame K (17%) but none between stevioside and cyclamate.[45]

There have been recent media announcements that rebaudioside C is a sweetness enhancer, enabling up to a 25% reduction in the use of carbohydrate sweeteners without impacting on perceived sweetness levels or other sensory properties. These announcements typically suggest that rebaudioside C is not sweet, although this claim is at odds with literature reports that rebaudioside C is approximately 30–40 times the sweetness of sucrose when measured at 2% sucrose equivalence[43] and 50–120 times the sweetness of sucrose, as reported in another publication.[46] Assuming rebaudioside C is, as reported, sweet, the claimed potentiation effects may be nothing more than sweetness additivity or synergy although it does not then follow that the sensory effects are without merit or commercial relevance.

9.2.2.3 Physical and chemical properties

Both stevioside and rebaudioside A appear to be stable sweeteners. One detailed study examining the stability of both sweeteners has been reported[47] in which their stability was monitored in carbonated beverages buffered with either phosphoric acid or citric acid. After 2 months storage at 37°C, some degradation was reported, but no significant changes were seen when formulated beverage products were stored at room temperature and below for 5 months. Some instability (20% loss) of rebaudioside A on exposure to UV light was reported following 1 week of exposure to sunshine, but stevioside was said to be completely stable under the same conditions. The authors conclude that these sweeteners are viable as commercial products in that they exhibit adequate hydrolytic stability.

The solubility of stevioside in water has been measured and found to be just less than 1.0% (w/v).[43] Rebaudioside A, being a more polar molecule than stevioside, is more soluble. Given the potencies of these sweeteners, their individual solubilities are not likely to be a limitation on their use potential. There are currently no published data on the solubility of rebaudioside C although, given its structural similarity to rebaudioside A, solubility may also be similar and therefore there are unlikely to be any technical issues due to limited solubility.

9.2.2.4 Physiological properties

There are data published on the metabolism (*in vitro* and *in vivo*) of stevioside and stevioside analogues such as rebaudioside A, the majority of which have been carried out in rats.

Steviol glycosides are not readily absorbed from the upper small intestine of the rat or human following oral administration. As human digestive enzymes do not hydrolyse β-glycosidic linkages, digestion in the small intestine is limited. Microbial fermentation occurs in the large intestine of both rat and human, releasing the aglycone steviol. Steviol is then absorbed, conjugated with glucuronic acid and excreted as steviol glucuronide, the primary route being in faeces for the rat[48] and urine for humans.[49] The study in rats also supported the conclusion that the metabolism of stevioside and rebaudioside A were essentially identical, thus leading the authors to conclude that toxicological studies carried out using stevioside could be used in the toxicological assessment of rebaudioside A.

9.2.2.5 Applications

Until the recent GRAS notification of rebaudioside A in the United States, the Japanese market for stevioside had been the most developed, with food processors using these sweeteners in a wide variety of applications. These include pickles, seafood, meat, fish, soy, chewing gum, beverages, table-top sweeteners and ice cream.[50] In Korea, stevioside is an accepted sweetener in baked products, table-top sweeteners, beverages and seasonings. Thus, steviol glycoside sweeteners have acceptance across the range of food and beverage applications normally associated with the use of high-potency sweeteners.

This practice has continued with GRAS approval for high content rebaudioside A extracts in the United States leading to product launches across many of the same food and beverage categories, including table-top sweeteners, beverages, desserts, dairy products and confectionery.

9.2.2.6 Safety

Although there is a long history of use for steviol glycosides, particularly in Japan where it is considered a 'food' because of its natural origin, attempts to gain regulatory approvals in Western countries have something of a chequered history. Much of this experience can be attributed not to concerns about the inherent safety of these sweeteners, but to the fact that many early safety studies suffered from the use of poorly defined and variable steviol glycoside content stevia-derived test materials. Crude and poorly characterised steviol glycoside containing products have been used in studies that have shown reproductive toxicity effects and, not surprisingly, the wide publication of these findings has frequently been cited as 'evidence' of safety concerns. This position was not helped by pronouncements from FDA and the Scientific Committee for Food (SCF) in Europe as to the then failure to demonstrate safety that were actually based on the inadequacy of material characterisation, rather than perceived safety issues. It is now accepted that whenever safety studies have been carried out using purified and fully characterised steviol glycosides, the reproductive safety of these sweetener materials has been fully demonstrated.[51,52]

A number of sub-chronic (13-week) toxicity studies have been completed in recent years, the results of which have been used to underpin regulatory approval procedures. These studies have all reported no statistically significant effects in the great majority of cases, although some impact on body-weight gains for groups receiving the highest doses tested were noted.[53] Other workers[54] also reported no compound-related alterations of blood, clinical chemistry or urinalysis parameters.

Although stevioside is not mutagenic,[55,56] its aglycone, steviol (13-hydroxy-ent-kaurenoic acid) has been shown to be mutagenic in some tests with *S. typhimurium* strains.[57] Matsui *et al.*[58] examined the genetic toxicity of stevioside and steviol in a range of mutagenicity tests with metabolic activation. Stevioside was not found to be mutagenic in any of the assays examined. Steviol, however, produced dose-related positive responses in some mutagenicity tests. However, *in vivo* tests found no genotoxicity for steviol[59] and the Joint Expert Committee on Food Additives concluded that 'the genotoxicity of steviol *in vitro* is not expressed *in vivo*'.[60]

Nutrition studies using rats have shown that stevioside itself does not pose a serious health issue, even when used in high concentrations over long periods. Yamada *et al.*[61] administered stevioside and rebaudioside A to rats for 2 years at a rate of 0.3–1% of the diet. Even at the highest dose rate, no differences were observed in biochemical, anatomical, pathological and carcinogenic tests on 41 organs following autopsy. Toskulkao *et al.*[62] found that stevioside at a dose of 15 g/kg body weight was not lethal to mice, rats or hamsters.

9.2.2.7 Regulatory status

JECFA reviewed stevioside in 2004 and granted a temporary 2 mg/kg body-weight ADI for steviol glycosides. JECFA asked that additional information regarding the pharmacological effects of stevioside in humans be provided by 2007. In order for the 'temporary' designation to be removed, further analytical data on steviol glycosides was also required. At that time, the Committee noted several shortcomings in the information available on stevioside. In some studies, the material tested (stevioside or steviol) was poorly specified or of variable quality, and no information was available on other constituents or contaminants. Furthermore, no studies of human metabolism of stevioside and steviol were then available.

Since then, there has been substantial regulatory progress. In July 2008 and on the review of additional safety data on defined and characterised materials, JECFA found steviol glycosides safe for use in food and beverages. Also, the US Food and Drug Administration (FDA) announced in December 2008 that it had no objection to the use of rebiana in food and beverages in the United States. In 2009, the French government was the first in the EU to approve the use of rebausioside A in food and beverages in France, an approval having an initial limit of 2 years. Finally, in 2010, the European Food Safety Authority published a Scientific Opinion confirming that steviol glycosides are safe for use in foods and beverages. Approval throughout the EU is anticipated, although the precise timing is currently uncertain.

Stevia rebaudiana extracts and the purified steviol glycoside sweeteners are approved as sweeteners in Japan, South Korea, China, Brazil and Paraguay.

9.2.3 Lo han guo (mogroside)

9.2.3.1 Introduction

The Chinese plant *Siraitia grosvenorii* is a perennial vine in the Cucurbitaceae (cucumber or melon) family that grows mainly in Guangxi Province, with most of the product from the mountains of Guilin. Siraitia fruits are used both inside and outside the People's Republic of China as a food, beverage, and traditional medicine. The sweet constituents of the plant are triterpene glycosides, known as mogrosides. Common names for the plant include: *lo han guo*, *lo han kuo*, Arhat fruit, Monk Fruit, *Fructus momordicae* and *Momordicae grosvenori* fructus.

The original botanical description of *S. grosvenorii* was published in 1941 by W.T. Swingle from plants collected in southern China. Swingle named the plant *Momordica grosvenorii* in honour of Dr. Gilbert Grosvenor, the president of the National Geographic Society, and sponsor of an expedition to collect Lo han guo in China. The plant was re-named *M. grosverorii* Swingle C. Jeffrey in 1979, but then new evidence indicated the plant should more correctly be named *S. grosvenorii* (Swingle) C Jeffrey.

Early work by Lee[15] established the structure of the sweet component to be a glycoside. Subsequent isolation of two sweet components (named mogrosides IV and V) was completed successfully[63] and their sensory properties described. Mogroside V is the most abundant component, occurring at around 1% in the dried fruits, and its triterpenoid glycoside structure is presented in Figure 9.5. Attached to the aglycone are a disaccharide composed of glucose-glucose linked β-1,6 and a trisaccharide of three glucose units linked β-1,2 and β-1,6. These hydrophilic moieties ensure mogroside V is a polar compound readily soluble in water and the β-linkages ensure that the glycoside is a stable compound.

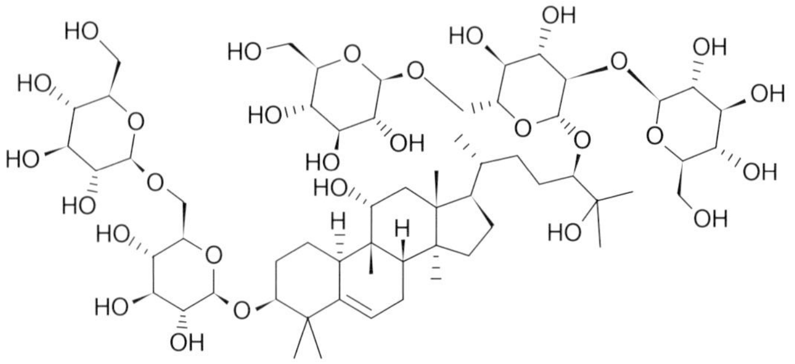

Fig. 9.5 Structure of mogroside. V

A process for making a useful sweetener from *lo han guo* was patented in 1995 by the Procter and Gamble Company (P&G).[64] As described in the patent application, the fruit itself, though sweet, has additional flavours that make it unsuitable for widespread use as a sweetener, so P&G developed a method for processing it to eliminate many of these undesired flavours. In the P&G process, the fresh fruit is picked before ripening and allowed to complete its ripening during storage so that processing begins with the just-ripe fruit. The peel and seeds are then removed, and the mashed fruit becomes the basis of a concentrated fruit juice or puree that can be used in food manufacturing. Further processing involves using solvents to remove volatile and off-flavour components.

The New Zealand-based company BioVittoria has successfully obtained GRAS status for their *lo han guo* extract that is marketed as Pure-Lo, or Monk fruit extract. Their production process involves simple infusion of the fruit in hot water followed by filtration, thus yielding an extract containing approximately 40% mogroside V. The balance of this preparation comprises other fruit constituents, some of which contribute colour and flavour. These flavour components, in particular, seem likely to limit the potential applications for the sweetener and it currently seems unlikely that mogroside V preparations will be suitable for use as sole sweeteners in most potential food and beverage applications. BioVittoria recently entered into a global marketing and distribution agreement with Tate & Lyle, thus providing T&L with a natural sweetener to complement their artificial high-potency sweetener sucralose.

9.2.3.2 Sensory properties

In his original work, Lee described the sweetness of mogroside as being 150 times as potent as sucrose, but no further details were provided. A subsequent report estimated the potency of mogroside V as approximately 250 times as sweet as sucrose at a 5% sucrose equivalent concentration.[65] Currently available commercial preparations are believed to be substantially less sweet, simply because of the mogroside V content, and are around 150 times the sweetness of sucrose at practical use levels.

Neither publication commented in detail on the flavour profile of mogroside, although the sweetener is known to deliver a taste profile that contains taste elements commonly seen in natural potent sweeteners, such as a slight delay to reaching maximum sweetness intensity and an aftertaste that contains liquorice and cooling elements.

9.2.3.3 Physical and chemical properties

Although there are no literature reports detailing the stability of mogroside, empirically it seems likely that it is a stable sweetener. Structurally, it resembles the steviol glycosides that are known to exhibit excellent stability. In addition, the indigenous use of the *lo han guo* fruit involves drying the fruit and then preparing an aqueous decoction that also indicates that the sweet principle is likely to be a relatively stable molecule. Aqueous solutions containing mogroside V are reported to be stable, even under boiling conditions,[66] and, as noted earlier, the β-linkages of the carbohydrate moieties are intrinsically resistant to hydrolysis.

Mogroside is said to be soluble in water and in a 50% aqueous ethanol solution, but no quantitative data are available.

9.2.3.4 Physiological properties

Extracts of *lo han guo* fruit have long been used indigenously to treat colds, sore throats and minor stomach and intestinal complaints.[67] Preparations made from *lo han guo* and even the dried fruit themselves have long been available in Chinese markets and medicinal stores in many countries, both in North America and in Europe. Recent reports suggest that the mogrosides may exhibit anti-cancer properties, possibly based on their reported anti-oxidant characteristics.[68,69]

9.2.3.5 Applications

The traditional use of the *lo han guo* fruit has been to prepare an aqueous extract that is then consumed as a tea or tonic drink. There is little experience of the use of this sweetener in applications other than this traditional use, although there have been some minor beverage products on the market in the United States that have contained 'lo han fruit extract' as a part of the overall sweetening system. In addition, there are now breakfast cereal products.

At this stage, it seems probable that the full technical potential of the sweetener remains to be defined, although it would seem obvious that because of its 'natural' status, applications such as table-top sweeteners will be a target.

9.2.3.6 Safety

Safety studies on mogrosides have not been extensive, although the long history of use in China suggests that the sweetener should be a safe compound. In safety studies that have been completed, it has been shown to be non-mutagenic in short-term predictive tests[70] and to produce no mortalities when administered to mice at doses up to 2 g/kg body weight.

9.2.3.7 Regulatory status

Lo han guo fruits and extracts are considered to be foods in China. A GRAS petition has been reviewed by FDA which issued a 'no objection' letter, thus affirming its GRAS status within the US market. A petition seeking approval in Europe is said to be in preparation.

9.2.4 Brazzein

9.2.4.1 Introduction

Brazzein is a sweet protein found in the fruits of a plant indigenous to tropical West Africa, *Pentadiplandra brazzeana* Baillon. Interestingly, it is not the only sweet protein found in the same plant, as the occurrence of another protein named pentadin was reported earlier.[5] The subsequent identification of a major protein isolated from fresh berries of the plant was reported[71] and its basic physical and sensory properties described.

The protein is located in the seed pulp and is present in ripe fruit at a concentration of approximately 1%, by weight. Brazzein is a small molecular weight protein composed of a single chain of 54 amino acid residues and has a molecular weight of 6473 Daltons.[72] In common with the sweet protein thaumatin, brazzein is stabilised by a number (four) of disulphide bonds intramolecularly.

9.2.4.2 Sensory properties

There are no literature reports that describe the detailed sensory properties of brazzein. Sweetness potencies are listed as ranging from 2000 times the sweetness of sucrose at 2% sucrose equivalence down to 500 times the sweetness of sucrose at 10% sucrose equivalence.[73] Its sweet perception is also described as being 'more similar to sucrose than that of thaumatin with a clean sweet taste with lingering aftertaste and with a slight delay longer than aspartame in an equi-sweet solution'.[74] However, given the extreme difference between the taste quality of sucrose and thaumatin, this qualitative description does not indicate that brazzein delivers a sucrose-like sweet taste profile. In addition to these taste characteristics, this author considers there to be some bitterness in the aftertaste and a cooling mouth sensation, not unlike that induced by neohesperidin dihydrochalcone, on tasting brazzein solutions.

9.2.4.3 Physical and chemical properties

Brazzein is similar to thaumatin in respect of its stability. As noted, it is stabilised by four disulphide bridges and these probably account for its stability to heat and low pH. Sweet taste is reported to remain, even following incubation at 98°C for 2 hours or 80°C for 4.5 hours in the pH range 2.5–8.0.[1] Ming and Hellekant[1] also speculate that the lack of free sulphydryl groups in the brazzein molecule may contribute to its excellent stability.

Brazzein is freely soluble in water (>50 mg/mL)[75] and acidic buffer solutions.

9.2.4.4 Physiological properties

Little has been reported on the metabolism of brazzein although since it is a protein composed of normal amino acids, metabolism through normal protein digestion routes is anticipated. Given the indigenous use of the sweetener in those West African countries in which the *Pentadiplandra brazzeana* plant grows, it is highly unlikely for there to be any health-related issues consequent on consumption of the sweetener.

9.2.4.5 Applications

Limited supply of sweetener material has restricted the independent evaluation of brazzein. Most of what is claimed about the sweetener and its suitability for use in mainstream foods and beverages is descriptive, rather than quantitative. Thus, it is said to taste purely sweet with no secondary tastes, although its slow onset of sweetness is acknowledged. There is also said to be no lingering sweetness character[1] and it is said to confer mouthfeel to low-calorie beverages and to function effectively in both citric acid and phosphoric acid buffer systems. The same authors report that there is both quantitative and qualitative synergy between brazzein and acesulfame K or aspartame, but no data are provided. Assuming the descriptive reports are supported by quantitative data, the rationale for considering a sweetener such as brazzein is clearly its 'natural' designation that would, of course, be negated through blending with artificial sweeteners. Blending with other natural sweeteners might be expected to confer sensory benefits simply because blends of high-potency sweeteners almost always taste better than the sweeteners individually.

Interesting variations on the usual ways of presenting or offering potent sweeteners have been demonstrated for brazzein with its expression in yeast[76] and in grains to be extracted as a sweetened flour, presumably with 'naturally pre-sweetened' breakfast cereals as a likely target market.[77]

9.2.4.6 Safety

No peer-review publications reviewing the safety of brazzein have been published. However, as noted, given that it is a small molecular weight protein comprising normal amino acids issues with its safety are not anticipated.

9.2.4.7 Regulatory

Brazzein currently has no regulatory approvals. There were media reports to the effect that brazzein would be commercial around 2009, but these have failed to materialise. At this stage, it is unclear whether the lack of approvals is due to limitations in a required package of safety data or whether the availability of raw material has been judged insufficient to justify market development.

9.2.5 Monatin

9.2.5.1 Introduction

Monatin is an indole derivative found in the root bark of a shrub native to South Africa called *Schlerochiton ilicifolius*. The structure of monatin was first elucidated by Vleggaar and co-workers,[7] using ^{1}H and ^{13}C NMR spectroscopy, to be 4-hydroxyl-4-(indol-3-ylmethyl) glutamic acid. It was given the common name monatin, derived from the indigenous language name *molomo monate*, or literally, 'mouth-nice'. The structure of monatin is presented in Figure 9.6.

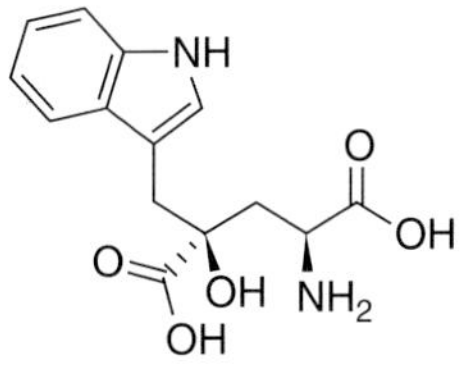

Fig. 9.6 Structure of monadic.

There are two asymmetric carbon atoms within the monatin structure, thus leading to four isomeric (diastereoisometric) forms of the molecule, named *R,R*; *S,S*; *R,S* and *S,R*. All four isomers have been shown to be present in a natural extract[78]), although the authors caution that isomerisation of the chiral centres during extraction and manipulation of samples might have been possible.

While there remains some uncertainty as to the relative sweetness of each individual isomer, it is clear that the *R,R*-isomer is the sweetest of the four monatin isomers. An early literature reference described monatin as being approximately 800 times as sweet as sucrose,[79] but the material assessed was an extract of the root bark that had undergone some purification and whose diastereoisomeric composition was then unknown. Now, it is generally accepted that the pure *R,R*-isomer is approximately 2000 times the sweetness of sucrose.

Extraction of monatin from its natural source for commercial purposes will not be a practical proposition. The concentration of monatin within the plant is extremely low ($<0.01\%$) and the plant itself only grows in rugged terrain in the Northern Transvaal of South Africa. Although the chemical synthesis of monatin has been demonstrated (e.g. see ref.[80]) such chemical processes are only of interest if the sweetener that is prepared using them can compete against the established synthetic sweeteners, purely on the basis of cost and technical performance. Thus, in order to have the possibility, in some markets, of retaining the cachet of being a 'natural' sweetener, monatin must be capable of being prepared at scale using commercially viable biosynthetic processes.

Biosynthesis of monatin has now been described (e.g. see ref.[81, 82]) that involve only the use of enzymes, thus raising the possibility that monatin may be capable of being synthesised at scale using a process that yields the sweetener in a sufficiently economic manner to be commercially attractive.

9.2.5.2 Sensory properties

The detailed sensory properties of monatin have not yet been described. Patent references[80] indicate that the relative sweetness of monatin is at least 1400 times that of sucrose with a highly acceptable sweet taste quality. Other reports indicate the relative sweetness of pure *R,R*-monatin is nearer to 2000 times sucrose.[83] Clearly, the level of interest in this sweetener strongly suggests that its sweet taste will prove to be of high quality.

9.2.5.3 Physical and chemical properties

Pure monatin is known to be relatively poorly soluble in water. It is for this reason that the mono-potassium salt has been the focus of development activity to produce crystalline monatin.[84] The potassium salt produced is said to have excellent water solubility. In addition, free monatin has also been crystallised and the resulting product claimed to exhibit much reduced levels of hygroscopicity.[85]

Details of the stability of monatin under food and beverage preparation and storage conditions have not been reported. There is evidence, however, of sensitivity of monatin (and other indole compounds) to the effects of exposure to UV light. For example, it has been suggested[86] that storing monatin or its salt in the presence of a free radical scavenger and/or in a UV light-shielding container is effective for preventing the decomposition of monatin.

Examples of free-radical scavengers that are claimed to improve the stability to UV light of monatin include classical anti-oxidants such as ascorbic acid, erythorbic acid, vitamin A, vitamin E, gallates and catechin.

9.2.5.4 Applications

Detailed descriptions of the applications potential of monatin are not yet available. However, it seems clear from what has been published that the sweetener delivers a high quality of sweet taste free from bitter or liquorice side/aftertastes that frequently are a feature of natural high-potency sweeteners. For these reasons, monatin seems likely to be considered as suitable for mainstream food and beverage applications, possibly even as a sole sweetener. This view is supported by the statement that of the natural high-potency sweeteners 'only monatin elicits a taste judged likely to have potential for use as a sole, natural source of sweetness in food and drink products'.[79]

9.2.5.5 Safety

Indigenous use of monatin for centuries in South Africa confirms that there is no acute toxic potential associated with monatin consumption. Now, safety tests involving standard toxicology testing protocols are clearly underway. The recent report of a successfully completed 90-day sub-chronic test in rats confirms that development of monatin is in progress.[82]

9.2.5.6 Regulatory status

Monatin currently has no regulatory status. No petitions seeking approval have yet been submitted for regulatory review.

9.3 CONCLUSIONS

Consumer interest in natural high-potency sweeteners has increased dramatically in recent years; interest fuelled largely by regulatory developments surrounding the sweetener rebaudioside-A and, to a much lesser extent, *lo han guo*. These sweeteners, along with thaumatin and brazzein, all present sensory challenges due to taste profiles that differ from that of sucrose or high-fructose syrups. However, in contrast to artificial sweeteners that are all amenable to structural modification in the hope that better tasting analogues may be discovered, natural sweeteners must be used 'as is', simply because any structural change introduced to a natural sweetener so as to improve its taste profile automatically destroys the 'natural' proposition and positioning. Therefore, although consumer interest levels are high, identification of a natural high-potency sweetener of the requisite sensory quality is not a trivial undertaking. It seems likely that monatin, alone of the current batch of natural sweeteners developed or in development, has the potential both to meet consumer demands for high-quality sweet taste and also to compete directly with the artificial sweeteners aspartame and sucralose.

REFERENCES

1. D. Ming and G. Hellekant. Brazzein, a new high potency thermostable sweet protein from *Pentadiplandra brazzeana* B., *FEBS Letters* 1994, 355, 106–108.
2. H. Yamashita, S. Theerasilp and Y. Kurihara. *Xth International Symposium on Olfaction and Taste, Oslo, Norway, 1989*. Abstract, Purification and partial structure characterisation of a new type of sweet protein having taste modifying action, Abstract 77.
3. Z. Hu and M. He. Studies on mabinlin, a sweet protein from the seeds of *Capparis masaikai*, I. extraction, purification and certain characteristics. *Acta Botanica Yunnanica*. 1983, 5, 207–212.
4. J.A. Morris and R.H. Cagan. Purification of monellin, sweet principle in *Discoreophyllum cuminsii*. *Biochimica Et Biophysica Acta* 1972, 261, 114–122.
5. H. van der Wel, G. Larson, A. Hladik, C.M. Hladik, G. Hellekant and D. Glaser. Isolation and characterisation of pentadin, the sweet principle of *Pentadiplandra brazzeana* Baillon. *Chemical Senses* 1989, 14, 75–79.
6. W.F. Daniell. Katemfe, or the miraculous fruit of the Sudan. *Pharmaceutical Journal* 1855, 14, 158.
7. R. Vleggar, L.G.J. Ackerman and P.S. Steyn. Structure elucidation of monatin, a high-intensity sweetener isolated from the plant *Schlerochiton ilicifolius*. *Journal of the Chemical Society, Perkin Transactions 1* 1992, 3095–3098.
8. Y-H. Choi, R.A. Hussain, J.M. Pezzuto, A.D. Kinghorn and J.F. Morton. Abrusosides A-D, four novel sweet tasting triterpene glycosides from the leaves of *Abrus precatorius*. *Journal of Natural Products* 1989, 52, 1118–1127.
9. M. Yoshikawa, T. Morikawa, K. Nakano, Y. Pongpiriyadacha, T. Murakami and H. Matsuda. Albiziasaponins A-E, five new triterpene saponins from the stems of *Albizia myriophylla*. *Journal of Natural Products* 2002, 65, 1638–1642.
10. H. Yamada and N. Mugio. Synthetic approach to intensely sweet glycosides: baiyunoside and osladin. *Studies in Plant Science* 1999, 6, 360–372.
11. P.J. Hyland and J. Kosugi. Bryonoside and bryoside – new triterpene glycosides from *Bryonia dioica*. *Phytochemistry* 1982, 21, 1379–1384.
12. M. Darise, K. Mizutani, R. Kasai, O. Tanaka, S. Kitahata, S. Okada, S. Ogawa, F. Murakami and F.H. Chen. Enzymic transglucosylation of rubusoside and the structure-sweetness relationship of steviolbisglycosides. *Agricultural and Biological Chemistry* 1984, 48, 2483–2488.
13. R.G. Shu, L.R. Xu, L.N. Li and Z.L. Yu. Cyclocariosides II and III: two secodammarane triterpenoid saponins from *Cyclocarya paliurus*. *Planta Medica* 1995, 61, 551–553.
14. G.E. DuBois. Non-nutritive sweeteners. The search for sucrose mimics. In: H-J. Hess (ed). *Annual Reports in Medicinal Chemistry*, Academic Press, New York; 1982, Vol. 17, 323–332.
15. C-H. Lee. Intense sweetener from lo han kuo (*Momordica grosvenorii*). *Experientia* 1975, 31, 533–534.
16. J.R. Sun, K.C. Cheng, T.Y. Pan and X.M. Si. A new acyclic sesquiterpene oligoglycoside from pericarps of *Sapindus mukurossi*. *Chinese Chemical Letters* 2002, 13, 555–556.
17. J. Jizba, L. Dolejs, V. Herout and F. Sorm. The structure of osladin – the sweet principle of the rhizomes of *Polypodium vulgare*. *Tetrahedron Letters* 1971, 18, 1329–1332
18. Y. Hashimoto, Y. Ohta, H. Ishizone, M. Kuriyama and M. Ogura. Periandrin, a novel sweet triterpene glycoside from *Periandra dulcis*. *Phytochemistry* 1982, 21, 2335–2337.
19. M. Katagiri, K. Ohtani, R. Kasai and O. Tanaka. Diterpenoid glycosyl esters from *Phlomis younghusbandii* and *P. Medicinalis*. *Phytochemistry* 1994, 35, 439–442.
20. J. Kim, J.M. Pezzuto, D.D. Soejarto, F.A. Lang and A.D. Kinghorn. Polypodoside A, an intensely sweet constituent of the rhizomes of *Polyupodium glycyrrhiza*. *Journal of Natural Products* 1988, 51, 1166–1172.
21. E.J. Kennelly, L. Cai, L. Long, L. Shannon, K. Zaw, B-N. Zhou, J.M. Pezzuto and A.D. Kinghorn. Novel highly sweet secodammarane glycosides from *Pterocarya paliurus*. *Journal of Agricultural and Food Chemistry* 1995, 43, 2602–2607.
22. J. Zhang, G-X. Chou, L-M. Liu and Z-J. Liu. Determination of rubusoside from *Rubus suavissimus*. *Journal of Chinese Mass Spectrometry Society* 2009, 30, 236–237
23. R. Kasai, N. Kaneda, O. Tanaka, K. Yamasaki, I. Sakamoto, K. Morimoto and S. Okada. Sweet diterpene glycosides of leaves of *Stevia rebaudiana* – synthesis and structure sweetness relationship. *Journal of The Chemical Society of Japan* 1981, 726–735.

24. V.D. Huan, K. Ohtari, R. Kasai, K. Yamasaki and N.V. Tuu. Sweet pregnane glycosides from *Telosma procumbens*. *Chemical and Pharmaceutical Bulletin* 2001, 49, 453–460.
25. N.I. Baek, M.S. Chung, L. Shamon, L.B. Kardone, S. Tsauri, K. Padmawinata, J.M. Pezzuto, D.D. Soejarto and A.D. Kinghorn. Selligueain A, a novel highly sweet proanthocyanidin from the rhizomes of *Selliguea feei*. *Journal of Natural Products* 1993, 56, 1532–1538.
26. C.M. Compadre, J.M. Pezzuto, A.D. Kinghorn and S.K. Kamath. Hernandulcin: an intensely sweet compound discovered by review of ancient literature. *Science* 1985, 227, 417–419.
27. R.L. Nie, T. Takashi, J. Zhou and O. Tanaka. Phlorizin and trilobatin, sweet dihydrochalcone glucosides from leaves of *Lithocarpus litseifolius*. *Agricultural and Biological Chemistry* 1982, 46, 1933–1934.
28. Y. Asahina and E. Ueno. Phylodulcin, a chemical constituent of Amacha. *Journal of the Pharmaceutical Society of Japan* 1916, 408
29. L. O'B. Nabors and G.E. Inglett. A review of various other alternative sweeteners. In: L.O'B. Nabors and R.C. Gelardi (eds). *Alternative Sweeteners*, Marcel Dekker, New York; 1986, pp. 309–323.
30. J.D. Higginbotham. Protein sweeteners. In: C.A.M. Hough, K.J. Parker, A.J. Vlitos (eds). *Developments in Sweeteners 1*, Applied Science, London; 1979, pp. 87–114.
31. B. Iyengar, P. Smits, F. van der Ouderaa, H. van der Wel and J. van Browersharen. Structure of the thaumatins. *European Journal of Biochemistry* 1979, 96, 193–204.
32. H. van der Wel, A. van der Heijden and H.G. Peer. Sweeteners. *Food Reviews International*, 1987, 3, 193–268.
33. J.D. Higginbotham. Talin protein (thaumatin). In: L.O'B. Nabors and R.C. Gelardi (eds). *Alternative Sweeteners*, Marcel Dekker, New York; 1986, pp. 103–134.
34. C.A.M. Hough and J.A. Edwardson. Antibodies to thaumatin as a model of the sweet taste receptor. *Nature*, 1978, 271, 381–383.
35. G.E. DuBois, D.E. Walters, S.S. Schiffman, Z.S. Warwick, B.J. Booth, S.D. Pecore, K. Gibes, B.T. Carr and L.M. Brands. Concentration-response relationships of sweeteners. In: D.E. Walters, F.T. Orthoefer and G.E. DuBois (eds). *Sweeteners, Discovery, Molecular Design and Chemoreception*, ACS Symposium Series 450, ACS, Washington, DC; 1991, pp. 261–276.
36. J.D. Higginbotham, M.G. Lindley and J.P. Stephens. Thaumatin. In: G. Charalambous and G.E. Inglett (eds). *The Quality of Foods and Beverages*, Academic Press, New York; 1981, pp. 91–113.
37. J.D. Higginbotham. Recent developments in non-nutritive sweeteners. In: T.H. Grenby, K.J. Parker and M.G. Lindley (eds.). *Developments in Sweeteners II*, Applied Science, London; 1983, pp. 119–155.
38. H. van der Wel and W.J. Bel. Enzymatic properties of the sweet-tasting proteins thaumatin and monellin after partial reduction. *European Journal of Biochemistry* 1980, 104, 413–418.
39. J.D. Higginbotham, D.J. Snodin, K.K. Eaton and J.W. Daniel. Safety evaluation of thaumatin. *Food and Chemical Toxicology* 1983, 21, 815–823.
40. C. Green. Thaumatin: A natural flavour ingredient, In: A. Corti (ed). *Low Calorie Sweeteners: Present and Future*, Karger, Basel; 1991, pp. 129–132.
41. A.D. Kinghorn and D.D. Soejarto. Stevioside. In: L.O'B. Nabors and R.C. Gelardi (eds). *Alternative Sweeteners*, Marcel Dekker, New York; Basel, Hong Kong; 1991, pp. 157–171.
42. A.D. Kinghorn and D.D. Soejarto. Current status of stevioside as a sweetening agent for human use. In: H. Wagner, H. Hikino and N.R. Farnsworth (eds). *Economic and Medicinal Plant Research*, Academic Press, London; 1985, Vol. 1, pp. 1–52.
43. S-H. Kim and G.E. DuBois. Natural high potency sweeteners. In: S. Marie and J.R. Piggott (eds). *Handbook of Sweeteners*, Blackie, Glasgow and London; 1991, pp. 116–185.
44. G.E. DuBois, P.S. Dietrich, J.F. Lee, G.V. McGarraugh and R.A. Stephenson. Diterpenoid sweeteners, synthesis and sensory evaluation of stevioside analogues non-degradable to steviol. *Journal of Medicinal Chemistry* 1981, 24, 408–428.
45. R.A. Frank, S.J.S. Mize and R. Carter. An assessment of binary mixture interactions for nine sweeteners. *Chemical Senses* 1989, 14, 621–632.
46. A.I. Bakal and L.O'B. Nabors. Stevioside. In: L.O'B. Nabors and R.C. Gelardi (eds). *Alternative Sweeteners*, Marcel Dekker, New York; 1986, pp. 295–307.
47. S.S. Chang and J.M. Cook. Stability studies of stevioside and rebaudioside A in carbonated beverages. *Journal of Agricultural and Food Chemistry* 1983, 31, 409–412.
48. A. Roberts and A.G. Renwick. Comparative toxicokinetics and metabolism of rebaudioside A, stevioside and steviol in rats. *Food and Chemical Toxicology* 2008, 46(Suppl. 7), S31–S39.

49. A. Wheeler, A. Boileau, P. Winkler, J. Compton, I. Prakash, X. Jiang and D. Mandarino. Pharmacokinetics of rebaudioside A and stevioside after single oral doses in healthy men. *Food and Chemical Toxicology* 2008, 46(7S), S54–S60.
50. S. Marie. Sweeteners. In: J. Smith (ed). *Food Additive Users Handbook*, Blackie, Glasgow/AVI, New York; 1991, pp. 47–74.
51. N. Mori, M. Sakanoue, M. Takeuchi, K. Simpo and T. Tanabe. Effect of stevioside on fertility in rats. *Shokuhin Eiseigaku Zasshi* 1981, 22, 409–414.
52. V. Yodyingyuad and S. Bunyawong. Effect of stevioside on growth and reproduction. *Human Reproduction* 1991, 6, 158–165.
53. L.L. Curry and A. Roberts. Acute and subchronic toxicity of rebaudioside A. *Food and Chemical Toxicology* 2008, 46(Suppl. 7), S11–S20.
54. A.I. Nikiforov and A.K. Eapen. A 90-day oral (dietary) toxicity study of rebaudioside A in Sprague-Dawley rats. Internat. *Journal of Toxicology* 2008, 27, 65–80.
55. J.M. Pezzuto, N.P. Nanayakkara, C.M. Compadre, S.M. Swanson, A.D. Kinghorn, T.M. Guenthner, V.L. Sparnins and L.K. Lam. Characterisation of bacterial mutagenicity mediated by 13-hydroxy-ent-kaurenoic acid (steviol) and several structurally related derivates and evaluation of potential to induce gluthathione S-transferase in mice. *Mutation Research* 1986, 169, 93–103.
56. M. Suttajit, U. Vinitketkaumnuen, U. Meevate and D. Buddhasukh. Mutagenicity and human chromosomal effect of stevioside, a sweetener from *Stevia rebaudiana* Bertoni. *Environmental Health Perspectives* 1993, 101(Suppl. 3), 53–56.
57. M. Matsui, K. Matsui, T. Nohmi, H. Mizusawa and M. Ishidate. Mutagenicy of steviol: an analytical approach using the Southern blotting system. *Eisei Shikenjo Hokoku* 1989, 300, 83–87.
58. M. Matsui, K. Matsui, Y. Kawasaki, Y. Oda, T. Nogushi, Y. Kitagawa, M. Sawada, M. Hayashi, T. Nohmi, K. Yoshihira, M. Jr. Ishidate and T. Sofuni. Evaluation of the genotoxicity of stevioside and steviol using six in vitro and one in vivo mutagenicity assays. *Mutagenesis* 1996, 11, 573–579.
59. Ha-Y. Oh, E-S. Han, D-W. Choi, J-W. Kim, M-O. Eom, I-W. Kang, M-J. Kang and K-W. Ha. *In vitro* and *in vivo* evaluation of genotoxicity of stevioside and steviol natural sweetener. *Yakhakhoe Chi* 1999, 43, 614–622.
60. JECFA. 2006. Safety evaluation of certain food additives. Sixty-third meeting of the Joint FAO/WHO Expert Committee on Food Additives, 8–17 June, 2004. Geneva, Switzerland. World Health Organisation (WHO), International Programme on Chemical Safety (IPCS), Geneva, Switzerland, WHO Food Additives Series, No. 54
61. A. Yamada, S. Ohgaki, T. Noda and M. Shimizu. Chronic toxicity study of dietary Stevia extracts in F344 rats. *Journal of the Food Hygienic Society of Japan* 1986, 27, 1–8.
62. C. Toskulkao, L. Chaturat, P. Temcharoen and T. Glinsukon. Acute toxicity of stevioside, a natural sweetener, and its metabolite, in several animal species. *Drug and Chemical Toxicology* 1997, 20(1–2), 31–44.
63. T. Takemoto, S. Arihara, T. Nakajima and M. Okuhira. Studies on the constituents of *Fructus momordicae*, III. Structure of mogrosides. *Yakugaku Zasshi* 1988, 103, 1155–1166.
64. G.E. Dawson. *Process and composition for sweet juice from Cucurbitaceae fruit*, US patent 5,411,755; 1995.
65. A.D. Kinghorn and D.D. Soejarto. Sweetening agents of plant origin. *CRC Critical Reviews in Plant Sciences* 1986, 4, 79–120.
66. L.O'B. Nabors and G.E. Inglett. A review of various other alternative sweeteners. In: L.O'B. Nabors and R.C. Gelardi (eds). *Alternative Sweeteners*, Marcel Dekker, New York; 1986, 309–323.
67. A.D. Kinghorn and D.D. Soejarto. Discovery of terpenoid and phenolic sweeteners from plants. *Pure and Applied Chemistry* 2002, 74, 1169–1179.
68. H. Shi. Antioxidant property of *Fructus momordicae* extract. *Biochemistry and Molecular Biology International* 1996, 40, 1111–1121.
69. T. Konoshima and M. Takasaki. Cancer-chemopreventive effects of natural sweeteners and related compounds. *Pure and Applied Chemistry* 2002, 74, 1309–1316.
70. A.D. Kinghorn and D.D. Soejarto. Intensely sweet compounds of natural origin. *Medicinal Research Reviews* 1989, 9, 91–115.
71. D. Ming and G. Hellekant. Brazzein, a new high potency thermostable sweet protein from *Pentadiplandra brazzeana* B. *FEBS Letters* 1994, 355, 106–108.
72. G. Hellekant and V. Danilova. Brazzein, a small sweet protein: discovery and physiological overview. *Chemical Senses* 2005, 30(Suppl 1), i88–i89.

73. H. Izawa, M. Ota, M. Kohmura and Y. Ariyoshi. Synthesis and characterization of the sweet protein brazzein. *Biopolymers* 1996, 39, 95–101.
74. J.F. Pfeiffer, R.B. Boulton and A.C. Noble. Modelling the sweetness response using time-intensity data. *Food Quality and Preference* 2000, 11, 129–138.
75. G.G. Birch. *Ingredients Handbook – Sweeteners (Ingredients Handbook Series)*. Leatherhead Food Research Association, UK; 2000.
76. Z. Guan, G. Hellekant and W. Yan. Expression of sweet protein brazzein by *Saccharomyces cerevisiae*. *Chemical Senses* 1995, 20, 701.
77. I. Faus. Recent developments in the characterization and biotechnological production of sweet-tasting proteins. *Applied Microbiology and Biotechnology* 2000, 53, 145–151.
78. A. Bassoli, G. Borgonovo, G. Busnelli, G. Morini and M.G.B. Drew. Monatin and its stereoisomers: chemoenzymatic synthesis and taste properties. *European Journal of Organic Chemistry* 2005, 1652–1658.
79. P.J. Van Wyk and L.G. Ackerman. 3-(1-amino-1,3-dicarboxy-3-hydroxy-but-4-yl)-indole compounds. US Patent 4,975,298; 1990.
80. E. Abushanab and S. Arumugam. Synthesis of monatin – A high intensity natural sweetener. US Patent 5,994,559; 1999.
81. P.M. Hicks, S.C. McFarlan, T.W. Abraham, D.C. Cameron, J.R. Millis, J. Rosazza, L. Zhao and D.P. Weiner. Polypeptides and biosynthetic pathways for the production of monatin and its precursors. US Patent 7,572,607; 2009.
82. A.B. Khare, B.H. Hilbert, C. Solheid, S.C. McFarlan, F.A. Sanchez-Riera, K. Floy and P.M. Hicks. Methods and systems for increasing production of equilibrium reactions. US Patent 7,888,081; 2011.
83. J. Hlywka, W.A. Brathwaite, A.I. Nikiforov, M.O. Rihner and A. Eapen. A 90-day dietary toxicity study of R,R-monatin salt, a natural high potency sweetener, in rats. Poster presentation Society of Toxicology Meeting, Washington, DC, March 2011.
84. K. Mori, E. Ono and T. Takemoto. Crystal of (2R,4R)-monatin potassium salt and sweetener composition containing same. US Patent 7,553,974; 2005.
85. Y. Amino, K. Hirasawa, K. Mori and T. Takemoto. Crystals of free (2R,4R)-monatin and use thereof. US Patent 7,935,377; 2011.
86. K. Mori. Sweetener composition. US Patent 7,781,005; 2010.

Summary Table for Part Two A brief summary of the characteristics of some high-potency (high-intensity) sweeteners.

The information given in this table is taken from published sources and/or from chapters in this book. The brief summaries are an attempt to provide the reader with a guide to important sweetener characteristics. It is only by using sweeteners in real applications will the reader appreciate the full potential of individual or blended sweeteners

	Acesulfame K	Aspartame	Aspartame–acesulfame salt	Cyclamate (sodium)	Saccharin (sodium)	Sucralose	Neotame
Year of discovery	1967	1965	1995	1937	1878	1976	1980s
Form	White odourless crystals	White odourless crystals	White odourless crystals	White practically odourless crystals	White odourless crystals	White practically odourless crystals	White odourless crystals
Relative sweetness[a] Sucrose = 1	130–200	180–200	350–400	30–50	300–500	400–800	7000–13000
Blends with other sweeteners	Qualitative and quantitative synergy with other sweeteners such as sodium cyclamate and sucralose. Used extensively with aspartame	Quantitative synergy with acesulfame K and/or saccharin, sodium cyclamate, stevia, glucose, fructose, sucrose and polyols	Already a blend can be used with other sweeteners	Quantitative synergy with acesulfame K aspartame, neohesperidine DC, saccharin and sucralose	Synergistic when used with other sweeteners such as aspartame, sodium cyclamate, sorbitol and mannitol	Synergy when used with other sweeteners such as acesulfame K, sodium cyclamate, saccharin, fructose	Limited synergy – some with saccharin. Some flavour-enhancing/modification properties – enhancing fruit, mint, cinnamic aldehyde and vanilla, modifying soy, vitamin and mineral aftertastes

Aqueous solubility	270 g/L @ 20°C	1% w/w @ 25°C	2.75% w/w @ 21°C	200 g/L @ 20°C	1200 g/L @ 20°C	28.2 g/100 mL @ 20°C	1.3% w/w @ 25°C
Melting (decomposition) range	>200°C (under decomposition)	Decomposes before melting	Decomposes before melting	169–170°C	>300°C	125°C (when heated from 115°C at 5°/min)	80.9–83.4°C
Storage and processing stability	Very good (under normal storage and processing conditions)	Very good stability in dry conditions (<8% moisture). Less stable in liquids, this is a function of pH, temperature and time	Good, as aspartame in solution	Good. No significant losses expected in common applications	Excellent. Stable to all the conditions to which it may be exposed in food applications	Very good. More stable than sucrose manifesting itself in its non-reactivity with other food components and in its resistance to hydrolysis under extremes of acid and heat	Excellent stability in dry conditions. Less stable in liquids, this is a function of pH, temperature and time

(Continued)

Summary Table for Part Two (Continued)

	Acesulfame K	Aspartame	Aspartame–acesulfame salt	Cyclamate (sodium)	Saccharin (sodium)	Sucralose	Neotame
	Possible product labelling associated with the use of high-intensity sweeteners include the following (not exhaustive): Diet, Light, Low carb, Reduced calorie, Low calorie, No-added sugar, Sugar free, Tooth-friendly, Reduced glycaemic response (diabetic suitability). It is advised that local expert opinion is obtained regarding the pertinent food regulations and labelling implications if these 'claims' are to be used on food products.						
Approval	Approved for use in over 100 countries	Approved for use in over 100 countries	Approved in countries where both aspartame and acesulfame are permitted to be used jointly	Approved for use in over 50 countries	Approved for use in over 90 countries	Approved for use in over 80 countries	Approved for use in over 25 countries
Regulatory status	EU[b] and US[c]	EU[b] and US[c]	EU: permitted as per amendment 2003/115/EC to the Sweetener Directive. US: GRAS status	EU[b] and under review for re-approval in the USA	EU[b]	EU[b], US[c], Australia, Japan and Russia	US[c] FDA approved use of neotame as a general purpose sweetener in July 2002. Approved in Australia and New Zealand. Not approved in EU

Labelling	Standard requirements	Standard requirements plus requirement to carry statements for PKU sufferers. EU: Contains source of phenylalanine. US: Phenylketonurics contains a source of phenylalanine.	As per aspartame	Standard requirements	Standard requirements	Standard requirements	Standard requirements

GRAS, Generally Recognized As Safe; PKU, phenylketonuria; ADI, acceptable daily intake; JECFA, Joint Expert Commission for Food Additives; FDA, Food and Drug Administration; GMP, Good Manufacturing Practice; DC, dihydrochalcone.

[a] Sweetness intensity depends on concentration, pH, temperature and the presence of other ingredients and is therefore application specific.

[b] EU – permitted for use in the EU by Directive 94/35/EC – known as the Sweetener Directive.

[c] US – approved by the FDA as a 'General Purpose Sweetener' allowing it to be used in most food and beverage applications. Also has GMP status, which means that there are no upper limits for its use in individual categories.

(*Continued*)

Summary Table for Part Two (Continued)

	Acesulfame K	Aspartame	Aspartame–acesulfame salt	Cyclamate (sodium)	Saccharin (sodium)	Sucralose	Neotame
E-number	E950	E951	E962	E952	E954	E955	
ADI	15 mg/kg of body weight (JECFA)	40 mg/kg of body weight (JECFA)	As per constituent parts	11 mg/kg of body weight (JECFA)	5 mg/kg of body weight	15 mg/kg of body weight (JECFA)	2 mg/kg of body weight (JECFA)

References

Soft Drinks World Supplement (Autumn 2004): Sweeteners.
T.H. Grenby. *Advances in Sweeteners*. Blackie Academic and Professional, Glasgow; 1996.
L. O'Brien Nabors. *Alternative Sweeteners*, third edition revised and expanded. Marcel Dekker, New York; 2001.
S. Marie and J.R. Piggott. *Handbook of Sweeteners*. Blackie and Son Ltd, Glasgow; 1991.

Part Three
Reduced-Calorie Bulk Sweeteners

10 Erythritol

Peter de Cock
Cargill, Vilvoorde, Belgium

10.1 INTRODUCTION

10.1.1 History

Erythritol is the sole non-caloric bulk sweetener and since it occurs naturally in many fruits and vegetables it has been consumed at low levels for as long as mankind has been eating such products. Erythritol was first isolated from the algae *Protococcus vulgaris* (now named Apatococcus lobatus) in 1852 by Lamy who named the substance phycit.[1] Later, erythritol was also isolated from the algae *Trentepohlia jolithus*.[2] In the early 1980s, when natural foods started to grow in popularity, Cerestar (now Cargill) initiated a research project to produce different types of polyols by fermentation since it was known that polyols like sorbitol, mannitol, xylitol, erythritol or glycerol could be produced via microbiological pathways.

During this research, a strain of yeast, which produced significant amounts of erythritol, was identified. Other research demonstrated that erythritol was essentially calorie-free and had a very high digestive tolerance, and in fact this was much higher compared with all other polyols. These unique properties led to the marketing of this new natural sweetener in Japan when it was approved there in the early 1990s.

Cerestar developed the fermentation process to use natural raw materials, improved yields, fermentation efficiencies and the purification process leading to high-purity erythritol.[3] Commercial production of erythritol began in 1993 for the rapidly developing Japanese market. Simultaneously, a programme was started to gain food approval in North America, Australia, Europe and other jurisdictions around the world and today, erythritol is approved for use in foods in more than 55 countries. This list and the range of applications of erythritol are still growing.

10.1.2 General characteristics

Erythritol is a white, anhydrous, non-hygroscopic, crystalline substance available in powdered or granular form with a mild sweetness and similar appearance to sucrose. It is a 4-carbon sugar alcohol or polyol and its small molecular size is responsible for many of erythritol's unique characteristics (Figure 10.1).

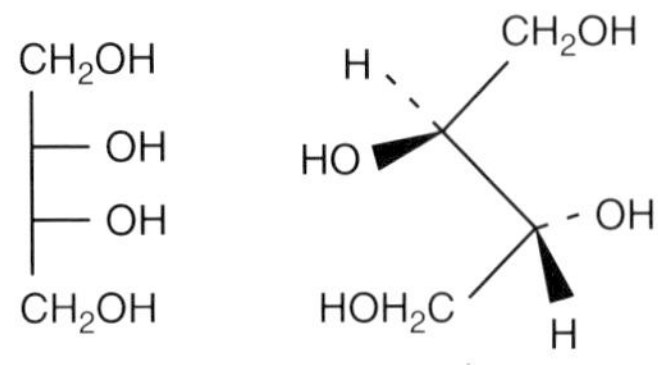

Chemical formula: $C_4H_{10}O_4$
Chemical names: 1,2,3,4,-butanetetrol, meso-erythritol
Molecular weight: 122.12

Fig. 10.1 Molecular structure and chemical formula of erythritol.

10.1.2.1 Non-caloric

Unlike other polyols, erythritol is non-caloric. This is a particularly useful benefit for a bulk sugar replacer. Owing to its small molecular size, 90% of ingested erythritol is absorbed in the small intestine but while it is well absorbed, it is not metabolised in the body. The kidneys remove erythritol from the bloodstream and it is excreted unchanged in the urine.[4] The small amount of erythritol not absorbed in the small intestine passes to the large intestine where it is excreted unchanged in the faeces. It is not fermented like other polyols[5] so there is no caloric contribution from the absorption of fermentation by-products such as volatile fatty acids (VFA). Consequently, erythritol contributes no energy to the body (0 cal/g).

10.1.2.2 No glycaemic or insulinaemic response

Since erythritol is not metabolised, it does not have any glycaemic or insulinaemic effect. This makes it a particularly useful sweetener for people wishing to reduce their post-prandial blood sugar levels.

10.1.2.3 Natural

Erythritol occurs naturally in a wide variety of fruits, vegetables and fermented foods. It is also present in the human body and in animals. The concentrations detected in various common foods are shown in Table 10.1. In the United States and Europe, no regulatory

Table 10.1 The natural occurrence of erythritol in various foods.

Foods	Erythritol content
Wine	130–300 mg/L
Sherry wine	70 mg/L
Sake	1550 mg/L
Soy sauce	910 mg/L
Miso bean paste	1310 mg/kg
Melons	22–47 mg/kg
Pears	0–40 mg/kg
Grapes	0–42 mg/kg

definition exists for use of the term 'natural' on food labels. If, for an ingredient, 'artificial' means a substance that does not occur in nature and 'synthetic' means a substance produced by chemical reaction, an argument can made that erythritol may be considered 'natural' in the US as the FDA has defined the terms artificial and synthetic. Erythritol is not considered artificial since both the finished product and the micro-organisms used to produce it are found in nature, and erythritol is not synthetic since it is produced by fermentation rather than chemical reaction. In Europe, fermentation is regulatory defined as a natural process.

10.1.2.4 High-digestive tolerance

Like beans and certain high-fiber foods, polyols can cause some undesirable gastrointestinal side effects if consumed in excessive amounts. Typical symptoms of overconsumption are flatulence and laxation. These symptoms are due to the poor absorption of low-digestible carbohydrates and consequent fermentation in the large intestine. As erythritol is well absorbed and not fermented, it can be consumed at relatively high levels without side effects. Clinical studies have shown erythritol to be the best tolerated polyol with no undesirable side effects at consumption levels two to four times higher than other polyols. Erythritol is better tolerated than many fibres such as inulin and better tolerated than sugars such as lactose and tagatose.

10.1.2.5 Non-cariogenic

All polyols are tooth-friendly. They do not promote tooth decay, as they cannot or can only be used poorly as a substrate by the oral bacteria, such as *Streptococcus mutans*, that cause dental caries. Like xylitol, erythritol can reduce dental plaque, thereby reducing the risk of developing dental caries.[6]

10.1.2.6 Anti-oxidant properties

Like other polyols, erythritol is a free radical scavenger. However, given the fact that erythritol is well absorbed and not metabolised, it has the ability to potentially exercise its anti-oxidant activity while circulating throughout the body before it is excreted in the urine.

10.1.3 Manufacturing process

Erythritol was the first polyol to be manufactured commercially by a natural fermentation process. The starting material is a dextrose- or sucrose-rich solution that is fermented by an osmophilic yeast such as *Moniliella pollinis* to yield a mixture of polyols containing mainly erythritol, with trace amounts of glycerol and ribitol. The yeast cells, other polyols and trace impurities are removed through a series of filtration and other separation steps. The erythritol is then crystallised from the concentrated liquor and dried to yield crystals with over 99% purity.

There are chemical routes to the synthesis of erythritol but they are complex and costly. Fermentation is a simpler process and is less expensive since the initial substrate is low in cost and readily available.

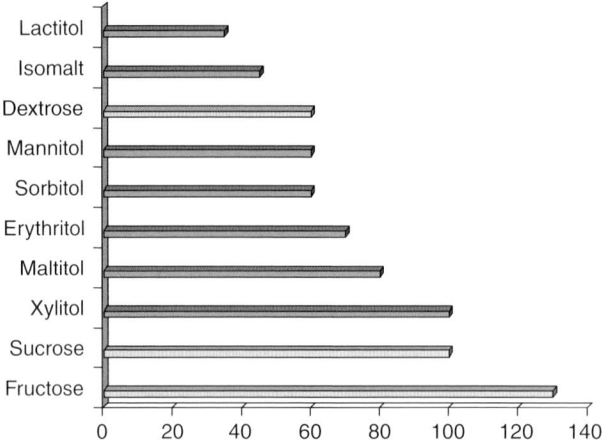

Fig. 10.2 Relative sweetness of polyols (10% aqueous solution).

10.2 ORGANOLEPTIC PROPERTIES

10.2.1 Sweetness intensity

Erythritol is a bulk sweetener, not an intense sweetener. As a bulk sweetener, erythritol provides volume, texture and microbiological stability similar to sucrose. Erythritol is 60–70% as sweet as sucrose, depending on the food or beverage formulation (Figure 10.2).

10.2.2 Sweetness profile

Erythritol has a temporal profile similar to sucrose. Quantitative descriptive analysis shows that erythritol solutions taste similar to sucrose (Figure 10.3).

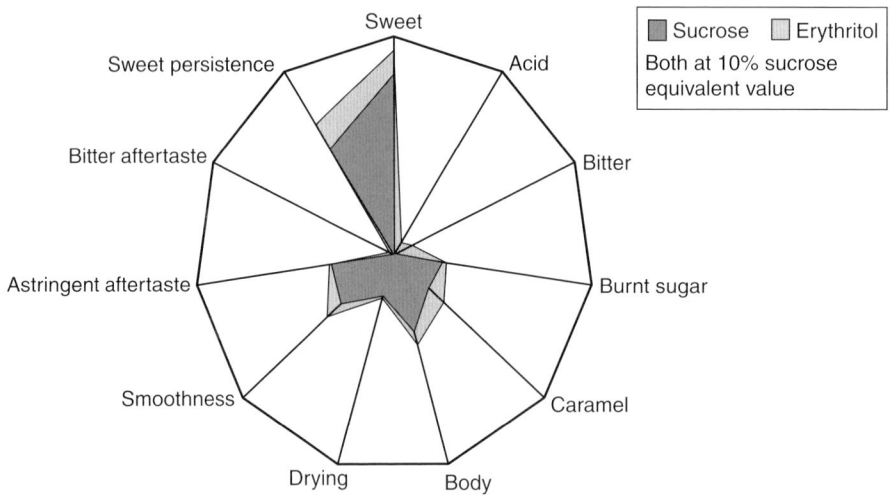

Fig. 10.3 Sweetness profile of erythritol and sucrose (10% sucrose solution and ISO-sweet erythritol solution).

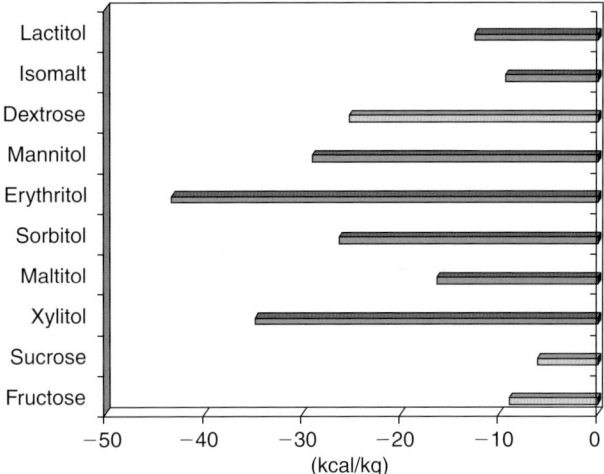

Fig. 10.4 Heat of solution of erythritol and other bulk sweeteners.

10.2.3 Cooling effect

When dissolved, erythritol exhibits a strong cooling effect owing to its high negative heat of solution. This characteristic is beneficial in some applications such as mints and candies, but it can be a challenge for product developers in other foods. Combining substances with a positive heat of solution with erythritol is an ideal means to offset this cooling effect in applications where it is not desirable. Combinations with soluble fibres such as inulin are often used (Figure 10.4). Patent pending technology has been developed to reduce the sensory cooling effect of polyols.[7]

10.2.4 Synergy with other sweeteners

Erythritol exhibits quantitative synergy with intense sweeteners as noted in Section 10.5.1. However, owing to the relatively high cost of sweetness of erythritol compared with intense sweeteners, erythritol would not be chosen for its quantitative sweetness synergy alone. More importantly, erythritol can improve mouthfeel and mask certain unwanted aftertastes such as astringency and the irritant effect of intense sweeteners (see Section 10.5.1). Qualitative sweetness improvements have been demonstrated with synthetic intense sweeteners such as saccharin, aspartame, acesulfame potassium and sucralose, and with natural intense sweeteners such as stevia, thaumatin and lo han guo.

10.3 PHYSICAL AND CHEMICAL PROPERTIES

10.3.1 Stability

Erythritol is very stable and does not decompose in either acid or alkaline environments. It does not contain a reducing end group so does not take part in Maillard-type browning reactions. As a reduced monosaccharide, it is also not subject to acid hydrolysis cleavage that would yield a reducing sugar. It shows excellent heat stability, even above 180°C, and is stable in heated concentrated solutions and when stored in a dry state. Heated, concentrated solutions of erythritol can be held for long periods of time without chemical breakdown or browning. The only means for decomposition of erythritol is microbial breakdown if the water activity increases to a suitable level.

10.3.2 Solubility

Erythritol is not as soluble as sucrose at room temperature, but approaches sucrose solubility at elevated temperatures. It is more soluble than mannitol or isomalt but less soluble than other polyols. It is often used in combination with maltitol in formulations where the erythritol solubility is a limiting factor, such as in a sugar-free pancake syrup (Figure 10.5a and b).

10.3.3 Melting point and other thermal characteristics

Erythritol melts at 121°C; see Figure 10.6 and Table 10.2 for comparisons with other polyols together with other thermal characteristics.

10.3.4 Viscosity

Owing to its small molecular size, erythritol develops a lower viscosity than sucrose or other polyols when equivalent concentrations are used (Table 10.3).

10.3.5 Hygroscopicity

Erythritol has a lower hygroscopicity than sucrose and other polyols. Erythritol does not begin to absorb water until the relative humidity reaches levels in excess of 90% (see Figure 10.7).

10.3.6 Boiling point elevation and freezing point depression

With erythritol's low molecular weight compared with sucrose and other polyols, it delivers greater boiling point elevation and freezing point depression per unit weight. Account must be taken of erythritol's lower solubility when used in frozen foods.

10.3.7 Water activity at various concentrations versus sucrose

As with freezing point depression and boiling point elevation, erythritol exhibits a greater reduction in water activity in solutions of equivalent concentration compared with sucrose and other polyols (Figure 10.8).

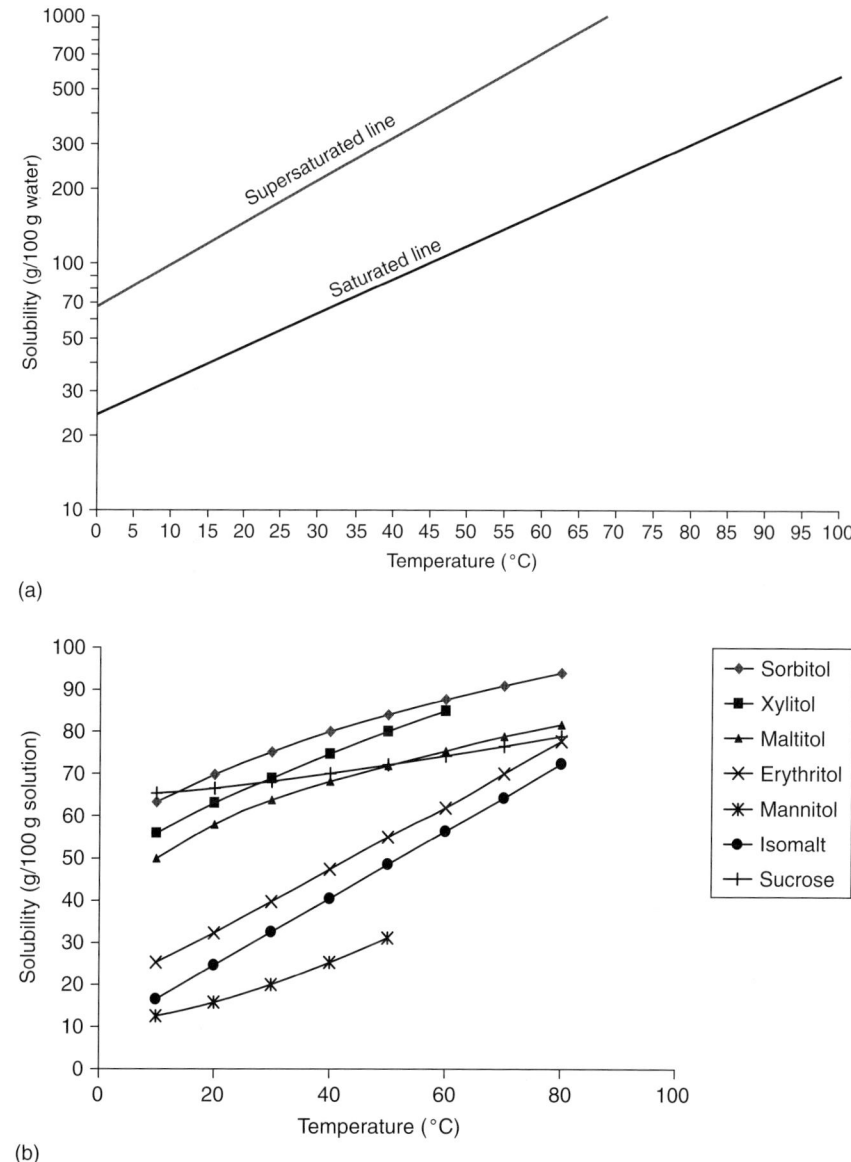

Fig. 10.5 (a) The solubility of erythritol. (b) Solubility of erythritol and other bulk sweeteners.

10.4 PHYSIOLOGICAL PROPERTIES AND HEALTH BENEFITS

10.4.1 Digestion of carbohydrates

The fate of dietary carbohydrates in the digestive system depends on their molecular size and chemical nature. Simple monosaccharides are directly absorbed through the mucosal cell layer of the intestine and their rate of absorption depends on active and passive absorption

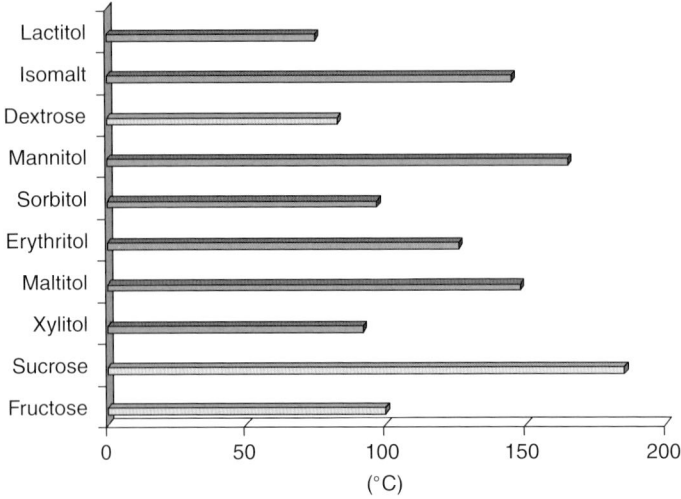

Fig. 10.6 Melting points of erythritol and other bulk sweeteners.

mechanisms. Glucose, for example, is actively – and therefore rapidly – absorbed compared with fructose that is absorbed through a carrier-mediated mechanism. Disaccharides first need to be hydrolysed by the intestinal enzymatic systems before they can be absorbed. Polysaccharides such as starch or glycogen owing to their complex, often branched, structures are only absorbed to the extent by which the digestive enzymatic system is capable of liberating the monosaccharide building blocks. Absorbed monosaccharides are oxidised to CO_2 to provide energy or are metabolised to other substances such as glycogen or fats for energy storage. The remaining unabsorbed polysaccharide structures or oligosaccharide residues move on to the colon where they are subjected to microbial fermentation. The resulting production and subsequent absorption of VFA may contribute additional energy.

10.4.2 Metabolic fate of erythritol

Erythritol is absorbed from the proximal intestine by passive diffusion in a manner similar to that of many low-molecular-weight organic molecules that do not have associated active transport systems. The rate of absorption of these types of molecules is related to their molecular size. Consequently, erythritol, a 4-carbon molecule, passes through the intestinal membranes

Table 10.2 Thermal characteristics of polyols (DSC).

Polyol	Tg (°C)	ΔCp (J/g°C)	Tm (°C)	ΔH (J/g)	Cryst.
Erythritol	−42	1.07	121	348	Yes
Mannitol	−39	1.75	167	291	Yes
Xylitol	−22	1.15	94	254	No
Sorbitol	−5	1.25	99	157	No
Isomalt	34	0.95	98/137	138	No
Maltitol	47	0.67	152	172	No

Tg, glass transition temperature (baseline shift); ΔCp, heat capacity at constant pressure; Tm, melting temperature (endothermic signal); ΔH, heat enthalpy; Cryst., static crystallisation (exothermic signal).

Table 10.3 Physical and chemical properties of erythritol and other bulk sweeteners.

	Erythritol	Xylitol	Mannitol	Sorbitol	Maltitol	Isomalt	Lactitol	Sucrose
Carbon (n°)	4	5	6	6	12	12	12	12
Molecular weight	122	152	182	182	344	344	344	342
Melting point (°C)	121	94	165	97	150	145–150	122	190
Heat of solution (cal/g)	−43	−36.5	−28.5	−26	−18.9	−9.4	−13.9	−4.3
Heat stability (°C)	60	60	60	60	60	60	60	160
Acid/alkaline stability (pH)	2–10	2–10	2–10	2–10	2–10	2–10	3	Hydrolyses
Viscosity	Very low	Very low	Low	Medium	Medium	High	Very low	Low
Hygroscopicity	Very low	High	Low	Medium	Medium	Low	Medium	Medium
Solubility (% w/w at 25°)	37	64	20	70	60	25	57	67

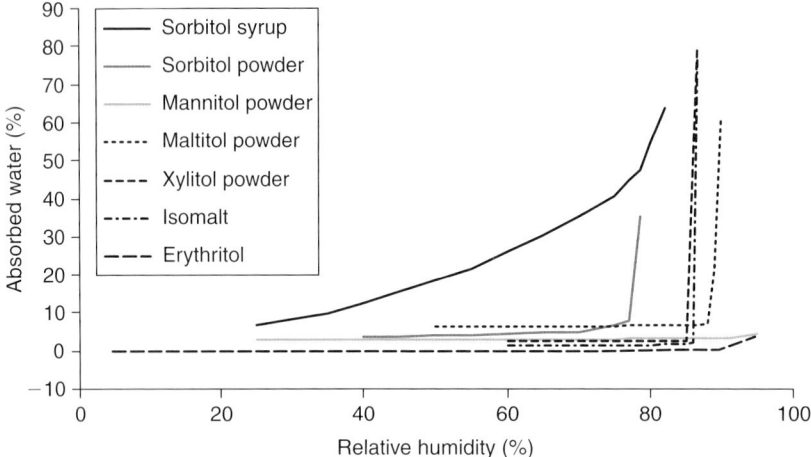

Fig. 10.7 Water absorption in erythritol in comparison with other bulk sweeteners.

at a faster rate than larger molecules such as xylitol (a 5-carbon molecule) and sorbitol (a 6-carbon molecule). The extent of the *in vivo* absorption of erythritol can be quantified easily since the absorbed fraction does not undergo systemic metabolism and is excreted unchanged in the urine. As a result, quantification of urinary excretion of erythritol is generally representative (excretion in the bile is comparatively small at less than 1% of the administered dose) of fractional absorption provided that the urinary collection period is of sufficient length (generally between 24 and 72 hour). In humans and animals, absorption ranges up to 90% or more have been reported.[8–10] Absorbed erythritol is rapidly distributed throughout the body with peak blood concentrations generally occurring within 1 hour of ingestion.

Studies with ^{13}C-erythritol[11] given to human subjects (at a dose of 25 g/day equivalent to a high end user) indicated that no $^{13}CO_2$ was expired confirming that erythritol was neither metabolised in the body nor fermented in the lower colon. The inability of faecal microflora to ferment erythritol was confirmed in *in vitro* incubation studies.[5,11] Both studies

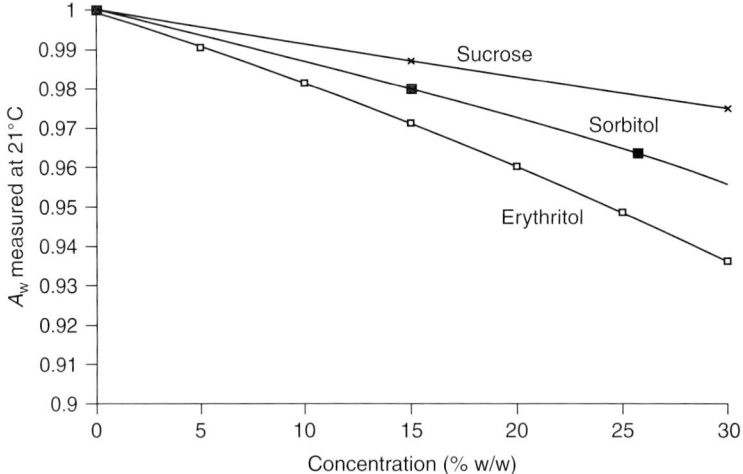

Fig. 10.8 Water activity of erythritol, sorbitol and sucrose at increasing concentrations.

used positive control substrates (e.g. lactitol and glucose or maltitol) that were readily fermented. The study of Arrigoni et al.[5] using the methodology outlined by Lebet et al.[12] extended the fermentation time to 24 hours to allow for the faecal micro-organisms to adapt to the substrate, erythritol. Taking all fermentation parameters into account, erythritol was completely resistant to bacterial degradation within 24 hours, thus excluding any adaptation within that period. The authors concluded that since under *in vivo* conditions more easily fermentable substrates enter the colon continuously, it seemed very unlikely that erythritol would be fermented *in vivo*.

10.4.3 Caloric value

Since erythritol is not systemically metabolised nor fermented in the colon, ingested erythritol provides no energy to the body and thus should be represented in nutrition labelling as 0 kcal/g. Therefore, erythritol is very useful as non-caloric bulk sweetener in weight management programmes. No other bulk sweetener is non-caloric.

10.4.4 Digestive tolerance

Consumption of excessive amounts of polyols can provoke undesirable intestinal side effects such as abdominal cramps, flatulence and laxation. Some of these symptoms are the result of osmotic effects, while others are the result of the fermentative degradation of these compounds in the colon. Numerous human tolerance studies have shown that the incidence and severity of these intestinal side effects and their threshold dose depend on the particular polyol consumed, the mode of ingestion, the existence of a previous adaptation period and the individual susceptibility for these kinds of effects.

Although erythritol chemically belongs to the group of polyols, it has a digestive tolerance that is much higher compared with all other polyols. Since 90% of the ingested erythritol is readily absorbed from the small intestine, minimal amounts reach the lower gut. Clinical studies show that gastrointestinal effects in adults ingesting erythritol at up to 1 g/kg body weight (up to 80 g per day) were not statistically different from those in persons ingesting sucrose at similar levels.[13,14] Consequently, erythritol does not cause laxation under the anticipated conditions of use.

Erythritol's high-digestive tolerance is particularly advantageous when using it in beverages since the intake from beverages can be relatively high. A use level of 1–3% is usually sufficient to generate the desired sensorial benefits. A concentration of 3.5% erythritol is isotonic.

Table 10.4[15–18] lists the maximum bolus dose not causing laxation for a number of polyols. These figures were generated in clinical studies that were conducted under very severe conditions where non-adapted test subjects consumed a single bolus dose, within 10 minutes, as a liquid or jelly on an empty stomach. Such test conditions allow better tolerance comparison and simulate worst intake conditions from beverages. Storey et al.[19] demonstrated that a single dose of 35 g erythritol consumed in a beverage within 15 minutes on an empty stomach (most severe conditions) was tolerated well without any symptoms whereas xylitol induced significant intestinal symptoms and laxation.

10.4.5 Glycaemic and insulinaemic response

Clinical studies demonstrate that consumption of erythritol does not raise plasma glucose or insulin levels. This makes erythritol a suitable sweetener for use by people who suffer from diabetes. In a study carried out by Bornet et al.,[8] mean plasma glucose and insulin levels,

Table 10.4 Maximum bolus dose not causing laxation of erythritol and other polyols.

Polyol	Maximum bolus dose not causing laxation (g/kg body weight)		Reference
	Male	Female	
Erythritol	0.66	0.80	Oku and Okazaki[15]
Sorbitol	0.17	0.24	Oku and Okazaki[15]
Maltitol	0.3	0.3	Koizumi et al.[16]
Isomalt	0.3	–	Oku[17]
Xylitol	0.3	0.3	BIO Clinica[18]

measured for up to 3 hours after ingestion of a single dose of 1 g erythritol/kg body weight, were unaffected by erythritol (Figure 10.9).

The importance of low-glycaemic foods is gaining more and more attention because of their potential health benefits. These include a lower risk of developing type-2 diabetes, a lower probable risk of a hypoglycaemic episode, long-term diabetic complications and coronary heart disease, and helping to manage obesity.

10.4.6 Dental health

The non-cariogenicity of erythritol has been demonstrated using various methods. *In vitro* incubation with a range of *Streptococcus* species has shown that neither lactic acid nor other organic acids are produced from erythritol.[20] Under the same test conditions, xylitol gave a similar response whereas maltitol, sorbitol, mannitol, lactitol and isomalt showed minor acid production. Streptococci are unable to metabolise erythritol and therefore cannot produce the glucosyltransferase that enables them to synthesise insoluble glucan plaque material. This was confirmed in separate *in vitro* tests showing that Streptococci did not produce any polymeric plaque material and subsequently could not adhere to a glass surface when incubated with erythritol, in contrast to the control group incubated with sucrose.

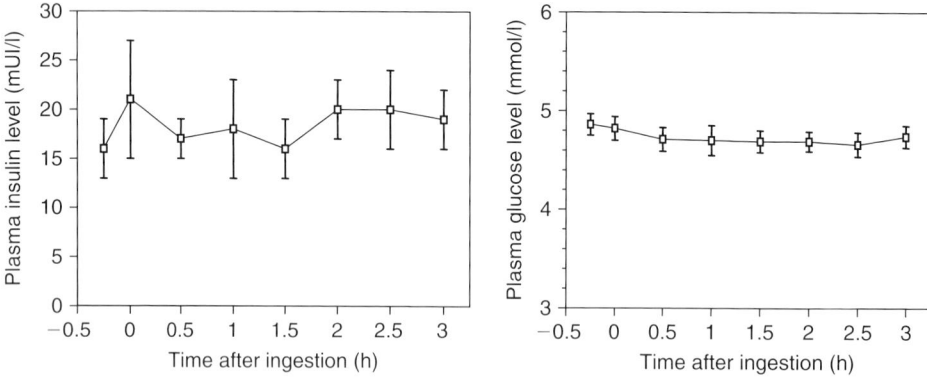

Fig. 10.9 Mean plasma glucose and insulin concentration after a 1 g/kg body weight single oral intake of erythritol by six healthy human volunteers. Error bars indicate standard error of the mean (SEM).

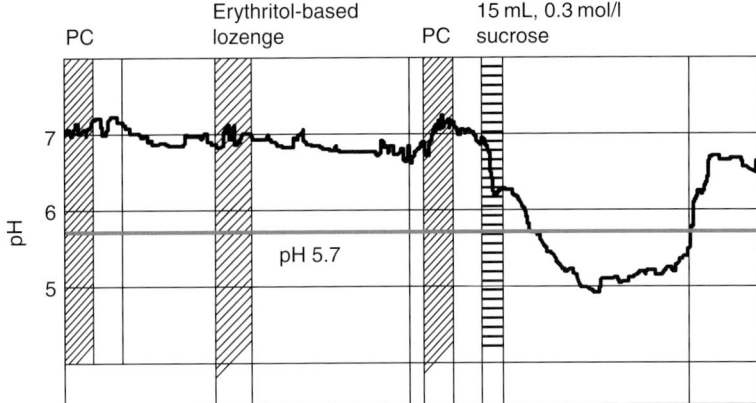

Fig. 10.10 Erythritol is safe for teeth as confirmed in the *in vivo* plaque-pH test. Example of a plaque-pH curve of interdental plaque in one volunteer during and for 30 minutes after consumption of an erythritol-based lozenge. A 0.3 mol/L (10%) sucrose rinse for 2 minutes was used as positive control (PC, paraffin chewing; id, age of interdental plaque (I) in days).

The *in vivo* plaque-pH test is commonly accepted as the reference method for measurement of the cariogenic potential of foods. This test is based on the formation of organic acids in the dental plaque after exposure to fermentable dietary carbohydrates, and measures the corresponding decrease of plaque-pH. The potential of a food to cause low plaque-pH values is generally agreed to be closely associated with the initiation of dental decay. If the plaque-pH value after ingestion of a food remains above the critical pH value of 5.7, the food is regarded as safe for teeth. Erythritol was studied in a plaque-pH test in the form of a crystallised tablet (lozenge). An example of a pH curve from the intra-plaque measurements during chewing of erythritol tablets is shown in Figure 10.10. Sucrose was used as a control. All tests consistently demonstrated the non-cariogenicity of erythritol.

In a 6-month clinical study, premonitory symptoms of dental caries (such as fresh plaque weight) were measured while using erythritol (lozenges and toothpaste) compared with xylitol, sorbitol and control.[6] The amount of dental plaque was significantly reduced in subjects receiving erythritol and xylitol, but not in the other experimental groups. The use of erythritol and xylitol was also associated with a statistically significant reduction in the plaque and saliva levels of mutans streptococci. The investigators conclude that erythritol and xylitol exert similar effects on the investigated risk factors of dental caries, although the biochemical mechanism of the effects may differ. These *in vivo* studies were supported by cultivation experiments in which xylitol, and especially erythritol, inhibited the growth of several strains of *mutans streptococci*.

10.4.7 Anti-oxidant properties

Polyols are excellent scavengers of hydroxyl radicals because of their many hydroxyl groups. However, this has only limited potential in animals and humans as polyols, with the exception of erythritol, are only poorly absorbed. Erythritol is highly bioavailable and, in addition, is not metabolised and circulates throughout the body where it potentially can exercise its radical scavenging activity.

Den Hartog *et al.*[21] evaluated the anti-oxidative and endothelium-protective properties of erythritol. Vascular health is strongly dependent on endothelial state, which regulates vascular tone and inflammatory responses. High glucose-induced endothelial cell apoptosis, likely via reactive oxygen species, has a pivotal role in diabetes-associated vascular diseases, including atherosclerosis. In competition assays, erythritol was shown to be an excellent hydroxyl radical scavenger (inert towards superoxide radicals) and inhibited chemical-induced haemolysis. Erythritol reacted with hydroxyl radicals forming erythrose and erythrulose by carbon-bound hydrogen abstraction. In streptozotocin-induced diabetic rats, erythritol had endothelium-protective effects and erythrose was found in urine. These endothelial protective effects of erythritol cannot, however, be explained by hydroxyl radical scavenging alone. More research is ongoing to elucidate the mechanism of action and to study the potential health benefits of these effects in humans.

10.5 APPLICATIONS

10.5.1 Table-top sweeteners

In table-top sweetener applications, erythritol is used at levels up to 99.9% as a non-caloric, non-cariogenic, non-glycaemic carrier for intense sweeteners. In these applications, the sensorial profile-modifying properties of erythritol are of great importance resulting in sweetness synergy, improved mouthfeel and masking of off-flavours. In addition, owing to erythritol's crystalline structure and density that is similar to sucrose, and its non-hygroscopic property, it offers excellent flowability and stability as a carrier.

10.5.1.1 Quantitative synergy with intense sweeteners

Mixtures of erythritol and intense sweeteners have been studied to determine if a synergistic effect is detectable. The approach consisted of developing and evaluating mixtures of sweeteners that, based on their individual concentration-sweetness response (C-R) functions, were expected to have equal sweetness intensity. Synergy between erythritol and an intense sweetener occurred in those cases where the sweetness intensity of the binary mixture was found to be statistically significantly higher than the additive sweetness intensity of the single solutions. Six ratios of erythritol–aspartame and of erythritol–acesulfame K were evaluated, in random order, for sweetness intensity using an unstructured line scale with anchors at 0% and 16% sucrose equivalent value (SEV) by comparing with sucrose controls (Table 10.5).

In general, mixtures in which erythritol was the major contributor to the sweetness, demonstrated significant synergism. Only minute amounts of aspartame or acesulfame K were necessary to boost the sweetness intensity of erythritol by about 30%.

Table 10.5 Percent synergy for binary mixtures of erythritol and intense sweeteners.

Sweetener combinations	Erythritol-intense sweetener ratio (sweetness contribution)					
	1–99	5–95	15–85	85–15	95–5	99–1
Expected sweetness (SEV)	10	10	10	10	10	10
Erythritol–aspartame	−3	−7	10	30[a]	25[a]	24[a]
Erythritol–acesulfame K	12	8	19[a]	32[a]	31[a]	27[a]

[a]Significant at $p < 0.05$.

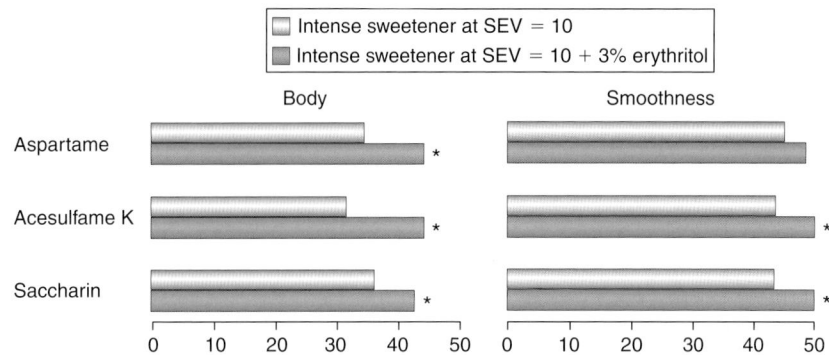

Fig. 10.11 The influence of erythritol on the mouthfeel attributes on body and smoothness (evaluated by addition of 3% (w/v) erythritol to intense sweetener solutions with a SEV of 10). *Significant difference ($p < 0.05$).

10.5.1.2 Improving mouthfeel and non-sweet flavour attributes of intense sweeteners

The influence of erythritol on the mouthfeel attributes, body and smoothness, was evaluated by addition of 3% (w/v) erythritol to intense sweetener solutions with a SEV of 10. Panellists assessed each attribute using an unstructured line scale equating to numerical scores of 0–100 (Figure 10.11).

As the results in Figure 10.11 show, scores for the mouthfeel attributes of body and smoothness are generally higher when erythritol is added to solutions of aspartame, acesulfame K and saccharin.

The impact of erythritol on non-sweet flavour and aftertaste attributes was studied by profiling pure intense sweetener solutions (SEV = 10) against binary mixtures of intense sweetener/erythritol at the same sweetness intensity. As an example, the impact of 1.5% erythritol on such attributes and the mouthfeel of sucralose is shown in Figure 10.12.

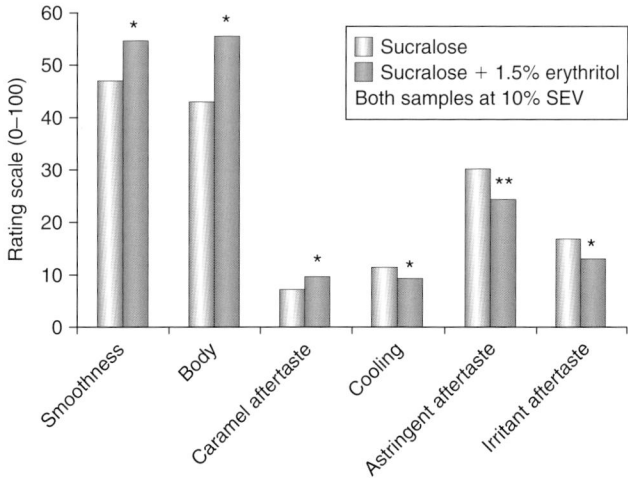

Fig. 10.12 The impact of erythritol on non-sweet flavour and aftertaste attributes (profiled using pure intense sweetener solutions (SEV = 10) against binary mixtures of intense sweetener/erythritol at the same sweetness intensity). *Significant difference ($p < 0.10$). **Significant difference ($p < 0.05$).

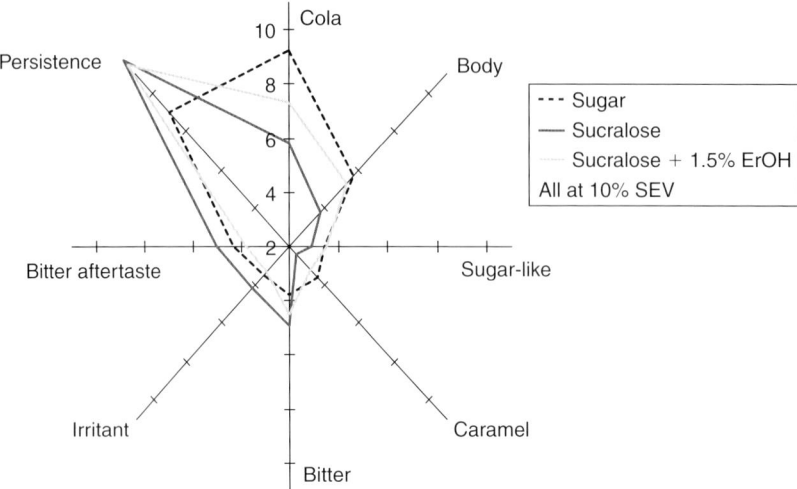

Fig. 10.13 Impact of 1.5% erythritol on taste profile of sucralose-sweetened diet cola at 10 SEV (scale 0–15).

10.5.2 Beverages

The quantitative and qualitative synergies that erythritol shows in combination with intense sweeteners, as illustrated in Section 10.5.1, are also very useful in low-calorie and diet beverages. The impact of 1.5% erythritol addition on the flavour profile of sucralose-sweetened diet cola is shown in Figure 10.13.

Addition of 1.5% erythritol moved the flavour profile of solely sucralose-sweetened diet cola much closer to the flavour profile of sugar-sweetened cola. In particular, the mouthfeel attribute, body and the bitter aftertaste improved significantly. No effect of erythritol on sweetness persistence was noted in this test.

Erythritol can also mask certain off-flavours in beverages like coffee, tea and grapefruit juice, as demonstrated in the following experiments.

Tea was prepared following a slightly modified British Standard Method (BS 6008: 1980). Leaf tea (0.85 g/100 mL) was mixed with freshly boiled mineral water, stirred initially, allowed to brew for 6 minutes and then stirred again. The tea was then sieved and added to an amount of erythritol (1% or 3% w/v). Coffee was prepared by adding 100 mL boiling water to 1 g instant coffee and the required amount of erythritol (1 or 3 g), and then stirred. Samples of grapefruit juice were prepared by thoroughly mixing 500 mL commercial grapefruit juice with the required amount of erythritol (5 or 15 g). All samples were profiled at room temperature (22°C) in triplicate (Figure 10.14).

In tea, the astringent aftertaste was significantly decreased when adding only 1% erythritol. When adding 3%, a significant increase in body and smoothness was observed, as well as a further decrease in astringency and bitterness. Also, the intensity of the tea flavour was decreased. Similar effects were observed for coffee, except that no significant difference was found for the attribute body. When adding 1% or 3% erythritol to grapefruit juice, no significant difference was observed in grapefruit flavour intensity or in sweetness, but smoothness was significantly increased and the astringent aftertaste reduced. Flavour profile modifications of this kind are very well appreciated by the vast majority of consumers of low-calorie foods, particularly beverages.

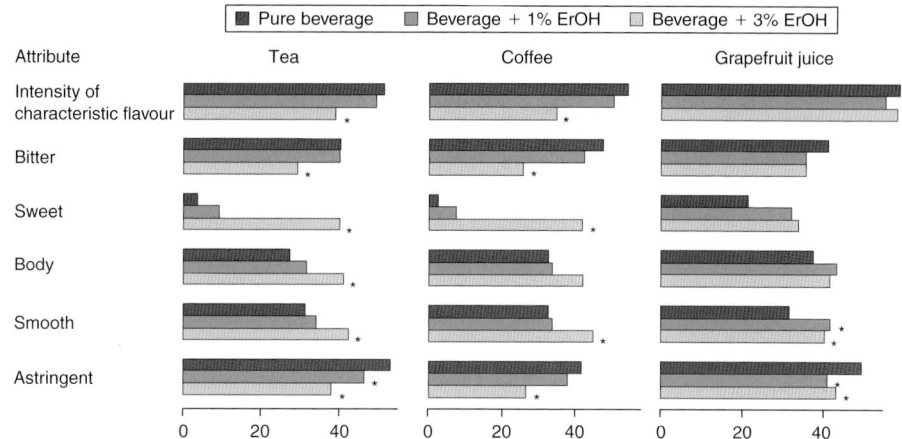

Fig. 10.14 Impact of erythritol on key attributes of tea, coffee and grapefruit juice (scale 0–100). *Significant difference ($p < 0.05$) compared to pure beverage.

In drinks sweetened with aspartame or aspartame–acesulfame K blends, erythritol can potentially mask off-flavours developed during storage as well as replace some of the sweetness of aspartame lost over time owing to its degradation. Hence, erythritol may improve the overall taste quality of aspartame-sweetened drinks throughout their shelf life.

10.5.2.1 Use of erythritol in non-caloric frozen carbonated beverages

A novel and unique use of erythritol is in non-caloric frozen carbonated beverages (FCBs) or slush beverages. Caloric products contain common sugars, such as sucrose or high-fructose corn syrup (HFCS), which are used as sweeteners at concentrations of about 10% w/v. These sugars play an important part in the freezing point depression of FCBs. Under normal operating conditions of FCB machines, the addition of caloric sweeteners depresses the freezing point of the product making them dispensable in a slush-like state. In contrast, a diet beverage, or non-caloric syrup contains no common sugars such as sucrose or HFCS, and thus lacks an effective freezing point depressant. Without a modified freezing point, diet syrup would freeze into blocks of ice in FCB machines rather than attaining the slush-like property found in caloric FCBs and necessary for proper dispensing. Approximately, 3.5% w/v erythritol can be used to achieve the same freezing point depression as 10% sucrose.

10.5.3 Chewing gum

Good-quality non-caloric and non-cariogenic chewing gum can be formulated using erythritol.

10.5.3.1 The cooling effect of erythritol

Erythritol crystals look very much the same as sugar, and have a clean sweet taste accompanied by a long and high cooling effect. This is due to the very high negative heat of solution of erythritol combined with its medium solubility. Dissolving 1 g of erythritol requires 43 calories, and this is the highest negative heat of solution of all bulk sweeteners. Only xylitol comes close at 36.5 cal/g. However, owing to the fact that xylitol dissolves more

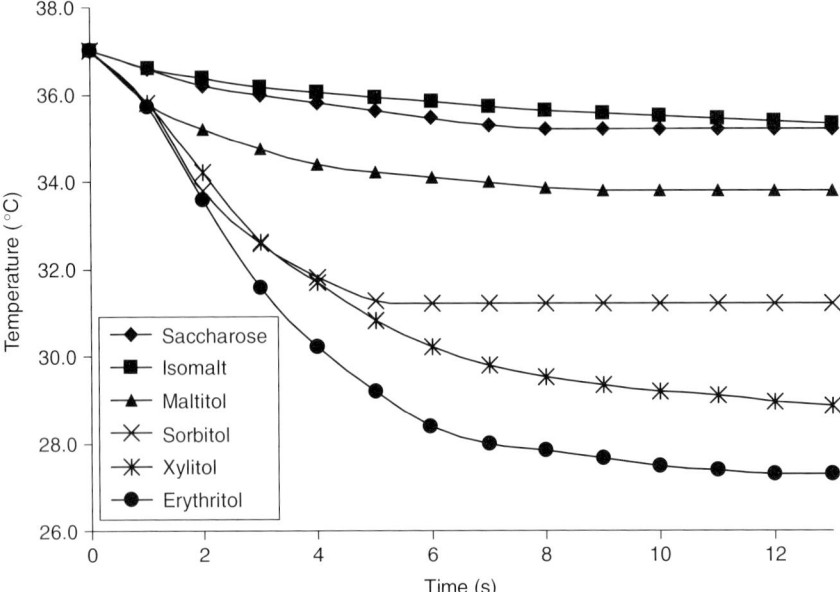

Fig. 10.15 Cooling effect of erythritol and other bulk sweeteners as a 30 g sample of sweetener dissolves in 100 g of water.

quickly, its cooling effect does not last as long as erythritol. Figure 10.15 shows the drop in temperature (cooling effect) over the first 13 seconds in a model system that simulates oral conditions comparing bulk sweeteners used in chewing gum manufacture.

10.5.3.2 Chewing gum sticks and centres

Erythritol is used in sticks or centres of traditional and sugar-free chewing gum to provide a softer texture and greater flexibility. This is done for two reasons. During processing, a greater flexibility reduces the risk that the chewing gum rope – after extrusion and cooling – will break, which would lead to machine downtime. The second reason for the use of erythritol in the base or centre is that it can improve shelf life since the soft and chewy texture is maintained over a longer period of time. This can be obtained in sugar-free chewing gum by partial replacement of other polyol(s) (such as sorbitol) with erythritol. A minimum of 5% erythritol (calculated on total sweetener level) is needed for a noticeable improvement, and a 20% erythritol level yields optimum results. A further increase in flexibility and a softer chewing gum can be obtained by a slightly coarser particle size. However, although erythritol powders show a range of particles, with the largest as high as 300 μm, it is generally preferred that all particles have a size less than that to avoid a grainy texture in the chewing gum. There is no need to change the manufacturing process when using erythritol (Figure 10.16).

10.5.3.3 Erythritol in chewing gum coating

The first few bites of a chewing gum are the most important to determine the level of consumer acceptance. For this reason, numerous studies have been carried out to evaluate the impact of polyols and polyol mixtures on the quality of sugar-free chewing gum coating. Parameters

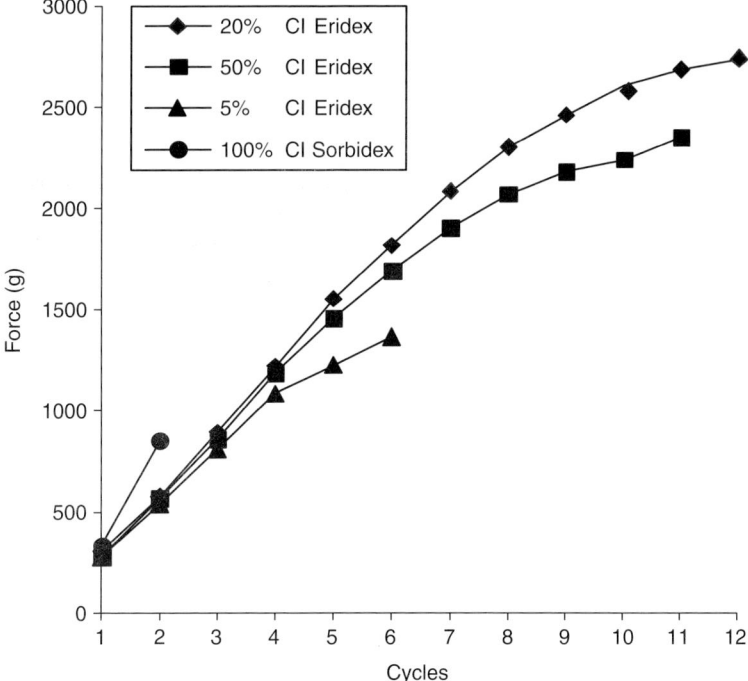

Fig. 10.16 Instron flexibility measurement of chewing gum after 2-month storage: comparison of 0%, 5%, 20% and 50% sorbitol (Sorbidex) replacement by erythritol (Eridex).

of key importance are crunchiness, coolness and stability (against moisture pick-up). Owing to its high speed of crystallisation, coatings based on solely erythritol have a rough surface texture. Therefore, erythritol is used in combination with another polyol such as sorbitol or maltitol, and the best results are obtained when such mixtures contain about 40% erythritol.

In terms of crunchiness and moisture stability, coatings based on high-purity maltitol score very highly and have become the reference for such products. However, maltitol does not contribute to the cool and fresh taste (similarly, isomalt). When 40% erythritol is used in combination with 60% maltitol, a clearly perceivable cooling effect is introduced and the stability against moisture pick-up is higher compared with maltitol alone (Table 10.6).

Coatings based on xylitol taste fresh with a strong cooling sensation. Comparable cooling can be obtained by using a 40:60 ratio of erythritol and sorbitol and although stability is slightly lower, the coating has the same crunchiness as xylitol. The coating time can also be shortened by about 30% owing to the fact that erythritol crystallises so easily and so fast. The

Table 10.6 Comparison of polyols on important chewing gum coating parameters.

	Sorbitol	Xylitol	Isomalt	Maltitol	Erythritol/sorbitol 40:60	Erythritol/maltitol 40:60
Crunchiness	3	2	4	6	4	6
Stability against moisture	3	2	5	6	1	7
Cooling effect	+	++	−	−	++	+

Scale 1–7 (1 = low, 7 = high).

erythritol/sorbitol mixture is, therefore, a cost-effective alternative to xylitol for improved coating. This blend has also been shown to give a better adhesion to the chewing gum centre compared with xylitol.

10.5.4 Chocolate

Chocolate is traditionally manufactured using a dry conching process at temperatures up to 80°C to obtain good flavour development and correct flow behaviour. However, use of most polyols necessitates a wet conching process at low temperature for a much longer time, resulting in less flavour development. The use of erythritol in chocolate compositions allows a dry conching process at high temperatures. Owing to the good heat stability and low-moisture pick-up of erythritol, it is possible to work at higher temperatures than traditionally used. This results in enhanced flavour development.

Erythritol-based chocolate is not only easy to produce but it can also be considered a true 'reduced-calorie' product. For example, 34% calorie reduction is obtained by using erythritol (at levels up to 50%) to replace sucrose in a classical formulation. Other polyols with a calorific value of 2.4 kcal/g (EU) do not provide the calorie reduction required for a 'light' chocolate claim by the simple replacement of sucrose, but necessitate important reformulation leading to an inferior quality chocolate.

Erythritol-based chocolate is non-hygroscopic, has excellent gloss, good breaking characteristics and has a pleasant cooling effect with good melting properties in the mouth (Figure 10.17).

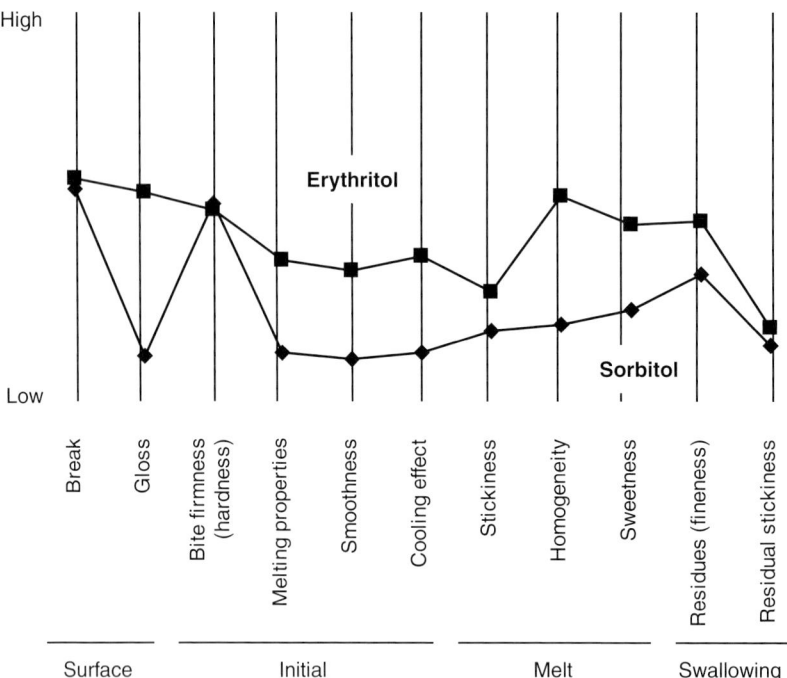

Fig. 10.17 Taste panel comparison between sorbitol and erythritol-based reduced-calorie and sugar-free chocolate.

Erythritol also partly inhibits the ability of oral bacteria to ferment lactose. Therefore, it allows the production of non-cariogenic milk chocolate using normal lactose-containing milk powder (confirmed by *in vivo* plaque-pH testing).

10.5.5 Candies

Fudge is a soft candy with a texture between that of a caramel and a fondant. It is a grained, medium-boiled confection containing milk solids and a high concentration of fat. Fudge contains both a solid and a liquid phase. Sugar-free fudge with texture and shelf-life properties equivalent to conventionally sweetened fudge can be produced using erythritol (up to 40%) in combination with maltitol syrup (75% maltitol) to control crystallisation.

Depending on the cooking temperature and the seeding level with erythritol, the texture can be varied from soft to hard. A typical formulation and manufacturing process for fudge containing erythritol is shown in the following table.

Formulation	%
Erythritol	22.5
Maltitol syrup (75% maltitol content)	45.5
Water	5.5
Unsweetened whole concentrated milk	18.7
Hydrogenated cocoa fat (including 0.2% GMS)	7.4
Lecithin	0.2
Sodium bicarbonate	0.2
Flavouring	As needed
Seeding blend	5–10
Erythritol	81
Water	19

Manufacturing process
- Mix the erythritol, maltitol syrup and water, and heat to 70°C.
- Add the milk, melted fat and other ingredients, and homogenise for 2 minutes.
- Heat the mixture to 140°C.
- Cool to about 100°C and seed with 5–10% of seeding blend.
- Cool and cut.

Erythritol can also be used in a range of other confectionery applications such as gelatin gums and hard candies. However, in these applications, the use of erythritol is limited owing to its crystallisation behaviour. In gelatin gums it can be used as a non-hygroscopic 'graining sugar', and in hard candies in its crystalline form at up to 50% w/w for its high cooling effect as a sherbet filling or 'sandwiched' between two layers of hard candy.

The latest innovation involves technology that broadens the melting range and the crystallisation peak of erythritol and shifts the melt crystallisation peak to a level low enough to allow depositing in moulds and control the formation of crystals.[22] The resulting candies have a smooth appearance, texture, and desired hardness. This novel technology allows the manufacture of deposited candies that are essentially free of calories, are safe for teeth, and have a novel texture and taste sensation.

10.5.6 Fondant

Fondants and creams are sugar confectionery products that contain mixed sugars and are made in two phases. The sugar crystals that constitute the solid phase are dispersed in a syrup containing higher sugars. Fondants and creams are similar in composition, though creams contain slightly higher residual water and a greater proportion of doctor solids.

Using erythritol, it is possible to obtain sugar-free, low-calorie, non-cariogenic fondant with identical technical properties to classical products, which was previously impossible with other low-calorie bulking agents. Pure erythritol induces too high a crystallisation, but 60% erythritol in combination with maltitol syrup as a liquid phase helps to control this phenomenon.

The preparation method for an erythritol-containing fondant is similar to a sucrose-based product. The erythritol crystals are dissolved in a specified amount of syrup phase and the mixture cooked to the required dry substance. After cooling the mass to 40°C, it is beaten to achieve the correct consistency and crystal size. The fondant is discharged into small containers and matured for 1 day before use.

Fondants with slightly different textures and fluidities can be obtained, depending on the beating temperature and time. Microscopic examination indicates that a good quality fondant has a majority of crystals between 5 and 10 μm. An erythritol fondant has a moisture content of about 6% and a low water activity (in the range of ± 0.5) that prevents microbial spoilage. A typical erythritol-based fondant recipe and manufacturing process is given in the following table.

Formulation	%
Erythritol	50
Maltitol syrup (75% maltitol content)	50

Manufacturing process
- Dissolve erythritol in the maltitol syrup.
- Cook the mix to 140°C.
- After cooling to 40–45°C, the mass is beaten for 5–10 minutes to obtain the required consistency and crystal size.
- The fondant is placed in containers and matured for 1 day.

10.5.7 Lozenges

Lozenges are a form of confectionery usually made from a finely milled sugar that is kneaded to a doughy consistency with water and a binding agent. A gum such as gum Arabic or gelatine is usually added to help retain the tablet shape. The malleable dough is cut into suitable shapes that are dried and hardened.

Lozenges may, of course, contain a suitable flavour, colour and/or acid, and should have a hard, brittle consistency. Polyols such as xylitol, sorbitol, mannitol, maltitol, isomalt and lactitol have been used to produce low-calorie/non-cariogenic lozenges. Although all these polyols can be used in confectionery, they are far from equivalent in all applications and the contrasting demands of the various type of products require careful polyol selection to find the correct combination of properties to replace part or all of the sugar.

Research has shown that lactitol, crystalline maltitol and erythritol are the only suitable materials to produce a lozenge with a lower calorie content than conventional sugar-based products but which are otherwise identical. In addition to the lower-calorie content and pronounced cooling effect, erythritol has the added advantage over lactitol and crystalline maltitol of requiring a shorter drying time to obtain the required texture at low residual moisture content. Lozenges based solely on erythritol (up to 99%) have excellent shelf-life properties, even when stored under high humidity conditions. A typical formulation and manufacturing process for erythritol-based lozenges is given in the following table.

Formulation	
Erythritol	1.5 kg
Gelatine solution (10% d.s.)	230 mL
Flavouring	As needed

Manufacturing process
- Prepare 10% gelatine (170 Bloom) solution by slowly adding the gelatine to warm water (50°C).
- Place the erythritol in a Z-blade mixer that has been preheated to 40–45°C.
- Slowly add the warm gelatine solution and flavouring while mixing.
- Mix for 10 minutes to obtain a smooth, homogeneous paste.
- Remove the paste from the kneader, roll out and cut in shape.
- Stove at 45°C for 8 hours.

10.5.8 Bakery (pastry) products

Bakery products with their high proportion of flour, butter, sugar and other ingredients are by definition difficult, but not impossible, products to manufacture in calorie-reduced form. However, since erythritol is non-caloric, sugar replacement gives some interesting results.

10.5.8.1 Bakery fillings

Conventional 'fat plus sweetener' compositions (fat creams) contain sucrose as the sweetener and are particularly used in butter cream fillings for cakes and biscuits. These compositions have a substantial calorie content and a typical fatty mouthfeel. When using erythritol as a sugar substitute, the cooling sensation of erythritol has the effect of masking the fatty mouthfeel, thereby giving a more refreshing and attractive product. In addition, the calorie content of the composition is also reduced. The best texture of the fat cream is obtained with erythritol at a use level of 60% and a particle size distribution below 300 μm. The manufacturing process and shelf-life properties of the erythritol-based fat cream are similar to those of conventional fat/sucrose compositions:

Formulation	%
Erythritol (<300 μm)	60
Shortening	40
Flavouring	As needed

Manufacturing process
- Gently mix all the ingredients for 5 minutes at full speed in a Hobart mixer.

Reduction of calories in bakery products is particularly difficult if only the sugar is replaced. Erythritol can reduce the caloric content by more than 30% in some sugar-rich applications without changing the organoleptic quality of the product significantly. Erythritol can be used successfully in cookies, biscuits and cakes, where it improves baking stability and shelf life at an addition level of about 7%.

Hard biscuit formulation	%
Margarine	10
Sucrose	7.65
Erythritol	7.65
Whole milk powder	1.23
Salt	0.46
Pyrophosphate	0.15
Vanillin	0.04
Biscuit flour	61.5
Sodium bicarbonate	0.3
Ammonium bicarbonate	0.15
Sulphite	0.07
Water	10.8

Hard biscuit manufacturing process
- Cream the margarine, sucrose and erythritol in the Hobart mixer for 5 minutes at high speed.
- Dissolve the baking agents in part of the water.
- Sieve the flour and milk powder into the blender, and add the salt and vanillin.
- Add the cream to the blender and mix for 30 minutes at high speed. Dough temperature should be 32–36°C.
- Add the dissolved products and the water during the mixing process.
- Let the dough rest for 30 minutes.
- Take out the dough with a biscuit-seal or with a biscuit machine.
- Bake the biscuit for 7 minutes at 250°C.

Good tasting, calorie-reduced creamed icings (full fat) with a moderate cooling effect can be produced using a combination of erythritol and maltitol syrup, and a calorie reduction of close to 50% can be achieved. In fruit fillings, the addition of erythritol can significantly enhance the characteristic flavour of the fruit.

In sponge cake, erythritol (in combination with maltitol) can provide improved shelf life and cake volume in comparison with the classical (sucrose-based) product.

Sponge cake formulation	%
Flour	11.44
Native wheat starch	14.64
Emulsifier	5.08
Glucono delta lactone	0.57
Trisodium phosphate	0.286
Colouring	0.011
Vanillin	0.034

Sodium bicarbonate	0.286
Aspartame	0.023
Eggs	34.3
Maltitol powder	14.85
Erythritol	9.9
Water	8.58

Sponge cake manufacturing process
- Measure the liquid phase into a bowl.
- Blend in the dry ingredients.
- Mix together at high speed for 6 minutes in the Hobart mixer.
- Measure 300 g of batter into a biscuit tin.
- Bake for 20 minutes at 190/200°C.

Generally, compared with sucrose, erythritol gives a different melting behaviour, more compact dough, softer end products, and less colour formation in bakery products.

10.6 SAFETY AND SPECIFICATIONS

The safety of erythritol is well documented.[4, 23–25] In June 1999, the Joint WHO/FAO Expert Committee on Food Additives (JECFA) assessed the safety of erythritol and assigned an acceptable daily intake 'not specified'; the highest safety rating that JECFA can give. The key features that contribute to the overall safety of erythritol are that it is readily absorbed, not systemically metabolised and excreted unchanged in the urine. In addition, erythritol occurs endogenously in various tissues of the body and it is present naturally in the diet. Moreover, both animal toxicological studies and clinical studies have consistently demonstrated the safety of erythritol.

Table 10.7 provides the specification tests, limiting values and reference test methods. The primary criterion is the assay for erythritol content, where a limit value of 'not less than 99.5%' has been established. Due to this very high level of purity, the only additional criteria for quality assurance are typical indicators, such as those for reducing sugars, loss on drying and residue on ignition. Finally, levels of trace elements are controlled by the test for heavy metals.

10.7 REGULATORY STATUS

Erythritol has been authorised for use in foods in more than 50 countries including Europe, the United States, Japan, Canada, Mexico, Brazil, Argentina, Turkey, Russia, China, India, Australia and New Zealand. Erythritol also has been awarded the valuable international endorsement of JECFA and is included in the General Standard for Food Additives (GSFA-list) of the Codex Alimentarius under INS number 968. In addition, erythritol is used as an excipient in pharmaceutical products and included in the European and Japanese Pharmacopoeia.

Table 10.7 The specification tests, limiting values and reference test methods.

Specification	Requirement	Test method
Erythritol assay	Not less than 99.5%	HPLC method
Ribitol + Glycerol	Not more than 0.1%	HPLC method
Reducing sugars	Not more than 0.3%	Method similar to FCC IV
Heavy metals (as lead)	Not more than 5 mg/kg	FCC IV method
Lead	Not more than 0.1 mg/kg	FCC IV atomic absorption spectrophotometric graphite furnace method I
Residue on ignition	Not more than 0.1%	FCC IV method (sulphated ash)
Loss on drying	Not more than 0.2%	FCC IV method

10.8 CONCLUSIONS

Erythritol belongs chemically to the group of polyols and shares some of their basic characteristics such as sweetness, very high acid and temperature stability, non-reducing, tooth-friendly and suitable for diabetics. However, there are significant differences between erythritol and other polyols. Erythritol is the only polyol produced commercially by a natural fermentation process, it has a unique metabolic profile that delivers no energy to the body and it has the highest digestive tolerance. These are the main reasons for the success of erythritol: the wide and continuously growing interest from consumers and the food and beverage industry to reduce sugar and calorie intake naturally without compromising taste.

REFERENCES

1. A. Lamy. Ueber einige Bestandtheile des *Protococcus vulgaris*. *Journal für Praktische Chemie* 1852, 57(1), 21–28.
2. M. Bamberger and A. Landsiedl. Erythritol in *Trentepohlia jolithus*. *Monatshefte fuer Chemie* 1900, 21, 571–573.
3. H. Röper and J. Goossens. Erythritol, a new raw material for food and non-food applications. *Starch/Stärke* 1993, 45, 400–405.
4. W.O. Bernt, J.F. Borzelleca, G. Flamm and I.C. Munro. Erythritol: a review of biological and toxicological studies. *Regulatory Toxicology and Pharmacology* 1996, 24, S191–S197.
5. E. Arrigoni, F. Brouns and R. Amado. Human gut microbiota does not ferment erythritol. *British Journal of Nutrition* 2005, 94(5), 643–646.
6. K.K. Mäkinen M. Saag, K.P. Isotupa, J. Olak, R. Nõmmela, E. Söderling and P.L. Mäkinen. Similarity of the effects of erythritol and xylitol on some risk factors of dental caries. *Caries Research* 2005, 39(3), 207–215.
7. R. Vercauteren. Reducing the sensory cooling effect of polyols. PCT patent application WO 2008/125344 Al, 2008.
8. F.R.J. Bornet, A. Blayo, F. Dauchy and G. Slama. Plasma and urine kinetics of erythritol after oral ingestion by healthy humans. *Regulatory Toxicology and Pharmacology* 1996, 24, S280–S286.
9. M. Ishikawa, M. Miyashita, Y. Kawashima, T., Nakamura, N. Saitou and J. Modderman. Effects of oral administration of erythritol on patients with diabetes. *Regulatory Toxicology and Pharmacology* 1996, 24, S303–S308.

10. K. Noda, K. Nakayama and T. Oku. Serum glucose and insulin levels and erythritol balance after oral administration of erythritol in healthy subjects. *European Journal of Clinical Nutrition* 1994, 48, 286–292.
11. M. Hiele, Y. Ghoos, P. Rutgeerts and G. Vantrappen. Metabolism of erythritol in humans: comparison with glucose and lactitol. *British Journal of Nutrition* 1993, 69, 169–176.
12. V. Lebet, E. Arrigoni and R. Amado. Measurement of fermentation products and substrate disappearance during incubation of dietary fibre sources with human faecal flora. *Lebensmittel-Weissenschaft Und-Technologie-Food Science and Technology* 1998, 31, 473–479.
13. W. Tetzloff, F. Dauchy, S. Medimagh, D. Carr and A. Bär. Tolerance to subchronic, high dose ingestion of erythritol in human volunteers. *Regulatory Toxicology and Pharmacology* 1996, 24, S286–S295.
14. F.R.J. Bornet, A. Blayo, F. Dauchy and G. Slama. Gastrointestinal response and plasma and urine determinations in human subjects given erythritol. *Regulatory Toxicology and Pharmacology* 1996, 24, S296–S302.
15. T. Oku and M. Okazaki. Laxative threshold of sugar alcohol erythritol in human subjects. *Nutrition Research*, 1996, 16(4), 577–589.
16. N. Koizumi, M. Fujii, R. Ninomiya, Y. Inoue, T. Kagawa and T. Tsukamoto. Study on transitory laxative effect of sorbitol and maltitol. Estimation of 50% effective dose and maximum non-effective dose. *Chemosphere* 1983, 12, 45–53.
17. T. Oku. Oligosaccharides with beneficial health effects: a Japanese perspective. *Nutrition Reviews* 1996, 54(11), S59–S66.
18. BIO Clinica. Estimated maximum dose not causing laxation of various sweeteners and their reference. Reported (in Japanese) in BIO Clinica, 1996, 11(11), (817) 31.
19. D.M. Storey, A. Lee, F. Bornet and F. Brouns. Gastrointestinal tolerance of erythritol and xylitol ingested in a liquid. *European Journal of Clinical Nutrition* 2007, 61, 349–354.
20. J. Kawanabe, M. Hirasawa, T. Takeuchi, T. Oda and T. Ikeda. Noncariogenicity of erythritol as a substrate. *Caries Research* 1992, 26, 258–362.
21. G.J.M. den Hartog, A.W. Boots, A. Adam-Perrot, F. Brouns, I.W.C.M. Verkooijen, A.R. Weseler, G.R.M.M. Haenen and A. Bast. Erythritol is a sweet antioxidant. *Journal of Nutrition* 2009, 26(4), 449–458.
22. M. Gonze and R. Vercauteren. Erythritol-based hard coatings. PCT patent application WO 2009/036954 Al, 2008.
23. I.C. Munro, W.O. Bernt, J.F. Borzella, G. Flamm, B.S. Lynch, E. Kennepohl, A. Bär and J. Modderman. Erythritol: an interpretive summary of biochemical, metabolic, toxicological and chemical data. *Food and Chemical Toxicology* 1998, 36, 1139–1174.
24. WHO. Erythritol. In Safety Evaluation of Certain Food Additives and Contaminants. WHO Food Additives Series: 44, pp. 15–70. IPCS International Program on Chemical Safety in cooperation with the Joint FAO/WHO Expert Committee on Food Additives (JECFA). World Health Organization, Geneva; 2000.
25. SCF. Opinion of the Scientific Committee on Food: Erythritol. Scientific Commitee on Food. European Commission, Health and Consumer Protection Directorate-General; 2003. SCF/CS/ADD/EDUL/215 Final. Opinion expressed 23 March, 2003.

11 Isomalt

Anke Sentko and Ingrid Willibald-Ettle

BENEO GmbH, Mannheim, Germany

11.1 INTRODUCTION

Isomalt is a polyol (synonym for sugar alcohol) made from sugar and used like sugar to replace sugars (like sucrose, high fructose corn syrup, glucose syrup and others) on a one for one basis. Like sugar, it has multiple functions in a product, in particular enabling the development of top-quality confectionery products with specific nutritional/functional characteristics, for example sugar-free products, products that have a low glycaemic or insulinaemic response, reduced calorie products and products that do not promote tooth decay.

The first steps in the development of isomalt were taken in 1957 when the bacterial fermentation of sugar to isomaltulose, brand name Palatinose™ (6-O-α-D-glucopyranosyl-fructofuranose) and the intermediate in isomalt production, was discovered by Südzucker.[1] Later, the manufacturing process for isomalt by the hydrogenation of isomaltulose was developed.[2,3]

The production process of isomalt involves two essential steps. In the first step, the α(1-2) glycosidic linkage between the glucose and the fructose molecule in sucrose is rearranged by an immobilised enzyme system to an α(1-6) linkage resulting in isomaltulose. Like sucrose, this is a disaccharide composed of glucose and fructose, although the α(1-6) linkage in isomaltulose is more stable than the α(1-2) linkage in sucrose. In the second step, hydrogenation of the fructose part of isomaltulose takes place resulting in a combination of two disaccharide alcohols, 6-O-α-D-glucopyranosyl-D-sorbitol (1,6-GPS) and 1-O-α-D-glucopyranosyl-D-mannitol dihydrate (1,1-GPM).[4] The generic name of the mixture of these disaccharide alcohols (isomers) is isomalt (synonym hydrogenated isomaltulose). Figures 11.1 and 11.2 show steps one and two of the isomalt production process and the chemical structure of isomalt.

On the basis of this process, isomalt is produced as white crystalline material and is available in a range of variants and types (particle size distribution), each tailored to specific application needs: ISOMALT ST, ISOMALT GS, ISOMALT DC and ISOMALT LM, as described in the following text. The difference between these products is based on the GPM/GPS ratio: ISOMALT ST is approximately 1:1 and ISOMALT GS is approximately 3:1. Additionally, GPS forms anhydrous crystals, whereas GPM has two mol of water of crystallisation. Additional drying of ISOMALT ST leads to ISOMALT LM, a low moisture

---

Sweeteners and Sugar Alternatives in Food Technology, Second Edition.
Edited by Dr Kay O'Donnell and Dr Malcolm W. Kearsley.
© 2012 John Wiley & Sons, Ltd. Published 2012 by John Wiley & Sons, Ltd.

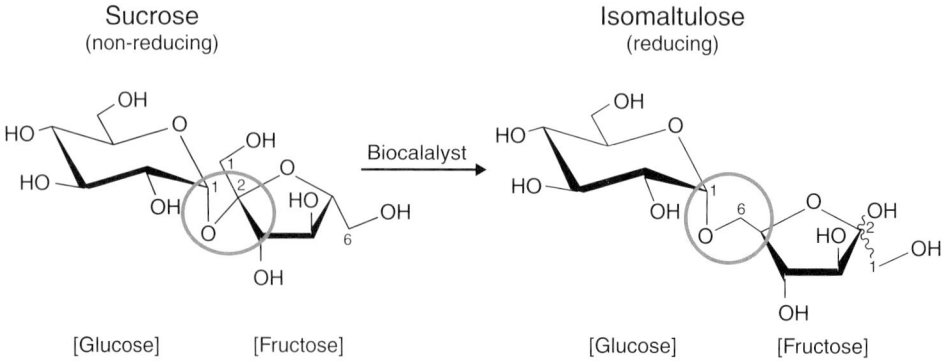

Fig. 11.1 Production step 1: Enzymatic conversion of sucrose to isomaltulose.
Source: Adapted from Schiweck.[4]

grade (<1% moisture). In a milling step, ISOMALT ST or GS can be milled and this is the starting material for ISOMALT DC, an agglomerated grade. All variants (except the DC variant) are available in different particle size distributions as granulates or as fine powders.

11.2 ORGANOLEPTIC PROPERTIES

11.2.1 Sweetening potency versus sucrose

The sweetening potency of isomalt, as shown in Figure 11.3, is between 0.45 and 0.6 compared with sucrose (=1.0). Investigations[5] have shown that sweetening potency is a function of concentration; that is, it increases at higher concentrations as shown in Figure 11.3.

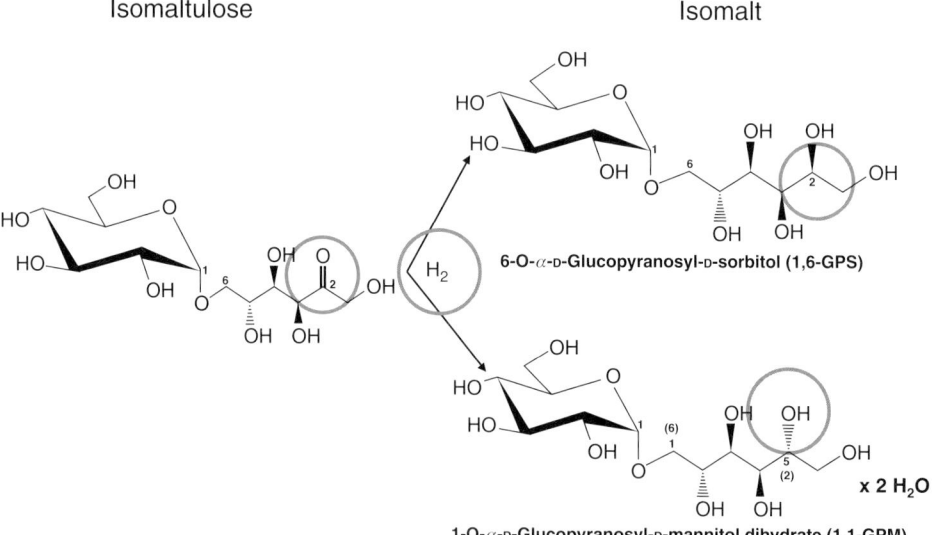

Fig. 11.2 Production step 2: Hydrogenation of isomaltulose to isomalt.
Source: Adapted from Schiweck.[4]

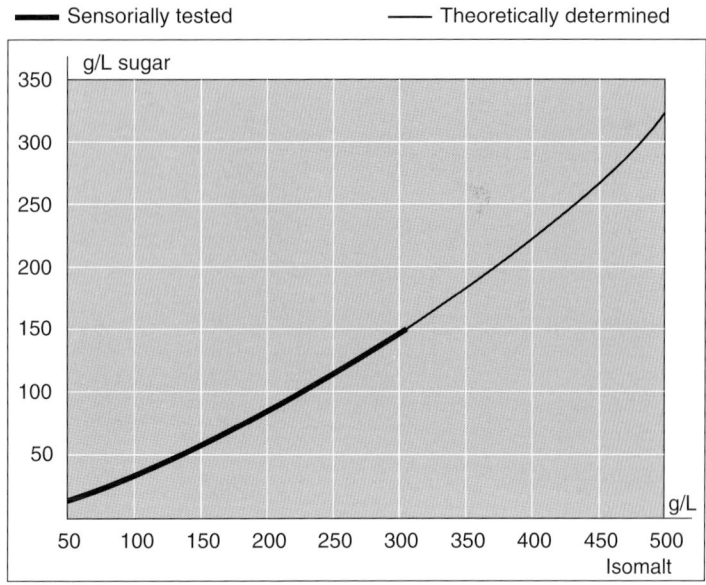

Fig. 11.3 Comparison of aqueous isosweet solutions of isomalt and sucrose at 20°C in g/L. *Source*: Adapted from Paulus and Fricker.[5]

11.2.2 Sweetening profile versus sucrose

The sweetening profile of isomalt can be described as a pure, sweet taste without any accompanying taste or aftertaste and is similar to that of sucrose.

11.2.3 Synergy and/or compatibility with other sweeteners

Since it enhances flavour transfer in foods, isomalt is often combined with both non-nutritive and nutritive sweeteners. Synergistic effects in sweetening power occur when isomalt is combined with other bulk sweeteners such as maltitol syrup (or hydrogenated starch hydrolysate syrup (HSH)), lactitol, sorbitol, mannitol or xylitol.

Sensory studies with isomalt[6] have shown that combinations of intense sweeteners with isomalt in finished products (e.g. boiled sweets) match the sweetness and taste profile obtained with sugar (Figure 11.4).

An additional advantage of combinations of sweeteners and isomalt is that the isomalt tends to mask the bitter aftertaste of some sweeteners and bulking agents.

11.3 PHYSICAL AND CHEMICAL PROPERTIES

11.3.1 Stability

Isomalt is extremely resistant to chemical degradation owing to its very stable glycosidic bonds.

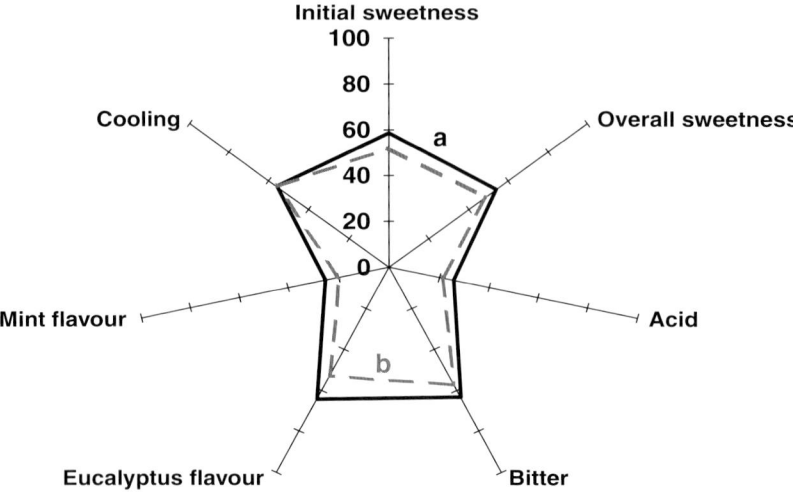

Fig. 11.4 Comparison of the taste profile of menthol-eucalyptus-flavoured candies (a, Isomalt and Aspartame; b, Sugar).
Source: Adapted from Schäppi.[6]

11.3.1.1 Acid and enzymatic hydrolysis

Isomalt is resistant to acid and enzymatic hydrolysis, because the glycosidic linkages in the two isomers of isomalt at the 1,1 and 1,6 positions have a lower dissociation energy than the glycosidic hydroxyl group linkage between the same two monosaccharides in sucrose. Isomalt takes more than 5 hours to completely hydrolyse in a 1% HCl solution at 100°C. Sucrose, in contrast, is completely hydrolysed in a 1% HCl solution after approximately 5 minutes (see Figure 11.5).[7] Enzymatic hydrolysis is not an issue in the production of foodstuffs and may be ignored. Since isomalt contains only a minimal percentage of free

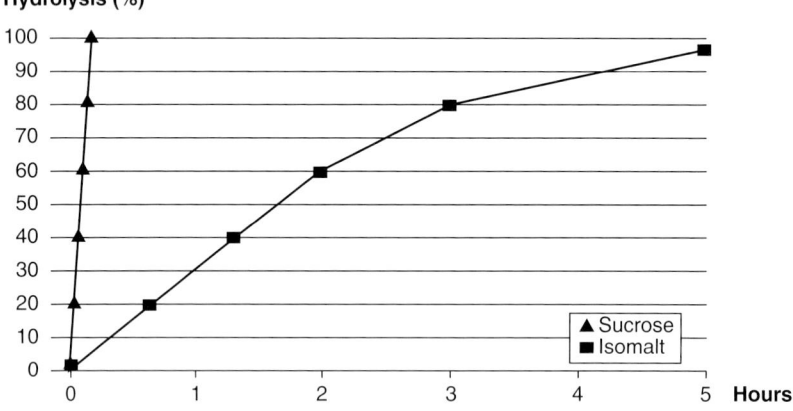

Fig. 11.5 Hydrolysis of isomalt and sucrose over time in a 1% HCl solution (pH ~ 1) at 100°C.
Source: Adapted from Südzucker AG, ZAFES [1987, Ref 7].

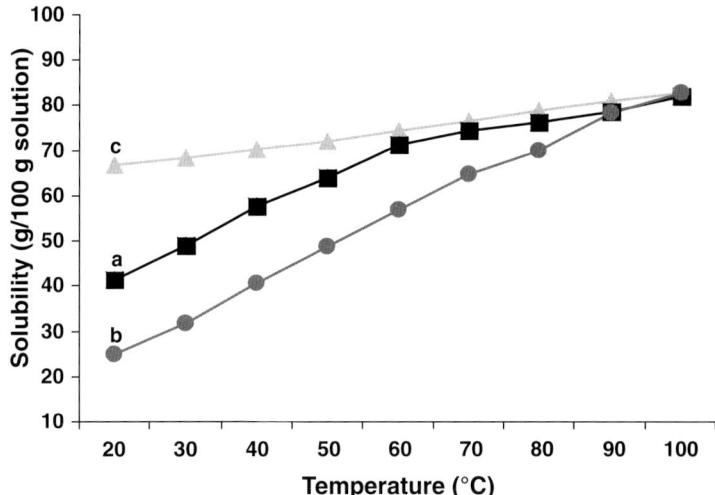

Fig. 11.6 Comparison of the solubility of isomalt variants versus sugar (a, ISOMALT GS; b, ISOMALT ST; c, Sucrose).
Source: Adapted from Südzucker AG, ZAFES (1999, Ref 7).

glucose, sorbitol and mannitol, there is no substrate for microbiological growth and similarly products manufactured with isomalt also display a high level of microbiological stability.

11.3.2 Solubility

At 20°C, ISOMALT ST has a solubility of *24.5 g/100 g* solution, although at most processing temperatures, the solubility is comparable with sucrose. The solubility of ISOMALT ST in water is adequate for most applications because the solubility increases at higher temperatures (Figure 11.6).[7]

ISOMALT GS has a solubility of *41.5 g/100 g* solution at 20°C and therefore a higher solubility than all other isomalt grades.

The pronounced tendency of isomalt to crystallise from super-saturated solutions is associated with its low solubility. This is a desirable attribute in pan coating.

11.3.3 Viscosity

The viscosity of aqueous isomalt solutions over the temperature range 60–90°C does not differ significantly from that of corresponding sucrose solutions. As far as production engineering is concerned, there are no special handling requirements (Figure 11.7).[7,8]

11.3.4 Heat of solution

A large number of sugar substitutes have a negative heat of solution, responsible for the so-called cooling effect associated with these products. This is experienced when the substance dissolves from the crystalline state. While this is a desirable feature in peppermint and

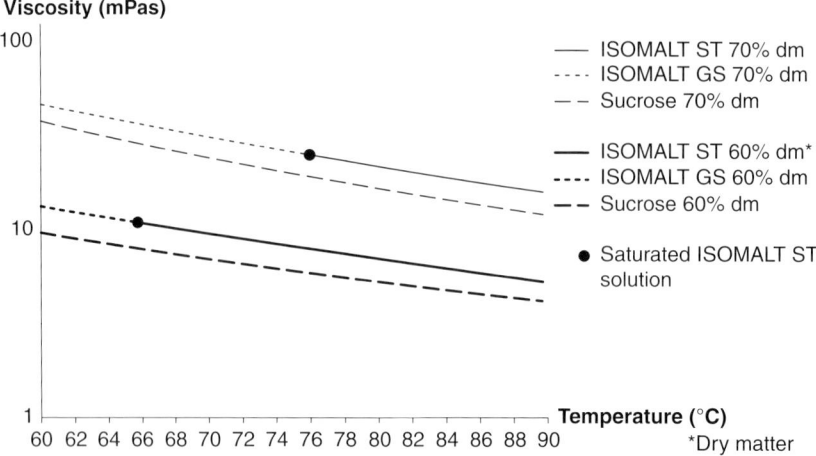

Fig. 11.7 Viscosity of isomalt and sucrose solutions.
Source: Adapted from Südzucker AG, ZAFES (2001, Ref 7); Sugar Technologists Manual (1995, Ref 14).

menthol products, it is frequently regarded as undesirable in a large number of other products, such as baked goods and chocolate. Compared with other bulk sweeteners, isomalt has a heat of solution, comparable with that of sucrose (Figure 11.8).[9,10]

The heat of solution of ISOMALT ST lies between the values of GPM and GPS and depends on the actual moisture content of the isomalt.

11.3.5 Boiling point elevation

The boiling point of highly concentrated isomalt solutions has been established experimentally. A higher boiling temperature must be maintained with isomalt solutions to obtain the same final water content compared with sugar/glucose solutions (see Figure 11.9).[11] This is

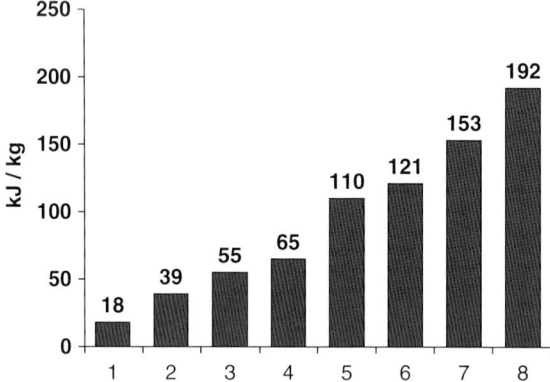

Fig. 11.8 Heat of solution of different substances (1, sucrose; 2, isomalt; 3, maltitol; 4, lactitol monohydrate; 5, sorbitol; 6, mannitol; 7, xylitol; 8, erythritol).
Source: Adapted from von Rymon-Lipinsky and Schiweck;[9] Cammenga and Gehrich.[10]

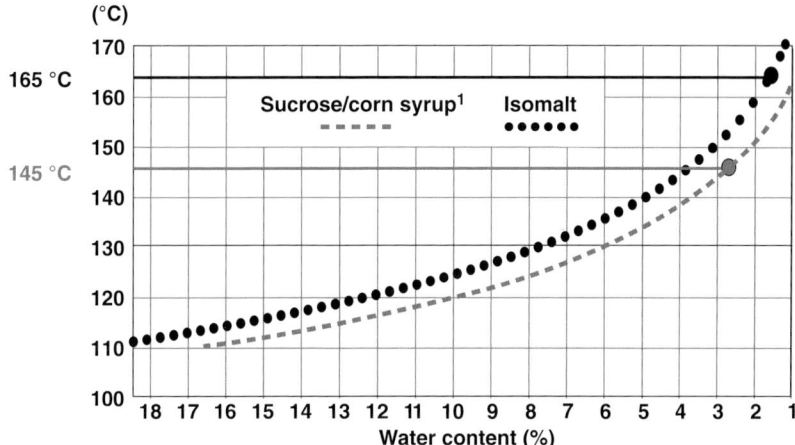

Fig. 11.9 Boiling temperature at atmospheric pressure of isomalt and sucrose/corn syrup solutions (superscript 1 represents ratio 100/100, dextrose equivalent (DE) of corn syrup: 39–42%). *Source*: Adapted from Mende.[11]

due to the fact that sucrose/glucose syrup contains oligosaccharides that lower the boiling point evaluation of sucrose/glucose mixtures. Generally, vacuum is applied in the cooking process of hard candies to reduce the required temperature.[12]

11.3.6 Melting range

Isomalt exhibits no distinct melting temperature, but a melting range between 140°C and 155°C depending on the specific ratio of GPS/GPM. The isomers GPS and GPM or GPM dihydrate form an eutectic system and the minimum melting temperature represents the eutectic mixture corresponding to a composition of 50% GPM and 50% GPS.[12]

11.3.7 Hygroscopicity – moisture uptake at various relative humidities

The sorption isotherms (see Figure 11.10)[7] show that, at a storage temperature of 25°C and a relative humidity (RH) up to approximately 85%, no isomalt variants absorb an appreciable amount of moisture. Any moisture absorption up to this level of humidity is incorporated into the GPM crystal as water of hydration, since isomalt crystallises with two mol of water of crystallisation. Under these conditions, no free water is available and therefore isomalt can be stored and distributed without any special care.

This key physical property explains why products exclusively or mainly based on isomalt (e.g. hard candies) do not tend to be sticky and have a long shelf life. The addition of isomalt to products that are mainly based on hygroscopic ingredients like HSH syrup, polydextrose or sorbitol syrup (e.g. low boilings, soft cakes or gum products) improves their quality by reducing stickiness and extending shelf life.

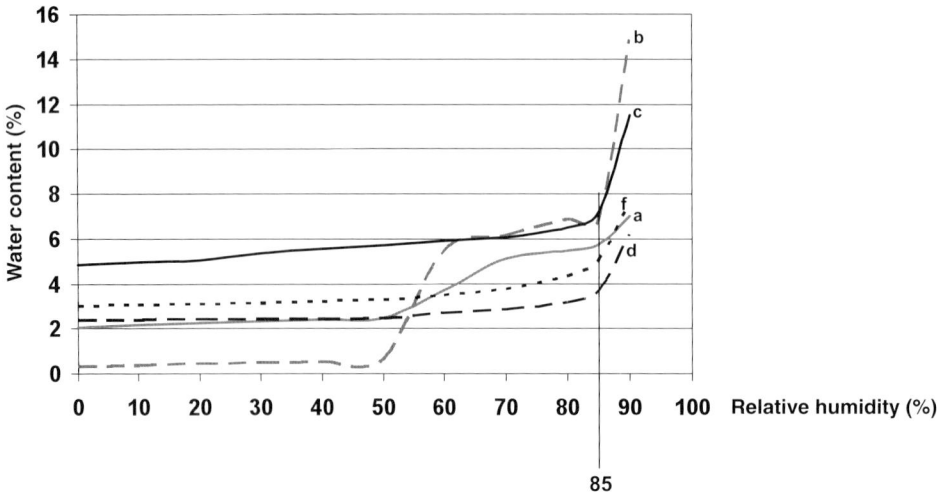

Fig. 11.10 Sorption isoterms of different isomalt variants (a, ISOMALT ST; b, ISOMALT LM; c, ISOMALT DC 100; d, ISOMALT DC 200; e, ISOMALT GS).
Source: Adapted from Südzucker AG, ZAFES (1999, Ref. 7)

11.3.8 Water activity at various concentrations versus sucrose

As shown in Figure 11.11, the water activity of isomalt solutions at various concentrations show they have a water activity similar to sucrose solutions at the same concentrations. This is due to the molecular nature and chemistry of the isomalt molecule as it has an affinity for water similar to sucrose.[13,14]

In Table 11.1, some physico-chemical properties of isomalt, GPS and GPM are listed and compared with sucrose.[7,13–19]

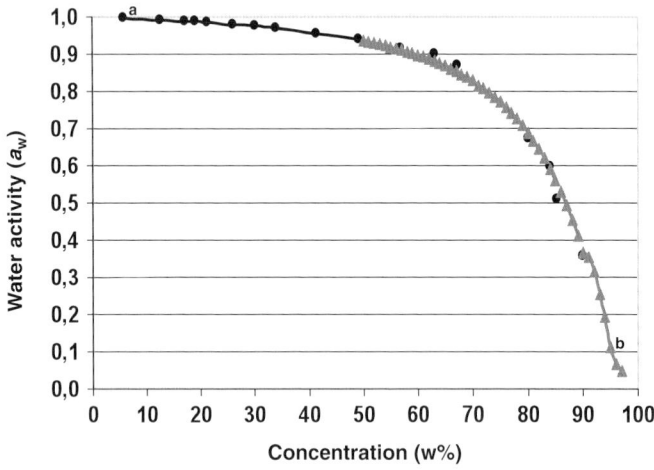

Fig. 11.11 Water activity of different substances with increasing concentration (a, isomalt; b, sucrose).
Source: a, Zielasko (1990); b, Bubnik (1995).

Table 11.1 Physico-chemical properties of isomalt compared to sucrose.

	Isomalt[a]	GPM	GPS	Sucrose
Molecular mass (g/mol)	—	344,324 anhydrous 380,356 dihydrate	344,324	342,303
Crystal group	—	Orthorhombic P2$_1$P2$_1$P2$_1$ [1]	Monoclinic P2$_1$ [2]	Monoclinic
Specific rotation	91.5° [3]	88.7° [3]	92.9° [3]	66.5°
Melting temperature/range (°C)	140…155 [6]	168 [4]	166 [4]	186 ± 4 [3]
Heat of fusion (kJ/mol)	46.5 [4]	55.0 [4]	56.36 [4]	46.41 [3]
Heat of solution (kJ/mol)	5.29 Isomalt ST, 0% water [4] 19.52 Isomalt ST, 5.2% water [6] 14.25 Isomalt GS, 2.7% water [6] −12.48 amorphous glass [4]	30.02 crystal dihydrate [4] 6.30 crystal anhydrous [4] −12.90 amorphous glass [4]	10.39 crystal [4] −10.99 amorphous glass [4]	6.22 crystal [7] −13,79 amorphous [7]
Glass transition temperature (°C)	63 [8]	65 [4]	50 [4]	65 [3]
Heat capacity (J/mol K)	480 crystal [5] 490 amorphous [5]	370 crystal anhydrous [5] 410 amorphous [5]	670 crystal [5] 610 amorphous [5]	425.8 crystal [3] 490.2 amorphous [3]

[1]: Lindner.[19]
[2]: Lichtenthaler.[16]
[3]: Bubník.[14]
[4]: Cammenga.[10]
[5]: Zielasko.[13]
[6]: Südzucker CRDS, internal data.[7]
[7]: Gehrich.[18]
[8]: Raudonus.[17]
[a]Physical data of isomalt depend on actual ratio of GPM/GPS and on water content.

11.4 PHYSIOLOGICAL PROPERTIES

Isomalt is a hydrogenated carbohydrate and more specifically, it is a disaccharide alcohol. As such, it must be hydrolysed into its monomers (glucose, sorbitol and mannitol) to be absorbed. Carbohydrases in the small intestine, the sucrase–isomaltase complex and the glucoamylase–maltase complex of the intestinal brush border hydrolyse isomalt much more slowly than maltose or sucrose.[20–22] *In vitro* experiments in humans show that the relative rates of cleavage of isomalt by intestinal mucosal enzymes were 100:25:2 for maltose:sucrose:isomalt.[21]

On the basis of a number of studies in humans and animals, it can be concluded that in humans, a maximum of 10% of isomalt intake is absorbed and about 90% is fermented in the large intestine.[23]

Blood glucose response after the intake of isomalt is very low. Among healthy people, type-1 and type-2 diabetics, results showed that the highest blood glucose response to isomalt was 12% that of glucose, while the lowest was 2%. The insulin response to isomalt is also low. Unlike other carbohydrates such as sucrose or glucose, no reactive hypoglycaemic effect occurs with isomalt.[24] Consumption of snack foods or milk chocolate made with isomalt results in significantly lower increments in post-prandial glycaemia and insulinaemia than would result from a similar food made with sucrose. An improvement of the glycaemic control, measured as glycosylated haemoglobin Hb_{A1C}, after intake of 24 g of isomalt/d over a period of 12 weeks was demonstrated.[23] In a long-term human intervention study with type-2 diabetic subjects ($n = 31$) receiving a diet with foods containing 30 g isomalt per day, significant reductions in glycosylated haemoglobin, fructosamine, fasting blood glucose, insulin, proinsulin, C-peptide, insulin resistance (HOMA-IR), and oxidised LDL were observed after 12 weeks. Routine blood measurements and blood lipids remained unchanged. This is a clear demonstration of a significant improvement in the metabolic control of diabetes.[25] In healthy volunteers and volunteers with hyperlipidaemia, no influence on blood lipid parameters was observed, as well after the intake of 30 g isomalt or sucrose per day over a 4-week period ($n = 19$, controlled diet, randomised double-blind cross-over design), apart from lower apolipoprotein for all subjects.[26]

The low glycaemic properties of isomalt and its ability to reduce the post-prandial glycaemic response of food when replacing sugars were confirmed by the European Food Safety Authority (EFSA) during the claims evaluation process.[27]

In the large intestine isomalt is completely fermented by the bacterial flora and an increase in total biomass has been reported. The bacterial fermentation products are short-chain fatty acids, CO_2, CH_4 and H_2. *In vitro* it was demonstrated, that isomalt was a good source for butyrate production. With respect to the bacterial composition, a particularly significant increase of bifidobacteria was shown in a double blind cross over study with 19 volunteers after intake of 30 g isomalt per day for a period of 28 days, demonstrating the prebiotic effect of isomalt as a significant growth stimulation for bifidobacteria. The total bifidobacteria cell counts increased by 47% and the proportion of bifidobacteria growth with isomalt compared with sucrose increased by 65%. The total activity of the bacterial β-glucosidase was reduced after the intake of isomalt. Water content of the faeces was slightly higher (not significantly) in the isomalt group (12%) and stool frequency was slightly, but significantly, increased (1.3 vs. 1.1) indicating another positive aspect of isomalt in gut health - the prevention of constipation. The mean transit time was unaffected and the diet was well tolerated in both cases.[28]

All low- or non-digestible carbohydrates lead to a higher water content of the chyme in the small intestine owing to the unabsorbed molecules binding water and thus making the chyme more fluid. The large intestine is the place where re-absorption of water takes place and the capacity of re-absorption influences the consistency of the faeces. If the capacity for re-absorption of this water in the large intestine is exceeded, watery stools might occur. This is the physiological background for the laxative effect that low digestible carbohydrates can have. If this is found to be uncomfortable, it is recommended that the intake of low-digestible carbohydrates is reduced for some time to give the body time to adapt and establish its tolerance level.[29]

Studies that were carried out to specifically investigate tolerance are of limited value with respect to a real life situation as they are often based on questionnaires that prompt volunteers to report sensations, which they normally might not have even noticed.[23,29]

From tolerance studies carried out with isomalt, it can be concluded that up to 50 g per day spread over the day, as would normally be the case when consuming sugar-free candies or chewing gums, for example, are well tolerated by most individuals.[30–38]

Owing to its limited digestion and metabolism, isomalt provides less energy or calories (joules) to the body than highly digestible carbohydrates such as sugar, glucose or starches. A large number of studies in animals and humans with various study designs are available to evaluate and establish how much energy isomalt provides. The US Federation of American Societies for Experimental Biology (FASEB), Life Science Research Office (LSRO), prepared an assessment of the energy values of certain polyols, including isomalt, used as food ingredients.[39] On the basis of the aforementioned studies they established an energy value of max. 2 kcal/g for isomalt; this being consistent with a value of 90–100% of isomalt energy escaping digestion and being fermented in the colon. The scientifically based value of 2 kcal/g can be used for food labelling purposes in a number of countries including the United States and Canada where the scientific basis was evaluated by Health Canada.[40] The European Union fixed one common energy conversion factor for polyols at 2.4 kcal/g or 10 kJ/g.

Another important property of isomalt is related to oral health and the prevention of dental caries. Isomalt's cariogenic potential was classified as 'none'.[41,42]

This is due to the fact, that almost no micro-organisms of the oral flora are capable of fermenting isomalt. This is demonstrated by measuring acid production after rinsing with an isomalt solution (intra-oral and intra-plaque). In addition, a number of experiments with animals, humans and *in vitro* were used to prove the non-cariogenic properties of isomalt.[43]

While the cariogenic potential based on acidogenesis is a passive mechanism and undoubtedly the major effect, there are some other minor active mechanisms to be considered for both isomalt and xylitol. For example, the bacteriostatic properties of xylitol, the inhibition of polysaccharide syntheses (plaque) for isomalt and the promotion of remineralisation, in particular for isomalt but also for xylitol.[23] As these effects are based on *in vitro* studies, their relevance in practice is not proven; therefore, a superiority claim for any one polyol is not justified.[44]

Isomalt received an approved health claim in the United States, as it is included in the Health Claim Regulations of the US FDA on dietary, non-cariogenic carbohydrate sweeteners and dental caries (21 CFR 101.80).[45] In the EU, EFSA evaluated isomalt in the context of their claims evaluation process related to claims on dental health. EFSA concluded that consumption of food containing isomalt instead of sugar may help maintenance of tooth mineralisation by decreasing tooth demineralisation.[27]

Both authorities, FDA and EFSA, made reference to the pH-telemetry methodology according to which the consumption of the food that is intended to bear the dental claim should not lower plaque pH below 5.7 during and up to 30 minutes after consumption. In Japan, isomalt and its dental properties are covered by the FOSHU regulation.[46]

11.5 APPLICATIONS

On the basis of its properties, the isomalt variants are ideal sugar replacers in many applications in the food and pharmaceutical industries.

Isomalt's outstanding properties make it the preferred choice, particularly in the field of sugar-free confectionery with regard to taste quality and shelf life. In terms of production engineering, existing processing equipment can be used for all applications without requiring major changes. In some cases, slight formula and process parameter modifications are recommended. In addition to panned goods, chewing gum and chocolate, sugar-free hard candy is an outstanding application for ISOMALT ST.

11.5.1 Hard candies

A wide range of sugar-free boiled sweets whether deposited, stamped or filled can be produced with ISOMALT ST. Hard candies with a longer shelf life will be obtained if the water content in the finished product is below 2%. These candies are stable with regard to water absorption. The production process is very similar to that for candies based on sucrose and glucose syrup with only minor changes to formulae and processing being required. During the manufacture of isomalt hard candies, either batch or continuous, the following characteristics, which differ from sucrose/glucose syrup, have to be taken into consideration:[47]

- Lower solubility
- Higher boiling point
- Lower viscosity of the melt
- Higher specific heat capacity

Table 11.2 shows a basic recipe for isomalt hard candies.[48]

In some cases, the preparation of a pre-solution of isomalt may be necessary and depending on the equipment used for the dissolution, the ratio between isomalt and water may need to be changed. Generally, a basic syrup with 75–80% dry matter at approximately 110°C should be prepared. When isomalt is blended with other bulk sweeteners (e.g. syrups) in a recipe, the water content of these ingredients should be taken into account.

Table 11.2 Basic recipe for isomalt hard candies.

Ingredients	%
ISOMALT ST-M	75.00
Water	24.00
Citric acid	0.80
Flavour	According to manufacturer's data
Colour solution	According to manufacturer's data
Intense sweetener (if required)	Depending on sweetening power
Total	100.00

Source: BENEO-Palatinit GmbH.[48]

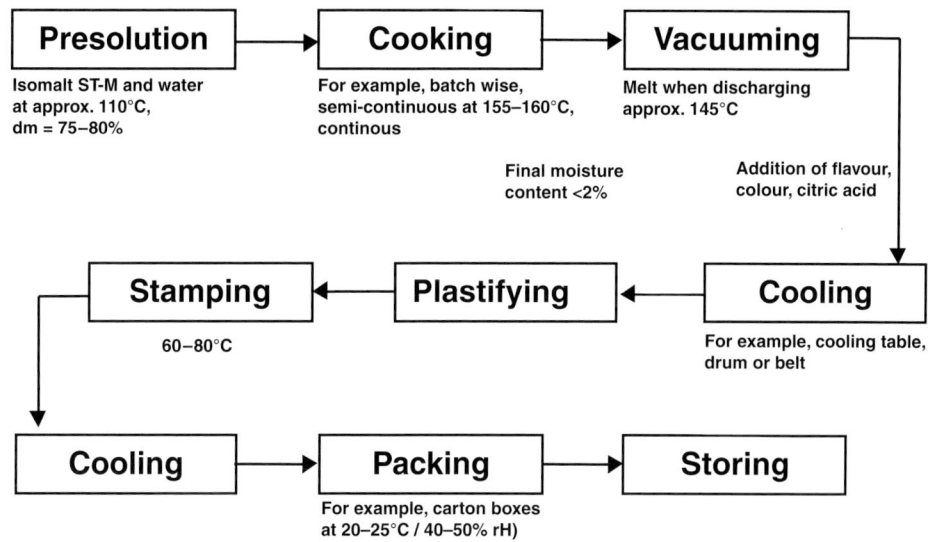

Fig. 11.12 Flow chart of an isomalt candy process. Source: Adapted from BENEO-Palatinit GmbH.[48]

11.5.1.1 Production process

A simplified flow chart showing the production of isomalt hard candies is given in Figure 11.12.

As with sucrose, it is important to dissolve all of the crystalline material to prevent uncontrolled crystallisation during further processing. Since isomalt has a lower solubility than sucrose, a higher temperature and – in the case of continuous dissolving – longer retention times are required for dissolution.

In practice, this means that the throughput, for example in a continuous or pressure dissolver, has to be reduced compared with sucrose/glucose syrup if it is not cooked under vacuum, and additionally, the pressure of the steam to achieve the dissolution has to be sufficient (>5 bar or 72.5 psig) to guarantee a minimum temperature of 110°C.

A clear crystal free solution with a dry matter of 75–80% can be obtained when the isomalt/water mixture (preferably, hot water pre-added) is heated to approximately 110°C. Experience has shown that a batch-wise system, for example an open, heated and agitated (double-jacketed) kettle, is one of the most suitable methods for completely dissolving isomalt. To guarantee continuous production, an evaporation tank between the dissolution vessel and the cooker has to be used.

11.5.1.2 Cooking

Isomalt hard candies can be produced in all the main types of batch, semi-continuous and continuous candy cookers. As previously described, a higher boiling temperature is necessary to achieve a water content of <2% in isomalt hard candy. Depending on the layout of the equipment used with respect to the applied vacuum, the following boiling temperatures are recommended:

- Cooking without applied vacuum: ~ 165°C (steam pressure: > 8 bar or 116 psig)
  For example,
  ○ Atmospheric batch cooker

- Cooking with vacuum applied after cooking: 155–160°C (steam pressure: ~ 8 bar or 116 psig)
  (vacuum: ~ 0.9 bar or 27″ Hg)
  For example,
  ○ Atmospheric/vacuum batch cooker
  ○ Semi-continuous cooker
  ○ Continuous coil cooker
  ○ Micro film heat exchanger/evaporator
  ○ Rotory cooker

- Cooking under vacuum: 135–140°C (steam pressure: ~ 5.5 bar or 80 psig)
  For example,
  ○ Vacuum batch cooker
  ○ Semi-continuous vacuum cooker
  ○ Continuous vacuum coil cooker
  ○ Vacuum micro film heat exchanger/evaporator

11.5.1.3 Cooking under vacuum

Owing to the higher boiling temperature, isomalt cooked masses show a low viscosity. The crucial part to the successful adaptation of *continuous processing* to isomalt consists of a temperature reduction during cooking, and this can be achieved by cooking under vacuum. This results in a subsequent viscosity increase for the cooked mass facilitating discharge of the mass from the vacuum chamber and further processing. To guarantee optimum turbulence and therefore mixing of the isomalt solution, the vacuum chamber should be filled to approximately one-third of the chamber volume. If cooking under high vacuum is not possible, a reduction of the mass throughput compared with the traditional formulation using sucrose and glucose syrup might be required to ensure the longest possible holding time of the melt in the vacuum chamber.

11.5.1.4 Discharging

Depending on the equipment used and the quality and time of applied vacuum, the discharge temperature of the isomalt melt varies between 130°C and 150°C. In the case of continuous coil cookers, the draw-off rollers or the extraction screw should not be heated as high as with traditional candy (steam pressure of approximately 0.5 bar or 7.25 psig versus a normal steam pressure of approximately 3 bar or 43.5 psig). This ensures a lower operating temperature that will improve the adhesion of the cooked mass and maintain a continuous discharge.

11.5.1.5 Additives

While the addition of additives can be achieved easily in a mixing screw (e.g. inline mixer), when processing continuously, a cooling of the melt is required depending on the discharge temperature required. Ideally, the temperature of the melt should not exceed 110°C to avoid too much flavour loss. While heat-stable Sucralose® and acesulfame K can be added during the dissolving process aspartame has to be added after vacuum cooking. To avoid an agglomeration in the hot melt and to guarantee a homogeneous distribution of the aspartame, it can be blended with other crystalline additives (e.g. acids, menthol) or dissolved in liquid citric acid, for example (this has then to be injected into the vacuum chamber).

11.5.1.6 Cooling/plastifying

Owing to the higher cooking temperature (except when cooking under vacuum) of an isomalt melt, the viscosity is lower than that of a corresponding sucrose/glucose syrup melt. In addition, the specific heat capacity of isomalt is approximately 17% higher than that of sucrose. Both have to be taken into consideration when cooling down the isomalt melt until a plastified mass is obtained and it will take longer to cool isomalt masses to a temperature of 100–110°C where the viscosity of the mass is comparable to that of sucrose/glucose syrup melt. The cooling until a plastified stampable mass is reached (60–80°C) is extended since isomalt has a high-heat capacity value (cp-value). In practice, when processing continuously, the throughput of the mass to be cooled is already reduced by the previous process steps as described. Therefore, the cooling capacity of the cooling belt (reduced speed) is sufficient in most cases.

11.5.1.7 Batch roller/stamping

No special requirements are needed to process on a batch roller and in the subsequent stamping process. Depending on the recipe, the ready-to-work-with temperature of the plastified isomalt candy mass can vary between 60°C and 80°C on the batch roller. During stamping, the temperature may vary between approximately 60°C and 70°C.

11.5.1.8 Cooling of stamped candies

In the cooling unit (conveyer belt/cooling tunnel), the stamped candies are cooled down to approximately 25–30°C until packed or stored.

11.5.1.9 Storage/packaging

While hard candies based on hygroscopic ingredients have to be packed immediately after cooling in moisture protective packaging materials, isomalt-based candies can be stored before packaging if required. This can be helpful with respect to the packaging capacity.

The candies might be slightly adhesive immediately after production. This property disappears after approximately 24-hour storage at 20–25°C and 40–50% RH. Isomalt candies keep their gloss, have very low hygroscopicity and remain stable for a long period of time. No twist-wrapping and, more important, no excessive packaging material such as aluminium-coated foils or single-flow-packs are required. Owing to the very low moisture absorption, a convenient flip-top carton box wrapped with polypropylene foil is sufficient as packaging. The most important characteristic is the water vapour permeability of the foil. It is recommended to be less than $1.0 \text{ g/m}^2/24 \text{ h}$ (DIN 53 380, at 23°C and 85% RH).

To guarantee smooth packaging of candies in bulk or carton boxes, the environmental conditions in the packaging room should be kept at the equilibrium RH of isomalt candies that is between 40% and 50% RH at 20–25°C, depending on the formulation and the aforementioned conditioning period.

11.5.1.10 Centre-filled and other speciality stamped hard candies

All kinds of filling material and filling methods (e.g. filling tube or co-extrusion) can be used to fill isomalt hard candies. As with the filling process for candies based on sucrose/glucose syrup, a temperature equilibrium between the candy mass and the filling is necessary, for example, molten isomalt chocolate, fruit fillings based on HSH syrup and fruit pulp or chewable masses. As an example for crystalline fillings, a pleasant flavour and mouth-cooling effect can be achieved by using blends of xylitol, edible acids, vitamins, sherbet powder (sodium bicarbonate/tartaric acid) or flavour in different compositions according to the desired taste.

Also, specialities like laminated (e.g. with no added sugar peanut paste), pulled, striped or layered isomalt hard candies can be manufactured using traditional technologies.

11.5.1.11 Deposited hard candies

A wide range of deposited isomalt hard candies can be produced using conventional depositing technology. Depending on the flexibility of the depositing system, centre-filled, striped or layered candies that are smooth, glossy and transparent can be manufactured. The recipe and production of the isomalt melt to be deposited is similar to that previously described, until the addition of additives. Preferably, the melt is deposited in Teflon moulds (e.g. an endless Teflon moulding belt) at a minimum depositing temperature of 130°C. Depending on the moulding belt length and the cooling capacity, the candies can be demoulded after approximately 15–20 minutes. When cooling the sweets, the specific heat capacity of isomalt, which is approximately 17% higher than sucrose, and the candy size have to be taken into consideration.

When a continuous belt is used, the belt speed may need to be decreased.

11.5.1.12 Pharmaceutical hard candies

High-quality pharmaceutical hard candies like cough drops can be produced with isomalt. Taking into consideration the process temperature, suitable medical additives like menthol are added to the isomalt melt in the same manner as other additives. To guarantee a homogeneous distribution of the actives, an appropriate mixer has to be used. Owing to the very low moisture absorption of isomalt hard candies, the shelf life with respect to, for example, oxidation or hydration of actives is extended. Another advantage of isomalt in pharmaceutical hard candies is its lower solubility. This physical property leads to a slower release of actives when sucking the candy, thereby extending the effect (see Figure 11.13).[49]

11.5.1.13 Shelf-life/storage properties

Isomalt-based products provide a similar taste and texture compared with traditional products. In addition, their shelf life is extended owing to the unique physical and chemical properties of isomalt. Hundred percent ISOMALT-ST-based candies have an excellent shelf life due to their very low water absorption. No stickiness or cold flow occurs during storage (see Figure 11.14).[7]

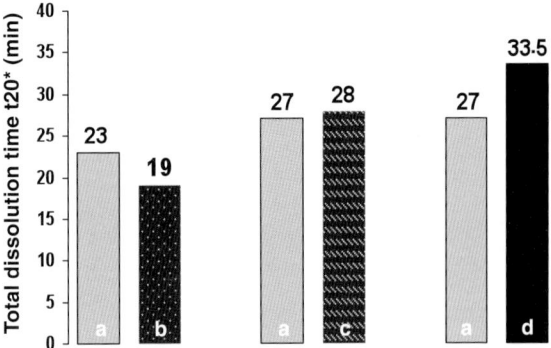

Fig. 11.13 Total dissolution time of three hard candy pairs (sugar/sugar-free). t20* is the required dissolution time to reach a concentration of 20% dry solids; Different values due to differences in size and surface of the three pairs (a, sucrose/glucose syrup; b, sorbitol syrup; c, maltitol syrup; d, ISOMALT ST). *Source*: Adapted from Dörr and Willibald-Ettle.[49]

11.5.2 Chocolates

ISOMALT LM chocolates already on the market show that their outstanding qualities, their sugar-like neutral taste, snap and melting behaviour are preferable to other sugar-free chocolates. No cooling effect is present when eating ISOMALT LM chocolate.

Furthermore, the very low glycaemic and insulinaemic effect of isomalt contributes to the formulation of low glycaemic products like chocolate-coated cereal bars, cookies or chocolate bars.

When establishing a recipe for an ISOMALT LM chocolate, the normal recipe for chocolate based on sucrose can be used. Depending on the refining process and specific requirements ISOMALT LM-E or ISOMALT LM-PF should be used.

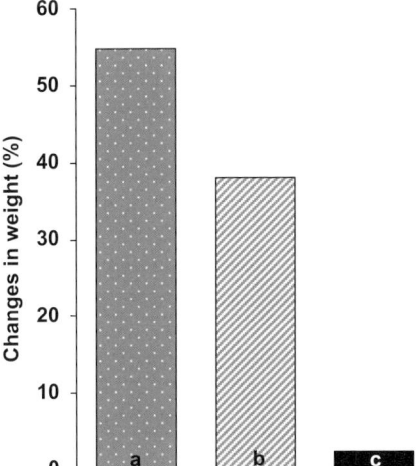

Fig. 11.14 Changes in weight by water absorption based on initial weight of hard candies. Storage test at 25°C, 80% RH after 7 days without packaging (a, sorbitol syrup; b, HSH; c, ISOMALT ST). *Source*: Adapted from Südzucker AG, ZAFES (2000, Ref 7).

Table 11.3 Basic recipe for chocolate.

Ingredients	Dark chocolate (%)	Milk chocolate (%)
ISOMALT LM-E or LM-PF[a]	45.60	42.00
Cocoa liquor	45.20	12.90
Cocoa butter[b]	8.50	20.00
Whole milk powder	—	21.40
Hazelnut paste	—	3.00
Lecithin	0.50	0.50
Flavour	0.10	0.10
Sweeteners[c]	Approximately 0.10	Approximately 0.10
Total	100.00	100.00

Source: BENEO-Palatinit GmbH.[48]
[a]ISOMALT LM-E has superior handling properties during pre-refining and refining due to its dust free quality. If, however, a two-step refining process cannot be realised in the production, it is advisable to use ISOMALT LM-PF to reach optimal refining properties and fineness.
[b]In case a softer texture is desired, up to 2% of cocoa butter can be replaced by purified butter fat.
[c]By addition of intensive sweeteners, the sweetness of the chocolate can be varied as you like. A 1:1 blend of acesulfame K and aspartame gives a superior sweetness profile. Alternatively, approximately 0.02% sucralose can be used to reach an iso-sweet product.

If a two-phase refining process is possible, ISOMALT LM-E has superior handling properties due to its dust-free quality. In the case of traditional processes with a one-step refining or production system like ball mills, ISOMALT LM-PF would be the preferable variant.

To obtain the same sweetness as sucrose chocolates, intense sweeteners like aspartame and/or acesulfame K and/or Sucralose may be added in small amounts. The concentration of the intense sweetener also depends on the flavour (e.g. vanilla) and its concentration.

In some formulations as shown in Table 11.3,[48] the sweetness of isomalt might be sufficient and addition of an intense sweetener would not be necessary. This depends very much on the overall recipe where components like milk powders or cocoa may contribute to the overall sweetness.

To achieve a further 'calorie-reduction' in chocolate, the total fat content can be reduced (e.g. to approximately 30%) and a part of ISOMALT LM (up to 50%) can be replaced by bulking agents like polydextrose (Litesse®) or the prebiotic fibre inulin (Orafti®).

For 'tooth-friendly' milk chocolate, milk powder with a low lactose content, that is milk proteins (e.g. sodium caseinates), have to be used in the formulation.

If a fibre and/or prebiotic claim is desired, ISOMALT LM can be combined with, for example 5–10% inulin (Orafti®).

11.5.2.1 Production by a conventional method

As shown in Figure 11.15,[48] a conventional production line for chocolate can be used for ISOMALT LM chocolates. Cocoa mass, milk powder, ISOMALT LM (type depending on the refiner system) and a part of the cocoa butter are blended in a kneader. The homogeneous mass is then refined until a particle size of about 20 μm is obtained. The refined chocolate powder (flake) is then transferred to the conche by pre-adding the residual cocoa butter.

As usual, the water content of all recipe ingredients, especially milk powder, should be as low as possible. This lowers the risk of agglomeration (e.g. of sugar/polyols, proteins and cocoa fibres) during the conching process that would give a rough mouthfeel to the chocolate.

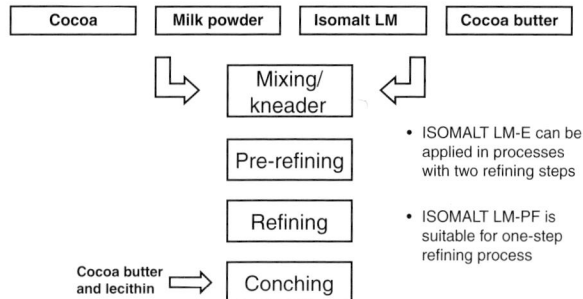

Fig. 11.15 Production of ISOMALT LM chocolate.
Source: Adapted from BENEO-Palatinit GmbH.[48]

ISOMALT LM has a very low moisture content (<1%). Owing to the very low water content and low hygroscopicity of ISOMALT LM (under normal chocolate production conditions <50% RH), the same conching temperature as in conventional sugar-based chocolate production can be used, that is for milk chocolate 70°C, for dark chocolate 80°C.

As in sugar chocolate production, the ISOMALT LM containing chocolate mass must have a plastic consistency in the conche (in the first hours of conching, wet conching is recommended). Since the crystal morphology of ISOMALT LM is different from that of sucrose (the specific surface area is larger), a slightly higher fat content might be required during refining and at the beginning of conching. At this stage, more fat is required to cover all the crystals with fat to give the correct plastic consistency to optimise the conching process.

The need for a higher fat level in the process depends on the desired final fat content of the chocolate and has to be adjusted depending on the individual formulation, as is usually done with sugar formulations. Depending on the type of conche used (dry- or wet-conching systems), fat content must be taken into consideration.

For chocolates with a very low fat content (e.g. below 29%), the addition of lecithin at the beginning (or even during the filling phase) and during conching can be useful to adjust the consistency for an optimal conching process.

Alternative chocolate production methods like McIntyre, IIoveras or Wiener ball mill systems are also suitable for the production of isomalt chocolates.

Liquid ISOMALT LM chocolate can be best stored at a temperature of approximately 45°C. Occasional agitation is recommended. When melting isomalt bulk chocolates, the temperature should not exceed 55°C.[50]

11.5.3 Low Boilings

Low-boiling technology can be used to produce chewable confectionery. In this context, sugar-free low boilings require a balance between the crystalline and the non-crystallisable phase, gelatine and fat to achieve a chewable product and to prevent stickiness. In sugar-free formulations, ISOMALT GS corresponds to sucrose, that is the crystallisation promoting component of the formula and hydrogenated starch hydrolysate (HSH or maltitol syrup corresponds to the glucose syrup component, i.e. the crystallisation impeding component of the formula).

- Mixing of water, ISOMALT GS, polyol syrup, e.g. maltitol syrup, gum arabic, vegetable fat, emulsifiers, Sucralose®
- Boiling to 125–130°C; target moisture content: approx. 7%
- Vacuuming from case to case
- Addition of citric acid and ISOMALT ST-PF, mix homogeneously
- Adding flavour and mix, empty the kettle
- Cooling down the mass to 44–46°C
- Pulling the cool mass for 5–10 minutes (temperature then 47–49°C)
- Alternatively the process stage pulling can be replaced by utilising a pressure whipping machine
- Cooling of the pulled mass until the optimal consistency for further processing has been reached
- Optional seeding with ISOMALT ST-PF (<10%)
- Forming/cutting

Fig. 11.16 Manufacturing of isomalt low boiling.
Source: Adapted from BENEO-Palatinit GmbH.[48]

The consistency and chewability of the final product depend on the crystallisation properties of the crystallisable component, and are also influenced by the final moisture content of the low boiling mass and other ingredients. In the case of sucrose/glucose syrup-based low boilings, the sucrose part re-crystallises quickly and is controlled at a water content of 6–10%. If the moisture content is too low, no crystallisation occurs and the candy hardens like a high boiling (amorphous-like). A water content which is too high would lead to a product which is not form stable (e.g. cold flow, sticky). The same procedure can be used for the sugar-free versions, with one major difference. Since all sugar substitutes do not crystallise as spontaneously as sucrose, an addition of seeding crystals as a crystallisation initiator is recommended.

It has been shown experimentally that by seeding the low boiling mass with powdered ISOMALT ST (ST-PF), after the boiling process and before the homogenisation or pulling step, an improvement in the control of the shape stability and chewing quality of the final product is possible. At the same time, there is an increase in the viscosity of the caramel mass, reduced stickiness and increased form stability. If no seeding is initiated, the dissolved isomalt would re-crystallise slowly. Since isomalt re-crystallises with two mol of water, the content of free water in the mass decreases over time until a value is reached where the product becomes hard. Therefore, it is very important to speed up the crystallisation by seeding with powdered isomalt while maintaining a constant water content (approximately 7%) during production. Owing to the fact that isomalt has a low solubility, part of the added isomalt powder will remain as crystallisation seeds for the dissolved isomalt and the finished quality will be obtained very quickly. The product remains unchanged during further prolonged storage.

The homogeneous distribution of the added powder has a decisive influence on the structure and storage stability of the low boiling. Distribution can be achieved in different ways: With a homogenising machine (e.g. Zehnder Homozenta) or by beating in the powder under pressure in a batch pressure beater (e.g. Ter Braak system); see Figure 11.16.[16,48]

Table 11.4 gives a formulation for a fruit flavoured low boiling with gum Arabic.[48]

Table 11.4 Basic recipe for isomalt chewy candies.

Ingredients	%
ISOMALT GS	24.40
ISOMALT ST-PF	8.40
Maltitol syrup	51.00
Water	7.85
Vegetable fat	5.64
Gum Arabic	0.95
Citric acid	0.90
Emulsifiers[a]	0.60
Flavour	0.25
Sucralose	0.01
Total	100.00

Source: BENEO-Palatinit GmbH.[48]
[a]For example, E 471 Grindsted HA 40 (Grindsted/DK) and E 473 Sisterna SB50 (Sisterna/NL) at a ratio 2:1

11.5.4 Chewing gum

For the production of chewing gum strips, pellets or balls, ISOMALT ST-PF is recommended. Owing to its low solubility and preferred sweetening profile, isomalt can be used as a functional ingredient in sugar-free chewing gum. The low solubility causes isomalt to remain in crystalline form in the chewing gum mass, which leads to less hardening of the chewing gum during storage. Recent findings show that the substitution of 100% of the mannitol and part of the sorbitol by isomalt leads to this improved stability (see Figure 11.17).[51] Research has shown that chewing gum containing 10–30% isomalt remain more flexible for a longer time than other sugar-free chewing gum. The machinability of the chewing gum mass is

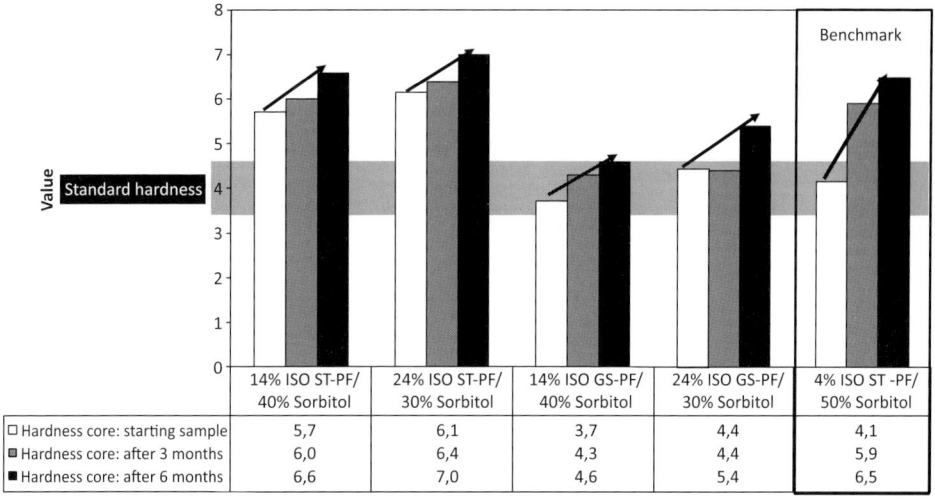

Fig. 11.17 Sensorial evaluation of isomalt containing chewing gum centres during shelf life at 30°C, 70% RH, 6 months, compared to sorbitol based chewing gum centres. Sensorial evaluation: scale: 0 (too soft) – 8 (too hard).
Source: Adapted from Südzucker AG, CRDS Offstein.[51]

improved by the addition of isomalt owing to less stickiness during process (extrusion, rolling and scoring). Less cleaning is necessary as less waste occurs during the cleaning of the Z-blade mixer because isomalt firms up the chewing gum mass making it much easier to discharge. This property leads to an improved filling and sealing of centre-filled chewing gum, especially during the extrusion process.

Sensory evaluation has shown that the low solubility of isomalt initiates an improved longer-lasting effect regarding the extended release of flavour and sweetness.

11.5.4.1 Isomalt flavour crystals

Owing to the porous surface of the isomalt crystal (granule), it can absorb a certain amount of flavour or colour. As the solubility of isomalt is low, dissolving during chewing is slower than sugar, so that its effectiveness for maintaining flavour lasts longer than sugar. Isomalt flavour crystals therefore give a pleasant flavour sensation and a 'gritty' effect to the chewing gum.

11.5.5 Pan coating with ISOMALT GS

ISOMALT GS is well suited for this application owing to its low solubility and enhanced tendency to crystallise.

Owing to its outstanding properties, ISOMALT GS coatings provide a variety of possibilities for the confectionery manufacturer who can benefit when using this material in sugar-free pan coated chewing gum where ISOMALT GS provides a good crunch and the coating has an excellent shelf life. This application is also important for the pharmaceutical industry, where active ingredients need to be protected. Centres containing pharmaceutical ingredients can be coated with ISOMALT GS to form a protective barrier.

The low hygroscopicity of ISOMALT GS-coated products means that no expensive packaging material such as blister packaging is required, a simple carton packaging such as a ZetKLIK-box is sufficient.

Hard coating with ISOMALT GS needs considerably shorter coating time compared to other polyols.

Another major advantage of ISOMALT GS in coating is its stability against the abrasion forces experienced during coating; this reduces damage and results in well-defined corners and edges of coated products.

11.5.5.1 ISOMALT GS pan coating

Coatings with ISOMALT GS can be easily applied with traditional coating equipment and processes and open as well as all closed automated coating systems can be used.

The coating procedure will depend on the type of coating equipment and on the type of centre required – whether it is a solid, a soft-centre or a chewing gum. Any type of convex centre can be coated with ISOMALT GS. In case the centre surfaces are rough, sticky or fatty, they should be pre-coated with a gum Arabic solution and a dusting powder (ISOMALT ST-PF) to bind moisture and form a firm, smooth surface. The amount of isomalt powder must be adjusted until the centres are suitable for final coating. In the case of chewing gum pellets, no pre-coating is normally required.

The improved solubility of ISOMALT GS allows a low working temperature of approximately 55°C at 65% dry solids of isomalt in the coating syrup. This is of advantage if heat sensitive ingredients are applied to the coating syrup.

11.5.5.2 Hard coating with ISOMALT GS

Hard coating follows the physical principles of crystallisation. The solution is applied to the rotating centres and dispersed on the surface. Water is evaporated with warm or cold air and thin crystalline layers develop on the surface. The air inlet temperature should be 10–40°C depending on centre type, and the humidity of the drying air should be as low as possible (maximum 40% RH).

During the coating process, the addition of solution should be adjusted to avoid overdosage and to guarantee a good distribution. At the beginning of a coating process, it is essential to limit the dosage of solution per layer to an amount sufficient to wet the centres completely. The ideal amount of solution per layer has to be adjusted in the course of the coating process depending on the character of the centres, the equipment used and on the coating conditions during the process. Some centres like chewing gum pellets or chewy candies tend to be sticky during the first three to four cycles. To overcome this period of stickiness, it is advisable to dry charge with ISOMALT ST-PF (in a 1:1 ratio with the coating syrup) as soon the solution is well distributed.

To achieve a smooth surface, it is recommended that the soluble solids content of the solution is lowered to decrease the dosage of syrup per wetting and to increase distribution times before drying with air. Colouring and polishing agents (e.g. carnauba wax) can be used as in traditional sugar coating. If white-coated products are required, the addition of titanium dioxide at up to 1% to the coating solution is recommended. This will improve the surface quality and appearance of the coated product.

The outstanding properties of ISOMALT GS lead to perfectly coloured coatings. Isomalt coatings become transparent on crystallisation, like glass incorporating the colour into its structure. This results in a brilliant appearance, especially for dark colours like artificial or natural reds.

ISOMALT GS gives excellent shelf-life performance. Based on its low tendency to absorb moisture, ISOMALT GS coated products are highly suitable for simple loose packaging (e.g. flip-top boxes).

The required coating time depends on the nature of the centres and the desired coating thickness and an extended coating time is needed compared with sugar coating. In comparison to other sugar replacers, (see Figure 11.18)[7] isomalt needs a significantly shorter coating time.

Table 11.5 shows a typical ISOMALT GS solution coating formulation.[48]

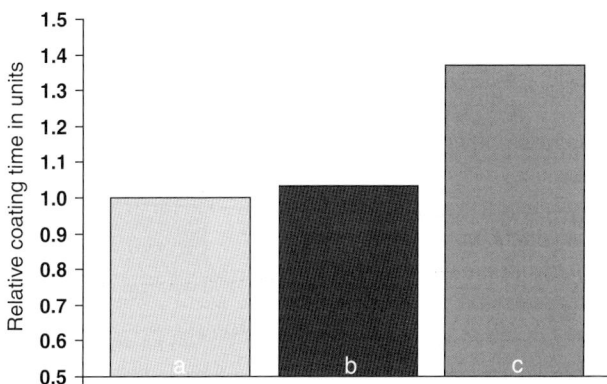

Fig. 11.18 Relative coating time of different bulk sweeteners. Method: Driacoater Vario 500/600 (a, sucrose; b, isomalt; c, maltitol).
Source: Adapted from Südzucker AG, ZAFES (2001, Ref 7).

Table 11.5 Basic recipe for ISOMALT GS solution coating.

Ingredients	%
ISOMALT GS	65.00
Water	29.80
Aspartame	0.05
Acesulfame K	0.05
Gum arabic 50%	4.10
TiO2	1.00
Total	100.00

Source: BENEO-Palatinit GmbH.[48]

11.5.6 Compressed tablets

ISOMALT DC was designed for direct compression to manufacture sugar-free confectionery or over-the-counter (OTC) tablets and it can be compressed on conventional machines (e.g. rotary presses).

It is essential to have a compressible powder with excellent flowability, with a narrow particle size distribution and without a tendency to become sticky.

ISOMALT DC is available in two variants. ISOMALT DC 100 is based on ISOMALT ST, and the ISOMALT DC 200 is based on ISOMALT GS.

The two variants have different properties when consumed (sucked). ISOMALT DC 200 results in a smoother surface with faster flavour and drug release, as it is more soluble than ISOMALT DC 100.

Both variants have a similar crushing strength.

As shown in Figure 11.19, ISOMALT DC is highly flexible and enables the production of tablets with almost any desired hardness and dissolution characteristics, be it lozenges, chewables or effervescents.[7]

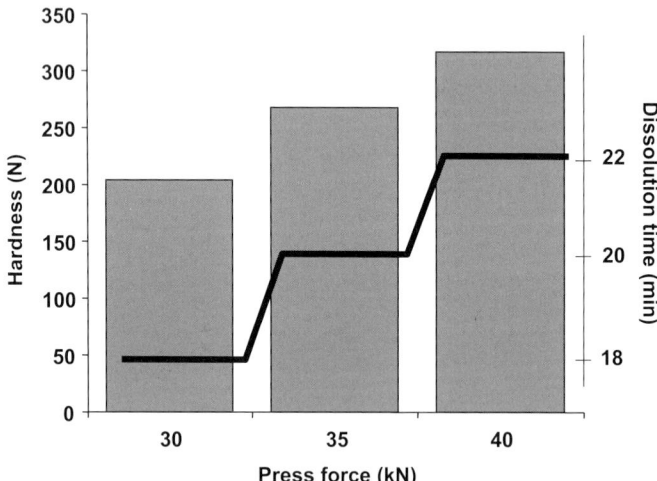

Fig. 11.19 Comparison of the dissolution time of isomalt tablets depending on applied press force. Tablet geometry = 18 mm; m = 1000 mg; with facet.
Source: Adapted from Südzucker AG, ZAFES (2000, Ref 7).

Table 11.6 Basic recipe for ISOMALT DC tablet formulation.

Ingredients	%
ISOMALT DC	98.40
Intense Sweetener	0.3
Citric acid	0.3
Flavour	0.5
Magnesium stearate	0.5

Source: Palatinit GmbH (2005).

Another outstanding feature of ISOMALT DC is that, in general, it does not produce undesired cooling effects and enhances flavours of all kinds.

Both grades exhibit a low hygroscopicity common to isomalt that allows less packaging material.

Depending on the other components of tablet formulations, it is normally not necessary to add a binding agent. In order to lubricate the compression, dyes 0.5–1% of calcium or magnesium stearate should be added. In some Asian countries, the use of stearates is not permitted but sucrose esters may be used instead. Finally, an ISOMALT DC tablet formulation can be as simple as shown in Table 11.6.[48]

11.5.7 Baked goods

The baking properties of isomalt were evaluated at the Institute for Cereal, Potato and Starch Technology in Detmold, Germany.[52] Since then, continuous research work has developed a wide variety of baked goods, different doughs and local specialities to meet all taste expectations.

Isomalt is an interesting ingredient in formulations with claims such as 'no sugar added', 'low glycaemic response' and 'light' baked goods.

By replacing the sugar in a formulation, a caloric reduction of between 10% and 20% can be achieved while still producing a tasty product.

The following properties of isomalt have to be taken into consideration when creating a recipe for baked goods:

- ISOMALT ST has a low solubility (24.5%/20°C).
- ISOMALT GS has a higher solubility (41.6%/20°C).
- No Maillard reaction occurs with isomalt.
- Mild sweetness without aftertaste.
- Isomalt has low hygroscopicity.

ISOMALT ST would be the preferred choice for cookies or biscuits and can be used alone to replace sucrose. Depending on the recipe, the isomalt content of the finished product varies between 10% and 20%. Owing to its low hygroscopicity, finished products have an excellent shelf life and biscuits, for example, stay crunchy over a longer period of time – a very desirable attribute. No, or little change to the manufacturing process is necessary when using isomalt.

In sponge cake, ISOMALT GS is recommended owing to its higher solubility. A combination with non-crystallising sugar replacers and humectants is recommended to increase the water-binding capacity of isomalt in products with a high moisture content.

11.5.8 Fruit spreads

The quality of ISOMALT GS fruit spreads is rated very positively with respect to taste, consistency, appearance and calorie content. Since ISOMALT GS remains stable in an acidic environment, the fruit spread possesses a better gelling behaviour than sucrose. During storage ISOMALT GS spreads show low syneresis. Syneresis describes the water separation on the surface of jam, resulting in an unpleasant appearance and poor spreading ability. The browning of ISOMALT GS spread is comparable with that of jam made with sugar. The taste of ISOMALT GS fruit spread is very fruity with no aftertaste. Also, where a low-dry substance (between 25% and 30%) is required in the finished product, a pleasant taste profile in combination with intense sweeteners can be obtained. Table 11.7 shows a basic recipe for fruit spreads.[48]

Table 11.7 Basic recipe for fruit spreads.

Ingredients	%
Fruits (strawberries)	66.00
ISOMALT GS	33.21
Sorbic acid	0.04
Citric acid	0.26
Pectin[a] (low esterified and amidated)	0.46
Acesulfame K	0.03
Total	100.00

Source: BENEO-Palatinit GmbH.[48]
[a]Fa. Herbstreith & Fox, AU 015, 125° SAG.

11.5.9 Breakfast cereals, cereal bars and muesli

Sugar-free and sugar-reduced products are becoming more and more important in nutrition.

With a high sugar content of about 10–20% and in some cases up to 40 %, breakfast cereals and corresponding products are prime targets for sugar-free or sugar-reduced variants.

A good taste and texture, that is the pleasure of consuming a product, is as important as its nutritional value. Most of the products on the market often just contain less sugar or some of the sugar has been replaced by fibre, for example. A simple reduction of sugar or replacement with ingredients that do not have the basic properties of sugar, results in texture and taste changes in the products and typical product qualities are lost.

The substitution of sugar by ISOMALT ST in breakfast cereals and other cereal products would result in similar or better products with respect to taste, bite, bowl-life and colour.

The most important property of ISOMALT ST in cereal products is that it can replace sugar on a weight for weight basis.

Additionally, ISOMALT ST's physical properties, particularly its low solubility, low hygroscopicity and stability against Maillard reaction, help to improve the shelf life and texture of cereal products.

11.5.9.1 Extruded breakfast cereals

In extruded cereals, sucrose can easily be replaced by ISOMALT ST-F or ST-PF. Typically, extruded cereals contain between 5% and 15% sucrose. 'No sugar added' versions can be

Table 11.8 Typical formulation for extruded cereals. Cereal shape: rings, thickness 5–6 mm.

Ingredients	%
Wheat flour type 405	69.30
Oat flour HK	15.00
ISOMALT ST-PF	12.00
Malt light	1.00
Malt dark	2.10
Salt	0.60
Total	100.00

Source: BENEO-Palatinit GmbH.[48]

formulated by a total replacement of sucrose with ISOMALT ST that is technically very similar to sucrose with respect to the processability and the final product's density and appearance.

ISOMALT ST works like a lubricant and helps to reduce the process forces in extruder technology.[53]

Extruded cereals with ISOMALT ST often have a crunchier texture and a harder bite than sucrose-based products. Owing to its low solubility, ISOMALT ST might also help to extend the crunchiness of breakfast cereals when combined with milk (extended bowl-life).

Table 11.8 gives a typical recipe for an extruded cereal product.[48]

11.5.9.2 Coated breakfast cereals

Coated cereals are usually made in traditional coating equipment like coating pans or drums. In comparison with the coating process for chewing gum, for example, for cereals, one coating cycle (application of engrossing syrup) is required.

Compared to sugar glazes or frostings, isomalt requires an adjustment of the drying parameters of the coated products before packaging. In general, the drying temperatures of the traditional production line have to be reduced. Process details and basic formulations for isomalt glazes and frostings have been developed (Ref. 48).

Depending on the isomalt variant, glazes or frostings can be produced. ISOMALT ST is most suitable for producing very shiny and stable glazed products.

An ISOMALT ST solution with a solids content of 80–82% for glazings and an ISOMALT GS solution with a solids content of 76–79% for frostings can be prepared at the appropriate temperature.

Owing to its very good crystallisation behaviour, ISOMALT GS is very suitable for the production of frosted coatings with a white, crystallised appearance.

In general, isomalt's low solubility and low hygroscopicity have a positive effect on the shelf life of finished products.

11.5.9.3 ISOMALT ST in cooked cereal products like corn flakes

ISOMALT ST has been used successfully to manufacture corn flakes. Basically, corn grits are cooked in an ISOMALT-ST-based syrup until the grits are saturated with respect to moisture. After a pre-drying process, the soaked grits are formed into flakes using flake rollers, which squeeze the soaked grits through a gap between two rollers. The resulting flakes are then toasted at high temperature for a short time.

Since ISOMALT ST does not undergo a Maillard reaction (browning), the colour of the final product is slightly lighter and the toasted flavour is slightly reduced.

Owing to its physical properties, corn flakes containing ISOMALT ST have a very good shelf life and a good crunch.

11.5.9.4 Soft cereal bars, crunchy bars and muesli clusters

ISOMALT ST can be used as a binder for crunchy or soft granola/cereal bars as well as for cereals. For soft textures, combinations with maltitol syrup are recommended. Maltitol syrup acts as a humectant owing to its non-crystallising and hygroscopic nature and isomalt rounds off the formulation by improving the taste profile and reducing its stickiness. Additionally, the product shelf life is affected positively owing to isomalt's low hygroscopicity.

The addition of inulin or oligofructose (Orafti®) allows further claims like 'prebiotic' or 'fibre enriched'.

For crunchy muesli bars or muesli clusters, ISOMALT ST can be used as the sole sugar replacer or can be combined with inulin/oligofructose, if a fibre or prebiotic claim is desired.

The basic process is first to produce a candy-like mass, which is then blended with muesli, granola or cereal base. Other processes use a highly concentrated ISOMALT ST solution that is blended with the cereal/muesli base and then dried and formed into clusters or bars.

Owing to the very low hygroscopicity of isomalt, crunchy bars and muesli clusters have an excellent shelf life and retain their crunch for a longer time.

11.5.10 Overview – further applications

- Bread spreads (e.g. hazelnut / chocolate spread)
- Nougat
- Croquant
- Marzipan
- Fondant
- Ice cream/sorbet
- Whipped masses for chocolate bars
- Table-tops

For all the aforementioned applications, formulations and process suggestions are available.[48]

11.6 SAFETY

The safety of isomalt has been evaluated and confirmed by a large number of health authorities worldwide as part of the approval process, resulting in approvals as a food or as a food additive depending on national legislation. Internationally, isomalt was evaluated by JECFA the WHO/FAO's Joint Expert Committee on Food Additives and advisory body of the UN's Codex Alimentarius. The assigned INS or E-number for isomalt is 953. JECFA assigned an

'ADI not specified' as the result of their safety evaluation, which is the best result possible since the committee did not see a need for a limiting numerical ADI (acceptable daily intake).[54]

Accordingly, isomalt is approved for use in food in general according to Good Manufacturing Practice in international food legislation (Table 3 of the Codex Alimentarius General Standard on Food Additives). For international trade, a specification was established by JECFA that included an HPLC method for determination of isomalt as such and in food.[55,56]

A monograph on isomalt was developed within Food Chemical Codex and published in 2009.[57]

11.7 REGULATORY STATUS: WORLDWIDE

Isomalt is approved as a food or as a food additive in more than 70 countries and in all major regions of the world, for example all countries in the European Union, NAFTA and MERCOSUR, Russia, China (new resource food status), Japan (food status), India, Australia/New Zealand and a number of Asian countries. On a global basis, it has a history of safe use for about 30 years, combined with nutritional benefits such as low-blood glucose response, good dental health properties and reduced caloric contribution compared with traditional sugars. As previously described, the scientific basis for the physiological properties of isomalt are well established and acknowledged by authoritative bodies in a number of countries, for example Europe, United States, Canada and Japan.[27,40,45,46] Internationally, isomalt is included in the General Standard on Food Additives (GSFA) of the Codex Alimentarius for use according to Good Manufacturing Practice.[56] Isomalt is also used as an excipient in pharmaceutical applications in an increasing number of countries. Isomalt is available in pharmaceutical grades according to the monograph on isomalt in the European Pharmacopoeia.[58]

11.8 CONCLUSIONS

Isomalt is the only bulk sugar substitute made exclusively from sucrose. It is a mixture of the two disaccharide alcohols 1,1-GPM and 1,6-GPS derived in a two-step production process, comprising an enzymatic conversion followed by hydrogenation. Owing to its physiological properties, it can be used as a basic component in a wide range of high-quality products that claim, for example, 'sugar free', 'calorie-reduced' and/or 'does not promote tooth decay'. In addition, isomalt products suit a low glycaemic lifestyle. Owing to its taste profile, which is similar to sucrose, isomalt-based products are barely distinguishable from their traditional counterparts.

In terms of production engineering, existing processing equipment can be used for all applications without requiring major changes. Applications like panned goods, chocolate, baked goods, cereal products and sugar-free hard candy illustrate the outstanding application of isomalt.

To summarise, isomalt is an ideal bulk sweetener that can be used to replace sugar in a wide range of applications. Its properties meet consumer demand for good-tasting healthier products with a long shelf life.

REFERENCES

1. R. Weidenhagen and S. Lorenz. Palatinose (6-O-alpha-D-glucopyranosyl-D-fructofuranose), a new product obtained from the bacterial fermentation of saccharose. *Zeitschr. Zuckerind.* 1957, 7, 533–534.
2. H. Schiweck. Palatinit – production, technological characteristics, and analytical study of foods containing Palatinit. *Alimenta* 1980, 19, 5–16.
3. H. Schiweck and M. Munir. Isomalt (Palatinit®), a versatile alternative sweetener – production, properties and uses. In: D. Albert Bartens (eds). *Carbohydrates in Industrial Synthesis*, Berlin, 1992; pp. 5–55.
4. H. Schiweck. *Isomalt, Ullmann's Encyclopedia of Industrial Chemistry*, Chemie, Weinheim; 1994, A25, pp. 426–429.
5. K. Paulus and A. Fricker. Zucker-Ersatzstoff: Anforderungen und Eigenschaften am Beispiel von Palatinit. *Lebensmitteltechnologie und Verfahrenstechnik* 1980, 31, 128–132.
6. D. Schäppi. Sensorische Untersuchungen an Zuckeraustauschstoffen, 3. Bericht, ETH Zürich, 1996 (unpublished).
7. Südzucker AG, ZAFES (CRDS), private communications, 1987, 1999, 2000, 2001, 2010.
8. Palatinit GmbH, Infopac, 2001.
9. G. von Rymon-Lipinsky and H. Schiweck (eds). *Handbuch Süßungsmittel – Eigenschaften und Anwendungen*, Behr's Verlag, Hamburg; 1990.
10. H.K. Cammenga and K. Gehrich. Eigenschaften und Rekristallisation amorpher Zucker und Zuckeraustauschstoffe, AiF Vorhaben Nr. 12097N, 2002.
11. K. Mende. Thermodynamische und rheologische Untersuchungen an Saccharidlösungen und –schmelzen, TU Berlin; 1990.
12. A. Sentko and J. Bernard. Isomalt. In: Lyn Nabors (ed.). *Alternative Sweeteners* 2012; CRC Press, New York, 275–297.
13. B. Zielasko. Ermittlung physikalisch-chemischer Daten von Isomalt und seinen Komponenten, Technische Universität Carolo-Wilhelmina zu Braunschweig, Dissertation; 1997.
14. Z. Bubnik, et al. *Sugar Technologists Manual*, Baertens, Berlin; 1995.
15. H.K. Cammenga, L.O. Figura and B. Zielasko. Thermal behaviour of some sugar alcohols. *Journal of Thermal Analysis* 1996, 47, 427–434.
16. F.W. Lichtenhalter and H.J. Lindner. The preferred conformations of glycosalditols. *Liebigs Annalen der Chemie* 1981, 2372–2383.
17. J. Raudonus, J. Bernard, H. Janßen, J. Kowalczyk and R. Carle. Effect of oligomeric or polymeric additives on glass transition, viscosity and crystallization of amorphous isomalt. *Food Research International* 2000, 33, 41–51.
18. K. Gehrich. Phasenverhalten einiger Zucker und Zuckeraustauschstoffe, PhD Thesis, TU Braunschweig, 2002.
19. H.J. Lindner and F.W. Lichtenthaler. Extended zigzag conformation of 1-O-D-alpha-glucopyranosyl-D-mannitol. *Carbohydrate Research* 1981, 93, 135–140.
20. U. Grupp and G. Siebert. Metabolism of hydrogenated palatinose, an equimolar mixture of alpha-D-glucopyranosido-1,6-sorbitol and alpha-D-glucopyranosido-1,6-mannitol. *Research in Experimental Medicine (Berlin)* 1978, 173, 261–278.
21. F. Heinz. Enzymatische Spaltung von Zuckeraustauschstoffen durch isolierte Enzyme und Enzymkomplexe der Dünndarmmukose, Med. Hochschule Hannover, Forschungsvorhaben Nr. 6539 Schlussbericht vom 30.09.1987.
22. S. Ziesenitz and G. Siebert; The metabolism and utilization of polyols and other bulk sweeteners compared with sugar. *Developments in Sweeteners*, Vol. 3, Elsevier Applied Science, London; 1987, pp. 109–149.
23. G. Livesey. Health potential of polyols as sugar replacers, with emphasis on low glycaemic properties. *Nutrition Research Reviews* 2003, 16, 163–191.
24. Sydney University's Glycaemic Index Research Service (SUGiRS), Glycaemic Index Report Isomalt, Sydney, Australia; 2002.
25. I. Holub, et al. Improved metabolic control after 12-week dietary intervention with low glycaemic isomalt in patients with type 2 diabetes mellitus. *Hormone and Metabolic Research* 2009, 41, 886–892.
26. A. Gostner et al. Effects of isomalt consumption on gastrointestinal and metabolic parameters in healthy volunteers. *British Journal of Nutrition* 2005, 94, 575–581.

27. EFSA; Scientific Opinion on the substantiation of health claims related to the sugar replacers xylitol, sorbitol, mannitol, maltitol, lactitol, isomalt, erythritol, D-tagatose, isomaltulose, sucralose and polydextrose and maintainance of tooth mineralisation by decreasing tooth demineralization (ID 463, 464, 563, 618, 647, 1182, 1591, 2907, 2921, 4300), and reduction of post-prandial glycaemic responses (ID 617, 619, 669, 1590, 1762, 2903, 2908, 2920) pursuant to Article 13(1) of Regulation (EC) No 1924/2006. *EFSA Journal* 2011, 9(4), 2076
28. A. Gostner, *et al.* Effects of isomalt consumption on faecal microflora and colonic metabolism in healthy volunteers. *British Journal of Nutrition* 2006, 95, 40–50.
29. A. Lee, *et al.* Consensus statements from participants of the international symposium on low digestible carbohydrates. *British Journal of Nutrition* 2001, 85(Suppl.1), 1–3.
30. J.M. Gee, *et al.* Effects of conventional sucrose-based, fructose-based and isomalt-based chocolates on postbrandial metabolism in non-insulin-dependant diabetics. *European Journal of Clinical Nutrition* 1991, 45, 561–566.
31. G.A. Koutsou, *et al.* Dose-related gastrointestinal response to the ingestion of either isomalt, lactitol or maltitol in milk chocolate. *European Journal of Clinical Nutrition* 1996, 50, 17–21.
32. A. Lee, *et al.* Breath hydrogen after ingestion of the bulk Sweeteners sorbitol, isomalt and sucrose in chocolate. *British Journal of Nutrition* 1994, 71, 731–737.
33. A. Lee, *et al.* The comparative gastrointestinal responses of children and adults following consumption of sweets formulated with sucrose, isomalt and lycasin HBC. *European Journal of Clinical Nutrition* 2002, 56, 755–764.
34. D. M. Paige, *et al.* Palatinit digestibility in children. *Nutrition Research* 1992, 12, 27–37.
35. D. Pometta, *et al.* Effects of a 12-week administration of isomalt on metabolic control in Type-II Diabetics. *Aktuelle Ernährung* 1985, 10, 174–177.
36. M. Spengler, *et al.* Tolerability acceptance and energetic conversion of isomalt in comparison with sucrose. *Aktuelle Ernährung* 1987, 12, 210–214.
37. D. Storey, *et al.* The comparative gastrointestinal response of young children to the ingestion of 25 g sweets containing sucrose or isomalt. *British Journal of Nutrition* 2002, 87, 291–297.
38. A. Zumbé. Comparative studies of gastrointestinal tolerance and acceptability of milk chocolate containing sucrose, isomalt or sorbitol in healthy consumers and type II diabetics. *Zeitschrift für Ernährungswissenschaft* 1992, 31, 40–48.
39. FASEB/LSRO (Federation of American societies for Experimental Biology/Life Science Research Office), The Evaluation of the Energy of Certain Sugar Alcohols used as Food Ingredients, June 1994, Bethesda, USA
40. Health Canada, Guide to Food Labelling and Advertising, Chapter 6, The Elements within the Nutrition Facts Table. http://www.inspection.gc.ca/english/fssa/labeti/guide/ch6e.pdf
41. J. Featherstone. Effects of isomalt sweetener on the caries process: A review. *Journal of Clinical Dentistry* 1994, 5(3), 82–85.
42. T. Imfeld. Efficacy of sweeteners and sugar substitutes in caries prevention. *Caries Research* 1993, 27(1), 50–55.
43. S. Ziesenitz. Basic structure and metabolism of isomalt. In: *Advances in Sweeteners*, Blackie Acad. & Prof.; 1996, pp. 109–133.
44. Scientific Committee on Medicinal Products 2002, Revision of the scientific opinion on the effects of xylitol and other polyols on caries development adopted by the SCMP; 2 June 1999.
45. US FDA; 21 CFR 101.80; http://www.accessdata.fda.gov/scripts/cdrh/cfdocs/cfcfr/CFRSearch.cfm?fr=101.80; 2 June 2011
46. T. Shimizu. Health claims on functional foods: the Japanese regulations and an international comparison. *Nutrition research Reviews* 2003, 16, 241–252.
47. B. Fritzsching. Isomalt in hard candy application. *The Manufacturing Confectioner*, November, 1995, 75(11), 65–73.
48. BENEO-Palatinit GmbH, 2011, Technical Application Sheets.
49. T. Dörr and I. Willibald-Ettle. Evaluation of kinetics of dissolution of tablets and lozenges consisting of saccharides and sugar alcohols. *Pharmazeutische Industrie* 1996, 58(10), 947–952.
50. BENEO-Palatinit GmbH, Chocolate makes people feel good – even without sugar. *Innovations in Food Technology* 2003, 21, 66–68.
51. Südzucker AG, CRDS, private communication, 2010/2011.
52. W. Seibel, *et al.* Backtechnische Wirkung des Zuckeraustauschstoffes PALATINIT (Isomalt), Getreide, Mehl und Brot, Detmold; 1986, pp. 8–10, pp. 239–242, pp. 269–274, pp. 302–306.

53. Palatinit GmbH. Sugar reduced breakfast cereals can now be enjoyed, *Food Marketing & Technology* 2002, 16(4), 6–10.
54. JECFA. Toxicological Evaluation of Certain Food Additives and Contaminants. *WHO Food Additive Series* 1987, 20, 205–237
55. JECFA. Compendium of Food Additive Specification, Addendum 4, FAO Food and Nutrition Paper 52, 1996, pp. 79—83.
56. FAO/JECFA. http://www.fao.org/ag/agn/jecfa-additives/details.html?id=622; 2 June 2011.
57. Food Chemical Codex (FCC). *Monograph on Isomalt*, 2009, 7th edn, 2010, The United States Pharmacopeial Convention, Rockville, USA.
58. European Pharmacopoeia. *Monograph on Isomalt*; 2005, 7th edn, Vol. 2, Published by: Directorate for the Quality of Medicines & HealthCare of the Council of Europe (EDQM), Strasbourg, 2010, pp. 1837–1839.

12 Lactitol

Christos Zacharis
Active Nutrition, DuPont Nutrition & Health, Surrey, UK

12.1 HISTORY

Lactitol is part of a group of alternative sweeteners known as sugar alcohols, also referred to as polyols and was discovered in 1920 by Senderens,[1] a French food chemist. He and Paul Sabatier are credited with founding the modern hydrogenation process, which they implemented in the production of lactitol and this method, catalytic hydrogenation, is still used. The first useful preparation of lactitol was made by Karrer and Buchi in 1937[2] and lactitol has been produced and marketed commercially since the 1980s.

Lactitol is a disaccharide composed of sorbitol and galactose, and is produced from lactose, a milk sugar, by catalytic hydrogenation using Raney nickel as the catalyst.[3] A 30–40% lactose solution is prepared and heated to approximately 100°C. The reaction is carried out in an autoclave under a hydrogen pressure of 40 bar or higher. On sedimentation of the catalyst, the hydrogenated solution is filtered and purified by means of ion-exchange resins and activated carbon. The purified lactitol solution is then concentrated and crystallised. The monohydrate, as well as the dihydrate and anhydrous forms can be prepared depending on the conditions of crystallisation.[4] Hydrogenation under more severe conditions (130°C, 90 bar) results in partial epimerisation to lactulose and partial hydrolysis to galactose and glucose followed by hydrogenation to the corresponding sugar alcohols lactitol, lactulitol, sorbitol and galactitol (dulcitol).[5]

Lactitol crystallises in several anhydrous and hydrous forms, but is usually commercially available either as the monohydrate or the anhydrous crystalline form. It is also known as lactit, lactositol and lactobiosit. The chemical abstract service (S) registry number is 585-86-4 and the molecular weight of lactitol monohydrate ($C_{12}H_{24}O_{11} \cdot H_2O$) is 362.34. The chemical structure of lactitol, 4-O-(β-D-galactopyranosyl)-D-glucitol, is shown in Figure 12.1.

12.2 ORGANOLEPTIC PROPERTIES

Sweetness is a key element in the perception of taste and quality. It plays an acknowledged role in food choice and represents an obvious factor in determining the consumer appeal of many products. Consumers demand the same sweet taste associated with sugar-sweetened products, without any discernable or unpleasant aftertastes.

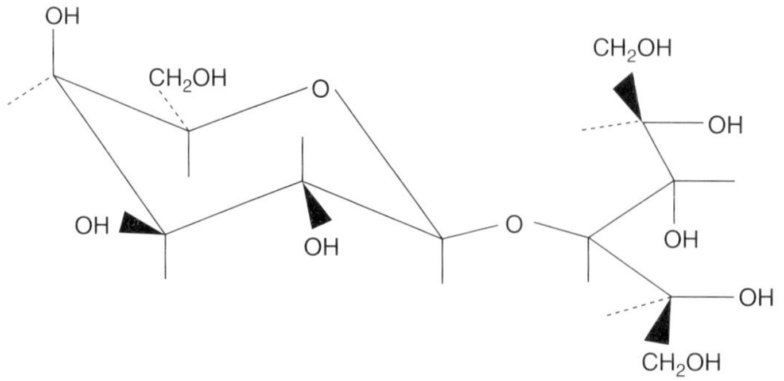

Fig. 12.1 The molecular structure of lactitol.

Lactitol is an odourless, white crystalline powder of very high purity and flow ability, with a mild clean sweet taste without any aftertaste. This closely resembles the taste profile of sucrose, although lactitol has only 40% of sucrose's sweetening power (Figure 12.2). Lactitol increases in relative sweetness when its concentration is increased (see Table 12.1). This can be of benefit in some food systems particularly in those foods that are considered 'too sweet', for example, fondants, jams, biscuits and surimi. In some products, the lower sweetness will allow the flavour of other ingredients to develop and be perceived more clearly.

This mild sweetness also renders it ideal for use as a bulk sweetener to partner with low calorie and intense sweeteners such as alitame, aspartame, acesulfame K or sucralose. This can bring the level of sweetness up to the same level as would be achieved with sucrose as the ingredient, but the product would contain fewer calories.

Lactitol exhibits very similar technical and handling properties to sucrose, hence it can be substituted for sucrose in almost any application.

12.3 PHYSICAL AND CHEMICAL PROPERTIES

12.3.1 Stability

Lactitol is a polyol with nine OH-groups that can, for example, be esterified with fatty acids to produce emulsifiers[3] or that can react with propylene oxide to produce polyurethanes.

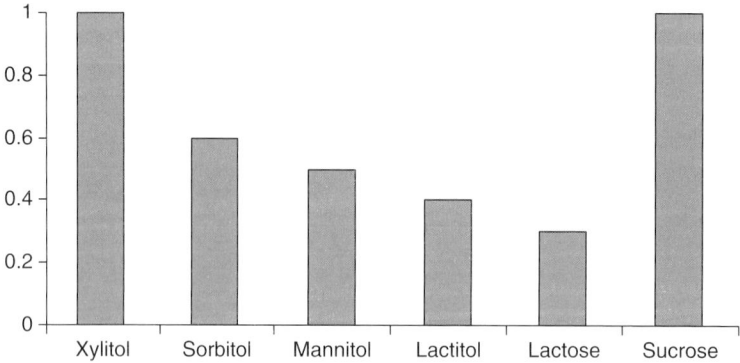

Fig. 12.2 The relative sweetness of some sweeteners.

Table 12.1 The relative sweetness of lactitol and sucrose solutions at 25°C.

Solution concentration (%w/w)	Relative sweetness of lactitol (sucrose = 1)
2	0.30
4	0.35
6	0.37
8	0.39
10	0.42

Owing to the absence of a carbonyl group, lactitol is chemically more stable than related disaccharides such as lactose. Lactitol's stability in the presence of alkali is much higher than that of lactose, whereas its stability in the presence of acid is very similar to that of lactose. The absence of a carbonyl group also means that lactitol does not take part in non-enzymatic browning (Maillard) reactions.

Lactitol solutions have excellent storage stability. A 10% lactitol solution in the pH range 3.0–7.5 at 60°C shows no decomposition after 1 month. After 2 months at pH 3.0, some decomposition (15%) is detected. At a higher pH, this does not occur. Hydrolytic decomposition of lactitol is observed with increasing temperature and especially with increasing acidity. Sorbitol and galactose are the main decomposition products. At high pH, lactitol is stable even at 105°C (Table 12.2). When heated to temperatures of 170–240°C, lactitol is partly converted into anhydrous derivatives (lactitan), sorbitol and lower polyols.

12.3.2 Solubility

The solubility of a polyol will greatly influence the perception of sweetness in the finished product. Highly soluble polyols result in finished products that exhibit rapid sweetness onset and impact, while poorly soluble polyols can provide a milder, more prolonged background sweetness. The solubility of all polyols increases with temperature. Lactitol has a reasonably good solubility when compared with sucrose (Figure 12.3). This means that modifications to production processes are minimised, and substituting the sugar in a product with lactitol is preferable to using other less-soluble polyols such as maltitol or isomalt. Solubility of the ingredients is important for many products (e.g. ice cream and hard-boiled sweets). A low solubility can make manufacturing some products more difficult (e.g. in hard-boiled sweets, most of the water will have to be evaporated and more water will have to be added initially when the solubility is low). This in turn takes more time and production costs would escalate. The solubility of lactitol at lower temperatures is less than that of sucrose but still good enough not to cause any inconvenience during processing. Lactitol can be dissolved at lower temperatures than sucrose, hence energy and processing costs are lowered. Lactitol

Table 12.2 Decomposition of a 10% lactitol solution under extreme conditions of pH and temperature.

pH	% Weight recovery at 24 h and 105°C			
	Lactitol	Galactose	Sorbitol	Not identified
2.0	42.9	28.1	21.4	7.6
10.0	98.0	N/A	N/A	2.0 (acids)
12.0	98.3	N/A	N/A	1.7 (acids).

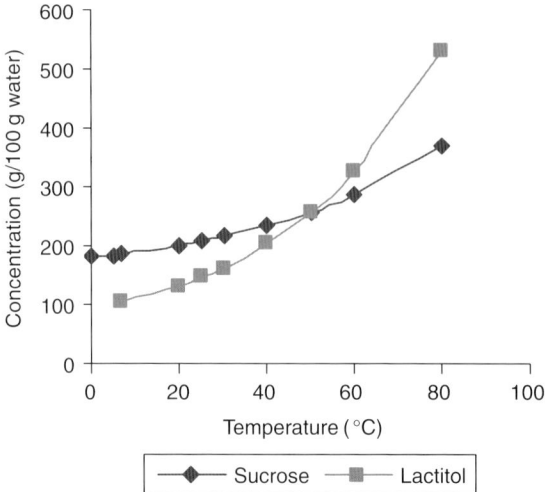

Fig. 12.3 Lactitol solubility g/100 g water.

is particularly advantageous in tablet formulations as it allows faster release of the active ingredient and/or flavour. The solubility of lactitol compared with other polyols is shown in Figure 12.4.

12.3.3 Viscosity

Sweetener viscosity can have a strong impact on a product's manufacturing characteristics, texture and mouthfeel. The viscosity of a polyol in solution is largely determined by its molecular weight. The viscosity of lactitol in aqueous solution when compared to sucrose is

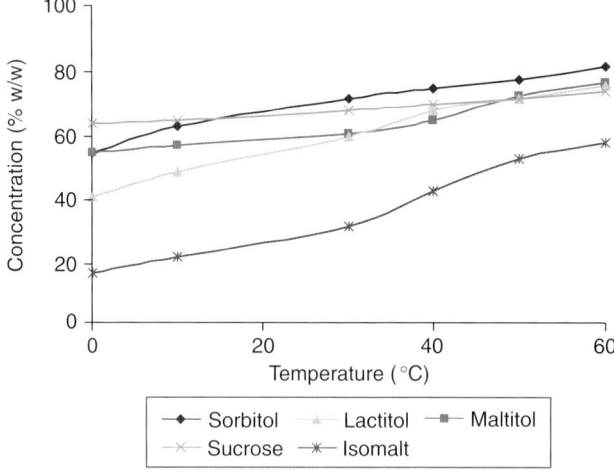

Fig. 12.4 Solubility of lactitol and other sweeteners at different temperatures.

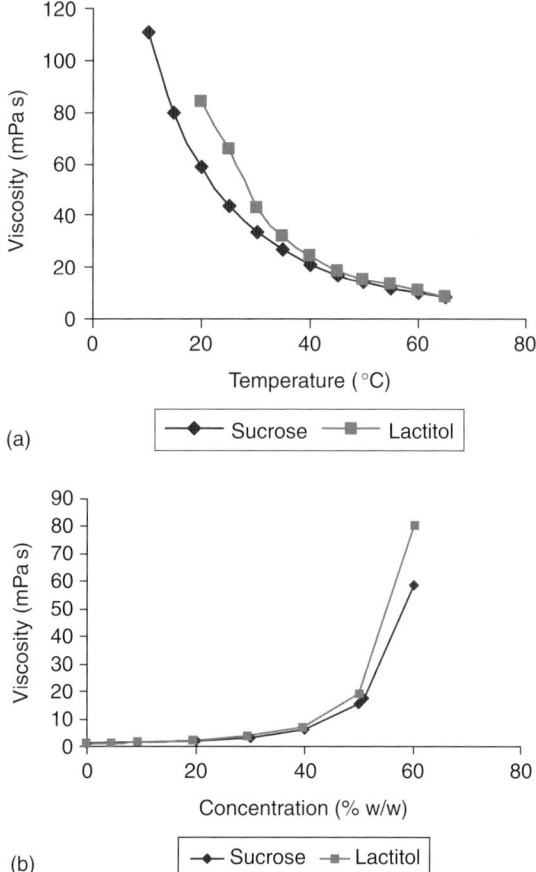

Fig. 12.5 Viscosity of lactitol solution (a) at 60% w/w concentration versus temperature, (b) at 20°C.

quite similar especially at low to moderate concentrations (Figure 12.5). At higher concentrations, the viscosity is slightly greater than in a sucrose solution, which can have a strong effect both on manufacturing characteristics and its organoleptic properties. For example, in the case of hard-boiled sweets, the viscosity of the lactitol melt is very important. In this case, a combination of lactitol and hydrogenated starch hydrolysate can be used (comparable to sucrose and glucose syrup). For this combination, a slightly lower viscosity is found than for its sucrose counterpart, but on cooling, the same viscosity (or plasticity) will be reached, so lactitol will also be suitable for moulded sweets.

12.3.4 Heat of solution

Compared with monosaccharide sugar alcohols, lactitol has less of a cooling effect and in applications such as chocolate production, it is much more desirable because a cooling effect would not be acceptable. In the anhydrous form, lactitol demonstrates even less cooling effect and has a similar value to sucrose. The anhydrous form is therefore the best option for chocolate applications (Table 12.3).

Table 12.3 Cooling effect measured at 37°C.

	Heat of solution (J/g)
Sucrose	−23
Anhydrous lactitol	−35
Monohydrate lactitol	−74

12.3.5 Boiling point elevation

Lactitol has a similar effect on the boiling point of water to sucrose (Figure 12.6). This is of great benefit when lactitol is used to provide bulk in foods such as hard candies and jams.

12.3.6 Hygroscopicity

Hygroscopicity is related to the amount of water absorbed by a product under certain environmental conditions. Lactitol is much less hygroscopic than sucrose and is the least hygroscopic of all the sugar alcohols with the exception of mannitol. Only under severe circumstances will lactitol absorb water. Lactitol is suitable for all applications in which water absorption is a critical parameter, like bakery products, tablets and panned confection. It is ideal as a dusting agent for chewing gum and lactitol-sweetened hard candies have much better resistance to moisture pick-up than their sugar equivalents. Crisp-baked goods will retain their freshness for much longer when made with lactitol but conversely, products such as cakes will require a humectant in the recipe to ensure a pleasant, lasting moist texture. With lactitol as an ingredient, there is no requirement at the manufacturing facility for expensive air conditioning or humidity control, again, saving on production costs.

12.3.7 Water activity

Water activity influences enzymatic activities, Maillard reactions, fat oxidation, microbial stability and texture, for example and these elements, combined, influence the shelf life of

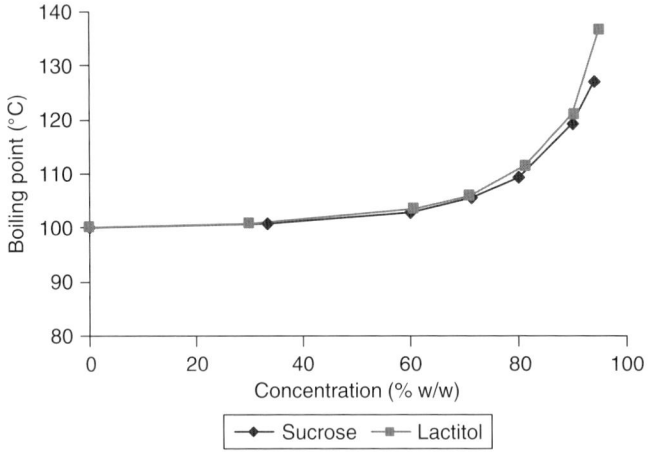

Fig. 12.6 The boiling point elevation of lactitol (at atmospheric pressure).

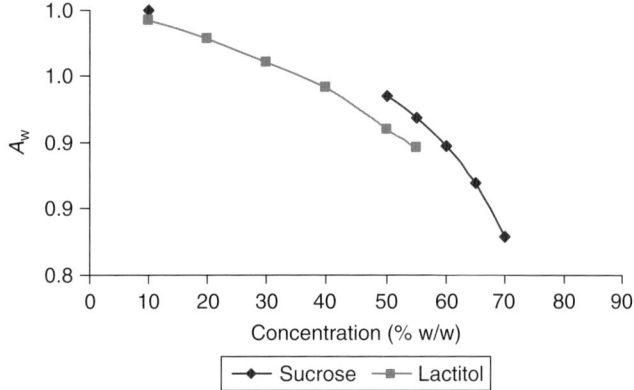

Fig. 12.7 Water activity of lactitol.

a product. The influence of lactitol on water activity is similar to that of sucrose (on a dry-solids basis) because the molecular weight of lactitol is almost identical to that of sucrose (lactitol monohydrate 362, sucrose 342). Therefore, Lactitol can directly replace sucrose in an application and have little effect on the shelf life of the finished product. Sorbitol, being a much smaller molecule, has a much larger effect on the water activity (Figure 12.7).

12.4 PHYSIOLOGICAL PROPERTIES

Society today is becoming increasingly health conscious with individuals taking more responsibility for their own health rather than relying on allopathic treatments. Disease prevention is a major focus. Nutrition plays a huge part in this and is an aspect of daily life over which individuals can exert some control.

Sugar has received rather a bad press over the past few years, with consumers being deluged with messages to purchase sugar-free, reduced-calorie and reduced-sugar products. However, many luxury or treat items have their main ingredient as sugar, hence fulfilling our desire for sweet foods. This is where lactitol and other sweeteners can be especially effective. Developments in sweetener technology and availability have enabled products to enter mainstream markets, which previously were associated with diabetic and slimming products. Polyols such as lactitol have fuelled this growth and allowed manufacturers to develop sugar-free products, which are of equal if not superior quality to traditional goods.

12.4.1 Metabolism

Like all foods, once ingested, lactitol passes to the small intestine. However, very little is absorbed here – approximately 2% by passive diffusion, and the remainder passes undigested to the colon; the distal part of the large intestine. Here, it becomes a substrate for the resident colonic microflora, is slowly fermented and converted into biomass, short-chain fatty acids (SCFAs), lactic acid, CO_2, and a small amount of H_2. Beneficial bacteria in the large intestine, such as *Bifidobacteria* and *Lactobacillus* spp., use lactitol as a substrate.

Lactitol consumption does not induce an increase in blood glucose or elicit the release of insulin, rendering it a desirable sugar substitute for diabetic patients. Since so little lactitol is absorbed by the small intestine,[6] this reduces the amount of calories available to the individual. It has been demonstrated that the nutritional caloric use of lactitol is half that of carbohydrates, with a metabolic energy value maximum of 2 kcal/g (8.4 kJ/g). However, for labelling purposes, the EC Sweeteners Directive assigns a value of 2.4 kcal/g to all polyols including lactitol. Legislation in the United States, Canada and Japan assign a caloric value of 2.0 kcal/g.

Karrer and Buchi[2] studied the action of galactosidase-containing enzyme preparations on the hydrolysis of lactitol into galactose and sorbitol. They found that lactitol was only hydrolysed very slowly by these enzyme preparations. Later studies reported in a German Patent[7] confirmed that lactitol is only slowly hydrolysed by enzymes at about a tenth the speed of lactose. On the basis of both these *in vitro* studies, a reduced-calorie value can be expected.

In two Hayashibara Patents,[8,9] it is claimed that lactitol has no caloric value because it is not digested or absorbed. At the Agricultural University of Wageningen in The Netherlands, a study was carried out[10] to determine the energy balance of eight volunteers on diets supplemented with either lactitol or sucrose. In this study, volunteers were kept for 4 days in a respiratory room; this was repeated: once with 49 g of sugar a day in a diet; and then with lactitol. The dosage of 50 g lactitol monohydrate was ingested in four to six portions during the day. Intakes of metabolisable energy (ME) were corrected, within subjects, to energy equilibrium and equal metabolic body weight. Further correction of ME intake was made towards equal actometer activity. With regard to the value of ME to supply energy for maintaining the body in energy equilibrium, the energy contribution to the body of lactitol monohydrate was 60% less than for sucrose.

As lactitol monohydrate contains 5% water and the results had an inaccuracy of standard error (SE) 10%, the final conclusion is justified that the metabolic energy of lactitol, on the basis of the dry substance, is at most 50% that of sucrose. A caloric value of lactitol in man at 2 kcal/g seems fully justified.

Lactitol is well tolerated, but as with all polyols, can cause laxation if consumed in excess. The Scientific Committee on Food to the EC (SCF-EC) evaluated lactitol and considered it to be a safe product; they stated that 'consumption of the order of 20 g per person per day of polyols is unlikely to cause undesirable laxative symptoms'.[11] The same statement also applied to isomalt, maltitol, mannitol, sorbitol and xylitol. The laxative effect varies between all the polyols and also depends on other factors such as mode and frequency of ingestion, diet, age and general gut health. Indeed, a therapeutic level is given as an effective medical treatment for chronic constipation.

12.5 HEALTH BENEFITS

12.5.1 Lactitol as a prebiotic

Both the biological and medical communities now recognise the significance of the role of the large intestine in both health and disease. Many scientific advances over the past 10–20 years have changed the perception that the large gut is simply an organ for the storage and secretion of waste matter with its main functional approach being the absorption of water and other nutrients from such material. Now, due to the profuse microbiota resident in the human colon, it is recognised that this organ can contribute significantly both nutritionally and metabolically.[12]

The perspective of the human colon in health and disease was actually suggested in seminal papers of Elie Metchnikoff at the start of the twentieth century, which indicated the clinical importance of host colonic microflora.[13] Rapid advances have been made owing to modern molecular-based techniques that characterise the composition of the bacterial mass and have lad to a greater understanding of the biological functions of this organ. The bacterial microflora of the human gut is widely accepted as an integral component of the functional food industry.

A prebiotic is a non-viable component of the diet that reaches the colon in an intact form and is selectively fermented by colonic bacteria.[14] The most recent definition regarding a prebiotic is 'A selectively fermented ingredient that allows specific changes, both in the composition and/or activity in the gastrointestinal microflora, that confer benefits on host well-being and health'.[15] The selectivity is by commensal bacteria already resident in the gut that are thought to have beneficial properties to the host regarding promotion of health. Examples of such colonies include *Lactobacilli* and *Bifidobacteria*, both of which are present in significant numbers. Health is improved by fortification of such selected bacteria. A schematic representation of the prebiotic concept is shown in Figure 12.8.

Most of the lactitol reaches the colon undigested where it is used as an energy source by intestinal microflora such as *Lactobacillus* and *Bifidobacteria*. The fermentation of lactitol favours the growth of saccharolytic (healthy) bacteria and decreases the amount of proteolytic

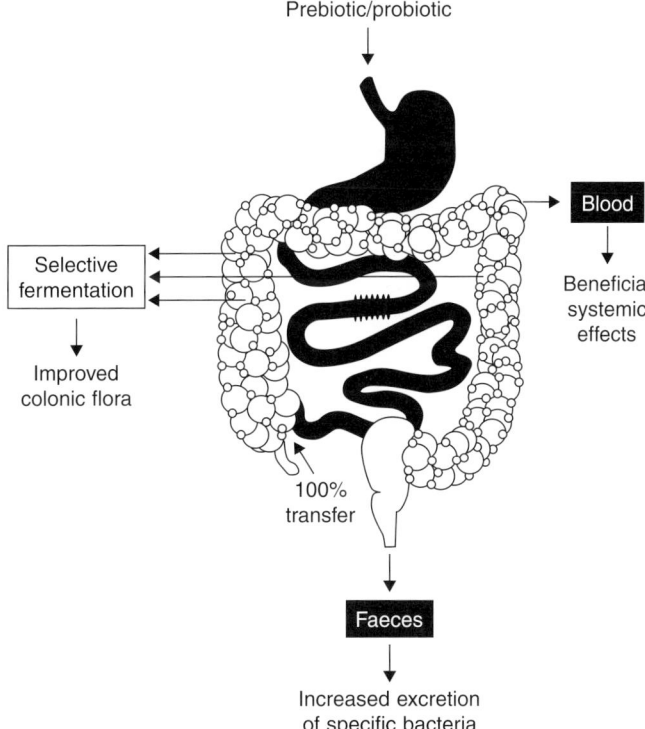

Fig. 12.8 Diagrammatic representation of the pre- and probiotic concepts. A prebiotic is a non-digestible food ingredient that beneficially affects the host by selectively stimulating the growth and/or activity of one or a limited number of bacteria in the large intestine.
Source: Reproduced from Leeds and Rowland.[8]

Table 12.4 Acidification and gas production by bacterial cultures with the addition of 10% lactitol.

Bacteria	Number of strains	Sugar-free control pH[a]	Lactitol pH	Gas
Staphylococcus aureus	4	6.5	6.3	N/A
Streptococcus faecalis	4	7.2	5.0	N/A
Klebsiella sp.	5	6.3	5.6	+
Escherichia coli	4	7.5	7.5	N/A
Enterobacter sp.	2	7.8	7.0	+
Citrobacter sp.	2	7.7	7.7	N/A
Bacteroides sp.	2	6.4	5.6	N/A
Clostridium perfringens	2	6.8	5.5	+++
Bifidobacteria	3	6.5	5.5	−
Lactobacillus sp.	3	6.8	5.3	−
L. acidophilus	3	6.6	4.7	−
Salmonella sp.	4	8.5	7.7	−
Yersinia enterocolitica	4	7.0	6.5	−
Helicobacter pylori	2	7.5	7.5	−

[a]The sugar-free produced no gas.
Source: Reproduced from Scevola et al.[20]

(unhealthy) bacteria such as gram negative *Bacteroides, Enterobacteria, Enterococci*, and various *Coliform* spp. by inhibiting adhesion of these bacteria to the epithelial cell walls[17–19] and creating a low pH (Tables 12.4 and 12.5). *Bifidobacteria* also produce organic acids such as acetic and lactic acid, which limit the growth of putrefactive and pathogenic bacteria. These bacteria are responsible for the production of undesirable enzymes such as β-glucuronidase, nitroreductase and azoreductase, and phenolic products, amines and endotoxins, all of which have been linked to various cancers and ulcerative colitis.[21] Reducing the colonies of such bacteria can only positively influence the health of humans and animals.

The fermentation of lactitol by *Bifidobacteria* and *Lactobacillus* spp. lowers the pH of the large intestine owing to the production of butyrate (butyric acid). It also generates SCFAs, which are thought to help protect and repair the intestinal wall, it is thought by stimulating cell division, which can mediate regeneration of the epithelial cells in the intestinal wall. The

Table 12.5 Bacterial growth with increasing doses of lactitol (inoculum at 10^4–10^8 CFU/mL.

Bacteria	Lactitol			
	1%	2%	5%	10%
Salmonella sp.	−	−	− −	− − −
Escherichia coli	−	−	− −	− − −
Staphylococcus aureus	0	−	− −	− − −
Shigella sp.	−	−	− −	− − −
Streptococcus faecalis	− −	− −	− − −	− − −
Klebsiella pneumoniae	0	0	0	− −
Yersinia enterocolitica	0	−	− −	− − −
Lactobacillus sp.	+	+	++	+++
L. Bifidus	+	+	++	+++
L. acidophilus	+	+	++	+++

Source: Reproduced from Scevola et al.[20]

colon epithelial cells derive their energy solely from SCFAs. Butyrate is a preferred nutrient and indirectly prevents colonic mucosal reduction, which can occur in cases of starvation. SCFAs are also thought to stimulate apoptosis which is literally programmed cell death and is the body's way of eliminating cells that are deleterious to the health of that area.

12.5.2 Lactitol to treat hepatic encephalopathy

Some studies have indicated the possibility of using lactitol to successfully treat cases of hepatic encephalopathy. It is an unabsorbed carbohydrate with a defined laxative threshold and superior taste properties and would make a good substitute for lactulose which is currently the medicine most widely used. Lactulose is excessively sweet and consequently is unpalatable to some patients. As lactitol is highly water soluble, less sweet than lactulose and not absorbed in the human intestine, therefore, it is thought that there is high potential for lactitol to be an alternative therapeutic agent to lactulose in the treatment of hepatic encephalopathy. A clinical trial supported this hypothesis showing lactitol was as effective as lactulose in treating patients with this condition.

12.5.3 Lactitol and diabetes

Owing to the unique metabolism of lactitol, there is negligible effect on blood sugar levels when consumed. It is not broken down to its component monosaccharides by any enzymic activity in the small intestine like some other carbohydrates but passes undigested to the colon, as previously described. Hence, there is no requirement for insulin in its metabolism, rendering it suitable for insulin-dependent diabetic patients (type 1) and non-insulin-dependent diabetic patients (type 2). The latter type can even reduce their diabetic status through diet management and lactitol can be instrumental in this respect owing to its lower calorie value. As with many of the polyols, lactitol exhibits a substantially lower glycaemic response (GR) value compared with sucrose or glucose and this can be used to reduce the overall glycaemic challenge of the diet by production of foods with a low GL.

12.5.4 Tooth-protective properties

Sugars are the major factor in the pathogenesis of dental caries. Oral bacteria convert sugars into polysaccharides that are deposited on the teeth; these plaque sugars are then fermented into acids. The acid demineralises the enamel and causes cavities. Lactitol is not fermented by these bacteria and as such is said to be non-cariogenic and cariostatic.

Three main types of experiments are used to evaluate the cariogenicity of foods and food ingredients:

1. Studies *in vitro*.
2. Experiments in laboratory animals.
3. Clinical trials and investigations in human subjects.

12.5.4.1 Studies in vitro

Among the earliest reports are those of Havenaar[22,23] on the use of lactitol by oral bacteria, with formation of acids. A number of plaque bacteria were found that could metabolise lactitol as a substrate, including *Streptococcus mutans* (which is known to possess cariogenic activity) and certain strains of *Streptococcus sanguis*, *Bifidobacteria* and *Lactobacillus*. These first experiments did not establish the speed of fermentation, which because of the limited time that sugars and sweeteners remain in the mouth is a determinant of their cariogenicity. It was later shown that the fall in pH was slow, leading to the conclusion that lactitol could be fermented slowly by these micro-organisms.

In addition to acid production, another important property of cariogenic bacteria is the capacity to synthesise extracellular polysaccharide from carbohydrate substrates, making dental plaque. No evidence for extracellular polysaccharide synthesis was found with lactitol.

A study on the cariogenic potential of lactitol was carried out by Grenby and Phillips[24] where lactitol was compared *in vitro* with five other bulk sweeteners. These were incubated with mixed cultures of human dental-plaque micro-organisms and acid development, insoluble polysaccharide synthesis, and the attack of the acid on enamel mineral was measured. The six different sweeteners fell into three groups. The highest acid generation, polysaccharide production and enamel demineralisation was found with glucose and sucrose, less from sorbitol and mannitol, and least of all from lactitol and xylitol.

12.5.4.2 Experiments in laboratory animals

The cariogenicity of lactitol was also assessed in rats.[25] Lactitol was incorporated in a powdered diet, consisting of 50% of a standard dietary formulation, 25% wheat flour and 25% of test substance. Lactitol was compared with sorbitol, xylitol, sucrose and a control consisting of 50% wheat flour and 50% usual diet. The rats were program-fed and none of the animals had diarrhoea. No significant adverse effects on general health were observed in any of the groups. The results showed the substitution of sucrose by lactitol-reduced caries significantly. The results for sorbitol and xylitol were in accordance with other studies in which these polyols were used.

In the second experiment, human food was fed to rats. Shortbread biscuits containing 16.6% sucrose or lactitol were incorporated at 66% in pulverised, blended diets fed to two groups of 21 rats, so that the final level of lactitol in the test diet was 11%. Again, after a period of 8 weeks, caries attack showed highly significant reduction when replacing the sucrose in the biscuits with lactitol.

The most recent studies on the dental properties of lactitol in rats assessed the lactitol at lower levels in the diet and in form of a finished human food product rather than as a raw ingredient in a blended animal diet.[25] At a level of 16% lactitol or 16% xylitol in the blended diet, the carious scores, lesion counts and severity of the lesions were so close on the two polyol regimens that they were indistinguishable but significantly beneficial compared with the 16% sucrose group.

12.5.4.3 Investigations in man

At the University of Zurich, a method has been developed for the *in vivo* determination of the plaque-pH in humans using an in-dwelling pH electrode. After consumption of chocolate or confections in which sucrose is replaced with lactitol, changes in plaque pH were detected by the electrode and transmitted electronically to a graph recorder.

The term 'zahnschonend' (safe for teeth/friendly to teeth) is used officially in Switzerland when the pH of dental plaque does not fall below 5.7 during a 30-minute period. Professor Muhlemann demonstrated in chocolates that lactitol is 'safe for teeth'.[26–28]

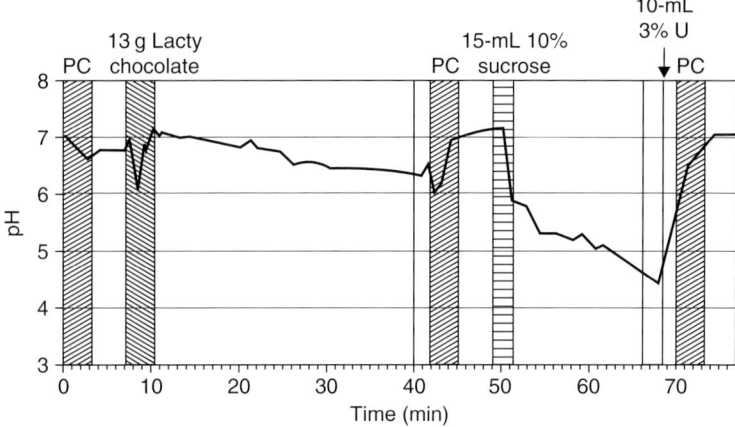

Fig. 12.9 Telemetrically recorded pH of 5 days interdental plaque in subject H.H. during and after consumption of 13 g lacty-plain chocolate. A 10% sucrose solution was used as a positive control (PC), 3-minute paraffin chewing gum (U), rinsing with 3% carbamide.

Figure 12.9 shows that during and after eating 13 g plain lactitol-containing chocolate, the plaque-pH does not fall below 5.7 whereas with a 15-mL 10% sucrose solution, the plaque-pH is reduced to about 4.5, which indicates that sucrose is being fermented into acids by the oral bacteria.

12.6 APPLICATIONS

Lactitol can be used in many applications as a direct substitute for sugar on a weight for weight basis. Where higher sweetness levels are required, these can be achieved by the addition of intense sweeteners with lactitol acting as the bulking agent.

12.6.1 Chocolate

Lactitol can be used successfully to produce sugar-free chocolate and both lactitol monohydrate and anhydrate can be used in this application. The difference in use between these two crystal forms during processing is reflected in the conching temperature. Lactitol monohydrate can be conched at higher temperatures; up to about 60°C as the water of crystallisation is tightly bound and not released during the process. Additionally, no atmospheric moisture is absorbed. Above this temperature, the water will be released and the viscosity of the chocolate mass will increase. The higher conching temperature means that the reduction in water content and removal of undesirable volatile components is accelerated and the overall flavour is improved.

Anhydrous lactitol is especially suited to chocolate manufacture. It has no water of crystallisation, and is therefore even more stable than the monohydrate form. It also demonstrates excellent stability in chocolate stored in liquid bulk form at elevated temperatures, showing no increase in viscosity even after 7 days (Figure 12.10). With anhydrous lactitol, the

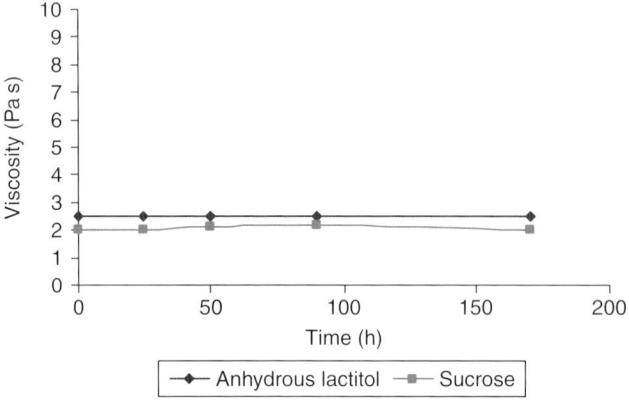

Fig. 12.10 Viscosity of chocolate at 50°C.

conching temperature can be as high as 80°C, allowing for stronger flavour development and improved operating efficiency. Another difference between the two forms is the cooling effect. The anhydrous lactitol has a lower cooling effect than the monohydrate. This will improve the taste of the chocolate, which should have a warm taste. Neither gives a scratching aftertaste, which can be a characteristic of other sweeteners in chocolate.

The sensory profiles of three types of 'no-added-sugar' chocolate (i.e. milk, white and dark) were evaluated and it was shown that there was a good synergistic effect between a lactitol/maltitol blend in milk chocolate. In particular, 'no added sugar' milk chocolate with the lactitol/maltitol blend showed an almost identical sensory profile to the standard (i.e. sugar) milk chocolate when considering key parameters, such as *sweetness* and *milk/caramel flavour*. Additionally, *cocoa taste* was significantly enhanced.

Lactitol was also shown to exert excellent creaminess and *sweet* taste owing to its interaction with the milk solids, in particular in the white chocolate, where the addition of high-potency sweetener was unnecessary.

In dark chocolate, lactitol was shown to promote the bitterness and the cocoa flavour, which are key parameters for this product.

As previously discussed, the physical properties of the anhydrous form of lactitol are such that the use of lactitol in chocolate applications guarantees a top quality end product, without introducing any process complications.

12.6.2 Baked goods

This area of application lends itself to the adoption of reduced-calorie and no-added-sugar recipes. Sugar can be replaced weight for weight by lactitol and an intense sweetener can be added if extra sweetness is required. Finished products generally have similar characteristics to sucrose equivalents. For some bakery products, for example biscuits, the crispiness of the product is one of the most important characteristics. The majority of bulk sweeteners have the tendency to be hygroscopic, but lactitol differs from these and demonstrates low hygroscopicity. This means crispness is assured and there is the added benefit of an extended shelf life. In soft bakery products, for example cakes, the low hygroscopic property of lactitol means that the addition of a humectant such as sorbitol is beneficial. This gives a smooth, moist feel without affecting the good structure and texture obtained with lactitol.

In addition to no-added-sugar bakery products, products reduced in calories or with functional claims can be developed. In these products, lactitol acts as an ideal bulk sweetener. In reduced-calorie products, lactitol is often combined with polydextrose (Litesse®). This combination is ideal to reduce the calorie content and maintain a high-quality product.

12.6.3 Chewing gum and confectionery

Sugar-free chewing gum is one of the most popular products in the confectionery market. All polyols exhibit a negative heat of solution, undergoing an endothermic reaction, absorbing heat from their surroundings as they dissolve. This is known as the 'cooling effect', as the dissolution of the crystals of polyol in the saliva in the mouth results in a perceivable cooling sensation. This property can be used to enhance and complement the delivery of mint flavours giving a refreshing and cooling finished product, ideal in chewing gum. All polyols can be used to create sugar-free chewing gum and lactitol can also be used as a bulk sweetener in this application, mainly as a replacement for sorbitol. The distinct advantage of lactitol over sorbitol is its low hygroscopicity nature, which requires no expensive air conditioning systems in the production area. Lactitol can be used as a sweetener in the gum base, as a rolling compound, as a dusting powder and in the panning layer for coated tablets. In the gum base, lactitol gives a more flexible structure than sorbitol and there is no need to increase the gum base level as is required in sorbitol-containing chewing gum.

The low hygroscopicity will increase shelf life, especially when stored at high temperatures and humidity. Furthermore, lactitol will improve mouthfeel and, compared with mannitol (which also has low hygroscopicity), has a better solubility that prevents sandiness in the chewing gum. Lactitol-sweetened gum exhibits a more pliable texture over a long period of time.

Marzipan provides an excellent application area for lactitol. Lactitol has a mild, sugar-like taste and its high solubility makes it preferable to other polyols or sweeteners. As it exhibits a low sweetness, the sweetness level can be enhanced by the addition of intense sweeteners if required.

Lactitol can also be used in a wide range of confectionery products including hard-boiled candies, chewy candies and toffee to provide reduced-calorie and diabetic-suitable products. In general, lactitol replaces the sucrose part of recipes, whereas glucose syrup needs to be replaced by hydrogenated starch hydrolysate, polydextrose solution or maltitol syrup. These latter products are used as crystallisation inhibitors and the optimum ratio with lactitol differs from product to product. For example, hard-boiled sweets need a 70/30 ratio (lactitol/hydrogenated starch hydrolysate). For best results, lactitol should be used in combination with other polyols or polydextrose, to prevent or control crystallisation. Such combinations result in products that are virtually identical to their sugar equivalents.

12.6.4 Ice cream and frozen desserts

Lactitol demonstrates a series of physical properties such as freezing-point depressing and high solubility that allows its use in ice cream. Its freezing-point depression is similar to that of sucrose and so can be used to directly replace it in both ice cream and other frozen desserts. This gives the same 'scoop' characteristics, but if a soft serve texture is required, a combination of monosaccharide sugars or sugar alcohols is recommended.

12.6.5 Preserves

Lactitol can be used in the preparation of jams and is especially useful in jams with less than 40% solids. The solution to reduced-calorie preserves is often achieved by use of a low dry solids content in the product. Such a jam, in combination with lactitol will result in an even larger reduction in calories, while maintaining good flavour and texture. Where high levels of bulk sweetener are required such as in high-solid jams (65% solids), lactitol in combination with other polyols or polydextrose is recommended to prevent crystallisation.

12.6.6 Tablets

Lactitol is ideal for use as an excipient in pharmaceutical preparations. It exhibits low reactivity when compared with sugars and other sugar alcohols. Granulated lactitol is especially suited to dry preparations due to the low hygroscopicity and this is especially important in tablet preparations. This low hygroscopicity prolongs shelf life and protects active agents against moisture. Figure 12.11 shows that lactitol tablets absorb hardly any water at 70% relative humidity. In pharmaceutical tablets, directly compressible sorbitol and mannitol are commonly used. The advantage of lactitol in this application is the low hygroscopicity compared with sorbitol and the improved solubility compared with mannitol. With respect to tablet hardness lactitol gives a hardness in between sorbitol and mannitol (Figure 12.12). This makes directly compressible lactitol a good alternative to either sorbitol or mannitol.

Lactitol is non-cariogenic and therefore especially suitable for children's products such as sugar-free vitamins.

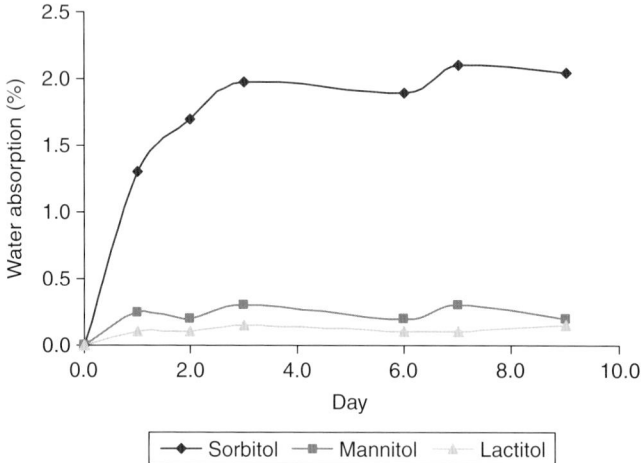

Fig. 12.11 Water absorption of tablets made with directly compressible polyols. Tablet hardness 120N; temperature 20°C; relative humidity 70%.

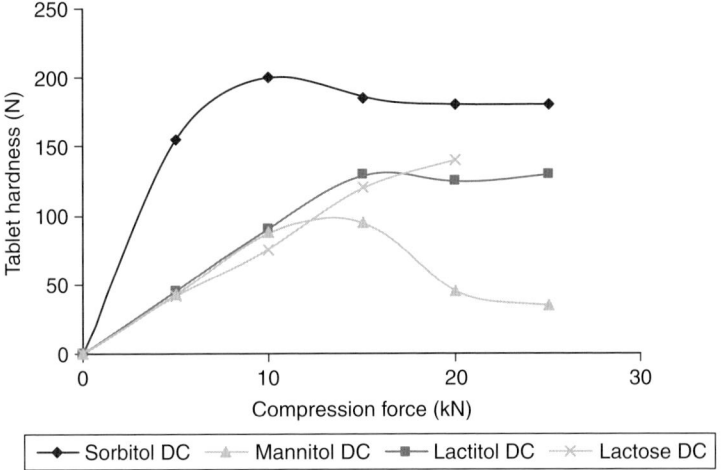

Fig. 12.12 The compression profile of directly compressible polyols on a rotary press.

12.7 REGULATORY STATUS

Argentina	Approved
Australia	Allowed
Brazil	Allowed
Canada	Allowed
EU	Lactitol is allowed as a sweetener in all EU countries regulated by the Sweeteners in Food Directive No. 94/35/EC up until the time that the aforementoned Directive will be fully replaced by Regulation (EC) 1333/2008 of 16 December 2008 regarding food additives.

Lactitol can also be used as an additive; this is regulated by the Food Additives other than Colours and Sweeteners No. 95/2/EC up until the time that the aforementioned Directive will be fully replaced by Regulation (EC) 1333/2008 of 16 December 2008 regarding food additives. The E-number for lactitol is E 966.

Israel	Allowed
Japan	Allowed
Norway	Allowed
Sweden	Allowed
Switzerland	Allowed
United States	Lactitol is self-affirmed generally recognized as safe (GRAS).

12.8 CONCLUSIONS

Lactitol is a commercially available bulk sweetener. Its physical properties guarantee optimal product performance during processing and storage of a food product. Being so similar to sucrose, its use facilitates the development of new sugar-free and light food products.

The Joint FAO/WHO Expert Committee of Food Additives (JECFA) approved lactitol in 1983. The Committee allocated an acceptable daily intake (ADI) 'not specified'.[11]

The SCF-EC evaluated lactitol and considered it to be a safe product; they stated that 'consumption of the order of 20 g/person/day of polyols is unlikely to cause undesirable laxative symptoms'.[29] The same statement also applied to isomalt, maltitol, mannitol, sorbitol and xylitol.

The safety data are not summarised here. Full data are described in CIVO-TNO (Zeist, The Netherlands) reports, which are published in the *Journal of the American College of Toxicology*.[30] These reports have been evaluated by JECFA, SCF-EC and the US Food and Drug Administration.

REFERENCES

1. J.B. Senderens. Catalytic hydrogenation of lactose. *Comptes Rendus* 1920, 170, 47–50.
2. P. Karrer and J. Buchi. Reduktionsprodukte von disacchariden: maltit, lactit, cellobit. *Helvectica Chimica Acta* 1937, 20, 86–90.
3. J.A. van Velthuijsen. Food additives derived from lactose: lactitol and lactitolpalmitate. *Journal of Agricultural and Food Chemistry* 1979, 27, 680–686.
4. C.F. Wijnman, J.A. van Velthuijsen and H. van den Berg. Lactitol monohydrate and a method for the production of crystalline lactitol. European Patent 39981; 1983.
5. T. Saijonmaa, *et al.* Preparation and characterization of milk sugar alcohol dihydrate (abstract). *Sixth European Crystallograph Meeting, Barcelona*; 1980, p. 26.
6. D.H. Patil, G.K. Grimble and D.B.A. Sil. Lactitol, a new hydrogenated lactose derivative; intestinal absorption and laxative threshold in normal human subjects. *British Journal of Nutrition* 1987, 57, 195–199.
7. Maizena GmbH. Verwendung van Lactit als Zuckeraustauschstoff. German Patent, 2,133,428; 1974.
8. K. Hayashibara. Improvements in and relating to the preparation of foodstuffs. UK Patent 1,253,300; 1971.
9. K. Hayashibara and K. Sugimoto. Containing lactitol as a sweetener. US Patent 3,973,050; 1975.
10. A.J.H. Van Es, L. de Groot and J.F. Vogt. Energy balances of eight volunteers fed on diets supplemented with either lactitol or saccharose. *British Journal of Nutrition* 1986, 56, 545–554.
11. Joint FAO/WHO Expert Committee on Food Additives. IPCS Toxicological evaluation of certain food additives and contaminants. *Lactitol. WHO Food Additives Series No. 27*, WHO, Geneva; 1983, pp. 82–94.
12. G.P. Macfarlane and G.R. Gibson. Metabolic activities of the normal colonic flora. In: S.A.W. Gibson (ed.). *Human Health: The Contribution of Micro-organisms*, Springer-Verlag, London; 1994, pp. 17–52.
13. E. Metchnikof. *The Prolongation of Life*, William Heinemann, London.
14. G.R. Gibson and M.B. Roberfroid. Dietary modulation of the human colonic microbiota. Introducing the concept of prebiotics. *Journal of Nutrition* 1995, 125, 1401–1412.
15. 6th International ISAPP Workshop, London Ontario Canada, 2008. In International Scientific Association of Probiotics and Prebiotics (ISAPP), Annual Report, 1 January to 31 December 2008.
16. AR Leeds, IR Rowland. *Gut Flora and Health: Past, Present and Future*. London: Royal Society of Medicine Press; 1996.
17. G. Lebek and S.P. Luginbuhl. Effects of lactulose and lactitol on human intestinal flora. In: H.O. Conn and J. Bircher (eds). *Hepatic Encephalopathy: Management with Lactulose and Related Compounds*, Michigan; 1989, pp. 271–282.
18. M. Finney, J. Smullen, H.A. Foster, S. Brokx and D.M. Storey. *European Journal of Nutrition* 2007, 46, 307–314.
19. A. Drakoularakou, O. Hasselwander, M. Edinburgh and A.C. Ouwehand. *Food Science and Technology Bulletin: Functional Foods*, 2007, 3(7) 71-80.
20. D. Scevola, G. Bottari, A. Franchini, A. Guanziroli, A. Faggi, V. Monzillo, L. Pervesi and L. Oberto. The role of Lactitol in the regulation of the intestinal micro flora in liver disease. *Giornale di Malattie Infettive e Parassitarie* 1993, 45(7–8), 906–918.

21. G.R. Gibson, J.H. Cummings and G.T. Macfarlane. Growth and activities of sulphate-reducing bacteria in gut contents of healthy subjects and patients with ulcerative colitis. *FEMS Microbiology Ecology* 1991, 86, 103–112.
22. R. Havenaar. *Microbial Investigations on the Canogenicity of the Sugar Substitute Lactitol*. Unpublished report from the University of Utrecht (The Netherlands), Department of Preventive Dentistry and Oral Microbiology, June 1976.
23. R. Havenaar, *et al*. Some bacteriological aspects of sugar substitutes. In: B. Guggenheim (ed.). Health and Sugar Substitutes. Proceedings of the ERGOB Conference, Karger, Geneva and Basel; 1978, pp. 99, 192–198.
24. T.H. Grenby and A. Phillips. Studies of the dental properties of lactitol compared with five other bulk sweeteners *in vitro*. *Caries Research* 1989, 23, 315–319.
25. J.S. van der Hoeven. Cariogenicity of lactitol in program-fed rats. *Caries Research* 1986, 20, 441–443.
26. T.H. Grenby and A. Phillips. Dental and metabolic observations on lactitol in laboratory rats. *British Journal of Nutrition* 1989, 61, 17–24.
27. H.R. Muhlemann. *Gutachten Uber Die Zahnschonenden Eigenschaften Von Lacty Schokolade und Lacty-Milch-schokolade Der Firma C*. Holland, unpublished report from Zahnarztliches Institut der Universitat Zunch, July 1977.
28. T.H. Grenby and T. Desai. A trial of lactitol in sweets and its effects on human dental plaque. *British Dental Journal* 1988, 164, 383–387.
29. The Scientific Committee for Food. Food Science and Techniques. Report on Sweeteners, EC-Document III/1316/84/CS/EDUL/27 rev., 1984.
30. R.M. Diener. *Journal of the American College of Toxicology* 1992, 11(2), 165–257.

13 Maltitol Powder

Malcolm W. Kearsley[1] and Ronald C. Deis[2]

[1] Reading, UK
[2] Corn Products International, Newark, DE, USA

13.1 INTRODUCTION

Maltitol powder is considered as a bulk sweetener and has been used to replace sucrose specifically and sugars generally in foods and related products for about 25 years in Europe and the United States and for a longer period in Japan.

There are strict legal purity requirements for all polyols including maltitol powder (and specifically the maltitol content of maltitol powder) depending on local/regional legislation. For example, the EU Sweeteners in Food Regulations specify a minimum 98% maltitol content (on a dry basis). Dry products containing less than 98% maltitol are defined as dried maltitol syrups.

In the United States, FCC and USP-NF guidelines require the maltitol content of maltitol powder to be not less than 92% and not more than 100.5% maltitol calculated on a dry basis.

Maltitol is considered to be safe for the teeth (non-cariogenic) and suitable for diabetics; low glycaemic index (GI) and reduced-calorie claims for foods can often be made to take advantage of these properties. In particular, maltitol can be used to make no–added-sugar chocolate and baked goods where it functions as probably the best replacement for sucrose because maltitol, like sucrose, is a dimer with many similar colligative properties. Maltitol is also used in coatings for tableted gum where it gives a highly desirable crunchy texture and a glossy surface. The laxative effect of polyols generally is often quoted as the major reason for their lack of market penetration (other than their cost vs. sugars), but fortunately maltitol is one of the better tolerated polyols. Sales of sugar-free/no-added-sugar foods are generally still strong, and maltitol will play a key role in meeting the demand for these products.

When considering replacement of the two main bulk carbohydrates in foods (sucrose and glucose syrups), maltitol offers the closest approximation to the properties of sucrose and the closely related maltitol syrups can replace glucose syrups thereby offering a complete 'sugar replacement' package.

There is an abundance of general and patent literature available on maltitol – production, properties and applications and the reader is directed specifically to two reviews by Kato and Moskowitz[1] and Kearsley and Deis.[2]

13.2 PRODUCTION

Maltitol powder is produced in the United States, Europe and Asia, and although production is highly patented, it is now almost a commodity ingredient for the food and related industries. Commercially, it is manufactured by the catalytic hydrogenation of either maltose or more usually very high maltose glucose syrup in which the free reducing aldehyde groups on the maltose are reacted with hydrogen to give stable alcohol groups. This change in functional group radically changes some of the properties of the maltose. For example, maltitol is more than twice as sweet as maltose and is safe for teeth (non-cariogenicity) yet retains many of the parent sugar's other desirable features including structure, bulk and functionality. Therefore, maltitol has many of the advantages of sucrose but without the associated disadvantages. This makes it an almost perfect sugar replacer.

The hydrogenation conditions to manufacture maltitol are similar to those used to make other polyols. The maltose or maltose syrup is converted readily to maltitol through reaction with hydrogen at high temperature and pressure (typically, 100–150°C/100–150 bar) in the presence of a suitable catalyst (e.g. nickel, molybdenum, palladium or platinum). Reaction times are typically in the order of 1–2 hours.

In common with other hydrogenation processes, the maltose or maltose syrup used to make maltitol must be highly purified to ensure long catalyst life and this is easily achieved using ion exchange and carbon treatment of the raw material. Any catalyst dissolved in the maltitol during the reaction is removed by further ion exchange treatment.

Development work is also continuing on higher performance catalysts, and for example, hydrogenation of maltose to maltitol using a cobalt–phosphorous–boron catalyst has been described as having greater selectivity than the usual Raney nickel catalyst.[3]

The maltose or maltose syrup is usually produced from either maize (corn) or tapioca starches by enzymic hydrolysis using established techniques. The starch is first converted to a very high maltose syrup, normally containing more than 85% maltose, and from this material, there are several routes to make maltitol depending on processing and patent issues. All routes involve either aqueous crystallisation or melt crystallisation. In the former, a saturated solution of maltitol (or maltose) is cooled and seeded with crystals of the solute. The maltitol crystallises out in pure form and is recovered by filtration or centrifugation and dried.

In melt crystallisation, water is progressively evaporated from the maltitol syrup until the solids content exceeds about 95%. The polyol is not in solution under these conditions but is in the form of a high solids melt. After seeding with maltitol powder, the molten maltitol solidifies as it leaves the cooker. Drying, milling and screening complete the process to give maltitol powder. Melt crystallisation does not result in purification of the maltitol.

13.2.1 Alternative methods of maltitol manufacture

1. Recover high-purity maltose by crystallisation from the maltose syrup and hydrogenate this to give maltitol syrup. Maltitol powder can then be recovered from this maltitol syrup by aqueous or melt crystallisation. In both cases, maltitol powder purity should be in excess of 99% with aqueous crystallisation giving a slightly purer product. There is no change in purity with melt crystallisation as what goes into the crystallisation unit comes out.

2. An alternative to crystallising maltose is to use industrial-scale chromatography to produce a purified maltose stream from the 85%+ maltose syrup that can then be hydrogenated and the maltitol recovered by aqueous or melt crystallisation.
3. The maltose syrup can be hydrogenated and a high-purity liquid maltitol recovered using liquid chromatography. Maltitol powder can be recovered from the purified maltitol stream by aqueous or melt crystallisation. Again, purity is in excess of 99%.
4. Produce a maltose syrup with a maltose content in excess of 92–93%; hydrogenate and recover the maltitol powder by melt crystallisation. This gives a relatively impure product but one which is suitable for many applications, for example chocolate and baked goods, but not where crystallisation of the maltitol is important, for example coating of gum. The balance of 7–8% polyols in such a maltitol powder is made up mainly of sorbitol and maltotriitol.

Owing to the additional processes that need to be carried out on the starting material (very high maltose syrup), maltitol is significantly more expensive than sucrose or glucose syrup and typically twice the price of sorbitol powder. Like other polyols, it is available in a range of packaging formats from bags to bulk.

Although the production processes for maltitol powder are well established, development work continues to refine both the process and the physical attributes of the product.

13.3 STRUCTURE

Maltose is a disaccharide consisting of two glucose (dextrose) molecules joined by an α(1-4) linkage leaving a residual reducing group available for hydrogenation. Hydrogenation opens the terminal glucose residue, converting it to sorbitol, and the specific structure produced and its spatial geometry are responsible for the unique characteristics of maltitol.

The structure of maltose and its subsequent conversion to maltitol is shown in Figure 13.1.

13.4 PHYSICAL AND CHEMICAL PROPERTIES

Manufacture of sugar-free or no-added-sugar foods normally involves replacement of sucrose and glucose syrup with appropriate, non-sugar bulking agents. Polyols generally meet this requirement but optimum results are only obtained when the sugars are replaced by polyols with identical or at least similar properties. For sucrose replacement, maltitol powder represents the best option and in many applications, for example chocolate, 'one for one' replacement of sucrose is possible.

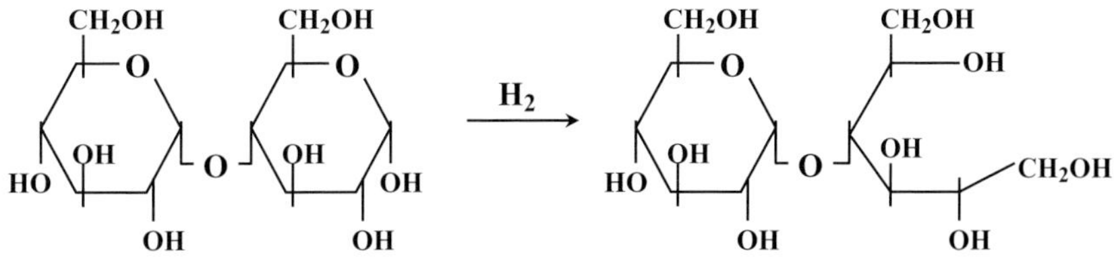

Fig. 13.1 Hydrogenation of maltose to maltitol.

13.4.1 Chemical reactivity

Maltitol does not take part in Maillard browning reactions and is relatively stable to high temperatures.[4] In most applications, this is a positive attribute but where browning is desirable in a food addition of caramel colour and flavour is usually necessary. This also extends to reaction with other ingredients in the food system such as colours, flavours, enzymes and vitamins where hydrogenation blocks reducing groups that would otherwise be more reactive over an extended shelf life.

13.4.2 Compressibility

Maltitol powder can be compressed directly to make tablets, but generally these are not as hard as sorbitol-based products, and owing to the high price of maltitol compared with sorbitol, sorbitol powder is usually preferred in this application. Since sorbitol is only about half as sweet as sucrose, sorbitol tablets require addition of high-potency sweeteners. Maltitol tablets require no such additions.

13.4.3 Cooling effect (heat of solution)

Although sucrose has a slight negative heat of solution of about -4 cal/g, it is not considered to have any appreciable cooling effect when consumed. Maltitol has a similar value to sucrose (-5.5 cal/g), again confirming maltitol as the optimum polyol to replace sucrose in foods. This is particularly important in 'no-added-sugar' chocolate where a cooling effect is undesirable.

13.4.4 Humectancy and hygroscopicity

Maltitol powder has little functionality as a humectant in no-added-sugar or sugar-free foods. It is also one of the least hygroscopic polyols and since it only absorbs moisture when the relative humidity exceeds 80%, maltitol contributes to making very stable food products. It is less hygroscopic than sucrose.

13.4.5 Molecular weight

The colligative properties of sugars and polyols are related to their molecular weight and the properties of importance in foods include viscosity, freezing point depression, boiling point elevation and osmotic pressure. These are mainly important in 'liquid' food systems (e.g. ice cream and hard candy) as opposed to solid systems (e.g. chocolate and biscuits), and therefore, relatively inexpensive maltitol syrup is often the polyol of choice where different composition products give a balance of sweetness combined with the properties of a high-molecular-weight ingredient.

Osmotic pressure of polyols is important when products containing these ingredients are consumed as any undigested/absorbed polyol can draw water from the body into the gut leading to osmotic diarrhoea. Pure maltitol (>98%) has a molecular weight of 344 and in fact, less pure versions (92–98% maltitol) are not dissimilar as the small amount of lower molecular weight sorbitol is balanced by the small amount of higher molecular weight maltotriitol (hydrogenated trisaccharide). Therefore, maltitol powder has almost double the molecular weight of the monosaccharide polyols sorbitol and mannitol and more than double the molecular weight of xylitol or erythritol. This gives a reduced osmotic pressure at the same concentration and potentially less severe digestive issues.

13.4.6 Solubility

Maltitol has a relatively high and similar solubility to sucrose – maltitol 175 g per 100 g of water at 25°C, sucrose 185 g. This is important specifically in those applications where maltitol is dissolved prior to use, for example when pan coating chewing gum and generally when foods containing maltitol such as chocolate and biscuits are eaten as the maltitol dissolves readily in the mouth, providing sweetness.

13.4.7 Sweetness

Sweet foods are considered highly desirable by the majority of consumers and are consumed far in excess of their value in alleviating hunger and thirst. Maltitol has a sweetness of about 90% that of sucrose and so can be used in most applications without addition of high-potency sweeteners. Different grades of maltitol powder with different maltitol content will vary slightly in sweetness, for example a maltitol powder containing in excess of 98% maltitol will be sweeter than a powder containing 92–93% maltitol. For most applications, these small differences are not important.

Hydrogenation of sugars usually results in an increase in their sweetness although the increase varies widely from sugar to sugar. Maltose is unique in this respect in that hydrogenation more than doubles its sweetness[5] and this behaviour has been attributed to the distinctive structure of maltitol.[6]

13.5 PHYSIOLOGICAL PROPERTIES

13.5.1 Calorific value

Owing to their different composition, it is most unlikely that all polyols would have the same calorific value. In the EU, however, maltitol and all polyols except erythritol have a calorific value of 2.4 kcal/g (10 kJ/g),[7] while in the United States, maltitol has a value 2.1 kcal/g,[8] and in Japan, an initial value of 1.8 kcal/g was proposed[9] but this was later amended to 2 kcal/g. In this respect, the values used in the United States and Japan are probably a more accurate reflection of the value for maltitol.

The different calorific values are not unique to maltitol and calculation of the calorific value of a food can therefore be very difficult depending on where it is being sold. It is predictable that different polyols will be digested to a greater or lesser extent, and in the EU, for example, some polyols will undoubtedly have a higher value than 2.4 kcal/g and some will be lower. The general lower values compared with sugars at 4 kcal/g are due to the relatively poor absorption of polyols in the small intestine. While some polyols will be hydrolysed and/or absorbed at this site, some will pass to the large intestine where they will be fermented to volatile fatty acids that are absorbed by the body and contribute about 2 kcal/g to the general calorie pool. If the gut is overloaded, polyols may draw water from the body into the gut and be excreted in very watery faeces.

Maltitol can be used to replace sugars in foods to make 'reduced-calorie' claims if appropriate reductions are made; currently, this is at least a 30% reduction in calories for such a claim in the EU and 25% in the United States although there are proposals to reduce the EU requirement to 25% in line with Codex and US recommendations.

This type of claim is already permitted in the United States.

In the United States, foods with 15% reduction in sugars may make the claim 'now contains 15% less sugars'. This claim is intended to encourage incremental sugar reduction and may only be made if the energy content of the reformulated product is equal to or less than the energy content of the original product. It may be used for a maximum of 1 year following the placing on the market of the reformulated product. This has also been discussed in the EU but to date not been adopted.

13.5.2 Dental aspects

Like other polyols, maltitol is either not fermented or fermented to a much lower extent than sugars by the oral bacteria. Fermentation of sugars leads to acid production and prolonged exposure of tooth enamel to acid conditions where the pH falls below 5.7 leads eventually to tooth decay. Since no acid is produced from polyols generally and maltitol specifically, maltitol is described as being non-cariogenic.[10] Maltitol can be used as a sweetener in foods that remain in the mouth for a considerable time, for example hard candy, chewing gum or chocolate, and therefore, provide a means of making such products safe for the teeth. This is the basis for a health claim in the United States – US Code of Federal Regulations, Title 21:101.80 (example: 'Frequent eating of foods high in sugars and starches as between-meal snacks can promote tooth decay. The sugar alcohol maltitol used to sweeten this food may reduce the risk of dental caries'). The US Food and Drug Administration (FDA) permits the claim 'does not promote tooth decay' and sugar-free foods containing polyols generally and maltitol specifically are included under this general heading.[11]

Even though a food may contain maltitol, when considering the cariogenic potential of the food the whole food should be taken into account as some components may be cariogenic. For example, chocolate contains milk that contains fermentable lactose, so even if the traditional sucrose is replaced with maltitol, the chocolate may still be cariogenic. The food can be tested at internationally recognised institutes to ensure they do not cause a lowering of the plaque pH and if they are 'safe for teeth'. Several studies have shown that maltitol *per se* is not cariogenic[12–14] and detailed information on the cariogenicity of foods generally is available in the literature.[15]

13.5.3 Diabetic suitability

Control of a subject's blood glucose and weight are critical factors in the management of diabetes and control of diet and maltitol can be used to advantage in both of these areas owing to its reduced GI and calorie content.[16,17] The metabolism of maltitol has been discussed extensively in the literature (e.g. by Dwivedi[18]) and in addition to giving reduced glucose levels after consumption, insulin levels are also reduced.

While diabetics are recommended to eat a normal diet and manage their carbohydrate intake, this has not prevented development of food products specifically for diabetics. Maltitol provides the same bulk as sucrose and can be used to make a range of high-quality diabetic foods, for example chocolate and biscuits.

13.5.4 Glycaemic index

The concept of GI was developed over 20 years ago to allow diabetics to manage their diets. GI is defined as the incremental area under the blood glucose response curve of a 50 g digestible carbohydrate portion of a test food, expressed as a percentage of the response to the same amount of carbohydrate from a standard food (typically, glucose) taken by the same subject.[19] Practically, it is method of ranking foods according to the extent to which they raise blood glucose levels after consumption. Foods containing carbohydrates that break down quickly after ingestion, giving a fast and high-blood glucose response, have the highest GI values, while foods containing carbohydrates that break down slowly after ingestion, giving a slow and low blood glucose response, have the lowest GI values. These are placed on a scale where glucose is given a GI of 100:

Low GI = 55 and below
Medium GI = 56–69
High GI = 70 and above

Maltitol is considered to have a low GI (about 35),[20] and low-GI diets (e.g. Atkins) have been promoted for weight loss and in improving the body's sensitivity to insulin.

13.5.5 Laxative effects

Although maltitol has been used in sugar-free/no-added-sugar confectionery and baked goods for many years for its calorie reducing and safe for teeth properties, and for its lowering effect on blood glucose, there is always a concern that excessive consumption may lead to laxative side effects. This concern is not restricted to maltitol, of course, but polyols generally and also fibre, a low-digestible carbohydrate, where overconsumption will have similar effects. Even overconsumption of simple sugars such as sucrose and fructose can produce the same reaction. For maltitol, the problem arises because we do not have the necessary enzymes in our bodies to fully metabolise maltitol in the small intestine, and maltitol is also described as a low-digestible carbohydrate. This follows logically from the non-cariogenic nature of maltitol where the bacteria in the mouth also do not possess the necessary enzymes to fully metabolise maltitol. Any maltitol not broken down in the upper gastrointestinal tract can, at certain concentrations, lead to an osmotic imbalance, and/or fermentation by bacteria in the lower gut, causing some digestive discomfort in the form of flatulence and diarrhoea. If the maltitol is broken down and too much sorbitol generated, then this could lead to a similar effect.

Many excellent papers have been published over the years concerning the laxative effects of polyols and related products (e.g. by Livesey[21]).

Although there is often no consistency in published data for the tolerance of maltitol, it is still possible to draw some conclusions and to offer general guidelines on the maximum level that can be consumed first as a single dose and secondly on a daily basis without causing significant digestive problems.

Disaccharide sugar alcohols such as maltitol are slowly digested by enzymes in the small intestine although in the short time they are in this region of the gut the process is usually incomplete depending on the amount ingested. Maltitol is hydrolysed by maltase enzyme into its component parts – glucose and sorbitol and the glucose produced is actively absorbed into the body while the sorbitol component is passively absorbed. Although maltase will hydrolyse both maltitol and maltose, hydrolysis of maltitol occurs at a much slower rate than maltose.[10]

It is relatively easy to overload this system by consuming large amounts of maltitol, and any maltitol not hydrolysed passes to the colon where it is fermented by bacteria producing gas and in larger quantities upsets the water balance producing osmotic laxation.[22] Similarly, with large doses of maltitol, more sorbitol is produced and this may exceed the amount that can be passively absorbed through the gut wall and into the blood stream. Unabsorbed sorbitol passes with the maltitol to the colon where it is also fermented. Generally, larger quantities of disaccharide polyols can be tolerated compared with monosaccharide polyols as the former are less osmotically active and, in the case of maltitol, half the molecule is, of course, glucose that is well tolerated in the body.[23]

The literature suggests a maximum of 30 g of maltitol can be consumed as an individual portion without laxation in the majority of adults and a maximum of 50 g of maltitol can be consumed per day. Therefore, maltitol is one of the best tolerated polyols. Generally, the greater the body weight, the less severe the laxative effect of maltitol and it has been reported that the maximum single dose in both men and women is 0.3 g/kg body weight (no laxative effect) with a daily maximum intake of 0.8 g/kg body weight.[24]

A more recent study[25] using chocolate containing maltitol showed that the product was well tolerated with no significant change in bowel habit or intestinal symptoms even at a daily dose of 45.6 g of non-digestible carbohydrate sweetener. This is of importance not only for giving manufacturers a sugar replacement that can reduce calorie intake but also for providing a well-tolerated means of delivering high levels of non-digestible carbohydrates into the colon, bringing about improvements in the biomarkers of gut health.

Ingestion of partially hydrolysed guar gum with maltitol has been reported as suppressing the transitory diarrhoea caused by maltitol. While fibre in large doses is often reported to have a laxative effect, it is interesting to note that in lower quantities fibre can reduce the laxative potential of non-digestible sugar substitutes.[26] Cellulose was also reported to have a similar effect.[27]

There is generally a lack of information regarding the laxative potential of maltitol (or in fact, any of the polyols) in children. A recent study[28] showed that maltitol was relatively well tolerated by children at up to 15 g per day and that its use could be extended to include new sugar-free confectionery and food products.

However, all values regarding daily intake and maximum dose levels should be treated only as guidelines when developing new products using maltitol as other factors such as gender, age, body weight and health can also affect tolerance. Maltitol in common with other polyols is usually better tolerated in solid foods and taken with other food ingredients such as fat or protein.

Prolonged consumption of maltitol (and all polyols, in fact) leads to adaptation with consequent improved tolerance,[21] and after 4–5 days on a relatively high intake of maltitol, for example, the laxative effects tend to be reduced and may in some cases disappear.

The Codex Alimentarius recommends that if a food provides a daily intake of sugar alcohols in excess of 20 g, there should be a statement on the label to the effect that the food may have a laxative effect. However, this is not mandatory, just a guideline and if polyols are used within sensible guidelines they can improve a formulation without causing problems.

13.6 APPLICATIONS IN FOODS

Although maltitol may be produced by melt or aqueous crystallisation and in different purity grades, the products can be used interchangeably in most applications. The grades are normally available in a range of particle sizes, and there will usually be an optimum particle size profile for each application. In some applications, minor processing changes may be required when using different purity materials, while in others, for example compressed tablet production (where the crystal morphology is important), the nature of the finished product will be different for different maltitol powders (>98% purity) or dried maltitol syrups (>about 93% purity). In some applications, for example pan coating, where crystallisation rate is important, only very high purity grades are acceptable to give a hard crunchy coating. In chocolate, where the milling characteristics of the different purity grades are different, changes to the chocolate-making process may be required.

Maltitol also crystallises in a similar way but at a slower rate than sucrose. In the cases where this is a problem, crystalline maltitol can be added with liquid maltitol as a seeding agent.

Maltitol is one of the most soluble of the polyols and again similar to sucrose. The texture and flavour release in a confection made using this product will be similar to a sucrose-based product, and this allows its use as an effective replacement for sucrose.

There is additionally some difference in sweetness between the grades, based on their maltitol content. The milling and screening processes used during manufacture of maltitol give a range of particle sizes in the powders, and these are used in different applications (like different particle size grades of sucrose in fact).

Maltitol can be used in 'diet', 'light' and 'reduced' (sugar or calorie) foodstuffs and in 'sugar-free' or 'no-added-sugar' products. It is widely used in bakery applications, pan coating of gum, as a sweetener in the gum itself and in chocolate although quantities used vary from category to category. Owing to the global rise in obesity and the demand for reduced-calorie foods, it seems likely that the demand for sugar-free confectionery will increase.[29]

While polyols have many similar properties – all are white, odourless and sweet (to a greater or lesser extent), for example – their other properties are usually quite different. Sucrose can be replaced directly in confectionery by several of the permitted polyols but best quality products are made when the polyol has properties most similar to sucrose. Of the disaccharide polyols, maltitol, lactitol and isomalt, maltitol has properties closest to sucrose and additionally is less hygroscopic. Like all polyols though, maltitol has found greatest application in specific products and is by no means used in every food application to replace sucrose. To avoid possible laxative effects, it is important to ensure the maltitol content of the foodstuff remains below about 20 g per serving.

13.6.1 The main food applications of maltitol

13.6.1.1 Chocolate

In theory, no-added-sugar/sugar-free chocolate can be made with a number of polyols although in practice and as in all applications, one specific polyol will tend to give the best product. In its anhydrous crystalline form, with low hygroscopicity, high melting point, low cooling effect, high sweetness (no additional high-potency sweetener needed) and stability, maltitol is the polyol of choice to replace the traditional sucrose in no-added-sugar chocolate. Additionally, substitution of sucrose with maltitol requires negligible processing changes and typically maltitol powder with a larger particle size is preferred to avoid the use of excess additional cocoa butter in the recipe.

Chocolate developed using a blend of polydextrose and maltitol not only had good sensory and rheological/processing characteristics but additionally contained 25% fewer calories than a full sugar product.[30]

Although not present in large volume, 'sugar-free' and 'no-added-sugar' chocolate products are available in the EU and the United States. In the EU, it is not permissible to partly replace the sugars in chocolate with maltitol but it is permissible to completely replace all added sugar with maltitol and call the product a 'no-added-sugar (milk) chocolate with sweeteners'. In the United States if any polyols are added to chocolate that has a standard of identity it may no longer be called 'chocolate' and would have to be called 'chocolate flavoured' or some other name.

13.6.1.2 Sugar-free panning

Soft panning: A combination of maltitol powder and maltitol syrup can be used to give a soft coating to products such as jelly beans.
Hard panning: The biggest application is in hard-coated gum where maltitol gives a hard, crunchy coating with minimal process changes. This application is highly patented.

The literature provides many references to sugar-free panning. For example:

- the attributes of panned products coated with maltitol together with the process involved to make a high-quality coating;[31]
- a low-cost method for producing non-chip coatings with maltitol.[32]

13.6.1.3 Dairy applications

Maltitol can be used to replace sucrose in ice cream.[33] It has approximately the same molecular weight as sucrose and gives the same freezing point depression as sucrose and hardness of the ice cream. Where sucrose and glucose syrup are used in the ice cream formulation, a blend of maltitol powder and maltitol syrup will give a product with a similar solids content and freezing point profile to the traditional product.[34] Maltitol provides the sweetness, while the higher molecular weight polymers in the maltitol syrup provide viscosity and air retention.

Maltitol or maltitol syrup can also be used in drinkable yogurts and flavoured milks.

Maltitol has also been shown to protect proteins during freeze drying[35] and may offer similar protection to milk proteins in ice cream, leading to higher quality finished products.

In addition to use of maltitol in dairy applications, it can also be used to formulate no-added-sugar, non-dairy desserts based on fruit or fruit puree.

13.6.1.4 Bakery applications

It is very difficult to make sugar-free baked goods as most contain flour that in turn contains small amounts of sugars. No-added-sugar and reduced sugar products are though possible. As in the majority of foods, sucrose and glucose syrups are the traditional sweeteners used in baked goods and the molecular weight and solubility of the sweetener system are similarly important with regard to the sweetness, starch gelatinisation temperature, water immobilisation, protein denaturation and the overall texture of the product. These are not only important in relation to the stickiness and hygroscopicity of the finished product but also control the biscuit size (spread during cooking), its volume and its baking characteristics. Maltitol can again replace sucrose directly in most baked goods and is generally considered to be the best sucrose replacer in this application. Maltitol syrups can replace glucose syrups and polyglycitols can replace maltodextrins. None of the polyols take part in Maillard browning reactions, and no-added-sugar baked products are normally less brown after baking. Again, a limit of about 20 g of maltitol per serving should be recommended to consumers although in products such as biscuits where large quantities are often consumed, this is easily exceeded.

Since maltitol does not take place in browning reactions, an additional advantage of using this material in biscuits is that less hydroxymethylfurfural (HMF) is produced during baking.[36] HMF has been implicated as potentially toxic and carcinogenic in rats.[37] No such correlations have been associated with intake in humans however.

A blend of polydextrose with maltitol (or lactitol) has been proposed as a replacement for sucrose in baked goods, giving a product with lower GI, while containing a prebiotic soluble fibre.[38] A similar effect on GI was observed when low-calorie muffins were produced using maltitol to replace the sugar and high-amylose corn starch to partly replace the wheat flour of a conventional recipe. This product also had a lower fat content than a traditional muffin and gave reduced insulin demand on consumption.[39]

It has also been reported that maltitol can be used to partially or totally replace fat in some baked goods, leading to additional calorie reduction[40]

With rising obesity levels, encouraging consumers to eat healthier foods choices is a big challenge. One option is to improve the nutritional value of traditional foods and 'healthier' cakes have been made using fructo-oligosaccharides to replace fat and maltitol to replace the sugar.[41]

13.6.1.5 Composite foods

Baked goods and dairy products often contain additional components such as caramel pieces, cream fillings, marshmallow, nougats and chocolate. It is important that any such additions do not contribute added sugars to the sugar-free or no-added-sugar finished product.

13.7 LABELLING CLAIMS

In addition to claims regarding the calorific value of a foodstuff, the replacement of traditional sweeteners such as sucrose and glucose syrups in foods with polyols is usually supported by appropriate labelling claims regarding the 'sugar' content of the product.

The following claims are those permitted in the EU and the United States, but similar claims are also used in other countries.

At the time of going to press these are:

Sugar-free
EU: The product contains less than 0.5 g of sugars per 100 g or 100 mL (where sugars are defined as mono- and disaccharides).
United States: Less than 0.5 g of sugars per reference amount and per labelled serving (21CFR101.60(c)).

Low sugar
EU: The product contains less than 5 g of sugars per 100 g for solids or 2.5 g of sugars per 100 mL for liquids.
The United States: This claim is not defined.

No-added sugar

EU: The product does not contain any added sugars (mono- or disaccharides) or any other food used for its sweetening properties. If sugars are naturally present in the food, the following should also appear on the label: 'contains naturally occurring sugars'.

The United States: The same claim is permitted.

Reduced sugar

EU: The reduction in sugars is at least 30% compared with the product for which no claims are made.

The United States: The reduction in sugars is at least 25% compared with the product for which no claims are made.

13.8 LEGAL STATUS

Extensive toxicological testing has shown that maltitol is safe for consumption and JECFA (the Joint FAO/WHO Expert Committee of Food Additives) has given it an acceptable daily intake 'not specified'.[42–44] It is permitted for food use in most countries, while some countries (EU, the United States) have their own specific legal requirements and purity criteria for this product others do not and they instead often adopt the Codex Alimentarius specification for maltitol. Legislation is designed to protect the consumer from fraud and poor-quality products and ensure production of food that is safe to eat. In the EU, polyols are considered as food additives and their use in foods is controlled by the Sweeteners in Food Regulations. Maltitol (and maltitol syrups) have been given the E Number E965 and are authorised for food use at *quantum satis* for the range of food products listed in the Regulations, typically confectionery, baked goods, ice cream, desserts and fruit preparations. Generally, polyols are not normally permitted in beverages (erythritol being an exception) owing to laxation issues from over consumption. In the EU, maltitol (and other polyols) cannot be used in foods in conjunction with sugars unless the polyol is present for a technological function other than sweetness or a 30% reduction in calories results from the combination. In the United States, the restrictions on the use of polyols in combination with sugars do not apply and polyols are considered either a food additive or 'Generally Recognized As Safe' (GRAS) by the FDA. Maltitol has self-affirmed GRAS status.

13.9 CONCLUSIONS

Of all the permitted polyols, maltitol offers unique opportunities to the new product formulator owing to its physical and chemical properties and similarity to sucrose. The main advantages and properties of maltitol when used in food products are as follows:

- A dimer with the same basic properties as sucrose.
- Bulk sweetener with a clean, sweet taste (about 90% of the sweetness of sucrose).
- Reduced calorie compared with sucrose and other sugars.
- Low-GI and low-insulin response (suitable for diabetics).
- Safe for teeth.
- Not hygroscopic.
- Heat stable.

REFERENCES

1. K. Kato and A.H. Moskowitz. Maltitol. In: L. O'Brien Nabors (ed.). *Alternative Sweeteners*, Marcel Dekker, New York; 2001.
2. M.W. Kearsley and R.C. Deis. Maltitol and maltitol syrups. In: H. Mitchell (ed.). *Sweeteners and Sugar Alternatives in Food Technology*, Blackwell Publishing Ltd., Oxford; 2006.
3. H. Li, P. Yang, D. Chu, H.L. Li, P. Yang, D. Chu and H. Li. Selective maltose hydrogenation to maltitol on a ternary Co–P–B amorphous catalyst and the synergistic effects of alloying B and P. *Applied Catalysis A: General* 2007, 325(1), 34–40.
4. M.W. Kearsley. The control of hygroscopicity, browning and fermentation in glucose syrups. *Journal of Food Technology* 1978, 13(4), 339–348.
5. M.W. Kearsley, S.Z. Dziedzic, G.G. Birch and P.D. Smith. The production and properties of glucose syrups. *Starke* 1980, 32(7), 244–247.
6. G.G. Birch and M.W. Kearsley. Some human physiological responses to the consumption of glucose syrups and related carbohydrates. *Die Starke* 1977, 29(10), 348–352.
7. European Economic Community Council. Council directive: nutrition labelling for foodstuffs. *Official Journal of the European Communities*, No. L276/41, 1990.
8. Life Sciences Research Office. In: J.M. Talbot, S.A. Anderson and K.D. Fisher (eds). *The Evaluation of the Energy of Certain Sugar Alcohols Used as Food Ingredients*. Prepared for the Calorie Control Council, Atlanta, Georgia, Federation of American Societies for Experimental Biology, Bethesda, MD, 1994.
9. Japanese Ministry of Health and Welfare. Evaluation of the energy value of indigestible carbohydrates in special nutritive foods, *Official Notice of Japanese Ministry of Health and Welfare*, no. Ei-shin 71, Tokyo, 1991.
10. S.C. Ziesentiz and G. Siebert. Polyols and other bulk sweeteners. In: T. Grenby (ed.). *Developments in Sweeteners 3*, Elsevier-Applied Science Publishers, London; 1987.
11. US Food and Drug Administration. Food Labelling: Health Claims: Sugar Alcohols and Dental Caries, Final Rule. *Federal Register*, 61, No. 165:43433, August 23, 1996.
12. R. Firestone, R. Schmid and H.R. Muhlemann. The effects of topical applications of sugar substitutes on the incidence of caries and bacterial agglomerate formation in rats. *Caries Research* 1980, 14, 324.
13. K. Matsuoka. On the possibility of maltitol and lactitol and SE-58 as non-cariogenic polysaccharides. *Nihon University Dental Journal* 1975, 49, 334.
14. J. Rundgen, T. Koulourides and T. Encson. Contribution of maltitol and Lycasin to experimental enamel demineralisation in the human mouth. *Caries Research* 1980, 14, 67.
15. Scientific Consensus Conference on the Methods for Assessment of the Cariogenic Potential of Foods. *Journal of Dental Research Special Issue*, 1986, 1473–1543.
16. T.M.S. Wolever. Dietary carbohydrates in the management of diabetes: importance of source and amount. *Endocrinology Rounds* 2002, 2(5), 2–5.
17. T.M.S. Wolever, A. Piekarz, M. Hollands and K. Younker. Sugar alcohols and diabetes: a review. *Canadian Journal of Diabetes* 2002, 26(4), 356–362.
18. B.K. Dwivedi. Polyalcohols: sorbitol, mannitol, maltitol and hydrogenated starch hydrolysates. In: L. O'Brien and R.C. Gelardi (eds). *Alternative Sweeteners*, Marcel Dekker, New York; 1986.
19. Carbohydrates in Human Nutrition. Report of the joint FAO/WHO Expert Consultation, Rome, 14–18 April, 1997.
20. G. Livesey. Health potential of polyols as sugar replacers, with emphasis on low glycemic properties. *Nutrition Research Reviews* 2003, 16, 163–191.
21. G. Livesey. Tolerance of low digestible carbohydrates – a general view. *British Journal of Nutrition* 2001, 85(Suppl.1), S7–S16.
22. H.A. Grabitske and J.L Slavin. Gastrointestinal effects of low-digestible carbohydrates. *Critical Reviews in Food Science and Nutrition* 2009, 49, 327–260.
23. P.J. Sicard and Y. Le Bot. From Lycasin to crystalline maltitol: a new series of versatile sweeteners. *Presented at the International Conference on Sweeteners – Carbohydrate and Low Calorie, 22–25 September 1988*, Los Angeles, CA. Sponsored by the American Chemical Society.
24. N. Koizumi, M. Fujii, R. Ninomiya, Y. Inoue, T. Kagawa and T. Tsukamoto. Studies on transient laxation effects of sorbitol and maltitol: estimation of 50% effective does and maximum non-effective dose. *Chemosphere* 1983, 12(1), 45.

25. E. Beards, K. Tuohy and G. Gibson. A human volunteer study to assess the impact of confectionery sweeteners on the gut microbiota composition. *British Journal of Nutrition* 2010, 104(5), 701–8.
26. S. Nakamura, R. Hongo, K. Moji and T. Oku. Suppressive effect of partially hydrolysed guar gum on transitory diarrhea induced by ingestion of maltitol and lactitol in healthy humans. *European Journal of Clinical Nutrition* 2007, 61(9), 1086–1093.
27. T. Oku, R. Hongo and S. Nakamura. Suppressive effect of cellulose on osmotic diarrhea caused by maltitol in healthy female subjects. *Journal of Nutritional Science And Vitaminology (Tokyo)* 2008, 54(4), 309–14.
28. C. Thabuis, M. Cazaubiel, M. Pichelin, D. Wils and L. Guerin-Deremaux. Short-term digestive tolerance of chocolate formulated with maltitol in children. *International Journal of Food Sciences and Nutrition* 2010, 61(7), 728–38.
29. A. Zumbe, A. Lee and D. Storey. Polyols in confectionery: the route to sugar free, reduced sugar and reduced calorie confectionery. *British Journal of Nutrition* 2001, 85(Suppl. 1), S31–S45.
30. C. Rodrigues Gomes, F. Zaratini Vissotto, A.L. Fadini, E. Vaz de Faria and A. Motta Luiz. Influence of different bulk agents in the rheological and sensory characteristics of diet and light chocolate. *Ciencia e Tecnologia de Alimentos* 2007, 27(3), 614–623.
31. B. Huzinec. Sugarless panning. *Manufacturing Confectioner* 2010, 90(6), 41–50.
32. S. Arai, S. Sakai and N. Imanishi. Method for providing coated product. United States Patent 2006/0286204 A1; 2006.
33. R.C. Deis, C.E. Kuenzle and B.W. Tharp. Ice cream and ice cream formulations containing maltitol. United States Patent 2007/0059404 A1; 2007.
34. P. Bordi, D. Cranage, J. Stokols, J. Palchak and L. Powell. Effect of polyols versus sugar on the acceptability of ice cream among a student and adult population. *Foodservice Research International* 2004, 15, 41–50.
35. S. Kadoya, K. Fujii, K. Izutsu, E. Yonemochi, K. Terada, C. Yomota and T. Kawanishi. Freeze-drying of proteins with glass-forming oligosaccharide-derived sugar alcohols. *International Journal of Pharmaceutics* 2010, 389(1/20), 107–113.
36. C. Delgado-Andrade, J.A. Rufian-Henares and F. Morales. Hydroxymethylfurfural in commercial biscuits marketed in Spain. *Journal of Food & Nutrition Research* 2009, 48(1), 14–19.
37. T. Husøy, M. Haugen, M. Murkovic, D. Jöbstl, L.H. Stølen, T. Bjellaas, C. Rønningborg and H. Glatt. Dietary exposure to 5-hydroxymethylfurfural from Norwegian food and correlations with urine metabolites of short-term exposure. *Food and Chemical Toxicology* 2008, 46, 3697.
38. M. Kweon, L. Slade and H. Levine. Exploration of low-glycemic-impact sugars and polyols in cookie baking, using SRC, DSC and RVA. In: *Dietary Fibre: New Frontiers for Food and Health*, Wageningen Academic Publishers, The Netherlands; 2010, pp. 513–528.
39. J. Quílez, M. Bulló and J. Salas-Salvadó. Improved postprandial response and feeling of satiety after consumption of low-calorie muffins with maltitol and high-amylose corn starch. *Journal of Food Science* 2007, 72(6), 407–411.
40. A.I. Bakal, S. Nanbu and T. Muraoka. Foodstuffs containing maltitol as sweetener or fat replacement. European Patent 039299 B1; 1993.
41. A. Wagner. Sugars reduced and fibre enriched products with Actilight® and Maltilite®: innovation in dairy products and snacks, *Wellness Foods Europe* 2008, May, 4–8.
42. Twenty Fourth Report of the Joint FAO/WHO Expert Committee on Food additives, Rome. WHO Technical Report Series No. 653; 1980.
43. Toxicological Evaluation of Certain Food additives and Contaminants. *Report Prepared by the 29th Meeting of the Joint FAO/WHO Expert Committee on Food Additives, 3–12 June 1985*, Geneva, Switzerland. Published by Cambridge University Press on behalf of WHO; 1987, pp. 179–206.
44. Reports of the Scientific Committee for Food Concerning Sweeteners. Commission of the European Communities. Sixteenth Series. Report EUR 10210 EN; 1980. Office of Official Publications of the European Communities, Luxembourg.

14 Maltitol Syrups

Michel Flambeau, Frédérique Respondek and
Anne Wagner
Tereos Syral, Marckolsheim, France

14.1 INTRODUCTION

Maltitol syrups and polyglycitols have been used in foods and primarily in confectionery for more than 20 years in Europe (the authorization is almost finalized for polyglycitol syrup), the United States and for a longer period in Japan; they were among the first 'sugar-free' ingredients to be used commercially.

They are now indispensable for the manufacture of high-quality 'sugar-free' and 'no-added-sugar (NAS)' products. Maltitol syrups and polyglycitols may be classified under the general heading of hydrogenated starch hydrolysates (HSH) or polyols or sugar alcohols. Theoretically, HSH could describe any hydrogenated starch hydrolysis product and could also include products containing sorbitol and mannitol as well as blends thereof, which may contain polymeric sugar alcohols, hence conferring new technological functionalities in terms of viscosity, texture and hygroscopicity. In practice, however, the term HSH is used to describe what are basically hydrogenated glucose syrups of any degree of hydrolysis.

Polyols are frequently marketed as being safe for the teeth, suitable for diabetics, low glycaemic response (GR) and reduced calorie, the last two categories are currently of specific interest to food manufacturers. Overweight and obesity are now a major cause of illness and death-related diseases in many developed and also developing countries. While the obsession with diets and dieting was once an age-related issue, it now affects every age group. Obesity among children is of particular concern owing to the sedentary lifestyle many of them now follow. Lack of exercise is seen as a particular problem in gaining weight. The media constantly debates what we should and should not be eating, what foods are good for us and what foods should be avoided. 'Diet', 'light' and 'reduced' labels on foodstuffs are commonplace, and it is now possible to buy a wide range of such items. Among these healthier products is a growing range of 'sugar-free' confectionery items including chewing gum, tableted mints and related products, hard candy, chocolate and chewy candy. It does seem likely that the manufacture of sugar-free confectionery will increase as the demand for 'low GI' and 'low carb' foods increases. Maltitol syrups and polyglycitols are set to play a key role in meeting this demand. A good example of this trend is the chewing gum market that has almost totally shifted to sugar-free, while soft-gum products aimed principally at children remain largely formulated from high-calorie sweeteners.

Sugar-free confectionery is almost exclusively based on the replacement of the traditional sweeteners sucrose and glucose with polyols or sugar alcohols that provide both bulk and sweetness. Of the permitted sugar alcohols, maltitol offers the closest approximation to the properties of sugar and different glucose syrups can be replaced by the appropriate maltitol syrup.

A more recently developing market for maltitol syrup is medicinal syrups, such as cough syrups, which are increasingly available in sugar-free form.

14.2 PRODUCTION

In common with the majority of polyols, maltitol syrups and polyglycitols are manufactured by the catalytic hydrogenation of the appropriate glucose syrup where reactive aldehyde groups are replaced by stable alcohol groups. By changing only the reactive reducing groups, the polyol retains much of the parent sugar's structure, bulk and function and simultaneously gains other beneficial properties. This makes them an excellent sugar replacer.

The raw material for the manufacture is starch and while this can be from any source, maize (corn), wheat and tapioca starches are most widely used commercially.

14.2.1 Maltitol syrups

Where legislation exists on the use of maltitol syrups in foods or pharmaceutical applications (see Section 14.7), they are usually defined as products containing a minimum of 50% maltitol on a dry basis. The hydrogenated carbohydrate balance is not defined in EU or US Food Regulations, but normally consists of sorbitol and hydrogenated gluco-oligosaccharides. The starting materials for maltitol syrup manufacture are, therefore, diverse and are typically maltose syrups containing from 55% to 95% maltose on a dry basis. With such wide-ranging composition, maltitol syrups possess a wide range of physical and chemical properties of use to the product developer. At high maltitol levels, crystallisation of the maltitol limits the solids level that can be achieved in these syrups. For example, the solids content of a syrup containing over 90% maltitol on a dry basis will need to be less than 70% to avoid maltitol crystallisation.

As with maltitol powder, the starch is first liquefied before addition of the specific saccharifying enzymes that give the required maltose syrup. Depending on the saccharifying enzyme(s) used, α-amylase, pullanase, isoamylase, maltogenase, maltose levels ranging from 45% to 75% can be achieved. The maltose syrup is then hydrogenated, submitted, if required, to a chromatography separation step to increase maltitol level, evaporated to the required solids and packaged. Typical packaging is in 275 kg drums, 1000 kg semi-bulk containers and bulk tanks (about 20 mt), with a solids content from 67% to 85% depending on the composition.

14.2.2 Polyglycitols

In addition to maltitol syrups, there exists a related group of products with equally diverse properties and these are the polyglycitols. Polyglycitols are HSH containing less than 50% maltitol and less than 20% sorbitol on a dry basis. The balance is hydrogenated gluco-oligosaccharides. As with maltitol syrups, the starting materials for the manufacture of polyglycitols are wide-ranging from maltodextrins up to mid-dextrose equivalent (DE) glucose syrups. Polyglycitols are available either as concentrated aqueous liquids or in powder form depending on their composition. In addition to sorbitol, maltitol and maltotriitol, polyglycitols contain a high percentage of hydrogenated gluco-oligosaccharides with more than three glucose units joined to a terminal sorbitol unit. This can sometimes lead to 'clouding' problems on storage of liquids as the high-molecular-weight components retrograde and become insoluble.

Polyglycitols do though extend the range of properties of maltitol syrups and provide sugar-free alternatives to the whole range of glucose syrups and maltodextrins normally used in foods.

14.3 HYDROGENATION

The hydrogenation conditions to manufacture maltitol syrups and polyglycitols are not dissimilar to those to make sorbitol and the other polyols. Typically, batch reactors are used in which maltose/glucose syrups react with hydrogen at high temperature (typically, 100–150°C) and high pressure (typically, 30–150 bar) in the presence of a catalyst like Raney nickel, supported nickel, ruthenium, molybdenum, palladium or platinum. Reactions normally take 1–3 hours, depending on the conditions (temperature, pressure, sugar concentration and catalyst type, concentration and age) and on the product.

The catalyst cost is an important part of the overall production cost. Therefore, it is important to recover and reuse the catalyst as much as possible. Recovery of catalyst is also important with respect to costs linked to the removal of catalyst residues from the reacted syrup.

Key for increasing the lifetime of the catalyst is to use highly refined raw materials in the process. Activated carbon treatment and several ion exchange steps are very common in achieving the required purity. Depending on the catalyst type, catalyst separation at the end of the reaction typically occurs by sedimentation, filtration or a combination of both these processes. A small amount of fresh catalyst is added to the reactor after each hydrogenation to compensate for the loss at this stage. Depending on the application, further refining steps are applied on the maltitol syrup for removal of degradation products (e.g. gluconic acid), traces of catalyst in solution and for removal of sugars left after hydrogenation (typically, 0.1–0.2 %).

Typically, maltitol syrups are sold at 70–85% solids with maltitol contents of 55–75% (based on dry solids). The other solids are sorbitol and polymeric sugar alcohols that contribute viscosity and texturising properties to the syrup. These products are stored at elevated temperature to facilitate loading and handling and to avoid potential crystallisation issues that could occur at room temperature.

For syrups containing less than 50% maltitol (on a dry basis), it is often not possible to achieve a high enough solids consistent with easy handling, so these products are usually spray dried to give microbiological stability.

While the hydrogenation conditions ensure sterility immediately after the reaction, there is always a risk for microbial contamination in the following process steps. This is especially true for all steps involving dilutions, for example the processes of sweetening-on or sweetening-off sweetener streams. Maintaining a high temperature in these process streams will prevent microbiological problems. This also applies for incidental dilution, for example in storage tank heads where condensed water vapour can run down into the syrup causing localised dilution. This can be avoided by heating the top of the tank or by introducing a sufficient flow of sterile air in the tank headspace.

14.4 STRUCTURE

The structures of maltitol and maltitol syrups/polyglycitols are shown in Figure 14.1. During hydrogenation, only the terminal glucose residue reacts with hydrogen and is converted to sorbitol. Polyglycitols have basically the same structure as maltitol syrups differing only in the content of each hydrogenated component. They contain less than 50% maltitol and less than 20% sorbitol and a proportionally greater content of hydrogenated higher molecular weight gluco-oligosaccharides.

14.5 PHYSICO-CHEMICAL CHARACTERISTICS

Polyols differ in molecular weight and configuration, leading to big differences in properties between the different products. It is important when replacing sucrose and/or glucose with polyols that the manufacturer selects a polyol whose properties most closely mimic the sugar(s) to be replaced. Some manufacturers replace sucrose with a combination of polyol(s) and other bulking agents such as polydextrose and inulin to compensate for the lack of performance of a particular polyol, but one for one replacement is of course much more straightforward and the label cleaner. It is important to be aware, that while sucrose and

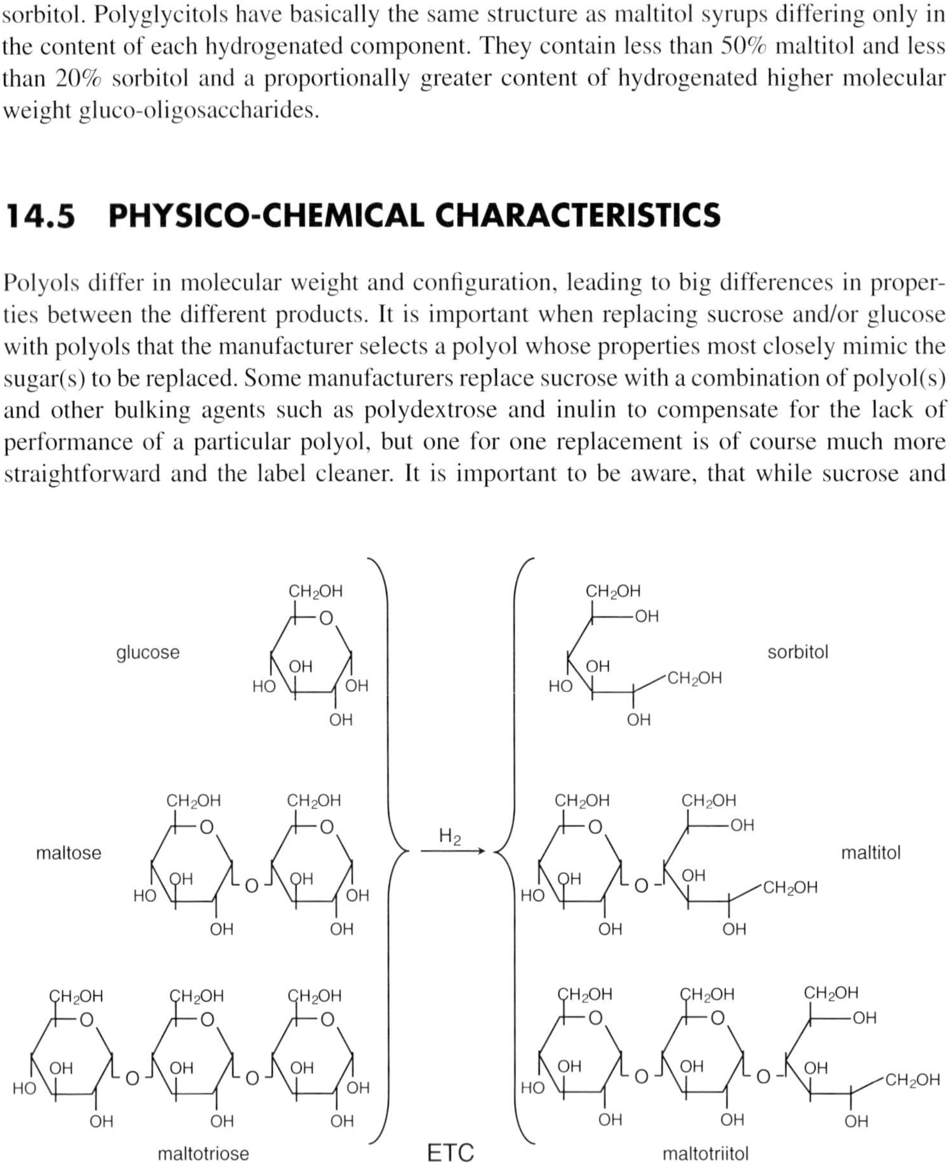

Fig. 14.1 Hydrogenation of maltose syrup to maltitol syrup.

glucose syrups can be replaced in foods with polyols, selection of an inappropriate replacement will lead to a poor quality finished product.

The various polyols have different sweetness, solubility, cooling effect, molecular weight and laxative effects and there can be wide variation across the permitted range of polyols. It is important that the formulator understands these differences so that they can select the correct polyol for a particular application. The days when poor products were bought because they were 'good for you' are long past and a successful product must now not only provide benefit to the consumer, but must also look good and taste good and cause no digestive discomfort, otherwise, it will fail.

14.5.1 Chemical reactivity

In common with other polyols, maltitol syrups and polyglycitols do not take part in Maillard browning reactions. Normally, this is a positive attribute for such products as browning is undesirable, but in some applications (e.g. caramel manufacture, biscuits), browning is an essential part of the manufacturing process. Sugar-free caramels require addition of caramel colour and flavour. It is often more appropriate to make a 'NAS' claim than a 'sugar-free' claim for caramels so that milk products containing lactose can remain in the formulation.

For biscuits, very often, a 1% addition of dextrose is made to permit good browning.

14.5.2 Cooling effect (heat of solution)

When the solid forms of some sugars or polyols are placed on the tongue, a distinct cooling effect may be noticed. This is more pronounced with some sweeteners than others. The effect, known as the 'heat of solution', is an exchange in energy that can either lower (in this example) or raise the temperature of a solution when a substance (sugar) is added to water (saliva). The smaller the particle size of the powder, the more quickly it will dissolve and the cooling effect is increased.

Most polyols in crystalline form have a cooling effect, with some being very significantly different from sucrose at -4.3 cal/g. Maltitol syrups and liquid polyglycitols have no cooling effect as such and in dried form the cooling effect depends on composition but is less pronounced than that of sucrose.

14.5.3 Humectancy

A good humectant is one that resists changes in moisture content as the humidity of its surroundings change. Maltitol syrups provide both sweetness and a degree of humectancy, which may be modified by varying the polymeric distribution in the product. Depending on their composition, polyglycitols are less effective humectants than maltitol syrups, and the high-molecular-weight polyglycitols in particular are, like maltodextrins, not good humectants.

14.5.4 Hygroscopicity

The hygroscopic tendencies of food ingredients can have very serious consequences during production and storage (shelf life) of the food itself and also may affect how the ingredient itself is stored. Maltitol is one of the least hygroscopic of the polyols and can be handled in most countries without the need for air conditioning in the manufacturing plant. In contrast, dried maltitol syrups tend to be very hygroscopic, while the moisture absorption properties of polyglycitols are relatively low.

14.5.5 Molecular weight

Polyols are available in a wide range of molecular weight from very low (erythritol at 122 g/mol) through to polyglycitols at over 1000 g/mol (with maltitol at 344 g/mol and maltitol syrups in the 400–600 g/mol range). Figure 14.2 gives the average molecular weights of the different polyols.

The molecular weight of sugars commonly used in confectionery range from high DE glucose syrups at around 250, through sugar at 342 to maltodextrins at over 1000.

Again, it is important to match the molecular weight of the polyol(s) with the sugar(s) being replaced to ensure the colligative properties remain unchanged.

Molecular weight can play a significant role in influencing the overall texture and functionality of the finished product. More importantly to the plant operator, it will determine how they will flow when in liquid or molten form during processing. Replacing a 43DE glucose syrup with sorbitol syrup in a foodstuff will completely change its processing characteristics and also the finished product. As a starting point, the 43DE glucose syrup must be replaced by a maltitol syrup or polyglycitol of similar molecular weight.

In addition to viscosity, molecular weight becomes important with regard to other properties – including freezing point depression, boiling point elevation and osmotic pressure.

Freezing point depression is an important factor when making ice cream, for example, where traditionally, sucrose, glucose syrups and maltodextrins are used for to provide functionality and sweetness. As in all applications, replacement of like with like will give the best results. Thus, sucrose can be replaced with maltitol powder, glucose syrup with maltitol syrup and maltodextrin with polyglycitol (if the ice cream is manufactured in the United States). In the EU, selection of a maltitol syrup with a proportion of high-molecular-weight hydrogenated saccharides would be the best option. However, knowing the proportion of

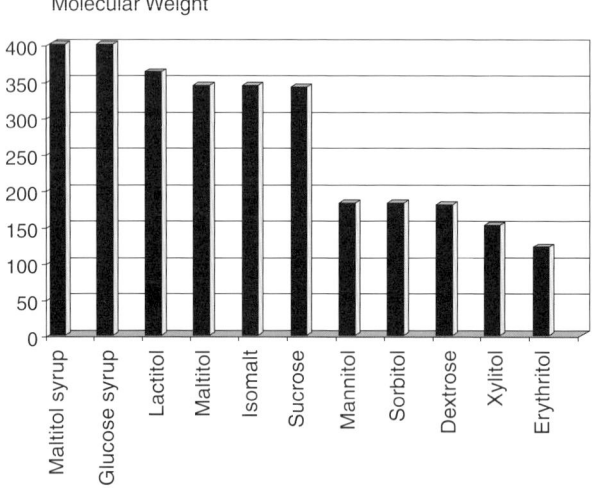

Fig. 14.2 Molecular weight of polyols.

sucrose, glucose and maltodextrin in a formulation, it may be possible to replace the three traditional products with one maltitol syrup with the same average molecular weight to give the same freezing point depression.

Osmotic pressure becomes important in the gut (see Chapters 1 and 3) and also with regard to the keeping qualities of a maltitol syrup. For example, it must be high enough to prevent microbial growth at the solids at which the syrup is sold.

Boiling point elevation is particularly important in hard candy manufacture as the product consists almost exclusively of sugars or polyols. Too high an elevation will increase the amount of energy required to evaporate water during the cooking stage.

14.5.6 Solubility

Solubility is generally defined as the amount of a solute (polyol) that can be dissolved in a solvent (water) at a given temperature before becoming saturated. As the temperature of the solution increases, more solid can be dissolved and if such a solution is cooled, the solute will crystallise out. This property is fundamental to the manufacture of almost all confections. Traditional confections such as caramels, marshmallow and hard candy in their most basic form are supersaturated sugar/glucose solutions, where crystallisation may or may not be required (grained or ungrained candy). Through many years of use, confectioners have utilised the solubility of sucrose to create the standards of appearance, taste, texture, shelf life and sweetness that we now come to expect from these types of products.

Polyols such as erythritol, isomalt and mannitol provide the lowest solubility and crystallise most readily. These may work well in applications that require low moisture, low hygroscopicity and rapid crystallisation, for example, nougats and grained marshmallows. Maltitol, lactitol and xylitol, on the other hand, are more soluble and of the polyols are most similar to sucrose. Although all crystallise relatively similarly to sucrose, lactitol is most similar followed by maltitol and then xylitol. The most soluble of the commonly used polyols is sorbitol but, like xylitol, it can form lumps during storage if conditions are not properly controlled.

Figure 14.3 gives the solubility of the different polyols.

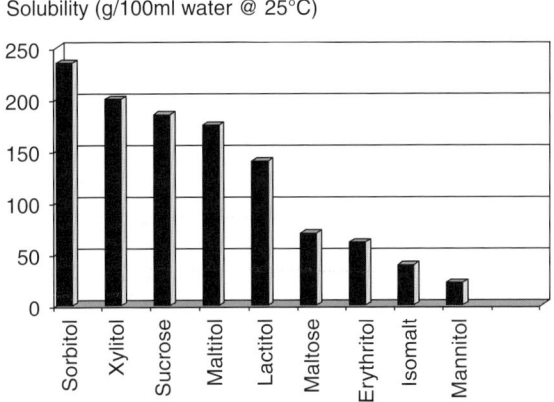

Fig. 14.3 Solubility of polyols.

Maltitol syrups and polyglycitol syrups behave more like glucose syrups and for the most part are very soluble. Only at very high maltitol levels is there a danger of maltitol crystallisation and mostly they can be supplied at 75% dry solids or higher without problems. They are typically available in a range of solids levels from 70% to 85% and are either used alone or in combination with other polyols – like erythritol, isomalt, maltitol and lactitol – to control crystallisation. These products therefore find a wide range of use in many sugar-free applications that would traditionally use glucose syrups.

14.5.7 Viscosity

Viscosity is related to molecular weight – the higher the molecular weight, the higher the viscosity. While hydrogenation increases the molecular weight of a reducing sugar, the effect is minimal and viscosity changes on hydrogenation are also minimal and will be virtually un-noticeable. For example, dextrose has a molecular weight of 180 and sorbitol 182: maltose 342 and maltitol 344. While maltitol has a low viscosity in solution, the range of related products has higher viscosity. Thus, maltitol syrups have higher viscosity than maltitol at the same solids content and finally some of the polyglycitols have very high viscosity in solution. Again, a range is found that can be used to advantage by the product formulator.

14.6 PHYSIOLOGICAL PROPERTIES

14.6.1 Calorific value

In the EU (in common with all other permitted polyols except erythritol), maltitol syrups are given the same average calorific value of 2.4 kcal/g (10 kJ/g). In practice, this makes calculation of the calorific value of a food very easy, but it is unlikely to be a true picture of what happens in reality. Owing to widely different compositional differences, it is most unlikely that they will be metabolised in the same way and have the same calorific value. Polyglycitols are a similar step away from maltitol syrups and while not currently permitted in the EU, it is likely they will also be allocated a calorific value of 2.4 kcal/g. Different polyols will be digested in a different way and while there may be some similarities between different polyols giving the same calorific value, it should not be assumed that this will be the case. Some polyols will have a higher value than 2.4 kcal/g and some will be lower. The lower value compared with sugars is due to the relatively poor absorption of polyols in the small intestine. While some polyols will be hydrolysed and/or absorbed at this site, what is not absorbed will find its way to the large intestine where it will be fermented to volatile fatty acids. These are then absorbed by the body and contribute about 2 kcal/g to the general calorie pool.

In many other countries, this difference is recognised and different polyols have different calorific values. This probably gives a truer picture of actual events. Table 14.1 gives the calorific value of polyols in the United States, Japan and the EU.

Reduced-calorie and low-calorie claims for foods are permitted in the EU. At the time of going to press, the EU food regulations require a 30% reduction in calories (compared with the product for which no claims are made) to make a 'reduced-calorie' claim.

Low-calorie foods must not be more than 40 kcal/100 g for solids or 20 kcal /100 mL for liquids.

Table 14.1 Calorific value of polyols.

	United States (kcal/g)	Japan (kcal/g)	EU (kcal/g)
Sucrose	4.0	4	4.0
Erythritol	0.2	0	0
Isomalt	2.0	2.0	2.4
Lactitol	2.0	2.0	2.4
Maltitol powder	2.1	2.0	2.4
Maltitol syrups	3.0	2.3–3.4	2.4
Mannitol	1.6	2.0	2.4
Sorbitol	2.6	3.0	2.4
Polyglycitols	3.0	2.3–3.4	2.4[b]
Xylitol	2.4	3.0	2.4

[b]Not yet permitted in EU but likely value when approved.

In the United States, reduced-calorie claims necessitate a 25% reduction in calories and low-calorie foods must meet the same requirements as in the EU. Polyols can be used to make reduced-calorie foods. This is straightforward for confectionery, but more difficult to achieve for compound foods containing protein, fat and carbohydrate.

14.6.2 Dental aspects

Dental caries is a disease affecting the hard tissue of teeth that can lead to progressive decay. Even if the prevalence has regularly decreased in developed countries over the last 30 years, tooth decay remains the third most prevalent disease in the world after heart disease and cancer. Both environmental and genetic factors can be involved in the development of tooth decay.

Bacteria in the plaque on the surface of the teeth are able to ferment carbohydrates in the foods we consume to produce acid, which causes a lowering of the pH at the tooth surface. If the plaque pH falls below 5.7, this can lead to demineralisation of the tooth enamel that is considered as a risk factor for tooth decay or dental caries. Polyols are either not fermented or fermented to a much lower extent by the oral bacteria and are described as non-cariogenic. While they may cause a slight fall in pH at the tooth surface, they do not normally cause a fall in plaque pH below 5.7. The use of polyols to replace sugars in those foods that remain in the mouth for a considerable time, for example confectionery, therefore provides a means of making such foods that are safe for teeth. A recent study has confirmed that maltitol was as effective as xylitol when used in sugar-free chewing gum to induce remineralisation of teeth in comparison to sugar gum.[1]

The effect of polyols on tooth mineralisation is the basis for a health claim in the United States – US Code of Federal Regulations, Title 21:101.80 (Example: 'Frequent eating of foods high in sugars and starches as between-meal snacks can promote tooth decay. The sugar alcohol maltitol used to sweeten this food may reduce the risk of dental caries'.). More recently, the European Food Safety Agency (EFSA) has also issued scientific opinions stating that replacing sugars in foods (which reduce plaque pH below 5.7) by different polyols including maltitol may maintain tooth mineralisation compared with sugar-containing foods, provided that such foods do not lead to dental erosion. The condition to bear the claim is that the considered foods do not lower plaque pH below 5.7 during and up to 30 minutes after consumption. While the polyols themselves are non-cariogenic, other components of the food may not be. For example, in chocolate manufacture, milk (containing lactose) may be used and maltodextrins are often used as carriers for flavours and colours. Therefore, it is important that the food as a whole be considered when deciding whether it is safe for teeth and not just the individual ingredients. It is possible to have finished products tested at certain recognised institutes to ensure they do not cause a lowering of the plaque pH.

14.6.3 Glycaemic index

Although this is a relatively new concept for foods and food ingredients, the concept of glycaemic index (GI) was developed over 20 years ago by David Jenkins at the University of Toronto as a tool to allow diabetics to better manage their diets.

GI is defined as the incremental area under the blood glucose response curve of a 50 g digestible carbohydrate portion of a test food, expressed as a percentage of the response to the same amount of carbohydrate from a standard food (typically, glucose) taken by the same subject.[2] It is a method of ranking foods according to the extent to which they raise blood glucose levels after consumption. Foods containing carbohydrates that break down quickly after ingestion, giving a fast and high blood glucose response, have the highest GI values, while foods containing carbohydrates that break down slowly after ingestion, giving a slow and low blood glucose response, have the lowest GI values. These are placed on a scale where glucose is given a GI of 100:

Low GI	55 and below
Medium GI	56–69
High GI	70 and above

In addition to being important in the control of diabetes, low-GI diets have also been promoted for weight loss (although more research is required in this case) and in improving the body's sensitivity to insulin.

The glycaemic response to the different polyols are well documented,[3] and Figure 14.4 shows the GI of different polyols with sucrose and glucose for comparison. Maltitol syrups are considered to have a low GI, and while some polyglycitols will also fall in this category, others will likely be classed as high GI. Specifically, these will be the higher molecular weight polyglycitols (molecular weight equivalent to maltodextrins) that will be easily hydrolysed in the upper gastrointestinal tract primarily to glucose, causing a significant rise in blood glucose. The metabolism of maltitol has been discussed extensively in the literature.[4]

14.6.3.1 Diabetic suitability

The American Diabetes Association has recognised that polyols produce a lower post-prandial glucose response than sucrose or glucose. They confirmed that their use appears to be safe including use by people with diabetes, even if no clear evidence has been provided that the amounts of polyols likely to be consumed will result in long-term improvement in glycaemia.[5,6]

While in Europe it was not considered that diabetics would require specific dietary recommendations in addition to those generally applicable to healthy people, EFSA has

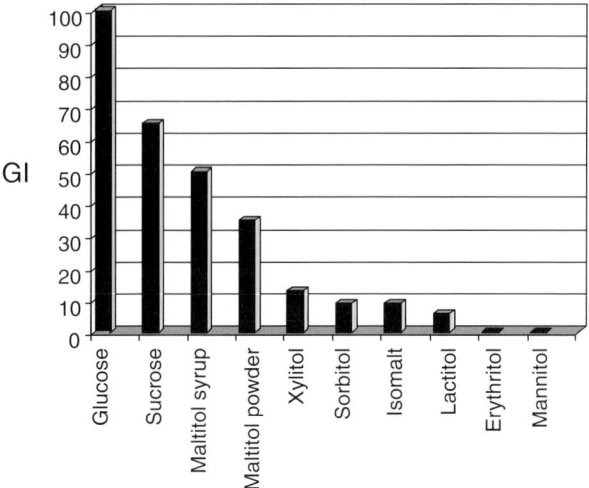

Fig. 14.4 Glycaemic index of polyols.

recently confirmed that reducing post-prandial blood glucose response (i.e. in comparison to a reference food) may be considered a beneficial effect providing that insulin response is not disproportionally increased. Experts have acknowledged that replacing sugars in total or to a significant amount by polyols, including maltitol will reduce post-prandial blood glucose response (Annex on nutrition claims of Regulation 1924/2006).

14.6.4 Toleration

Polyols have been an important feature of sugar-free confectionery for many years and have proved extremely useful in reducing calories, lowering blood glucose response and formulating tooth-friendly products. Polyols are low-digestible carbohydrates in so far that they cannot generally be completely hydrolysed by the enzymes present in the upper digestive human tract. What is not absorbed in the upper gastrointestinal tract can, at certain levels, lead to an osmotic imbalance and/or fermentation by bacteria in the lower gut – causing some digestive discomfort in the form of flatulence and diarrhoea. Polyols are not unique in this respect, as fibre also falls into the same category of low-digestible carbohydrates and overconsumption of fibre will have similar effects. Many excellent papers have been published over the years concerning the laxative effects of polyols and related products.[7]

Just as the physical properties of polyols are different, so the laxative properties of polyols differ. Figure 14.5 shows the laxation thresholds for various polyols. These values represent the average total amount of polyol that can be typically consumed daily before a laxative effect is observed but an important inter individual variability has to be taken into account. Generally, an individual can consume larger doses of maltitol per day than other polyols without discomfort and larger still doses of maltitol syrups and polyglycitols. The high-molecular-weight components of the syrups not only moderate the osmotic effects in the gut but are also hydrolysed more readily in the upper gut. Some polyglycitols can be tolerated at over 150 g per day without problems. Generally, the higher the molecular weight of the hydrogenated carbohydrate, the less the laxative effect. In the brush border

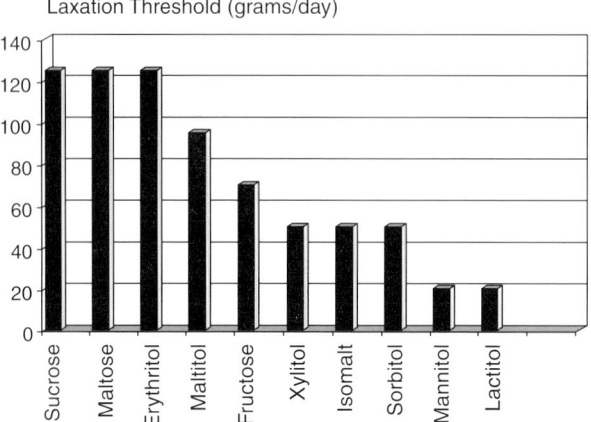

Fig. 14.5 Laxation of polyols.

region of the upper gut, amylase enzyme quickly reduces the molecular weight of the hydrogenated gluco-oligosaccharides, producing mainly glucose and maltitol and the glucose is actively and rapidly absorbed by the body. While the maltase enzyme located in this region hydrolyses maltose very quickly, it acts much more slowly on maltitol. While the glucose from any maltitol hydrolysis is again rapidly absorbed, the sorbitol produced is only passively absorbed. If large doses of maltitol are ingested, the system becomes overloaded, both in terms of maltitol hydrolysis by maltase and the amount of sorbitol being too much for the passive diffusion through the gut wall and into the blood stream. The maltase enzyme does not have time to hydrolyse all the maltitol and excess passes into the lower gut with any unabsorbed sorbitol – where they are fermented producing gas (and in larger quantities they upset the water balance producing osmotic laxation). The higher the molecular weight, the less maltitol (and thus sorbitol) are produced on hydrolysis, which explains why polyglycitols are better tolerated than maltitol syrups which in turn are better tolerated than maltitol. Factors such as individual response, age, colonic microflora, gender, psyche, health, diet, any drugs or antibiotics being taken and other foods consumed can all have an effect on the laxative effect of the polyol. Laxation is also dictated by chemical factors such as molecular weight (larger molecules will cause less osmotic imbalance), solubility and whether or not the substance is broken down and absorbed in the small intestine. Additionally, prolonged consumption of polyols leads to some adaption and improved tolerance. After 4–5 days on a relatively high intake of maltitol and related products, the laxative effects tend to be reduced and may in some cases disappear.

Currently, in the United States, only products containing sorbitol, mannitol and polydextrose are required to carry a warning label that 'excessive consumption can cause a laxative effect'.

In the EU, a warning label is required when the polyol comprises 10% or more of the food product.

In the Codex Alimentarius, if the food provides a daily intake of sugar alcohols in excess of 20 g, there should be a statement on the label to the effect that the food may have a laxative effect. This is not mandatory however, just a guideline.

Consumption of any polyol at levels more than 20–30 g per serving is not recommended even though warning labels may not be required. While individually a single food or confectionery serving will not cause discomfort, it is all too easy to overload the digestive system by consuming more than one portion of the product. While the objective is to make the quality of a sugar-free/NAS food the same or better than the traditional product, there is always the danger (as with any food) that it tastes so good it is overconsumed. Considering that many individuals will consume more than one serving size per meal, it would be wise to adhere to the following limits:

- For monosaccharides, less than 10 g per serving.
- For disaccharides, less than 15 g per serving.
- For polysaccharides, less than 20 g per serving.

If polyols are used within sensible guidelines, they can improve a formulation without causing problems and a recent study has confirmed that a single dose of 15 g of maltitol introduced in milk chocolate was well tolerated by children with an average body weight of 32 kg.[8]

When using maltitol syrups and polyglycitols in foods, there has inevitably to be a trade-off among calorific value, laxative effect and GI in addition to the other properties of the products. Thus, while maltitol has a lower calorific value and GI than a polyglycitol, it will be more laxative.

14.6.5 Sweetness

Sweetness is one of the most important properties of any sugar or sugar replacer. We generally like sweet foods and, despite manufacturers making foods with reduced sweetness, the demand for the sweet taste is still one of the most sought after. Polyols vary in sweetness from about 30% to 95% the sweetness of sugar but any lack of sweetness in the polyol can, of course, be compensated for by addition of a high-potency sweetener. However, just because a polyol might deliver the same sweetness as sucrose, it does not mean it will function like sucrose in the chosen application. Sweetness intensity is just one of the important properties when selecting the correct polyol(s) for a particular application, the sweetness profile as well as synergestic effects with other sweeteners and flavours need to be taken into account.

Sweetness of a carbohydrate is related to its molecular weight and structure and the chemical conversion of a reducing sugar to its corresponding sugar alcohol by the process of hydrogenation normally results in an increase in sweetness. It is interesting to note among the disaccharides the changes in sweetness that occur when the reducing sugar are hydrogenated. While maltose with its $\alpha(1\text{-}4)$-linked glucose residues increases in sweetness after hydrogenation, cellobiose with its $\beta(1\text{-}4)$-linked glucose residues and isomalt, a mixture of the two dimers glucose:sorbitol $\alpha(1\text{-}6)$ and glucose:mannitol $\alpha(1\text{-}1)$ decrease in sweetness after hydrogenation. Lactose also increases in sweetness on hydrogenation but to a much lower extent than maltose.[9]

It is immediately evident that maltitol is unique among the permitted polyols in that hydrogenation of maltose more than doubles the sweetness of the reducing sugar and this unusual high sweetness as been attributed to its unique structure.[10]

Maltitol syrups have a range of sweetness depending on their maltitol content – the more maltitol they contain, the sweeter they are and proportionally more sweet than the corresponding maltitol syrup starting material. Polyglycitols containing from 49% to less

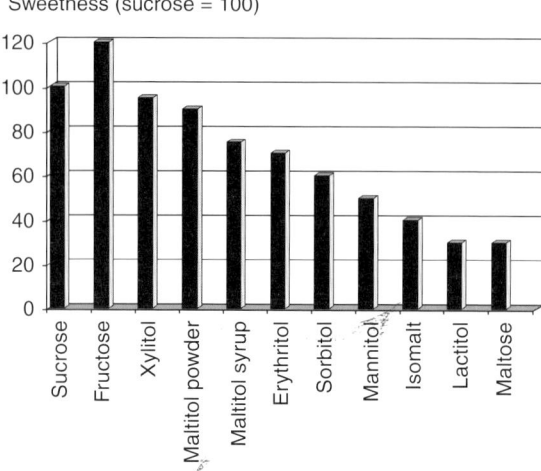

Fig. 14.6 Sweetness of polyols.

than 1% maltitol, range in sweetness from about 40% to 50% as sweet as sucrose to less than 10% as sweet.

Figure 14.6 shows the sweetness of a range of polyols.

14.6.6 Conclusions

Maltitol syrups and polyglycitols offer unique opportunities to the formulator of new food products owing to their wide-ranging physical and chemical properties.

The main advantages and properties of these products when used in food products are given as follow.

14.6.6.1 *Maltitol syrups*

- Bulk sweeteners with a clean, sweet taste, 60–90% of the sweetness of sucrose.
- Reduced calorie compared with traditional sugars.
- Suitable for diabetics.
- Safe for teeth.
- Low GR.
- The polyols of choice for NAS/sugar-free ice cream and confectionery.

14.6.6.2 *Polyglycitols*

- Bulk sweeteners with a clean, sweet taste, 10–60% of the sweetness of sucrose.
- Reduced calorie compared with traditional sugars.
- Mostly suitable for diabetics.
- Mostly safe for teeth.
- Low/medium GR.
- The polyols of choice for stability in hard candy, carriers for flavours, colours and enzymes.

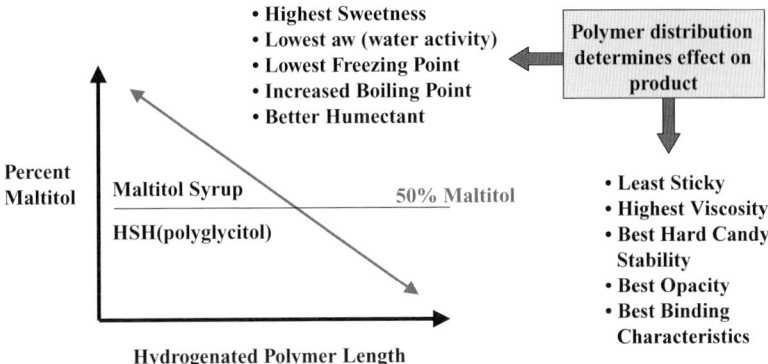

Fig. 14.7 Properties of maltitol, maltitol syrups and polyglycitols.

Figure 14.7 summarises the properties of maltitol syrups and polyglycitols.

14.7 APPLICATIONS IN FOODS

Obesity is major contributor to illness and death-related diseases and the obsession with diets and dieting among the population appears to be unlimited. We are constantly advised about what we should and should not be eating, what foods are good for us and what foods should be avoided. 'Diet', 'light' and 'reduced' sugar are widely used on labels for foodstuffs and it is now possible to buy a wide range of such items to meet the demands of an increasingly overweight and obese population.

Included in the number of healthier products is a growing range of 'sugar-free' confectionery items that, although increasing, is still limited compared with traditional sugar or sugar/glucose-based products. It does seem likely though that the manufacture of sugar-free confectionery will increase as the demand for 'low-GI' and 'reduced sugar' foods increases.[11] Similarly, NAS-baked goods are now well established although small in volume.

Sugar-free confectionery is almost exclusively based on the replacement of sugar and glucose with polyols, and they have been produced for over 20 years.

While maltitol syrups and polyglycitols are colourless and odourless (and the dried products are white and odourless), different products have different functional and technological properties. Sucrose and glucose syrups can be replaced directly in confectionery by several of the permitted polyols. However, poor selection of the replacements can lead to changes in reactivity, appearance, flavour, process and preparation, mouthfeel, texture, storage and handling, so it must not be undertaken lightly. However, optimum results will only be obtained when the disaccharide sucrose is replaced by a disaccharide polyol whose properties are most similar to sucrose, for example, maltitol powder. Similarly, the polymeric glucose syrup must be replaced by a hydrogenated equivalent and selection of the appropriate maltitol syrup (EU) or maltitol syrup/polyglycitol (United States) will depend on the DE, viscosity and composition of the glucose syrup being replaced.

It is important that the formulator understands the differences between different polyols so that the correct polyol can be selected for a particular application. For example, solubility is fundamental to the manufacture of almost all confections. Traditional confections such as caramels, marshmallow and hard candy in their most basic form are supersaturated sugar/glucose solutions where either crystallisation (grained) is required to some degree or not at all (ungrained). Through many years of use, confectioners have utilised the solubility of sucrose to create the standards of appearance, taste, texture, shelf life and sweetness that we now come to expect from these types of products.

Maltitol syrups and polyglycitol syrups behave like glucose syrups and for the most part are very soluble and are typically available as syrups in a range of solids levels from 70% to 85% or in some cases as spray dried powders. These products therefore find a wide range of use in many sugar-free applications that would traditionally use glucose syrups.

Molecular weight can play a significant role in influencing the overall texture and functionality of the finished product. Viscosity, freezing point depression, glass transition temperature (T_g) and boiling point elevation are all directly related to molecular weight. For the plant operator, it will determine how the products will flow when in liquid or molten form during processing. As a starting point, the 43DE glucose syrup must be replaced by a maltitol syrup or polyglycitol of similar average molecular weight.

14.7.1 Hard candy

Hard candy is essentially a sugar syrup that has been heated to reduce the moisture content to a very low level such that on cooling the product remains in a glassy state.

Sugar-free hard candy has been available for many years.[12]

Isomalt is often the preferred sweetening agent in hard candy as a result of its unique stability and manufacturers are prepared to tolerate the lack of sweetness and poor solubility to take advantage of this. Replacement of 15–30% of the isomalt with an appropriate maltitol syrup increases the overall solubility of the candy and also increases viscosity. Less flavour needs to be added and there will be a slight reduction in the amount of high-potency sweetener required in the product as the maltitol syrup will contribute more sweetness than the isomalt. No appreciable loss of stability should be found and there will be a noticeable improvement in quality, taste and longer term stability. There are now also maltitol syrups in the market that can be used alone to formulate sugar-free hard candy giving stability equivalent to sugar/glucose products. Rather than form a crystalline piece, a stable amorphous glass with a high T_g is formed. In this case, control of finished moisture to less than 1%, and preferably below 0.5%, is important owing to the plasticising effect of water. This process forms a very good wrapped candy with excellent colour and flavour release.

14.7.2 Aerated confectionery

Interest for the full range of confectionery products in sugar-free variant has developed. More recently, marshmallow formulated with maltitol syrup and maltitol powder as a replacement of glucose syrup and sucrose has been successfully developed. Use of maltitol syrups containing 55–65% maltitol gives good products and they form the major component of such formulations (up to 70% of the dry solids). Other ingredients in the formulation like gelatin or flavour and colour are mainly unchanged and any changes, if necessary, are relatively minor.

Manufacturing of sugar-free marshmallow can be made on standard production equipment and with process parameters close to sugar products. For the dusting powder applied to the external surface, a fine granulometry maltitol powder gives good and stable products and is preferred to other polyols.

The overall texture of a traditional sugar product can be achieved in sugar-free form and important parameters of the product such as density, chewiness and stickiness are almost identical. Depending on the sugar reference targeted, some change in the maltitol syrup to maltitol powder ratio may be necessary.

14.7.3 Caramels

In addition to being a product in their own right, caramels are also often used in combination with other confections such as chocolate, creams, nougats and hard candy. Additionally, they also find application in other food products such as nutritional bars, baked goods and frozen desserts.

Caramels consist essentially of fat, milk solids, sucrose and/or glucose syrup. It is the reaction between the reducing sugars and the milk proteins during heating (Maillard reaction) that gives caramels their unique brown colour and flavour. Caramels can vary greatly in texture with a structure ranging from a hard candy type to a free-flowing liquid depending on the final moisture content of the product, the amount of protein in the formulation, the ratio of sucrose to glucose syrup and the melting point of the fat. The same considerations apply when making 'NAS' or 'sugar-free' caramels where polyol(s) replace the sucrose and glucose syrup in the formulation.

When making caramels, the main difference between sugars and polyols is that, since polyols do not participate in the Maillard reaction, the formulator must use a caramel colour and flavour to achieve the desired colour and taste.

Local regulations with respect to specific labelling claims for 'sugar-free' dictate what ingredients in addition to the polyols are to be used in the formulation. Typically, in Europe, sugar-free products must contain less than 0.5 g sugars/100 g where sugars are defined as all mono- and disaccharides. Dairy ingredients such as sweetened condensed milk, condensed whole milk, dried whole milk powder and non-fat dry milk are often used in the manufacture of caramels. These can contain either sucrose or lactose in sufficient quantities to make it almost impossible to keep the sugars in the finished product below the required legal level. Proteins such as sodium caseinate, which contain little or no sugar, provide alternative materials to potentially enable such claims to be made. For these reasons, the formulation of sugar-free caramels is one of the most challenging of applications.

NAS caramels are more flexible and easier to formulate because they can use more traditional dairy ingredients, for example condensed whole milk, dried whole milk powder and whey concentrates. These ingredients bring no additional sugars during manufacture hence allowing such caramels to be considered 'NAS'. It may also be possible when the sugar content of these ingredients is very low and when combined with other 'sugar-free' products' such as hard candy and chocolate to create a finished product to which the term 'sugar-free' can be applied.

Where reduced-sugar claims are permitted, reduced-sugar caramels are relatively easy to formulate with little or no change in processing, sweetness or product functionality. In the United States, a reduction of 25% of the sugars with reference to a standard product will enable this claim to be made. 'Reduced-sugar' caramel does not need added caramel colour or flavour since there would be more than enough reducing sugars available for the Maillard reaction to occur.

The choice of polyol to make the caramel depends on the texture and structure of the caramel required and the ratio of sucrose to glucose syrup in the traditional product. For instance, as the ratio of sucrose increases over the glucose syrup, the more crystallisation or graining of the caramel will be apparent. This gives a product a soft, short texture but with limited shelf life owing to crystallisation depending on how much sucrose is used and the final moisture content. If the glucose syrup content increases, the caramel will become more sticky, chewy, firmer and stringy with limited crystallisation and an extended shelf life. The crystallising polyols such as erythritol, isomalt, lactitol, maltitol, sorbitol and xylitol will take the role of the sugar while maltitol and polyglycitol syrups take the role of the glucose syrup component of the caramel and control crystallisation as well as other textural properties. A grained caramel can be made using erythritol and maltitol syrup to give a short soft texture with a moderate shelf life. Alternatively, a polyglycitol syrup used alone would give a firm, chewy caramel that is stringy. It is, of course, not always easy to do in practice and other factors such as cooling effect, cost, laxation, legislation and solubility of the polyols must also be considered.

14.7.4 Sugar-free panning

Soft panning involves the coating of centres by applying successive layers of a non crystallising syrup and crystalline powder in a rotating drum or pan and is the type of coating applied to such products as jelly beans, where a combination of glucose syrup and crystalline sugar are used. To achieve the characteristic texture, it is important that the layers are not fully crystallised.

Both soft and hard panned sugar-free products can be manufactured.[13]

Soft panning can be adapted to sugar-free products by replacing the glucose syrup with a non-crystallising maltitol syrup (e.g. 55–65% maltitol) and replacing crystalline sucrose with maltitol powder. The relatively low content of maltitol in the syrup prevents crystallisation of the coating. Sugar-free jelly beans made with maltitol powder and maltitol syrups deliver similar characteristics to sugar-based products in terms of taste and texture but contain 40% less calories.

Hard panning involves the coating of centres with successive application of sucrose syrup layers, which crystallise and creates a hard and crunchy layer. Crystallisation occurs as water is evaporated and removed using dry/hot air in a rotating pan or drum. Typically, the coating will form about 30% w/w of the product. The contrast in texture with a softer centre (chewing gum, aerated confectionery, chocolate lentils) or brittle centre (nuts) creates a unique eating experience.

Replacement of sucrose syrup with a high-purity maltitol syrup (above 95% maltitol) gives a hard coating of high quality and one which is very similar to sucrose in many aspects including whiteness, crunchiness, low friability, good sweetness and stability to moisture uptake. In many aspects, maltitol is superior to other polyols like sorbitol or xylitol that could be used as an alternative in this application.

In terms of the coating process, maltitol can be used in coating equipment similar to those used for sugar coating, and coating conditions are similar to those used for sucrose coating. A good coating could be applied at a temperature around 30°C. Hard coating can be achieved using a high-purity crystallisable syrup or from a syrup obtained by dissolution of a high purity maltitol powder.

Blending different polyols will allow the creation of a wide range of coating textures by controlling the crystallisation speed. There are, however, many application patents covering coating with maltitol and blends thereof and the users should ensure that the process they use does not infringe these.

14.7.5 Chewing gum

Maltitol syrups are also used in the formulation of sugar-free chewing gums. Maltitol syrup helps prevent crystallisation of the polyols added in large quantity to the gum base. It contributes to give a soft and chewy texture to the chewing gum.

For chewing gum in dragees, coating with maltitol syrup gives a smooth but crunchy texture and a good resistance even in moist environment due to its low hygroscopicity (see Section 14.7.4).

14.7.6 Other sugar-free or reduced-sugar confectionery[14]

14.7.6.1 Sugar-free jellies

Maltitol syrup with a maltitol content in the range 55–65% enables production of jelly products with a chewy but short texture, good transparency and stability similar to sugar products. A high maltitol content should be avoided to prevent recrystallisation over shelf life. The syrup represents more than 90% of the composition of the jelly, other ingredients being gelatine, flavour and colour a higher dry solids in the syrup will enable reduced cooking time. In addition to a sugar-free claim, the calorie content of the jelly is reduced by more than 30% compared with a sugar reference.

14.7.6.2 Sugar-free chewy candies

Sugar-free products can be made using a combination of maltitol syrup and maltitol powder in a 2:3 to 1:3 ratio and they can, therefore, replace the sucrose and the glucose syrups typically used in these products. Candies can be made with similar elasticity and stickiness, but tend to be slightly less hard at first bite. Sugar-free products are available commercially, and they have gained a significant market share in several countries.

14.7.6.3 Reduced sugar wine gums

A combination of maltitol syrup, maltitol powder and maltodextrins can replace glucose syrup and sucrose to achieve a sugar reduction of 30%. Reformulated products can be made with similar properties to full sugar products in terms of elasticity and hardness, two important parameters for these types of products.

14.7.7 Dairy applications

In a conventional full-sugar ice cream, the sweetener system – typically, sucrose and glucose syrup – accounts for almost half of the total dry solids in the finished product. In addition to sweetness, the 'sweetener system' also provides some other essential technological functions, including freezing point depression (which affects machinability, dippability and shelf life), water retention and crystal size, bulk and shape retention. In the ideal situation, the NAS sweetening system should replace not only the sweetness but also these other functions.[15] In the past, typical NAS sweetening systems have developed using blends of ingredients including sorbitol, polydextrose, fructo-oligosaccharides, high-potency sweetener and maltodextrin. The sorbitol is typically added to replace the sucrose and the polydextrose to replace the glucose syrup solids. However, as both sorbitol and polydextrose lack sweetness, a high-potency sweetener needs to be added. Since sorbitol has half the molecular weight of sucrose and polydextrose (depending on the grade) may have a lower average molecular weight than the traditional ice-cream glucose syrup, a further ingredient, namely maltodextrin, may be required to adjust the freezing point depression. Recent work has demonstrated that this entire sweetening system can be replaced by an appropriately formulated maltitol syrup. Maltitol provides the sweetness as well as a molecular weight equivalent to sucrose and the higher molecular weight polyols present in the syrup replace the glucose syrup. Calorie reductions greater than 30% are possible using such systems. Therefore, the NAS ice cream has a solids content and freezing point profile equivalent to the full-sugar ice cream, as well as a similar sweetness profile.[16]

Drinkable yoghurts and flavoured milks have increased in popularity in recent years as an alternative to high-calorie beverages and as delivery systems for probiotics and prebiotics. The traditional route to sugar and caloric reduction in these beverages has been the addition of high-potency sweeteners and hydrocolloid stabilisers. The drawback to this type of approach is usually a change in flavour and texture compared with the full sugar version. A more practical approach to consider is the addition of maltitol and/or maltitol syrup to replace the sugar solids as the maltitol would contribute significantly to the overall sweetness of the product. To avoid possible laxative effects, it is important to ensure that the overall polyol content of the product remains below 20 g per serving. With maltitol the finished product will have a better texture owing to the more equivalent dry solids as well as a better sweetness profile.

14.7.8 Bakery applications

In bakery applications, sucrose and glucose syrups are the traditionally sweeteners, so the consumer has been very much conditioned to these products as the gold standards with respect to product quality. As in confectionery and dairy applications, the molecular weight and solubility of the sweetener system is also very important in bakery applications, because these properties not only affect the sweetness but also other factors including the starch gelatinisation temperature, water retention, protein denaturation and the bulk of the product. These characteristics, in turn, affect the spread, the volume, the baking characteristics and the stickiness of the biscuit. Many of the lessons learned from working with confectionery can be extended to baked goods (and also dairy products) because confectionery items such as caramels, creams, marshmallow, nougats and chocolate are also components of many baked goods. Since many of the physical and chemical characteristics of maltitol are similar to sucrose, maltitol can replace sucrose directly in most baked goods applications and maltitol syrups or polyglycitols are available to replace glucose syrups or maltodextrins. Maltitol syrups and polyglycitols do not participate in Maillard browning, and the products will therefore brown less during processing. To overcome this drawback, a low concentration of dextrose usually 1% is added to the formulation. Sweetness may also need to be adjusted, but the textural properties of the finished product remain unchanged. In very sweet products, close attention should be given to the grams of polyol per serving because consumers are quite likely to consume more than one serving. A limit of 20 g per serving should ideally be used. For this reason, reduced-sugar-baked goods should be considered as a healthy alternative to full sugar-baked goods rather than attempting to replace all of the sugar.

14.7.9 Ketchup

Sugar-free ketchup based on tomato paste, modified starches and glucose syrup can be developed in sugar-free form by replacing the glucose syrup with a maltitol syrup containing

55–65% maltitol. Maltitol syrup will represent about one-third of the formula. As this type of product is very sensitive to the sweet/acidic balance, an intense sweetener will often be used to fine tune the taste.

14.8 LEGAL STATUS

Maltitol syrups are permitted for food and pharmaceutical use in most countries. Some countries (EU, the United States) have their own legal requirements and purity criteria for these products, but where these do not exist, the corresponding Codex Alimentarius specifications are usually adopted. All are essentially similar and have the same objective – to present the consumer with high-quality food ingredients. In the EU, polyols are considered as food additives and their use in foods is controlled by the Food Improvement Agents Package and more specifically by the regulation on Food Additives (1333/2008/CE). Maltitol has been given the E-number E965 (i) for powder and (ii) for syrup, when both are used in a formulation, the single name of maltitol is used. They are authorised for food use at *quantum satis* for the range of food products listed in the Regulations. Depending on the country, this range typically includes confectionery, baked goods, ice cream and desserts and fruit preparations. Beverages are not usually permitted to contain maltitol and its derivatives as well as other polyols owing to possible overconsumption leading to laxative side effects. They are also not authorised to be used in foods in conjunction with caloric sugars unless the polyol is present for a technological function other than sweetness (e.g. sorbitol is often used as a humectant in baked goods), or unless at least a 30% reduction in calories results from the combination.

In the United States, the restrictions on the use of polyols in combination with sugars do not apply. Polyols are considered either a food additive or 'generally regarded as safe' (GRAS) by the Food and Drug Administration (FDA). Maltitol syrups and polyglycitols are both self-affirmed GRAS.

Generally, polyglycitols are less widely permitted than maltitol syrups. Currently, they are permitted in the United States (self-affirmed GRAS), Canada and Japan. In Europe, polyglycitols should soon be authorised as food additives as the European Commission has already approved and assigned the E-Number E964 to these products.

14.9 SAFETY

Extensive toxicological testing has shown that maltitol syrups and polyglycitols are safe for consumption and the Joint FAO/WHO Expert Committee of Food Additives (JECFA) has given all these products an acceptable daily intake (ADI) 'not specified' (WHO, Geneva 1998).

14.10 CONCLUSIONS

In food applications, selection of the most appropriate polyol(s) can be based on an understanding of the functionality of sugar and/or glucose syrup in a particular product and matching the properties of the polyol to the sugar(s) they are to replace. Optimum results will be obtained when these properties are most similar. Because consumers are likely to consume multiple servings, precautions should be taken to ensure that the product is a healthy alternative with no adverse effects.

REFERENCES

1. E.J. Lee, B.H. Jin, D.I. Paik and I.K. Hwang. Preventive effect of sugar-free chewing gum containing maltitol on dental caries in situ. *Food Science and Bitotechnology* 2009, 18, 432–435.
2. FAO/WHO. Carbohydrates in human nutrition. Report of the Joint FAO/WHO Expert Consultation, Rome, 14–18 April 1997.
3. G. Livesey. Health potential of polyols as sugar replacers, with emphasis on low glycemic properties. *Nutrition Research Reviews* 2003, 16, 163–191.
4. B.K. Dwivedi. Polyalcohols: sorbitol, mannitol, maltitol and hydrogenated starch hydrolysates. In: L. O'Brien and R.C. Gelardi (eds). *Alternative Sweeteners*, Marcel Dekker, New York; 1986, pp. 165–183.
5. T.M.S. Wolever. Dietary carbohydrates in the management of diabetes: importance of source and amount. *Endocrinology Rounds* 2002, 2(5).
6. T.M.S. Wolever, A. Piekarz, M. Hollands and K. Younker. Sugar alcohols and diabetes: a review. *Canadian Journal of Diabetes* 2002, 26(4), 356–362.
7. G. Livesey. Tolerance of low digestible carbohydrates – a general view. *British Journal of Nutrition* 2001, 85(Suppl 1), S7–S16.
8. C. Thabuis, M. Cazaubiel, M. Pichelin, D. Wils and L. Guerin-Dereumaux. Short-term digestive tolerance of chocolate formulated with maltitol in children. *International Journal of Food Science* 2010 (online).
9. M.W. Kearsley, S.Z. Dziedzic, G.G. Birch and P.D. Smith. The production and properties of glucose syrups. *Starke* 1980, 32(7), 244–247.
10. G.G. Birch and M.W. Kearsley. Some human physiological responses to the consumption of glucose syrups and related carbohydrates. *Die Starke* 1977, 29(10), 348–352.
11. A. Zumbe, A. Lee, and D. Storey. Polyols in confectionery: the route to sugar free, reduced sugar and reduced calorie confectionery. *British Journal of Nutrition* 2001, 85(Suppl 1), S31–S45.
12. C.E. Seaman, J.A. Bower, A. March. Sensory characteristics of sugar free and sugar based boiled sweets. *International Journal of Food Science and Nutrition* 1997, 48(5), 329–337.
13. B. Huzinec. Sugarless panning. The Manufacturing Confectioner, 2010, pp. 41–50. Presented at PMCA Conference 2010.
14. S. Roelle and N. Camuel. A step forwards for sugar free confectionery. *Innovations in Food Technology* 2010, February, pp. 22–23.
15. A. Wagner. Sugars reduced and fibre enriched production with Actilight and Maltilite: innovation in dairy products and snacks. *Wellness Foods Europe* 2008, May, pp. 4–8.
16. P. Bordi, D. Cranage, J. Stokols, J. Palchak and L. Powell. Effect of polyols versus sugar on the acceptability of ice cream among a student and adult population. *Foodservice Research International* 2004, 15, 41–50.

15 Sorbitol and Mannitol

Ronald C. Deis[1] and Malcolm W. Kearsley[2]

[1]Corn Products International, Newark, DE, USA
[2]Reading, UK

15.1 INTRODUCTION

Sorbitol and mannitol are the only polyols found naturally in any appreciable quantity; both have been used for over 50 years in foods and related products and were among the first 'sugar-free' ingredients. In common with all other polyols (except erythritol), they are produced by chemically reacting the appropriate reducing sugar with hydrogen.

As one of the first polyols to become commercially available, sorbitol was used in a wide range of products including fruit preserves, baked goods and confectionery. In particular, it was used in foods for diabetics since it does not cause an increase in blood glucose on ingestion. As the sugar-free bulk sweetener market matured, sorbitol, like other polyols, established itself in specific use areas, and sorbitol (and mannitol) now occupy what may be considered a small corner of the polyols market in terms of applications. However, in terms of volume consumed annually, the use of sorbitol far exceeds the use of all other polyols combined. The main applications are not in food but in toothpaste and mouthwash where sorbitol syrup provides both sweetness and the required 'sugar-free' formulation. Although limited in application, sorbitol and mannitol are indispensable in one sector of the food industry and that is in the manufacture of high-quality 'sugar-free' gum. Both are also used in sugar-free tabletted products in both the food and pharmaceutical industries. Sugar-free gum and tabletted products can be described as 'diet' or 'light'; however, they are generally not considered to be major contributors to calories even in traditional full sugar form as the volumes consumed are normally not large. The greatest benefit of the sugar-free products is that they are safe for teeth. Sugar-free confections based on sorbitol and mannitol also have a very low glycaemic index (GI).

15.2 PRODUCTION

Sorbitol was 'discovered' in 1872 in the berries of the Mountain Ash and is now known to occur naturally in a wide range of fruits and berries. Mannitol is also found naturally in marine algae, mushrooms and certain exudates from trees. Although found widely in nature, in common with most other polyols, their commercial production is currently built around the catalytic hydrogenation of the appropriate reducing sugar where reactive aldehyde and ketone groups are replaced by stable alcohol groups. The simple modification of hydrogenation enables the retention of much of the parent sugar's structure, bulk, and function while simultaneously giving the polyol other beneficial properties. Mannitol extraction from seaweed is a commercial operation in China and appears to be cost effective when compared with the traditional chemical route. The EU Sweeteners in Food Regulations also include mannitol produced by fermentation.

The usual raw materials for the manufacture of both sorbitol and mannitol are sugar, starch or glucose syrups, depending on price and availability. While the starch or syrups can be from any source, maize (corn) and tapioca starches are most widely used commercially.

Like maltitol, sorbitol is sold in both liquid and solid forms with several variations on the liquid theme to meet the demands of different applications. Owing to its low solubility, mannitol is available only as a crystalline solid.

15.2.1 Sorbitol powder

Sorbitol is a white, crystalline compound with the formula $C_6H_{14}O_6$. Crystalline sorbitol has been available commercially for more than 70 years. Dextrose (glucose) is the starting material for production, and although several dextrose-containing raw materials could be used to make sorbitol, the most common is dextrose derived from starch. Using conventional starch processing, enzyme technologies yield a high-dextrose content syrup, containing typically 94–96% dextrose on a dry basis, and this is a relatively inexpensive process and product. The dextrose is then crystallised out as the monohydrate to increase its purity, redissolved in water and hydrogenated. Sorbitol is the most soluble of the polyols and although it will crystallise at high solids and low temperatures, the crystals are small and not easily recovered from the mother liquor. Therefore, conventional aqueous crystallisation techniques to recover the sorbitol are not very efficient and other options have been developed. The sorbitol liquid can either be spray dried to produce the powder or the sorbitol liquid can be converted to a solid by melt crystallisation. This latter process involves progressive evaporation of water from the sorbitol syrup by application of high temperature and vacuum until the solids content exceeds about 95%. The sorbitol is not in solution under these conditions but is effectively a high solids melt. After seeding, the molten sorbitol is then bled from the cooker and solidified. Drying, milling and screening complete the process to give sorbitol powder.

Neither spray drying nor melt crystallisation of sorbitol result in purification with the same purity product leaving the drier as entered. Some of the properties of the sorbitol powder, particularly those related to crystal morphology (e.g. compressibility) can be manipulated by changing the manufacturing conditions. The milling and screening processes give a range of particle size in the powders and these are used in different applications. Packaging is typically in 25-kg bags, 500- and 1000-kg big bags and bulk tanks (about 20 ton).

The EU Sweeteners in Food Regulations specify a purity criteria for sorbitol powder (E420 i) of not less than 91% sorbitol on a dry weight basis, and therefore, it is possible to hydrogenate a 94–96% dextrose syrup directly and still meet this requirement for food use. For EU, pharmaceutical use sorbitol powders must conform to the requirements of the European Pharmacopoeia (EP). This specification is somewhat more restrictive, and a purity criteria of not less than 97% sorbitol on a dry basis is required. Since most sorbitol powder manufacturers produce both food and pharmaceutical grades of product, it is usually manufactured to the highest required standard so the product can be used for either application. Thus, the dextrose crystallisation route is preferred.

In the United States, for food use, sorbitol powder must comply with the Code of Federal Regulations (CFR) Title 21 Part 184.1835 that specifies that the product must comply with the Food Chemicals Codex. This states that the assay must be not less than 91.0% and not more than 100.5% D-sorbitol calculated on an anhydrous basis. Sorbitol powder is Generally Recognized As Safe (GRAS) by the US Food and Drug Administration (FDA). These same requirements also apply in the US Pharmacopoeia (USP).

15.2.2 Sorbitol syrups

Two basic sorbitol syrup types are manufactured – a crystallising grade and a non-crystallising grade. There are several variations around these two themes with a number of different syrups being available commercially. The crystallising grade is basically a pure sorbitol syrup, and it is used in those applications where powder would be dissolved in water before use, for example in pan coating. As the name suggests, it crystallises relatively easily especially if the storage temperature is reduced and/or concentration increased. The non-crystallising grade is essentially a hydrogenated high-dextrose equivalent glucose syrup, typically 85-90DE, and in addition to sorbitol, it also contains maltitol and maltotriitol with small amounts of higher molecular weight hydrogenated gluco-oligosaccharides. Depending on the manufacturer, it may also contain mannitol. Production of the non-crystallising grade is often linked to dextrose production (and hence sorbitol powder) where the mother liquor after dextrose crystallisation can be used as the base for the sorbitol syrup. The main application of the non-crystallising product is in toothpaste and mouthwash, but it also finds use as a humectant in baked goods. Typical packaging is in 275-kg drums, 1000-kg semi-bulk containers and in bulk tanks.

Like sorbitol powders, sorbitol syrups are manufactured for both food and pharmaceutical use, so products are normally manufactured to the highest required standard.

In the EU, for food use, sorbitol syrups (E420 ii) must comply with the Sweeteners in Food Regulations that specify a minimum of 50% sorbitol on a dry basis and a minimum of 69% total solids in the liquid, so a wide range of products can potentially be produced.

For pharmaceutical use, products must meet EP requirements. The non-crystallising grade must contain between 68% and 72% dry solids and a sorbitol content of 72.0–92.0%, while the crystallising grade must contain between 68% and 72% dry solids and a sorbitol content of 92.0–101.0%.

In the United States, for food use sorbitol syrup must comply with the Code of Federal Regulations (CFR) Title 21 Part 184.1835 that specifies that sorbitol must comply with the Food Chemicals Codex. This states that the assay must be not less than 64.0% on a dry basis. Sorbitol is GRAS by the US FDA. These same requirements also apply in the USP.

15.2.3 Mannitol

Mannitol is a white, crystalline compound with the formula $C_6H_{14}O_6$ and was originally isolated from the secretions of the flowering ash, called manna after their resemblance to the Biblical food. It is found in a wide variety of natural products, including almost all plants,[1,2] in concentrations from 20% (seaweeds) to 90% (tree exudates).

There are several routes that can be used to make mannitol.

The simplest is direct extraction from certain species of seaweed – this method is used commercially in China and linked to production of alginates from the same source.

In theory, it is also possible to make mannitol by hydrogenation of mannose although mannose is not readily available in large quantities.

The most common route to make mannitol is by hydrogenation of fructose (derived from either starch or sugar depending on price). When an aldo-sugar is hydrogenated, a single hydrogenated product is formed, for example dextrose (glucose) is converted into sorbitol and maltose into maltitol, but when a keto-sugar such as fructose is hydrogenated two products are formed. This is due to the two possible orientations of the hydroxyl groups on C2 in the fructose when the hydrogenation reaction occurs. Therefore, the C=O is converted to either H-C-OH (sorbitol) or HO-C-H (mannitol), and in fact under normal reaction conditions, the two are formed in equal amounts. By changing the pH of the hydrogenation system to slightly alkaline, however, it is possible to achieve slightly higher mannitol yields. Sorbitol and mannitol differ only in the position of the hydroxyl group on carbon 2 in the molecule and are therefore isomers.

Although production of mannitol from starch may be more cost-effective than from sugar, the process is slightly more complicated. The first part is the same as for sorbitol powder manufacturing that is the production of a 94–96% dextrose syrup. This syrup is then isomerised using immobilised glucose isomerase enzyme to give a high-fructose glucose syrup with a composition of about 42% fructose, 52% dextrose and 6% maltose. While this could be hydrogenated, it would yield only about 21–22% mannitol at best (only half the fructose being converted to mannitol), so it is normally subject to industrial chromatography to increase the fructose content to 90–95% before hydrogenation. The 'waste' stream from the chromatography can be utilised in a number of ways, for example in fermentation, blended with other starch hydrolysates or hydrogenated to give a non-crystallising sorbitol syrup.

When using sucrose, the first stage is hydrolysis to invert sugar. This can be hydrogenated directly but since invert sugar contains only about 50% fructose, and only about half of this is converted to mannitol, the theoretical yield of mannitol is only about 25%. In practice, it is often less. More usually, the fructose content of the invert sugar is enriched to 90–95% by industrial chromatography before hydrogenation giving mannitol yields approaching 50% or slightly higher. The 'waste' stream from the chromatography is almost pure dextrose that can be crystallised and sold as such or hydrogenated to give a high-purity sorbitol syrup and then sorbitol powder. The 'sugar route' to make mannitol is very straightforward and is a relatively clean process as sugar is a very high-purity raw material. Since 4 tons of sugar are required to make 1 ton of mannitol, it does, however, require a cheap source of sucrose to be an economic process. Fortunately, such sources are currently readily available.

In the EU, the Sweeteners in Food Regulations also permit the manufacture of mannitol by fermentation although the cost effectiveness of this process is currently debatable compared with conventional means. With regard to purity criteria, the regulations specify a purity criteria of 96–98% on a dry solids basis for mannitol produced by catalytic hydrogenation and not less than 99% on dry solids basis for mannitol produced by fermentation. It is interesting to note that the regulations do not include mannitol extracted from seaweed and such a product would presumably not currently be permitted in the EU.

The EP specifies a mannitol content of 98–102% of dry solids for mannitol powder.

In the United States, for food use, mannitol must comply with the Code of Federal Regulations (CFR) Title 21 Part 180.25 that specifies that mannitol must comply with the Food Chemicals Codex. The FCC states that the assay must be 96.0–101.5% on a dry basis. The USP has the same requirements.

Whatever process is used, the hydrogenated mannitol rich stream is then concentrated by evaporation, seeded with mannitol crystals and cooled. The mannitol then readily crystallises out and can be recovered by filtration. After washing it is dried, screened and packed. Typical packaging is 25-kg paper sacks and 500-kg bulk bags.

The mother liquor after mannitol crystallisation is normally blended off to give non-crystallising sorbitol syrup, hence the mannitol content of some of these products.

15.3 HYDROGENATION

The hydrogenation conditions to manufacture sorbitol and mannitol are basically similar to those for other polyols. The dextrose or glucose syrup or fructose is reacted with hydrogen at high temperature (typically, 100–150°C) and high pressure (typically, 100–150 bar) in the presence of a suitable catalyst. Reaction times are in the order of 1–2 hours depending on the conditions and the product being manufactured. The raw material for hydrogenation must be of the highest purity to prevent the catalyst becoming poisoned, to achieve the necessary purity, ion exchange, carbon treatment and/or crystallisation would normally have been used at some stage in its manufacture. Depending on the catalyst used in the process, the hydrogenated product may then be subject to a further ion exchange treatment to remove dissolved catalyst.

Handling and storage after hydrogenation must be carried out according to good manufacturing practice to ensure microbiological problems are avoided.

15.4 STORAGE

Sorbitol and mannitol powders are easily compressed, and therefore, good storage is critical to ensure the products remain free flowing. Pallets of product should not be stacked but rather a shelved storage system should be used. Warehouses should, of course, be clean and dry. Ambient temperatures are quite acceptable.

Sorbitol syrups are easily handled owing to their low molecular weight, so storage at 20–25°C is quite acceptable for non-bulk deliveries although higher storage temperatures are often used to facilitate handling. Bulk storage is normally at 50–60°C, again, to facilitate handling and prevent crystallisation of the higher sorbitol content syrups.

15.5 STRUCTURE

The structures of sorbitol and mannitol are shown in Figures 15.1 and 15.2 (the non-crystallising grades of sorbitol syrup also contain mannitol, maltitol and hydrogenated gluco-oligosaccharides (see also Chapter 14)).

Sorbitol and mannitol can exist in different polymeric forms. For mannitol, the most stable is the beta form. Sorbitol is reported to have four such forms (alpha, α; beta, β; gamma, γ; and delta, δ) and additionally a glass transition form (E). The different forms have different properties including solubility, melting range and stability. The gamma form of sorbitol is the most stable and modern manufacturing techniques produce sorbitol powder predominantly in this form. This is to avoid changes in the food or pharmaceutical product in which the

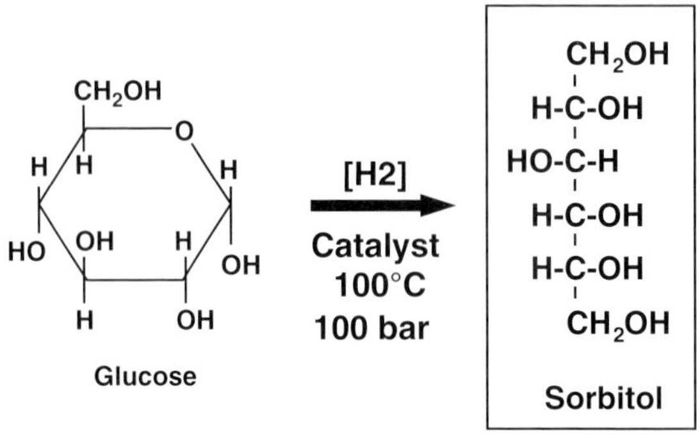

Fig. 15.1 Hydrogenation of dextrose to sorbitol.

powder is included, as during processing and storage, the tendency is for the unstable forms to change to the stable form with subsequent changes to the food or pharmaceutical product. While these changes occur slowly, they are accelerated by high temperatures and all the polymorphs will eventually assume the more stable gamma form.[3]

15.6 SAFETY

The Scientific Committee for Food in the EU published a comprehensive assessment of sweeteners in 1985 and concluded that sorbitol and mannitol are acceptable for use at levels 'not specified' meaning no limits are placed on their use.[4] Additionally, there are many other studies reaching the same conclusion. To affirm the GRAS status of sorbitol in the United States, safety data were evaluated by the Select Committee on GRAS Substances selected by the Life Sciences Office of the Federation of American Societies for Experimental Biology (*FASEB*). In the opinion of this Select Committee, there was no evidence demonstrating a hazard where sorbitol was used at current levels or at levels that might be expected in the future.

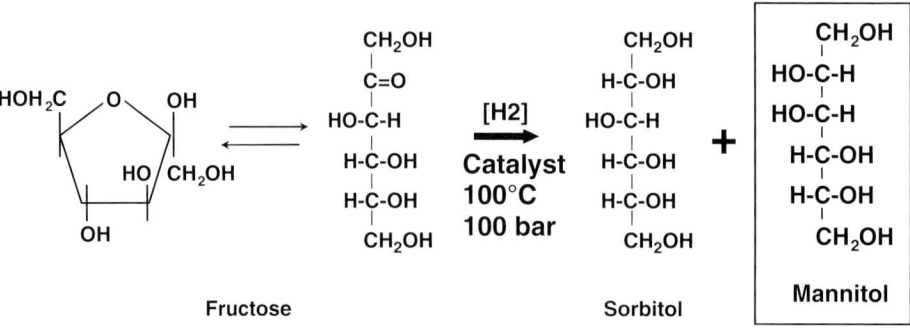

Fig. 15.2 Hydrogenation of fructose to sorbitol and mannitol.

The use of mannitol in food in the United States is broadly permitted by FDA food additive regulations (21 CFR 180.25).

The Joint Food and Agriculture Organization/World Health Organization Expert Committee on Food Additives (JECFA) also reviewed the safety data and concluded that sorbitol was safe.[5] JECFA also established an acceptable daily intake for sorbitol of 'not specified', the safest category in which JECFA can place a food ingredient. JECFA's decisions are often adopted by many countries that do not have their own agencies to review food additive safety.

For mannitol, JECFA reviewed the safety data and concluded that mannitol is safe.[6] JECFA has allocated a temporary Acceptable Dietary Intake of 0–50 mg/kg.

As with all food ingredients, if the consumer has any doubts about consuming a particular product, then medical advice should be taken.

15.7 PHYSICO-CHEMICAL CHARACTERISTICS

Although structurally sorbitol and mannitol are very similar products, differing only in the position of a single hydroxyl group in the molecule, they do in many cases have different properties that have led to quite specific applications for each of the products.

15.7.1 Chemical reactivity

In common with other polyols, sorbitol and mannitol do not take part in Maillard browning reactions. Normally, this is a desirable attribute although in some products the reactions of reducing sugars with proteins and amino acids generates desirable flavour and colours. On the other hand, if non-reactivity is the goal (protection of colours, flavours, enzymes, etc.), then polyols will work well in these applications.

15.7.2 Compressibility

The relatively low cost of sorbitol powder makes this the polyol of choice in direct compression applications to make tablets where it produces a harder tablet compared with other polyols. Although mannitol also gives hard tablets, its high price compared with sorbitol limits its food use. Mannitol is much less hygroscopic than sorbitol, and tablets made with mannitol are also less hygroscopic than sorbitol tablets. In some respects, mannitol is, therefore, easier to handle than sorbitol. Mannitol finds particular application in pharmaceutical tablets where the generally higher active cost outweighs the excipient cost. Where moisture sensitive actives are being used, the relatively lower hygroscopicity of mannitol in the finished product may be important.

It is reported that sorbitol produced by spray drying is easier to compress than melt crystallised material and results in less wear on the die punches in the tabletting press.

Different particle size grades of sorbitol will give different tablet hardness, texture (surface roughness in the mouth) and dissolution time.

15.7.3 Cooling effect

Sorbitol has a very pronounced cooling effect or negative 'heat of solution' when placed in water or in the mouth. The smaller the particle size of the powder and the faster the solubility, the more noticeable the effect. In practice, mannitol apparently has less of a cooling effect than sorbitol. This though is due to its low solubility compared with sorbitol and in fact mannitol gives a slightly greater cooling effect than sorbitol when completely dissolved (sorbitol -26.5 cal/g vs. mannitol -28.9 cal/g).

For directly compressible tableted confectionery where a strong cooling sensation coupled with good solubility are desirable, for example in mint or menthol based formulations, sorbitol is the polyol of choice in this application.

15.7.4 Humectancy

A good humectant is one which resists changes in moisture content as the humidity of its surroundings change, and it thereby limits the exchange of water between products.[7] Sorbitol is one of the best humectants available for use in foods and in syrup form finds application in baked goods. It is also used in shredded coconut for the same purpose. Sorbitol has the ability to gain or lose moisture slowly as the relative humidity of its environment changes. Thus, foods in which it is included are more shelf stable and retain their freshness for longer periods. Mannitol has no humectant properties.

15.7.5 Hygroscopicity

While sorbitol is considered very hygroscopic, mannitol is considered non-hygroscopic and this property illustrates one of the greatest differences between the products. The hygroscopic tendencies of food ingredients can have very serious consequences during production and storage (shelf life) of the food itself and also may affect how the ingredient itself is stored. Mannitol is the least hygroscopic of the polyols and does not start to absorb moisture until the relative humidity is over 90%. Sorbitol, however, has a greater affinity for water and is much more hygroscopic and starts to absorb moisture when the relative humidity reaches about 65%. This can cause problems when sorbitol is used in foods and especially in tabletting where the absorbed moisture can prevent the presses running. In Europe, where the typical relative humidity in the air is about 65–75%, air conditioning and humidity control are essential in factories handling sorbitol. Similarly, finished products must be well wrapped to prevent softening due to moisture absorption.

15.7.6 Molecular weight

Sorbitol and mannitol have the same molecular weight of about 182.

Viscosity, freezing point depression, boiling point elevation and osmotic pressure are all directly related to molecular weight, and it can therefore play a significant role in influencing the overall texture and functionality of a finished product.

In ice cream, freezing point depression is an important factor. Sorbitol has been used to lower the freezing point of ice cream to enable it to be served directly from the freezer as well as contributing to the overall sweetness of the product. Mannitol with its limited solubility has not found the same wide application.

Osmotic pressure becomes important in digestion (see Chapters 1 and 3 for further information). It is also important with regard to the keeping qualities of a sorbitol syrup, for example, and it must be high enough to prevent microbial growth at the solids at which the syrup is sold.

Boiling point elevation is particularly important in hard-candy manufacture but as neither sorbitol nor mannitol is widely used in this application, it is not a serious issue.

15.7.7 Solubility

Solubility is generally defined as the amount of a solute (polyol) that can be dissolved in a solvent (water) at a given temperature before becoming saturated. As the temperature of the solution increases, more solids can be dissolved and if such a solution is cooled, the solute will crystallise out.

With sorbitol and mannitol, we have the two extremes of the polyol solubility range. Mannitol is one of the least soluble polyols with only 22 g being dissolvable in 100 g water at 25°C (beta polymorph), while for sorbitol this figure is 235 g/100 g of water (gamma polymorph).

15.7.8 Viscosity

The viscosity of solutions of mannitol and sorbitol are little different from dextrose or fructose at the same solids content. Non-crystallising sorbitol syrups have a slightly higher viscosity owing to the slightly higher molecular weight material they contain.

15.8 PHYSIOLOGICAL PROPERTIES

15.8.1 Calorific value

In the EU, sorbitol and mannitol (in common with all other permitted polyols) are given the same calorific value of 2.4 kcal/g (10 kJ/g). This makes calculation of the calorific value of a food very easy, but whether it is the true value for these products is debatable. In the United States, for example, sorbitol is given a calorific value of 2.6 kcal/g and mannitol 1.6 kcal/g, while in Japan these values become 3.0 and 2.0 kcal/g, respectively. Although not specified, the calorific values of the non-crystallising grades of sorbitol syrup should be slightly higher depending on the proportion of glucose they release after breakdown in the gut.

Reduced calorie and low-calorie claims for foods are permitted in the EU and the United States and both sorbitol and mannitol can be used to make reduced calorie foods. It is, of course, important to consider the calorific value of other food ingredients when making such products.

In the EU, a 'reduced-calorie' claim necessitates a 30% reduction in calories compared with the product for which no claims are made and low-calorie foods must not be more than 40 kcal/100 g (167 kJ/100 g) or 100 mL of food.

In the United States, a reduced-calorie claims necessitate a 25% reduction in calories, light products must have a one-third reduction in calories and low-calorie foods must meet the same requirements as in the EU.[8]

15.8.2 Dental aspects

While tooth decay may no longer be the major problem it once was in terms of pain, suffering (and occasional death) and time off work for dental treatment, it continues to be a major health and economic issue.

Bacteria in the plaque on the surface of the teeth are able to ferment simple sugars in the foods we consume to produce acid that causes a lowering of the pH at the tooth surface. If the pH falls below 5.7 this can lead to decalcification of the tooth enamel and eventually to tooth decay or dental caries. Sorbitol and mannitol are not fermented by the oral bacteria and are described as non-cariogenic. The use of these products to replace sugars in those foods that remain in the mouth for a considerable time, for example tableted items and gum, therefore provides a means of making such foods that are safe for teeth. While sorbitol and mannitol are non-cariogenic, other components of the food may, of course, not be. Colours and flavours often use maltodextrins as carriers, and although such products may be sugar-free, they may not be safe for the teeth.

In the United States, the FDA have approved use of the statement 'does not promote tooth decay' on labels of sugar-free foods containing sorbitol and mannitol.[9]

15.8.3 Diabetic suitability

The various national Diabetic Associations normally recommend that diabetics consume a normal diet but take account of the sugars in the foods they consume. This has not stopped the development of foods specifically for diabetics and foods containing sorbitol and mannitol can be used by diabetics to manage their blood glucose levels since neither requires insulin in their metabolism. Products sweetened with sorbitol and/or mannitol may have a role in providing a wider variety of reduced-calorie and sugar-free food products to diabetics.[7,10]

15.8.4 Glycaemic response

The concept of GI was developed over 20 years ago by David Jenkins at the University of Toronto as a tool to allow diabetics to better manage their diets.

GI is defined as the incremental area under the blood glucose response curve of a 50 g carbohydrate portion of a test food expressed as a percentage of the response to the same amount of carbohydrate from a standard food (typically, glucose) taken by the same subject (FAO/WHO 1997). It is a method of ranking foods according to the extent to which they raise blood glucose levels after consumption. Foods containing carbohydrates that break down quickly after ingestion, giving a fast and high blood glucose response, have the highest GI values, while foods containing carbohydrates that break down slowly after ingestion, giving a slow and low blood glucose response, have the lowest GI values:

- *Low GI*: 55 and below.
- *Medium GI*: 56–69.
- *High GI*: 70 and above.

In addition to being important in the control of diabetes, a low GI diet has also been shown to be important in weight maintenance and potentially in weight reduction (so important to combat the current rising trends in obesity) and in improving the body's sensitivity to insulin.

Not unexpectedly, there is a wide range of GI values for the polyols. While it is unusual to eat only a polyol in isolation, in the case of sorbitol tableted confectionery, the polyol may constitute over 98% of the product. Therefore, GI should be measured on the finished food product and not on individual ingredients.

The GI of both sorbitol and mannitol are very low at about 10 and 0, respectively.

15.8.5 Tolerance

A person's response to low-digestible carbohydrates varies depending on individual factors such as amount and frequency of consumption. While sorbitol and mannitol have proved extremely useful in weight maintenance, lowering blood glucose response and formulating tooth-friendly products, they are not without disadvantages. Neither sorbitol nor mannitol are actively absorbed by the body from the gut but rather they pass into the blood stream by passive absorption. They are essentially low-digestible carbohydrates and if they are not absorbed in the upper gastrointestinal tract, they can, at certain levels, cause problems in the lower gut. This can take the form of an osmotic imbalance, where water is drawn into the lower gut, leading to osmotic diarrhoea and/or fermentation of the carbohydrate by bacteria in the lower gut leading to flatulence. Sorbitol and mannitol are not unique in this respect as fibre also falls into the same category of low-digestible carbohydrates and over consumption of fibre will have similar effects.

Mannitol is probably the least well-tolerated polyol and it is possible to consume on average only about 20 g per day and not more than 10 g per single serving before problems are seen. This is generally not too great an issue as mannitol finds only limited use in foods. With sorbitol, this number increases to about 40 g per day but again not more than 10 g per single serving. This guideline takes into account the tendency for the consumer to ingest more than one (sometimes many) serving of product. Manufacturers should be aware of this when designing packaging and pack size, but if sorbitol and mannitol are used within sensible guidelines, they can improve a formulation without causing problems. As with all polyols, prolonged consumption of sorbitol and mannitol leads to some adaptation and improved tolerance. The non-crystallising sorbitol syrups are slightly better tolerated than pure sorbitol owing to their maltitol and hydrogenated gluco-oligosaccharide content but the difference is negligible.

In the United States, products containing sorbitol and mannitol are required to carry a warning label that 'excess consumption can cause a laxative effect' if reasonably foreseeable consumption could result in daily ingestion of 50 g and 20 g of each polyol, respectively.

In the EU, a warning label is required when the polyol content of the food is 10% or higher.

In the Codex Alimentarius, if the food provides a daily intake of sugar alcohols in excess of 20 g, there should be a statement on the label to the effect that the food may have a laxative effect. However, this is not mandatory just a guideline.

15.8.6 Sweetness

Sweetness is one of the most important properties of any sugar replacer and both sorbitol and mannitol are much less sweet than sugar (sorbitol 60% as sweet and mannitol 50% as sweet). In foods formulated with these products, addition of a high-potency sweetener is normally necessary to increase the sweetness of the finished product to an acceptable

Table 15.1 Properties of sorbitol and mannitol.

Property	Sorbitol	Mannitol
Sweetness (sucrose = 100)	60	50
Cooling effect (cal/g)	−26.5	−28.9
Calorific value		
EU	2.4	2.4
United States	2.6	1.6
Japan	3.0	2.0
Maximum recommended intake (g/day)	40	20
Solubility at 25°C (g/100 g water)	235	22
GI	9	0
Hygroscopicity	High	Very low
Regulatory approval		
EU	Yes	Yes
United States	Yes	Yes
Japan	Yes	Yes

level. Hydrogenation of a reducing sugar normally results in either an increase in sweetness (dextrose to sorbitol), no real change in sweetness or occasionally a slight decrease in sweetness (cellobiose to cellobiitol). Hydrogenation of fructose to produce sorbitol and mannitol produces a very dramatic decrease in sweetness however. Perhaps though this is not a fair comparison and we should be looking at changes in sweetness in dextrose and mannose when considering sorbitol and mannitol rather than a hybrid molecule like fructose. Dextrose increases in sweetness and the conversion of mannose to mannitol results in little change in sweetness following the usual trend.

Table 15.1 summarises the main properties of sorbitol and mannitol.

15.9 APPLICATIONS IN FOODS

There is increasing consumer demand for good-tasting products with fewer calories (and fat) and the use of low-calorie sweeteners, bulking agents, fat replacers and other low-calorie ingredients help meet this consumer demand. Sorbitol and mannitol combine well with other ingredients and may be synergistic with other sweeteners.

15.9.1 Gum

Both sorbitol and mannitol are used in gum but for different purposes. Sorbitol is used in crystalline form in the gum itself to provide bulk and sweetness and typically the gum may contain 50–55% of sorbitol in this form. Additionally, sorbitol syrup may be added to the gum, sometimes in combination with maltitol syrup where it functions as a plasticiser and prevents the gum drying out. Since sorbitol has only about 60% of the sweetness of sugar, a high-potency sweetener need also be added.

Mannitol may also be included in the gum to reduce its hygroscopic tendencies, but it is used primarily as a dusting agent owing to its outstandingly low-moisture absorption properties. It is used to prevent the gum sticking to the rollers as the sheeted gum is produced and in the finished product to prevent the gum sticking to the wrapping paper.

15.9.2 Hard candy

While hard candy can be made from sorbitol, second- and third-generation polyols such as maltitol syrup and isomalt give higher quality products. Typically, the sorbitol will be cooked to very high solids (in effect a glass), seeded, deposited and allowed to crystallise. This is a difficult process and one reason why manufacture of sorbitol candy has slowly disappeared. This is not a hard candy in the traditional sense but a crystallised solid.

Although mannitol is now not used in hard candy, some of the first sugar-free products used a combination of maltitol syrup, mannitol and gum Arabic in their manufacture.

15.9.3 Tabletting

Sorbitol is widely used as a directly compressible tabletting excipient in the food industry where it provides bulk and sweetness to the product. A high-potency sweetener needs to be used in addition to the sorbitol to increase the sweetness of the product to an acceptable level. Typically, the sorbitol powder, flavour and lubricant are mixed and then fed to the tabletting machine. The mixing and tabletting must be carried out in air-conditioned areas, as sorbitol is hygroscopic. Sorbitol powder manufacturers usually make several grades of sorbitol powder of different particle size and all can be used to make tablets. However, there has to be a trade-off between product texture and flowability of product to the tabletting dies. The large particle size grades flow easily and fill the dies but give a coarse mouthfeel, while the finer grades flow less readily but give a smoother finished product. The combination of the cooling effect of the sorbitol and mint flavour makes directly compressed 'sugar-free' mints one of the major application areas for sorbitol powder. While all commercially manufactured sorbitol is the gamma polymorph this alone does not guarantee a hard tablet, the physical state of the sorbitol powder is equally important. For example, under the same tabletting conditions, an open-structured sorbitol powder will give a harder tablet than a powder with a more dense structure.

Mannitol is also used in tabletting but more so for pharmaceutical than food products. Fine powder grades are typically used in wet-granulated formulations and larger particle sizes in directly compressible products where good product flow is essential to fill the tabletting dies. Mannitol is also used in directly compressed chewable tablets where its slower solubility is utilised to provide sweetness for the duration the product is in the mouth.

15.9.4 Surimi

Reconstituted fish protein is a large and growing market in which the fish protein or muscle is minced and reformed into a range of products before freezing. Sugar was traditionally used as a cryoprotectant for the protein during the freezing step, and while it functioned effectively in this role, it made the product too sweet. Sorbitol has now largely replaced sugar in this application. During the freezing process, the sorbitol protects the protein from denaturation that would otherwise reduce its gel-forming ability and gel strength. The end result of this is reduction in its water-holding capacity and a less succulent product.

15.9.5 Cooked sausages

Sorbitol is used in cooked meat products to improve the flavour and to prevent charring during the cooking process. In the United States, this is permitted under USDA Regulation

9CFR318.7. When sausages are grilled, sugar and glucose syrups normally used in the product caramelise and char, whereas sorbitol will not.

15.9.6 Baked goods

Sorbitol has been used for many years as a humectant in baked products, nutritional bars and granola bars, extending shelf life by maintaining the moisture content of the product. Sorbitol syrup was the sugar alcohol initially used in many sugar-free biscuits and cakes for diabetics although much of this has now been replaced by other polyols. However, crystalline sorbitol continues to be used in these products owing to its lower cost.

15.9.7 Panning

Sugar-free panning with sorbitol is now a well-established process and follows many of the principles of sugar panning.[11,12] Sorbitol-coated products include tableted gum, chocolate products and jellybeans. A high-purity sorbitol solution is used to coat the centres and sorbitol powder is added to provide crystallisation nuclei.

15.9.8 Over-the-counter products

Non-crystallising sorbitol syrup can be used in children's cough medicines and pain relief products although its use has been largely superseded by the use of maltitol syrups that are sweeter and better tolerated.

15.9.9 Chocolate

Owing to its relative non-hygroscopic nature, mannitol was used in sugar-free/no-added-sugar chocolate-flavoured coatings although it has now been largely replaced with maltitol powder. While the low cost of crystalline sorbitol would seem to make it attractive in this application, its hygroscopicity leads to unacceptable moisture increases during refining and its powerful cooling effect is not desirable in chocolate.

15.10 NON-FOOD APPLICATIONS

15.10.1 Sorbitol

15.10.1.1 Toothpaste and mouthwash

This is the largest applications area for sorbitol and particularly the non-crystallising grade of the syrup, where it provides sweetness and acts as a humectant to prevent the toothpaste drying out. In mouthwash, sorbitol acts as a non-cariogenic sweetener and viscosifying agent.

15.10.1.2 Vitamin C manufacture

Sorbitol is used as an intermediate in the manufacture of vitamin C.

15.10.1.3 *Pharmaceutical laxative*

It has been known for many years that eating some types of fruit (e.g. pears and prunes) produced a laxative effect. This is due to the sorbitol content of the fruit, and although these products could be classed as 'natural' laxatives, sorbitol is now sold specifically for this purpose as its intake can be more closely controlled.

15.10.1.4 *Pharmaceutical tabletting*

Clinical
Sorbitol solutions are used for irrigation during surgery.

15.10.2 Mannitol

15.10.2.1 *Drugs*

Mannitol has reportedly been used to dilute illegal drugs before sale.

15.10.2.2 *Pharmaceutical tabletting:*

Clinical
Mannitol solutions are used for:

- irrigation during surgery;
- cerebral dehydration to decrease elevated intracranial pressure;
- renal protection to protect against renal failure.

15.10.2.3 *Medicinal*

Mannitol can also be used to facilitate transport of pharmaceuticals directly into the brain.

15.11 LEGAL STATUS

Sorbitol powder and syrups and mannitol are permitted for food and pharmaceutical use in most countries. Some countries (EU, the US) have their own legal requirements and purity criteria for these products but where these do not exist the appropriate Codex Alimentarius specifications are usually adopted. All are essentially similar and have the same objective – to present the consumer with safe, high-quality food ingredients. In the EU, sorbitol and mannitol are considered as food additives and their use in foods is controlled by the Sweeteners in Food Regulations. Sorbitol powder has the E number 420 (i) sorbitol syrups E 420 (ii) and mannitol E 421. They all are authorised for food use at *quantum satis* for the range of food products listed in the Regulations. They cannot be used in foods in conjunction with sugars unless the polyol is present for a technological function other than sweetness (e.g. where sorbitol is used as a humectant in baked goods) or a 30% reduction in calories results from the combination. In the United States, the restrictions on the use of polyols in combination with sugars do not apply.

In the United States, polyols are considered either a food additive or GRAS by the FDA. Sorbitol is considered GRAS (21CFR184.1835) with limitations under good manufacturing practices outlined in the citation. Mannitol is a food additive (21CFR180.25) allowed in pressed mints (max. 98%), hard candy and cough drops (max. 5%), chewing gum (max. 31%), soft candy (max. 40%), confections and frostings (max. 8%), non-standardised jams and jellies (max. 15%) and in all other foods at a maximum level of 2.5%.

15.12 CONCLUSIONS

Although sorbitol and mannitol may no longer be the first choice polyols for many sugar-free food applications, they still have a range of properties of interest to the formulator of new food products. Despite being first-generation products, they should not be discounted from new product development.

REFERENCES

1. P. Lawson. *Mannitol*. Blackwell Publishing Ltd.; 2007, pp. 219–225.
2. S. Song and C. Vieille. Recent advances in the biological production of mannitol. *Applied Microbiology and Biotechnology* 2009, 84, 55–62.
3. J.W. Du Ross. Modification of the crystalline structure of sorbitol and its effects on tabletting characteristics. *Pharmaceutical Technology* 1984, 8, 50–56.
4. Commission of the European Communities. Reports of the Scientific Committee for Food concerning sweeteners. Sixteenth Series. Report EUR 10210 EN. Office for Official Publications of the European Communities, Luxembourg; 1985.
5. Joint FAO/WHO Expert Committee on Food Additives. Toxicological evaluation of certain food additives: sorbitol. Twenty-sixth report. WHO Technical Report Series 683, pp. 218–228. Geneva, 1982.
6. Joint FAO/WHO Expert Committee on Food Additives. Toxicological evaluation of certain food additives: mannitol. Twenty-ninth report. WHO Technical Report Series 733, 35. Geneva, 1982.
7. Y. Le Bot and P.A. Gouy. Polyols from starch. In: M.W. Kearsley and S.Z. Dziedzic (eds). *Handbook of Starch Hydrolysis Products and Their Derivatives*, Blackie Academic and Professional, Glasgow; 1995, pp. 155–177.
8. Federation of American Societies for Experimental Biology. The evaluation of the energy of certain polyols used as food ingredients; June 1994.
9. American Dental Association. Position Statement on the Role of Sugar-Free Foods and Medications in Maintaining Good Oral Health. Adopted October 1998.
10. A.S. Le and K.B. Mulderrig. Sorbitol and mannitol. In: L. O'Brien Nabors (ed.). *Alternative Sweeteners*, 3rd edn, Marcel Dekker, New York; 2001, pp. 317–334.
11. J. Bogusz. Sucrose hard panning. 58th PM Production Conference, Hershey, PA; 2004, pp. 29–36.
12. R.F. Boutin. Sugarless panning procedures and techniques. 46th PM Production Conference, Hershey, PA; 1992, pp. 129–134.

16 Xylitol

Christos Zacharis

Active Nutrition, DuPont Nutrition & Health, Surrey, UK

16.1 DESCRIPTION

Xylitol is a sugar alcohol (or polyol) that has been used as a food additive and sweetening agent since the 1960s. It is a natural constituent of many fruit and vegetables[1,2] and although the level found is usually less than 1% (Table 16.1), it has always been a natural component of modern man's diet. The human body also produces 5–15 g xylitol per day during normal carbohydrate metabolism in the liver.[3] It is a five carbon polyol (Figure 16.1), or pentitol, which was first discovered and reported in 1891 by the German Nobel Prize winning chemist Emil Fischer.[4] Fischer named the new compound *xylit* (the German word for xylitol), which is believed to stem from the Greek word *xylon*, whose English translation is *xylem*. Xylem constitutes the woody element of higher plants, such as the beech wood from which Fischer first isolated xylitol.

At this time, xylitol's physiological role in the body was yet to be elucidated, and few could have speculated on the considerable potential that it would offer in various dietary, nutritional and medicinal applications. However, during the 1950s, it was discovered that xylitol is included in the intermediary metabolism of carbohydrates in the mammalian liver.[3,5] This discovery generated a wealth of new research that in turn revealed novel biochemical information regarding xylitol.

The development of a commercial production method for xylitol began in the 1940s, when researchers commenced work on the isolation of xylitol from xylose, and by the 1960s, an economically feasible manufacturing procedure for this process had been achieved. Prior to this development, xylitol had largely been considered a costly research chemical, used in specialised medical applications and research laboratories. This original manufacturing process was based on the use of Finnish birch trees as a raw material, and for this reason, xylitol was often referred to as 'birch sugar'. Xylitol can be manufactured from a variety of natural plant sources that contain the polysaccharide xylan,[6,7] a polymer of xylose. Although xylitol occurs freely in nature,[1,2] it is only economical to manufacture xylitol from these xylan-rich plant materials using a number of relatively straightforward chemical steps. Industrially, the most commonly used raw materials are hemicellulose sources, such as wood chips and corn-cobs (both sustainable methods). These materials typically contain 20–35% xylan, which is readily converted to xylose (wood sugar) by hydrolysis. Xylose is subsequently converted to

---

Sweeteners and Sugar Alternatives in Food Technology, Second Edition.
Edited by Dr Kay O'Donnell and Dr Malcolm W. Kearsley.
© 2012 John Wiley & Sons, Ltd. Published 2012 by John Wiley & Sons, Ltd.

Table 16.1 The natural occurrence of xylitol in various fruit and vegetables.

Product	Xylitol content (mg/100 g dry solids)
Yellow plums (*Prunus domestica* ssp. *italia*)	935
Strawberry (*Fragaria* var.)	362
Cauliflower (*Brassica oleracea* var. *botrytis*)	300
Raspberries (*Rubus idaeus*)	268
Endives (*Cichorium endivia*)	258
Bilberry (*Hippophae rhamnoides*)	213
Aubergine (*Solanum melongena*)	180
Lettuce (*Lactuca sativa*)	131
Spinach (*Spinacia oleracea*)	107
Onions (*Allium cepa*)	89
Carrot (*Daucus carota*)	86

xylitol via catalytic hydrogenation (reduction). Following the hydrogenation step, there are a number of separation and purification steps that ultimately yield high-purity xylitol crystals. This process is summarised in Figure 16.2. While other methods have been reported for the commercial manufacture of xylitol, these are not commonly used, and include conversion of glucose (dextrose) to xylose followed by hydrogenation to xylitol, and the microbiological conversion of xylose to xylitol.

Simultaneously, with developments in the industrial production of xylitol, researchers at the Dental School of the University of Turku, Finland, began to investigate the use of xylitol as a potential sugar substitute. Since it had long been known that sugar consumption was intimately associated with dental caries,[8,9] it was hypothesised that the substitution of dietary sugar with xylitol may reduce the problem.[10] The exceptionally positive results of this intensive research programme[11] generated significant interest in the potential dental benefits of xylitol all over the world, spawning further widespread research into the caries-inhibitory effects of xylitol.[12–21] These studies confirmed xylitol's unique dental benefits and established xylitol as the sugar substitute of choice for the promotion of improved oral health. At the same time, xylitol was confirmed as being safe for human consumption, and a wide variety of potential applications areas, such as its use as a sweetener in the diabetic diet, were highlighted and exploited.[22–26]

16.2 ORGANOLEPTIC PROPERTIES

16.2.1 Sweetness

Sweetness is a major factor in determining the appeal of any sugar-free product, and it is regarded as a major factor in the consumers' perception of product taste and quality.

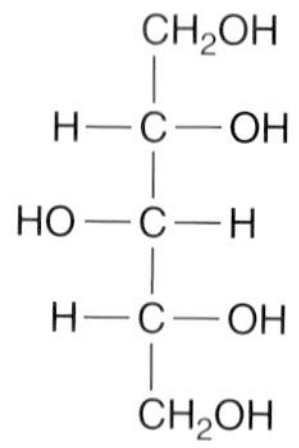

$$\begin{array}{c} CH_2OH \\ | \\ H-C-OH \\ | \\ HO-C-H \\ | \\ H-C-OH \\ | \\ CH_2OH \end{array}$$

Fig. 16.1 Chemical structure of xylitol.

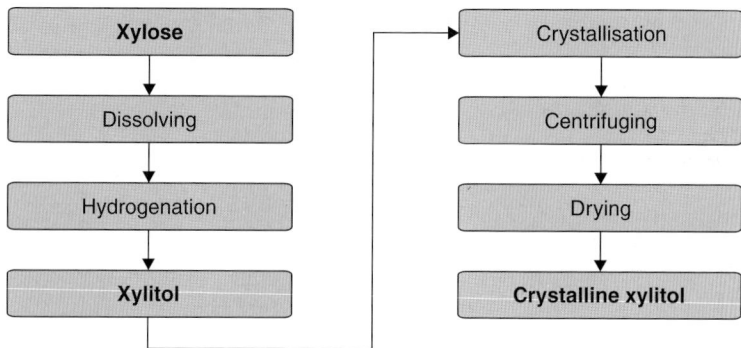

Fig. 16.2 Outline method of manufacture for xylitol.

Xylitol is the sweetest of all of the polyols (Figure 16.3),[27–31] being the only polyol to exhibit a sweetness intensity equivalent to that of sucrose.[32,33] Xylitol is equisweet with sucrose at concentrations of 10% solids, and is reported to be 20% sweeter than sucrose at a concentration of 20%.[33] Xylitol also exhibits a very similar sweetness time-intensity profile to sucrose.

16.2.2 Sweetness synergy

Xylitol can be combined with other polyols to produce significant sweetness synergy. For example, a 60:40 sweetener ratio of xylitol and sorbitol in chewing gum or an 80:20 sweetener ratio of maltitol and xylitol in chocolate will produce sugar-free products that are isosweet to their sugar-sweetened counterparts.

Xylitol has also found application in combination with high-intensity sweeteners, where it helps to provide a cleaner, more rounded and altogether more 'sugar-like' sweetness. The sweetness of xylitol can be utilised to compensate for the sweetness lost in systems using

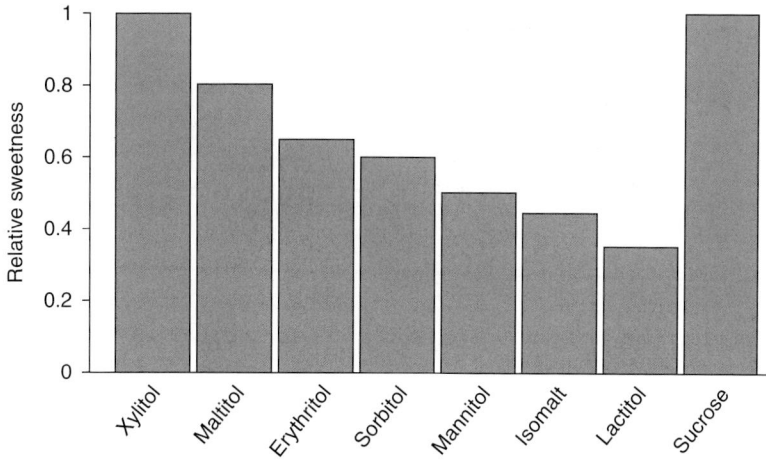

Fig. 16.3 Relative sweetness of polyol sweeteners.

aspartame in combination with aldehyde-based flavour systems (aldehyde groups have been shown to interact with aspartame, resulting in a diminished perception of sweetness).

16.3 PHYSICAL AND CHEMICAL PROPERTIES

16.3.1 Heat of solution

Xylitol has a very high negative heat of solution, meaning that it absorbs energy from its environment as it dissolves, causing a measurable drop in the temperature of the surroundings. This means that as xylitol dissolves in the mouth, a pleasant cooling effect is experienced. The perceived cooling effect of any polyol is the result of a combination of its negative heat of solution and solubility, as solubility is a major factor in the onset and perception of this effect. Xylitol's high solubility combines with its negative heat of solution to result in a significant perception of cooling upon consumption, which provides it with a clear taste advantage over both sugar and other polyol sweeteners in many applications. The cooling effect of xylitol is, for example, nearly 40% greater than that of sorbitol (Figure 16.4). As this is a physical effect, caused by the heat of solution as the crystal dissolves, it results in a natural cooling effect that is very different to the astringent cooling that agents such as menthol can produce in the finished product. Xylitol will, therefore, enhance and complement the delivery of mint flavours, providing the flavour with added impact, and producing a finished product with a refreshing cooling sensation immediately upon consumption. Fruit and berry flavours can also benefit from this effect, which enhances the freshness and juiciness of these more delicate flavours.

As the cooling effect is a result of the dissolution of the xylitol crystals, it will only be apparent in product formats where the xylitol is not already dissolved. Therefore, products in liquid form (e.g. syrups, toothpaste) or amorphous form (e.g. jellies, hard-boiled candies) will not typically exhibit a polyol-associated cooling effect upon consumption.

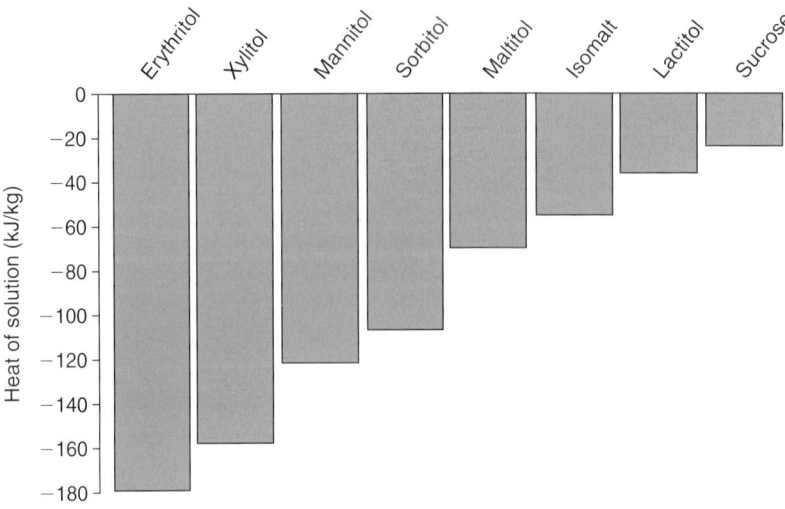

Fig. 16.4 Heat of solution of polyol sweeteners at 37°C.

16.3.2 Stability

The absence of a reducing group in the structure of xylitol means that, in common with other polyol sweeteners, it can be considered to be a non-reactive compound. As it contains no reducing groups, it will not take part in Maillard reactions. The boiling point of xylitol is 216°C, and caramelisation of xylitol will only occur if it is heated to temperatures near the boiling point for several minutes. Therefore, in applications that require caramelisation or non-enzymatic browning, the addition of a small quantity of reducing sugars or colour may be required. The stability of xylitol is not affected by pH, and it can therefore be used across a broad pH range (1–11).

16.3.3 Solubility

The solubility of a sweetener clearly exerts a significant influence over the perception and onset of sweetness and cooling (in the case of sweeteners with a high negative heat of solution). The xylitol molecule contains five hydroxyl groups and has a strong affinity for water, making it readily soluble in aqueous solutions (a property shared by many simple carbohydrates). The solubility of xylitol is similar to sucrose at ambient temperatures and higher than sucrose at elevated temperatures (Figure 16.5). This also allows for the formation of very high solids content solutions at elevated temperatures, a property that is particularly beneficial in hard-coating procedures, where high solids coating solutions can be used to significantly reduce process times.

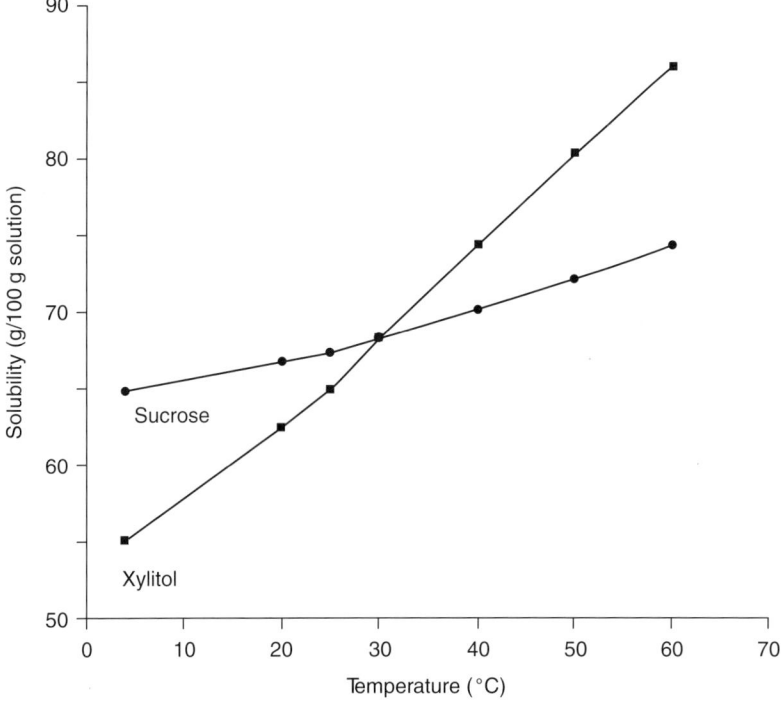

Fig. 16.5 Solubility of xylitol versus sucrose (g/100 g of solution).

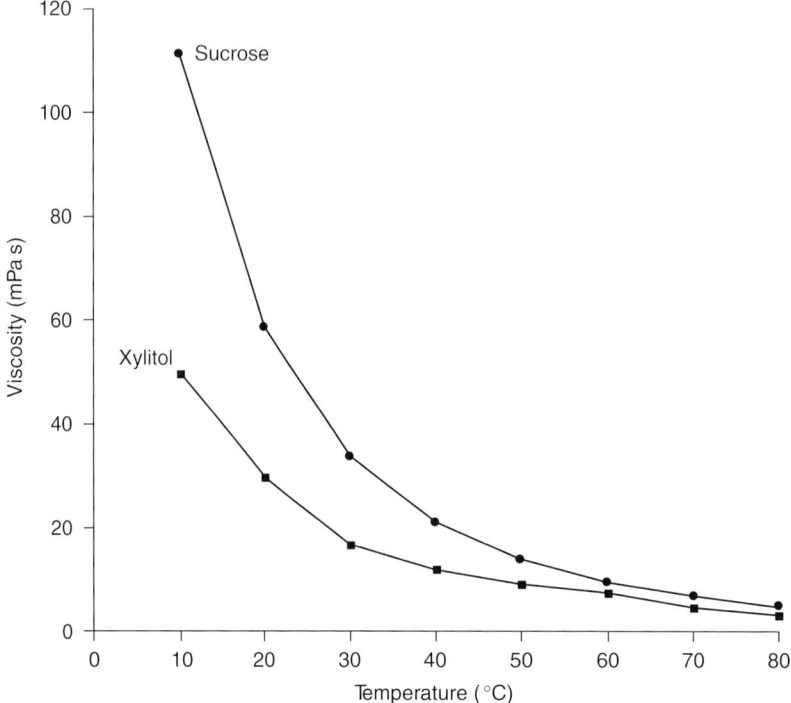

Fig. 16.6 Viscosity variation of xylitol and sucrose solutions with temperature (60% w/w solutions).

The solubility of xylitol in other solvents varies from only slightly soluble (e.g. ether at 0.4 g/100 g of solution) to soluble (e.g. methanol at 6.0 g/100 g of solution). Xylitol is only sparingly soluble in ethanol (1.2 g/100 g of solution). As with aqueous solutions, solubility is increased at elevated temperatures, for example xylitol is freely soluble in ethanol and methanol at a temperature of 50°C (14.0 and 16.0 g/100 g of solution, respectively).

16.3.4 Viscosity

The viscosity of a sweetener can have a strong impact on a product's manufacturing characteristics, texture and mouthfeel. As a monosaccharide sugar alcohol, xylitol has a lower viscosity in solution than sucrose at any given temperature (Figure 16.6) or concentration (Figure 16.7). It may, therefore, be necessary to compensate for this lower viscosity in certain applications by reducing the moisture content and/or increasing the quantity of high-molecular-weight components or gelling agents. Conversely, the low viscosity of xylitol can help to reduce the viscosity of confections containing high-molecular-weight components.

16.3.5 Boiling point elevation

The boiling point elevation of xylitol differs substantially from that of sucrose. Higher boiling temperatures are required to achieve any given concentration in the final product (Figure 16.8).

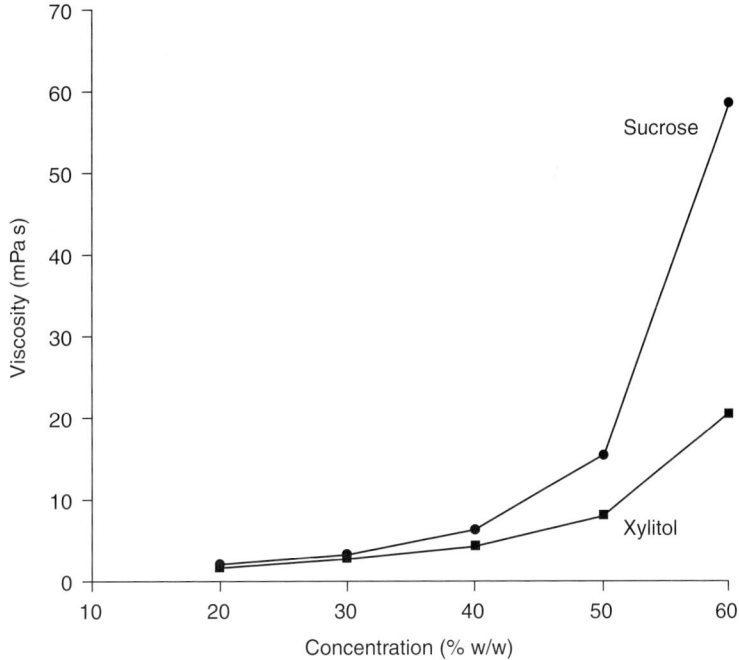

Fig. 16.7 Viscosity of xylitol and sucrose solutions at 20°C (variation with increasing concentration).

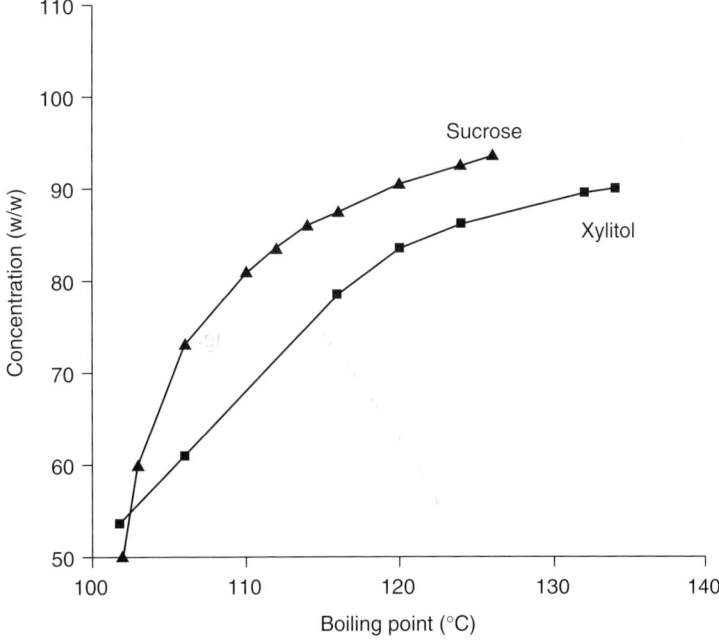

Fig. 16.8 Boiling point elevation effect of various concentrations of xylitol and sucrose.

16.3.6 Water activity

The water activity of a bulk sweetener can influence product microbial stability and freshness. Owing to its low molecular weight, xylitol exerts a higher osmotic pressure and, therefore, provides a lower water activity than equivalent solutions of sucrose (meaning that it effectively exerts a greater preservative effect in solution than sucrose). This makes xylitol a particularly useful sweetener to increase the solids and, therefore, the microbial stability of liquids.

16.3.7 Hygroscopicity

Xylitol is less hygroscopic than sorbitol, and marginally more hygroscopic than sucrose.

16.4 PHYSIOLOGICAL PROPERTIES

16.4.1 Metabolism

Xylitol, in common with the majority of other polyols, is only slowly absorbed from the digestive tract, largely owing to the lack of a specific transport system across the intestinal mucosa. After ingestion of xylitol, only a portion of the original dose is absorbed from the digestive tract. The actual amount of the ingested dose that is absorbed from the small intestine may vary from 25% to 50%, depending upon a variety of factors, not least of which is the total dose of xylitol. Upon entering the hepatic metabolic system, xylitol is readily sequestered in the liver, where it is further metabolised via the glucuronic acid–pentose phosphate shunt of the pentose phosphate pathway.[3,5,34,35] Figure 16.9 shows how this direct metabolism of xylitol fits into the process of normal carbohydrate metabolism. The conversion of L-xylulose to xylitol, and thence to D-xylulose, links the oxidative and non-oxidative branches of the glucuronate–pentose phosphate pathway, and yields glyceraldehyde-3-phosphate,

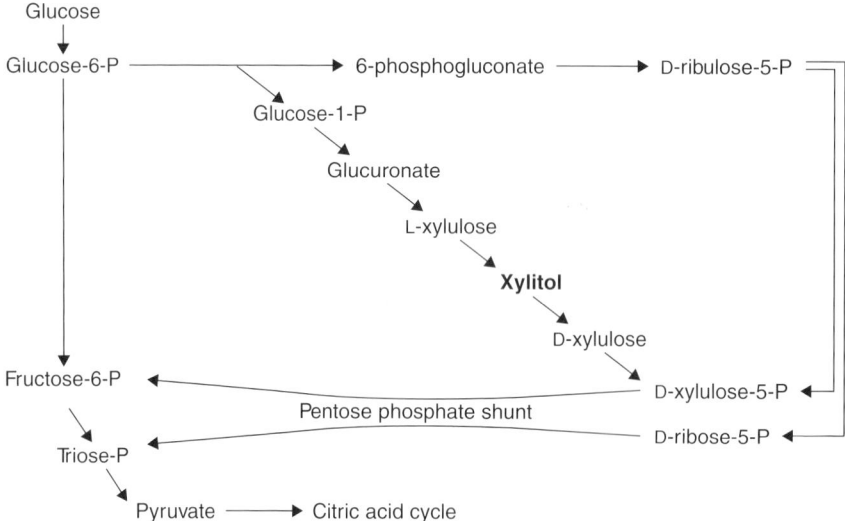

Fig. 16.9 Glucuronic acid–pentose phosphate shunt.

fructose-6-phosphate and ribose-5-phosphate, which are utilised in ribonucleotide biosynthesis.[36–38] Therefore, the pentose phosphate pathway facilitates the transformation of xylitol into intermediates of the glycolytic pathway, allowing for further degradation, or ultimately, transformation into glycogen.[3,34–36]

The portion of the ingested xylitol dose that is not absorbed from the small intestine (approximately 50–75%) passes to the distal parts of the gut, where it becomes a substrate for fermentation by the intestinal flora (i.e. it exhibits a prebiotic effect). The results of this fermentation process are predominantly short-chain fatty acids (SCFAs) (e.g. acetate, propionate and butyrate), and the development of small amounts of gas (H_2, CH_4 and CO_2).[39–41] The SCFAs so produced are subsequently absorbed from the gut and are further metabolised via the normal metabolic pathways of the host.[42] Acetate and butyrate are further metabolised in the liver, where they are utilised in the production of acetyl-CoA. Propionate is also almost completely metabolised in the liver, where it is utilised in the production of propionyl-CoA.[43,44]

SCFA production is a normal stage in the digestive process of mammals. In addition to polyols, the consumption of dietary fibres (celluloses, hemicelluloses, pectins, gums) also give rise to the production of SCFAs, as the prerequisite hydrolysing enzymes for these fibres are either absent from the small intestine or are inefficient in their operation. The majority of the SCFAs produced as a result of the bacterial fermentation of these fibres in the gut are absorbed and utilised by existing metabolic pathways in the host.[41,45] This indirect route of metabolism clearly plays an essential role in the utilisation of all fibres and low-digestible carbohydrates (including the polyols), and its contribution to the total energy intake from the diet appears to be significant.[46,47]

16.4.2 Suitability for diabetics

Following the ingestion of xylitol, the blood glucose and serum insulin responses are significantly lower than those following glucose or sucrose ingestion. This finding is linked to xylitol's relatively poor absorption and metabolism, and means that xylitol can be regarded as a suitable sweetener for use in diabetic and carbohydrate-controlled diets.[48–50] These observations clearly indicate that the conversion of xylitol to glucose, which has been reported to take place in the liver, occurs only slowly and, therefore, does not impact blood glucose concentration to any significant extent.

It was initially hypothesised that xylitol's lack of glucose and insulin responses would not be apparent when ingested in the context of a complex meal. In order to evaluate this hypothesis, a study was initiated which used 30 g of xylitol or sucrose to substitute 30 g of starch in a standardised meal as part of a diabetic diet regimen. The result of this study demonstrated that xylitol continued to exhibit significantly lower blood glucose and serum insulin responses than sucrose, even in the context of the complex meal.[51] Further studies in diabetic patients demonstrated that the incorporation of xylitol in the diet at levels of 30–60 g per day resulted in no adverse effects on the patients, particularly with respect to carbohydrate and lipid metabolism.[52,53]

As the control of blood glucose, lipids and weight are seen as the three main goals of diabetes management, xylitol represents an ideal sweetener for use in diabetic diets. Not only does xylitol have minimal impact on blood glucose, but it also provides fewer calories and a number of additional health-related benefits, without any reported negative impact on metabolic condition.

16.4.3 Tolerance

As with all polyols and slowly metabolised carbohydrates (e.g. lactose), the consumption of large doses of xylitol can cause certain gastrointestinal side effects. The factors affecting the tolerance of xylitol are its limited absorption and digestibility in the small intestine, which give rise to symptoms that are also commonly associated with malabsorption, including flatulence, accelerated intestinal transit times, bloating, borborygmi and in the most severe of cases, diarrhoea (laxation).[54] These gastrointestinal symptoms are transient and are readily reversible when consumption ceases. The gastrointestinal tolerance for xylitol will vary greatly between individuals, and depends upon a number of factors, including individual sensitivity, mode of ingestion, daily diet and previous adaptation to xylitol. Tolerance will also typically increase after repeated exposure to xylitol, a process termed adaptation. Adaptation has been observed with most polyols, and means that after a suitable period of consumption, doses of polyols can be increased without the risk of undesirable gastrointestinal symptoms.

Comparative studies on the gastrointestinal tolerance of different polyols are scarce and, therefore, it is difficult to accurately rank the polyols according to their laxative threshold. However, the gastrointestinal tolerance of xylitol and sorbitol has been directly compared in at least one early study.[55] In this study, 26 healthy adult volunteers received xylitol or sorbitol at a dose of 5 g on day 1, followed by subsequent increments of 5 g per day until the final dose of 75 g per day was reached on day 14. Diarrhoea was observed in nine subjects during the sorbitol period and in six subjects during the xylitol treatment. Repeated flatulence was recorded by 18 participants while consuming sorbitol, but by only 8 persons during the ingestion of xylitol. Overall, 21 out of 26 volunteers considered that xylitol was better tolerated than sorbitol.

In investigations on the use of xylitol and sorbitol in diabetic children, it was found that a daily dose of 30 g xylitol was well tolerated (except by one child), while the corresponding study with 30 g sorbitol had to be discontinued because of unacceptable side effects.[56] The good tolerance of xylitol in children has also been confirmed by another study, in which 13 children received increasing doses of 10, 25, 45, 65 and 80 g of xylitol for successive 10-day periods. No diarrhoea was reported at levels below 65 g per day, however, diarrhoea occurred at doses of 65 g per day in four children and 80 g per day in one child. Even in these cases, this side effect was noted only on one of the 10 days of the treatment.[57]

In summary, the gastrointestinal tolerance of humans to high oral doses of xylitol has been tested in both adults and children, with healthy volunteers and diabetic patients.[55–64] The results of these studies have invariably demonstrated that a daily dose of xylitol of the order of 30–40 g is unlikely to cause undesirable intestinal symptoms, except perhaps in particularly sensitive persons. In fact, there is good evidence from many studies that most people will tolerate levels of up to 50–70 g per day without any laxative effects. Expressed in an alternative way, a daily dose of xylitol of 0.5–1.0 g/kg body weight has been shown to be well tolerated by the large majority of individuals, and is unlikely to cause undesirable gastrointestinal side effects. This quantity can usually be increased after a suitable period of adaptation.[65]

16.4.4 Caloric value

A number of methods may be used to determine the caloric value of polyols. These include energy balance studies using indirect calorimetry, recovery measurements of polyols and their constituents at the lower end of the small intestine, breath hydrogen analysis, growth rate studies and carcass composition studies. Essential factors in the ability to estimate the caloric value of polyols are: how much of the ingested dose is absorbed, how much is available for fermentation by the colonic microflora and how much of the ingested dose is excreted unchanged in the faeces or urine. Caloric estimates can then be made on the basis of how the absorbed dose is further metabolised, the extent of microbial fermentation in the colon and the degree of absorption and metabolic utilisation of the resulting fermentation products.

The absorption of xylitol from the gastrointestinal tract is via a process of passive diffusion, and on the basis of *in vitro* and *in vivo* experiments, it has been estimated that only 25–50% of the ingested dose of xylitol is actually absorbed.[66] The *absorbed* portion of xylitol is fully energetically available to the host at approximately 4 kcal/g, while the *unabsorbed* portion (50–75%) is almost completely fermented by the intestinal flora, with approximately 58% becoming available to the host through the absorption of SCFAs. This value correlates well with the estimated 50% energy salvage proposed by one expert group.[67]

It is therefore possible to estimate a metabolisable energy value for xylitol of approximately 2.8–2.9 kcal/g, a value that is supported by the findings of an *in vivo* study in which xylitol was found to be only approximately 60% as effective as glucose in promoting growth.[68]

The use of indirect calorimetry allows a more accurate determination of the caloric value of polyols, and this method has also been used to study the caloric value of xylitol in healthy human volunteers.[69] This study observed an overall increase in carbohydrate oxidation of only 25% of that for glucose, together with an increase in metabolic rate of only 48% of that observed for glucose. This clearly demonstrates that the caloric value of xylitol should be considered to be approximately half that of glucose, namely 2.0 kcal/g. However, the design of this study may not have allowed for the complete absorption and metabolic utilisation of the SCFAs generated by bacterial fermentation of xylitol in the colon, and it is therefore most likely that the caloric value for xylitol would be somewhat higher than this. Therefore, the Federation of American Societies for Experimental Biology (FASEB) have determined the net energy value of xylitol to be 2.4 kcal/g.[66]

The reduced caloric value of xylitol and other polyols is widely acknowledged, and the catalogue of scientific data supporting these values continues to increase. The European Union (EU) currently allocates a caloric value of 2.4 kcal/g for all polyols (including xylitol), and the United States Food and Drug Administration (US FDA) has acknowledged the caloric value of xylitol as 2.4 kcal/g.

16.4.5 Health benefits

As already discussed, xylitol is regarded as a sugar-free bulk sweetener that can be used to replace sucrose and other fermentable carbohydrates in the diet. The use of xylitol as a replacement for sucrose offers numerous advantages, as it is not only sugar free and reduced calorie, but also exhibits a number of established health benefits including unique dental properties, satiating and prebiotic effects, as well as being suitable for diabetics and exhibiting a low glycaemic response.

16.4.5.1 Dental benefits

Nowhere is the correlation between health and diet more firmly established than in the case of dental caries (tooth decay). The relationship between dietary carbohydrates and dental caries has been recognised for over a century, and it is now generally accepted that the consumption of sugars and other fermentable carbohydrates in our diets is one of the main causative factors of tooth decay.[8,9,70–72] The bacteria that make up the oral flora are able to rapidly ferment these carbohydrates, and in doing so produce and secrete acid waste products. Accumulation of these acidic waste products results in a fall in the plaque pH, together with an associated drop in the pH of the oral cavity as a whole.[8,9,70–72] If the pH of the plaque falls below the recognised 'critical pH' of 5.7, then the acids initiate a process of demineralisation of the dental enamel, effectively dissolving the calcium and phosphate ions from the tooth surface. As calcium and phosphate are essential to the integrity of the tooth enamel, over time, this significantly weakens the tooth surface. Eventually, the tooth surface will collapse, resulting in what is recognised as a cavity, and if left untreated, to tooth loss. If fermentable carbohydrates are removed from the diet, then the bacteria in the oral cavity do not produce the acid waste products that initiate demineralisation of the tooth enamel, and the cycle is broken. In fact, numerous studies have clearly demonstrated that if the consumption of fermentable carbohydrates is reduced, the risk of dental caries decreases significantly.[71]

Clearly, a process of complete substitution of fermentable carbohydrates in the diet would be unfeasible, as well as undesirable in the context of overall health. Fortunately, such a drastic approach is not necessary, as numerous studies have demonstrated that reducing the consumption of foods rich in fermentable carbohydrates and limiting their intake to alongside regular meal times, or replacing fermentable carbohydrates with non-fermentable alternatives in snacks and confectionery products only, provides a significant benefit with regard to decreasing caries incidence.[9,71,73,74] Therefore, xylitol, and polyols in general, are increasingly utilised as replacements for fermentable carbohydrates in a wide range of foods, where their inherent non-cariogenic nature means that they do not contribute towards the development of dental caries. The usefulness of polyols as alternatives to sugars and as part of a comprehensive programme to reduce dental caries, including proper dental hygiene, has been recognised by the American Dental Association. The US FDA has also approved the use of a 'does not promote tooth decay' health claim in labelling for sugar-free foods that contain xylitol or other polyols. The following sections shall identify and explore some of the more specific dental health properties ascribed to xylitol.

Acidogenicity

Non-fermentable carbohydrates and associated sugar substitutes are not fermented to any great degree by the oral bacteria. Therefore, their consumption does not give rise to the acid production, and subsequent pH drop in the oral cavity, which is the initiator of the caries process.[74,75] For this reason, they are often referred to as being 'non-acidogenic' (i.e. they do not give rise to acid production by the oral bacteria) or 'non-cariogenic' (i.e. they do not contribute to the caries process).

Xylitol has been the subject of numerous acidogenicity studies, all of which have consistently demonstrated that xylitol is not fermented by the vast majority of oral micro-organisms, and that consumption of xylitol does not cause a fall in the pH of the plaque.[75–77] In fact, it is widely acknowledged that the consumption of xylitol results in the lowest level of acid production by the oral bacteria of any of the polyols.[78] Perhaps of equal importance is the fact that the oral bacteria do not appear to develop the ability to ferment xylitol (i.e. they do not adapt to ferment xylitol), even following several years of regular xylitol exposure.[79–81] Obviously, in the absence of acid production by the oral bacteria, the initiation of the dental caries process is prevented.

Efficacy of xylitol in caries prevention: animal studies
Xylitol can, therefore, be described as being non-cariogenic, and this trait has been studied in a number of studies, which have utilised rats as a caries model. In these studies, it was shown that when xylitol was used to replace fermentable carbohydrates in the daily diet of the rats, the rats did not develop caries; thereby demonstrating that xylitol has little, if any, caries potential.[75,82] The ability of xylitol to halt the progression of caries (e.g. to exhibit a cariostatic action) was also first investigated using a rat caries model. In these studies, the rats were provided with xylitol in conjunction with a normal diet containing fermentable (cariogenic) carbohydrates.[75] These studies demonstrated that xylitol is able to reduce the overall cariogenicity of the diet. When xylitol meals were administered alternately with fermentable carbohydrate meals, the caries scores of the animals were 35% lower than those in the control group.[75] These early rat studies also indicated that the inclusion of xylitol in the diet is able to promote the remineralisation of early caries lesions.

Efficacy of xylitol in caries prevention: human studies
The ability of xylitol to inhibit the development of caries has been demonstrated in numerous clinical and field studies. These studies have consistently shown that when xylitol-containing confectionery or chewing gum is consumed as part of a normal daily diet, in conjunction with accepted oral hygiene practices, new caries incidence is significantly reduced (typically by 40–80%).[11–21] The significant potential of xylitol as a component in caries prevention programmes was first recognised as part of the Turku Sugar Studies in the early 1970s.[83] These studies found that substituting sucrose in the diet with xylitol resulted in a dramatic reduction in the occurrence of caries. A short-term chewing gum study was also carried out following this initial study, which demonstrated that the regular daily use of xylitol-sweetened chewing gum also reduced caries incidence to a similar degree.[12] This stimulated a great deal of research throughout the subsequent three decades, and a number of long-term clinical studies have since observed reductions in caries incidence of between 40% and 100% for subjects regularly consuming xylitol-sweetened confectionery (Figure 16.10).[12–15,18,20,21] Perhaps the most significant of these studies was the 40-month Belize study.[20] This study represents the most comprehensive study to date of the caries-preventive properties of sugar-free chewing gum, and in particular xylitol-sweetened chewing gum. Nine treatment groups were included: one control group (no chewing gum), four xylitol groups (with a range of xylitol consumption from 4.3 to 9.0 g per day), two xylitol/sorbitol groups (with a total polyol consumption from 8.0 to 9.7 g per day), one sorbitol group (9.0 g per day) and one sucrose group (9.0 g per day). Compared with the no-gum group, sucrose gum usage resulted in a marginal increase in caries rate (relative caries risk 1.20), while sorbitol gum reduced the caries rate (relative caries risk 0.74). The four xylitol gums were most effective in reducing caries rates (relative caries risks from 0.48 to 0.27). The most effective product was a 100% xylitol pellet gum (relative caries risk 0.27). The xylitol–sorbitol combinations were less effective than xylitol, but reduced the caries rates significantly compared to the no-gum or sorbitol-gum groups. These results suggest that the regular usage of polyol-based chewing gum reduces caries rates in young subjects, with xylitol gums being most effective.

The potential of xylitol as a caries-preventative agent is not limited to chewing gum. Xylitol has also been demonstrated to be effective in helping to reduce caries when administered in the form of other confectioneries (tablets, chewy candy and hard candy), and even in traditional oral hygiene products such as toothpaste and mouthwash. In fact, all of the

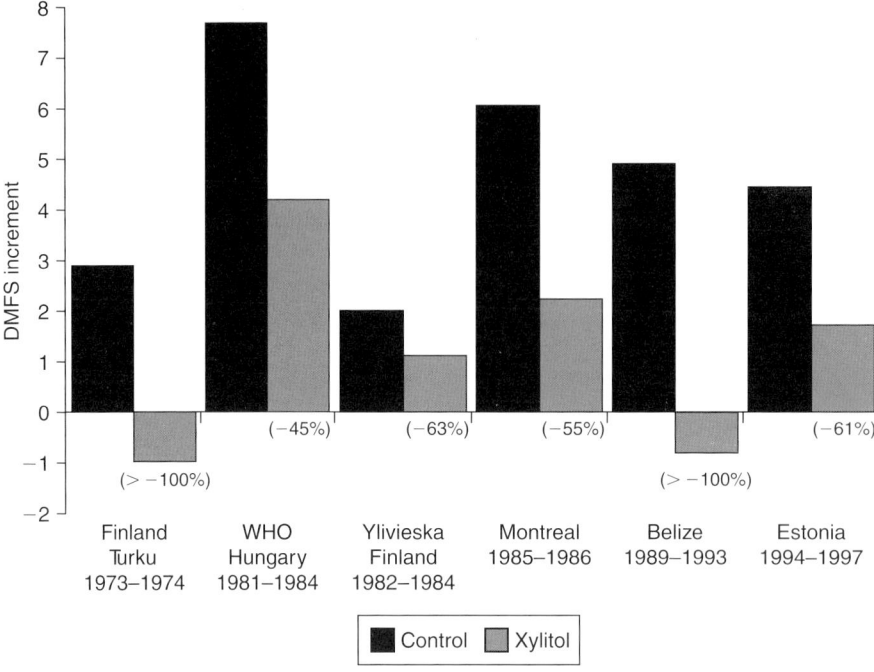

Fig. 16.10 Significant xylitol dental studies (values in parenthesis represent the percentage reduction in decayed, missing and filles surfaces (DMFS) increment observed in the xylitol group, compared to the control).

dental benefits observed from the regular use of xylitol in chewing gum (reduced caries, reduced plaque and specific inhibition of mutans streptococci) have also been observed in these other dosage forms.[13,16,21,83–93] As an example, one 3-year clinical study was conducted with 2630 children,[85] which evaluated the efficacy of a sodium fluoride/silica/10% xylitol dentifrice when compared to a sodium fluoride/silica dentifrice that contained no xylitol. After 3 years, subjects using the xylitol-containing dentifrice had a statistically significant reduction in decayed and filled dental surfaces (12.3% reduction; $p < 0.001$). A subsequent study by the same group,[86] using a sodium monofluorophosphate/dicalcium phosphate/10% xylitol dentifrice compared to a sodium monofluorophosphate/dicalcium phosphate dentifrice that contained no xylitol, also demonstrated that subjects using the xylitol-containing dentifrice had a statistically significant reduction in decayed and filled teeth after 30 months (14.8% reduction; $p < 0.05$). These studies would appear to support the theory that xylitol and fluoride act synergistically to enhance the efficacy of oral hygiene products.

The long-term caries-preventive potential of xylitol has been studied through the use of follow-up examinations in both the Ylivieska[17] and Belize[94] studies. Re-examination of the subjects of the Ylivieska study 2 or 3 years after discontinuation of the use of xylitol revealed a continued reduction in caries increment in the years following the use of xylitol of approximately 55%.[17] In teeth erupting during the first year of the use of xylitol chewing gum, the long-term caries-preventative effect was over 70%. Similarly, in a follow-up to the Belize study, subjects were found to have a highly significant reduction in caries risk 5 years after the discontinuation of xylitol consumption, with particular benefits observed in teeth erupting during and after the period of xylitol consumption.[94] These results suggest that the value of xylitol may be highest during periods of high dental activity such as eruption of new teeth.

Effect on dental plaque
The ability of xylitol to reduce total plaque levels via the inhibition of bacterial growth has been proposed as one of its main mechanisms of action in reducing caries.[95] Numerous *in vitro* and *in vivo* studies have reported a xylitol-mediated inhibition in plaque growth and/or re-growth, as measured using either a plaque index, plaque wet weight, plaque dry weight or plaque accumulation/re-growth method.[96–107] These studies have demonstrated that xylitol consumption is able to inhibit plaque growth by up to 50%.

It appears to be generally accepted that the oral bacteria are unable to ferment xylitol to acid end products, or to utilise it as a sole carbon source. While this would potentially impact plaque growth, it is unlikely to be the main mode of action in xylitol's reported effect on plaque growth. It has the been suggested that the dramatic reduction in plaque quantity observed for xylitol may be linked to the complete inhibition of mutans streptococci, through an inhibition of glucose metabolism[108] (see Section 'Specific effects versus mutans streptococci'). Again, it is unlikely that this is the main mechanism of plaque reduction, as it has also been demonstrated that tolerance to xylitol can develop in mutans streptococci, allowing these bacteria to continue fermenting glucose and other carbohydrates in the presence of xylitol.[95,109,110] It is important to note that this adaptation (tachyphylaxis) does not result in these bacteria developing the ability to actually ferment xylitol itself, but merely to ferment other dietary carbohydrates in its presence.[95,109–111] The plaque-reducing and non-cariogenic effects of xylitol in subjects with plaque containing large proportions of xylitol-tolerant mutans streptococci is still apparent.[105,107] Research has also clearly shown that xylitol-tolerant mutans streptococci have altered virulence factors in comparison to the xylitol-sensitive organism. This reduction in virulence has been attributed to decreased production of adhesive macromolecules (i.e. reduced adhesion properties),[110] increased susceptibility to agglutination by lysozyme and saliva, reduced adhesive properties and reduced cell-to-cell aggregation.[103] As mutans streptococci form such a large proportion of the dental plaque, and have a key role in maintaining the overall adhesivity of the plaque matrix, this means that the plaque as a whole becomes less adhesive.[95,111] This net reduction in plaque adhesivity could explain reductions in total plaque quantity, via an enhancement in plaque clearance from the tooth surface by saliva (a process enhanced by chewing).

In summary, convincing data exist demonstrating a clear dental plaque-reducing effect for xylitol. In plaque with an abundance of xylitol-sensitive oral bacteria, this reduction would appear to be related directly to the inhibition of growth, particularly of the mutans streptococci. However, as the proportion of xylitol-tolerant bacteria increases, there appears to be a concurrent reduction in virulence. This reduction in virulence facilitates a more efficient removal of the plaque by the host's defence and clearance mechanisms, chewing and oral hygiene practices. The plaque-reducing effect of xylitol has been shown to be superior to that of sorbitol in a number of studies and, the magnitude of this superiority (20–50%) suggests that the effect is not only due to xylitol's specific inhibition of the mutans streptococci, but is more likely to be a result of a combination of the factors outlined previously. Blends of xylitol and sorbitol in chewing gum have also been shown to be more effective in inhibiting plaque growth than sorbitol alone, but less effective than pure xylitol products.[96,101]

Specific effects versus mutans streptococci

Extensive microbiological investigation of the caries process has demonstrated that certain species of bacteria are associated with the initiation and progression of dental caries in both animals and humans.[95,112] One particular family of bacteria associated with dental caries is the mutans streptococci group of bacteria, which are generally considered as among the most virulent of the cariogenic bacteria and are frequently implicated as the main cause of dental caries in the developed world.[95,112,113] *Streptococcus mutans* itself is only one of the particular species belonging to this family of bacteria, which also includes *Streptococcus salivarius* and *Streptococcus sanguis*, among several other species and serotypes.

As previously highlighted, one of the key mechanisms of xylitol's plaque-reducing, and ultimately caries-inhibitory, effects is believed to be its effect against the mutans streptococci.[95] Xylitol is unique among the polyols as it appears to specifically inhibit the growth of mutans streptococci, resulting in reduced numbers of these highly cariogenic bacteria in the plaque. Xylitol also reportedly decreases the amount of insoluble plaque polysaccharides produced by these bacteria, therefore making the plaque as a whole less adhesive and easier to remove.[95,103,110] Prolonged use of xylitol has been shown to select for a population of *S. mutans*, which are less virulent (exhibiting reduced acidogenic potential and adhesive properties), and are therefore less likely to contribute towards further caries.[95,114,115]

This particular aspect of xylitol's dental benefits have been extensively studied[88,90,93,101,106,110,116–123] and reviewed.[95,108,124] As with all clinical data, a wide variety of results have been obtained from these studies, however, the use of xylitol has consistently been associated with significant reductions in both plaque and salivary mutans streptococci levels. These reductions are typically in excess of 20% of the baseline or control group levels studied, and appear to occur rapidly following the commencement of the individual's consumption of xylitol, with significant reductions observed following weeks rather than months of regular consumption. The dose of xylitol required to induce this effect has been observed to be as low as 1 g xylitol per day, in dosage forms as diverse as chewing gums, mints, mouth rinses and toothpastes.

The specific mechanisms of xylitol's effect against the mutans streptococci have been the subject of much debate over the last two decades, and the consensus of opinion on the subject appears to be that xylitol exhibits a combination of inhibitory effects, which can be divided into four main modes of action:[124]

1. Non-fermentability, therefore not encouraging bacterial growth.
2. When mutans streptococci in the plaque are exposed to xylitol, the proportion of xylitol-tolerant strains increases, which are less virulent (these tolerant strains are less acidogenic, and produce less of the adhesive exopolysaccharides that form the plaque matrix).[114,125]
3. The uptake and conversion of xylitol to xylitol-5-phosphate has an inhibitory effect on metabolism within the cell.[95] In some cases, this mechanism can result in the formation of intracellular vacuoles, degraded cell membranes and may even result in autolysis.[125–128]
4. Xylitol can participate in a 'futile metabolic cycle' in certain streptococci, wherein xylitol is taken up by the cell and converted to xylitol-5-phosphate. This is then split by sugar–phosphate phosphatases and the xylitol expelled from the cell. This represents an energy-wasting cycle and ultimately limits the metabolic efficiency of the organism.[95,129–132]

In summary, it has been established that the regular use of xylitol is able to significantly reduce the numbers and proportion of mutans streptococci resident in the oral cavity. The mechanisms by which xylitol brings about this reduction appear to be specific to mutans streptococci. However, as this family of bacteria typically represent a relatively large proportion of the oral flora, the impact of xylitol on these specific bacteria results in a significant change in the oral flora as a whole.

Other dental benefits associated with xylitol
(a) Enhanced remineralisation In addition to the prevention of demineralisation, the regular consumption of xylitol is also associated with the enhanced remineralisation of teeth already affected by caries. This has been demonstrated in several studies, most recently during the extensive clinical evaluations in Belize. Concurrent studies in primary dentition (children initially 6 years old) and permanent dentition (children initially 10 years old) both showed that chewing xylitol-containing chewing gum is more frequently associated with the arrest of caries than sorbitol or sucrose chewing gums, and has the effect of actually hardening existing caries lesions.[133]

(b) Prevention of mother to child transmission of cariogenic bacteria A newly identified benefit associated with the regular use of xylitol is that it can inhibit the transfer of cariogenic mutans streptococci bacteria from mothers to their children.[134] Several studies have demonstrated that mothers are the primary source of infection of mutans streptococci in the mouths of their children[135–137] and that preventing or delaying colonisation by these bacteria in early childhood can lead to significant reductions in tooth decay later in life.[138–140] In an innovative study,[134] children whose mothers regularly consumed xylitol chewing gum for the 2-year period following the birth of the child, who exhibited significantly reduced mutans streptococci colonisation at the age of 2 years, were compared to children whose mothers had not consumed xylitol chewing gum, but who had been treated with fluoride or chlorhexidine varnish treatments. At the age of 5 years, the children whose mothers had consumed xylitol chewing gum in the original study had significantly less dental caries (71–74% decayed, missing or filled teeth) than the children in the other two groups.[141] Interestingly, although the children of mother's in the chlorhexidine group exhibited reduced colonisation by mutans streptococci after 2 years, this did not translate into a reduction in caries incidence after five years. This further supports the theory that xylitol does not exert an antibacterial effect on the oral bacteria, but rather a 'modulating' effect on the oral flora as a whole, resulting in a shift towards a less virulent/cariogenic flora. There was even a further follow up whereby the effects of xylitol were evaluated in children at the age of 10 years (10-year follow up).[142] It was observed that the primary teeth of the children belonging to the xylitol group were maintained completely cavity free. This has been attributed to the fact that xylitol has influenced the bacterial composition of the bacterial strain transmitted from mother to child. In addition, children from high-risk mothers who used 100% xylitol containing chewing gum during the course of the study, when under the age of 2 years, appeared with their first cavities at the age of 8 years. These children also indicated 40% less caries when compared to the rest of the groups. In particular, the teeth of the children with mutans streptococci colonisation had developed caries as early as the age of 5 years. Perhaps the most remarkable finding of this study is that this impressive reduction in caries was achieved solely through the treatment of the mothers, as the children received no treatment during this period. In addition, the follow-up 10-year study is a proof of the long-lasting effects of xylitol, which are beneficial in this unique way.

16.4.6 Other health benefits associated with xylitol

16.4.6.1 Low glycaemic response

An added facet to xylitol's suitability as a diabetic sweetener is that it exhibits a very low relative glycaemic response (RGR) of 8, compared to glucose with a value of 100.[143] The RGR of a substance is measured in almost exactly the same way as the glycaemic index (GI), but takes into account the effect of both available and unavailable carbohydrates in the product (whereas the measurement of GI is only made in relation to the available carbohydrate in the product). The significance of the GI value system is that low-GI and low-RGR carbohydrates are broken down very slowly following ingestion, therefore releasing glucose gradually into the bloodstream. Conversely, high-GI and high-RGR carbohydrates are rapidly broken down, causing an immediate rise in blood glucose and concomitant insulin response. Nutritionists are increasingly recommending that individuals should reduce the overall glycaemic load of their diet as a whole, to decrease the risk of developing type-2 diabetes and to improve general health *per se*.

As xylitol has a substantially lower RGR value than sucrose or glucose, it can be used to reduce the overall glycaemic load of the diet through the manufacture of foods with a low glycaemic load, and therefore an improved nutritional profile.

16.4.6.2 Prebiotic effect

Xylitol, in common with most polyols, is a low-digestible carbohydrate. It is only partially absorbed in the upper intestine, with a minimum of 50% of the ingested dose passing to the lower intestine, where it is available for microbial fermentation. Colonic fermentation by the micro-organisms that inhabit the gastrointestinal tract can significantly influence the health and well-being of the host. Low-digestible carbohydrates are fermented in the colon by beneficial commensal bacteria, such as *Bifidobacterium* and *Lactobacillus* species, generating SCFAs. The net result of this fermentation is a reduced colonic pH and increased butyrate production. The dual effect of pH decrease and production of SCFAs has been shown to inhibit the growth of harmful gram-negative bacteria such as *Bacterioides* and coliforms. This process ultimately causes a shift in the microbial community of the gut and enhances the growth of the beneficial bacteria, in particular, *Bifidobacterium*.[144–146]

The reduction in intestinal pH associated with this shift in the gut microflora has been closely linked to improvements in epithelial function, maintenance of a healthy colonic epithelium and, ultimately, to a decrease in the risk of diseases of the colon. Reduced intestinal pH is also reported to aid the absorption of calcium ions[147,148] and decrease the formation and absorption of harmful ammonia in the colon. Intestinal microflora studies have indicated that ingested xylitol is able to influence the colonic and faecal microflora in animals and in humans.[149,150] The observed changes result in decreased faecal pH and an increase in the numbers of gram-positive bacteria in the faeces. Therefore, the portion of xylitol that passes to the colon following ingestion can be expected to exert a prebiotic effect.

16.4.6.3 Satiety effects

It is widely acknowledged that excess weight and obesity is driven by an imbalance between energy intake and energy expenditure. Energy intake is largely determined by satiety (the condition of feeling full) and satiation (the process of satisfying completely). Therefore, an important opportunity exists to help consumers feel fuller while consuming fewer calories. Studies have demonstrated that foods with a low caloric density are more satiating.[151,152] Foods with a low GI have also been shown to be more satiating, with one review of the area concluding that the 'consumption of high-GI carbohydrates may increase hunger and promote overeating relative to consumption of items with a lower GI'.[153] It is obvious from these data that xylitol can be used in the manufacture of foods that are less calorically dense and have a lower GI. However, several human clinical studies have also shown that consumption of xylitol is able to reduce subsequent food intake by enhancing satiety. Caloric reductions have been observed in the range of 5–25%, through a mechanism that may be related to prolonged gastric emptying.

Shafer et al.[154] studied the effects of the pre-meal consumption of simple solutions of xylitol on the subsequent gastric emptying of a solid-food component of a complex meal. After ingestion of 25 g xylitol, gastric emptying was markedly prolonged ($T_{1/2}$ 58 ± 5 minutes control versus 91 ± 7 minutes after xylitol ($p < 0.01$)). Total food intake after oral pre-loading with 25 g xylitol in water led to a caloric intake of 690 ± 45 kcal versus 920 ± 60 kcal ($p < 0.01$), a 25% reduction in the calories consumed. In the same study, glucose, fructose and sucrose failed to significantly suppress food intake. Although only the 25 g data were statistically significant, a dose response appeared to exist.

Blundell et al.[155] studied xylitol consumption in 20 male subjects; 25 g of xylitol consumed in a yoghurt 1 hour before a controlled lunch suppressed calorie intake by 10–20% compared to the control products, such as sucrose or fructose. Xylitol also suppressed appetite for another 4–5 hours after the lunch. Xylitol's satiating effect appears to be enhanced in combinations with polydextrose. King et al.[156] studied xylitol and polydextrose on 15 human subjects over a period of 10 days. Xylitol (25 g), xylitol:polydextrose (50:50; 25 g) or polydextrose (25 g) ingested in yoghurt 90 minutes prior to a controlled lunch, suppressed combined calorie intake by 13–17% versus the sucrose control product. This effect remained consistent from day 1 through day 10. Lunch energy intake was 7–12% lower for the test groups than for the control. Adding in the caloric content of the pre-loads strengthens this effect, resulting in a total reduction of energy intake of 13–17%. These studies demonstrate that the consumption of xylitol, alone or in combination with polydextrose (50:50), can induce satiety and thereby reduce caloric intake.

16.4.6.4 Prevention of acute otitis media

Acute otitis media is an inflammation of the area behind the eardrum (the tympanic membrane). This area is commonly known as the 'middle ear'. The inflammation is caused by a bacterial infection, causing an associated build-up of fluid behind the eardrum, which in severe cases may cause the eardrum to rupture. These symptoms are frequently associated with, or preceded by, signs of an upper respiratory infection, such as runny or stuffy nose, or cough. The occurrence of acute otitis media is most common in young children, with two out of three children experiencing at least one episode of acute otitis media before the age of 3 years. As antibiotic therapy is the traditionally prescribed treatment for this condition, it represents a significant cost to the healthcare system and a major use of antibiotics in children. Even after effective antibiotic treatment, 40% of children may retain non-infected residual fluid in the middle ear that can cause some temporary hearing loss. This may last for 3–6 weeks after the initial antibiotic therapy.

Two significant clinical trials have demonstrated that the regular consumption of products containing xylitol (particularly chewing gum) can reduce the occurrence of acute otitis media in children by as much as 40%.[157,158] While these studies showed that xylitol administered in chewing gum was the most effective route of administration, xylitol syrups and lozenges also exhibited significant reductions in both the occurrence, and recurrence, of acute otitis media. The mechanism of xylitol's preventative action against acute otitis media has been reported as the growth inhibition of various otopathogenic bacteria, but most markedly *Streptococcus pneumoniae*.[159] The mechanism of this growth inhibition appears to mimic that reported for xylitol against *S. mutans* and other caries-related organisms.[160] Xylitol has also been shown to significantly inhibit the attachment of *S. pneumoniae*, *Haemophilus influenzae* and *Moraxella catarrhalis* (all recognised otopathogenic bacteria) to the nasopharyngeal cells, thus also inhibiting the pathogenesis of the condition.[161] Studies carried out in the United States also support the aforementioned evidence[162] pointing out xylitol's good tolerability of 5 g three times per day or 7.5 g per day.

16.5 APPLICATIONS

Xylitol is a versatile sweetener, which has found application in almost all sectors of the food industry, both on its own and in combination with other bulk sweeteners. Xylitol can also be used as a household or 'table-top' sweetener, where it can be used in general domestic cookery and baking and for sweetening of beverages. While this represents an increasing market for the use of xylitol, by far the majority of the global usage of xylitol remains firmly in the commercial manufacture of prepared foods.

16.5.1 Confectionery

As confectionery is generally regarded as an indulgence or comfort product, consumers demand that confectionery provide quality and satisfaction. Sugar-free confectionery obviously faces the same challenge, and it must therefore offer the sweetness, flavour and mouthfeel of a sugar-sweetened counterpart, without any unpleasant aftertaste. Xylitol is an ideal sweetener for developing high-quality sugar-free products, as its sweetness and cooling effect create excellent tasting sugar-free confectionery with added health benefits. Xylitol can be effectively used in a diverse range of confectionery applications, either as the sole sweetener or in combination with other sugar-free sweeteners or bulking agents. Commercially, examples of xylitol-containing products exist for most confectionery types, including chewy candy, hard-boiled candy, gum Arabic pastilles, gelatine jellies, pectin jellies, starch jellies, toffees, caramels fondant, fudge, cast lozenges, compressed tablets and mini-mints. Crystalline xylitol can also be used as a sugar-free sanding and engrossing material for various types of confectionery.

While xylitol can be used alone to form hard candies, it is more commonly used in combination with other polyols or sugar-free bulking agents. As hard candy is an amorphous 'glass', xylitol's cooling effect is not apparent when the product is consumed (as the xylitol is effective in solution). In order to utilise xylitol's unique cooling effect in hard candies, it is necessary to incorporate the xylitol into the product in the form of a centre filling. This approach creates a pronounced cooling sensation from the centre of the confection when consumed, and can be achieved in a number of ways.

The use of xylitol in confectionery is not necessarily strictly limited to the production of sugar-free products. Increasingly, xylitol is being used for its functional benefits in sugar-based confectionery, where its excellent cooling and flavour-enhancing qualities can be used to provide improved flavour impact and added freshness to a wide variety of products.

16.5.2 Chewing gum

Sugar-free chewing gum is the most common commercial application of xylitol throughout the world, where it is used for its dental, technological and organoleptic properties. Xylitol's intense and distinctive cooling effect, in combination with its equal sweetness to sugar, makes it the perfect ingredient for use in sugar-free chewing gum formulations (both stick and pellet or dragee formats). Chewing gums produced exclusively with xylitol do not necessarily require the addition of intense sweeteners, although intense sweeteners are often added in order to enhance flavour and sweetness longevity. Xylitol can also be used to create novel textures in chewing gum and bubble gum products, as it generally forms a softer, more flexible product than other polyols.

16.5.3 Hard coating applications

Xylitol excels in sugar-free hard coating applications due to a combination of its excellent solubility and controllable crystallisation, and hard coating of sugar-free chewing gum pellets represents one of the major commercial applications of xylitol. Xylitol's functional properties make it ideal for this application, with numerous advantages over other polyols (Table 16.2). These factors mean that xylitol hard coating can be carried out significantly faster than with other sugar-free coating materials, while contributing its greater sweetness and pronounced cooling effect, allowing for a superior tasting, sugar-free hard coating.

Table 16.2 The advantages of xylitol in hard coating applications.

Property	Comments
High solubility	The high solubility of xylitol permits the production of very high solids and supersaturated coating solutions.
Low viscosity	Xylitol has a low viscosity in solution, meaning that high solids, supersaturated coating solutions are still of sufficiently low viscosity to pump and atomise during coating processes.
Controlled crystallisation	Crystallisation of xylitol in its most stable form can be controlled through normal process parameters.
Speed of processing	Xylitol's high solubility and controlled crystallisation result in rapid processing, with industrial processing times for xylitol coatings 30–60% lower than those obtained for the equivalent processes with other polyols.
Resistance to chipping	Xylitol hard coatings are less friable than coatings produced with other polyols, meaning that xylitol coated products exhibit a strong resistance to chipping during processing, packaging and transit.
Cooling effect	Xylitol's high negative heat of solution manifests itself as an intense cooling perception upon dissolution of the crystal. It exhibits the strongest and most noticeable cooling effect of all polyols, with a rapid cooling onset.
Sweetness	Xylitol is equisweet to sucrose, removing the requirement the need for intense sweeteners.

16.5.4 Chocolate

Xylitol can be used as a replacement for sucrose to produce great tasting no-sugar-added chocolate, although high levels of xylitol may be linked with a slightly 'scratchy' aftertaste (a phenomenon frequently associated with the use of monosaccharide sugars in chocolate). Despite this, there are a number of commercial products utilising xylitol as a sweetener in chocolate. One major advantage that xylitol has over most other polyols in the manufacture of chocolate is that the use of xylitol negates the need for the addition of intense sweeteners. The conching process for xylitol-containing chocolate masses does not present any significant variations over sugar-sweetened chocolates, although due to the moderately hygroscopic nature of xylitol, the conching temperature is best held at a temperature of 50°C or less, and the relative humidity of the manufacturing environment should be controlled at less than 60%.

16.5.5 Dairy products and frozen desserts

Xylitol can be used to replace sucrose in the manufacture of many dairy products and frozen desserts, such as yoghurt, mousse, ice cream and sorbet. The resulting products are capable of making a variety of health claims, including reduced calorie, no added sugar, sugar free (in certain products), suitable for diabetics, reduced glycaemic load and even tooth-friendly.

The freezing point of a frozen dessert depends on the quantity of dissolved solids. The higher the level of dissolved solids, the lower the freezing point of the product. The absolute freezing point of the mixture is determined by the molar concentration. The depressive effect on the freezing point produced by a single sugar or sweetener is referred to as the freezing point depression factor (FPDF). Sucrose has been chosen as the datum point and given a FPDF value of 1.0. Xylitol having a lower molecular weight than sucrose consequently has a higher FPDF value of approximately 2.2.

The texture of a frozen dessert is also influenced by the FPDF. The higher the FPDF of the mixture as a whole, the softer and more 'scoopable' the finished frozen dessert will be. Xylitol will increase the FPDF of a mixture more effectively than sucrose, and will therefore produce a softer and more scoopable finished product.

16.5.6 Baked goods

Xylitol is suitable for use in most forms of commercial- and domestic-baked goods applications, including biscuits, breads and cakes. The use of polyols in these applications permits the development of no-sugar-added and reduced sugar products, with at least some of the added health benefits discussed previously. Xylitol's most obvious advantage over other polyols and sugar-free bulk sweeteners in baked goods is its sucrose-like sweetness intensity and sweetness profile, meaning that the addition of intense sweeteners is rarely required if xylitol is used to replace sucrose. In fact, it has been reported that xylitol can be used as a direct replacement for sucrose in sponge cakes and other baked goods without adversely affecting the quality of the product, and while requiring only minimal, if any, adjustments to the original recipe.[163,164] Obviously, as with all polyols, the absence of reducing groups in the chemical structure of xylitol means that it will not participate in Maillard browning reactions, and therefore some reducing sugars may be required in the recipe in order to develop a deeper colour in the finished product.

Xylitol acts as an effective humectant in cakes and muffins. By binding the moisture within the product, xylitol has been shown to improve both the texture and shelf life of baked products. Specific tests in sponge cakes have shown that xylitol can significantly improve the shelf life of the product. These tests evaluated a full range of humectants, and samples were monitored for mould growth and deterioration of organoleptic properties over the test period. The sucrose containing control product exhibited mould growth after 15 days, while the product supplemented with 5% xylitol did not begin to exhibit mould growth until day 29. The organoleptic properties of this cake were also improved, with a moister texture, improved structure and improved surface appearance.

Xylitol products typically exhibit a slightly lower physical volume than conventional sucrose-based products, which is a result of xylitol's lower molecular weight, and hence viscosity in solution. This can be advantageous in certain applications, where xylitol can be used to improve the texture of the product. For example, in cereal bars, xylitol's lower viscosity can be used to soften the texture of hard bars, making them easier to bite and chew.

Xylitol's high negative heat of solution can also be used to provide a pleasant cooling effect in biscuit creams, fillings, fondants and icings. This cooling effect can enhance the flavour of the products, and adds increased freshness and 'juiciness' to fruit-flavoured creams and fillings.

16.5.7 Non-food applications

Xylitol has also found extensive use in non-food applications, particularly within the pharmaceutical, and cosmetic and toiletries industries. In pharmaceutical products, xylitol is utilised as a sugar-free sweetener, an inert excipient and as a parenteral energy source. In cosmetics and toiletries, xylitol is most frequently used as a humectant and skin-moisturising agent, although some therapeutic applications have also been reported.

16.6 SAFETY

Xylitol has been shown to have a very low order of toxicity via all routes of administration. Conventional tests for embryotoxicity, teratogenicity and reproductive toxicity have consistently yielded negative results, as have *in vitro* and *in vivo* tests for mutagenicity and clastogenicity.[25, 26, 165, 166]

In 1983, the Joint Expert Committee on Food Additives (JECFA), a prestigious scientific advisory body to the World Health Organisation and the Food and Agricultural Organisation of the United Nations, conducted a thorough evaluation of the available toxicological and safety data pertaining to xylitol,[167] and recommended: (1) an unlimited accepted daily intake (ADI) based on the safety of xylitol (this is the safest category into which this committee can place a food additive) and (2) that no additional toxicological studies were requested or recommended. According to JECFA, the term 'ADI – Not Specified' means that:

> [O]n the basis of the available data (chemical, biochemical, toxicological and other), the total daily intake of the substance, arising from its use at the levels necessary to achieve the desired effect and from its acceptable background in foods, does not, in the opinion of the Committee, represent a hazard to health. For this reason, and for the reasons stated in the individual evaluations, the establishment of an ADI is not deemed necessary.

This statement clearly indicates JECFA's confidence that xylitol can be safely used, without the need to restrict levels of inclusion, in food applications.

The Scientific Committee for Food (SCF) of the EU also determined that xylitol is 'acceptable for dietary uses', in its report of 1985.[168] This led to the approval of xylitol for use as a sweetener and food additive in most food applications in the EU at *quantum satis* (i.e. no maximum usage level is specified).

In 1986, the FASEB was commissioned by the US FDA to review all relevant safety and toxicological data concerning xylitol and other polyols.[169] The scientific conclusions of the FASEB report indicated that the use of xylitol in humans is safe. The report stated that xylitol appears to have a safety profile equivalent to that of sorbitol and mannitol, which are both widely used in confectionery, food and pharmaceutical products. The report also reaffirmed xylitol's acceptability as an approved food additive for use in food.

The numerous studies, reviews and independent evaluations concerning the safety of xylitol, have clearly established that xylitol is a safe and acceptable food additive for human consumption.

16.7 REGULATORY STATUS

Xylitol is approved for use in food in over 50 countries worldwide, and has been safely used in the food industry for over 30 years. Xylitol is also extensively approved for use in oral hygiene products, pharmaceuticals, cosmetics and toiletries.

The US FDA has approved the use of xylitol in 'foods for special dietary purposes' (21 CFR 172.395). Such approval extends to all sugar-free and no-sugar-added applications, as well as products making other specific dietary claims.

Within the EU, the use of xylitol as a sweetener is governed by the Sweeteners Directive (94/35/EC) and for non-sweetening applications, for example functional and technological uses, by the Miscellaneous Additives Directive (95/2/EC). The two aforementioned Directives will continue to be in force up until the time that they are fully replaced by Regulation (EC) 1333/2008 of 16 December 2008 regarding food additives.

The Japanese Ministry of Health and Welfare (MOHW) approved the use of xylitol for food applications at the end of 1997.

Specifications pertaining to the quality of xylitol are set forth in the current monographs of the Food Chemicals Codex, United States Pharmacopoeia/National Formulary, European Pharmacopoeia, Japanese Pharmacopoeia and the JECFA Compendium of food additive specifications.

The European Food Safety Authority (EFSA) issued with a positive opinion regarding children and chewing gum that is sweetened with 100% xylitol as part of Article 14 of the Regulation (EC) 1924/2006 regarding health claims. In particular, the proposed wording for such a claim was published as follows:

> Chewing gum sweetened with 100% xylitol has been shown to reduce dental plaque. High content/level of dental plaque is a risk factor in the development of caries in children (Commission Regulation 1024/2009).

In April 2011, the EFSA Panel on Dietetic Products, Nutrition and Allergies (NDA) issued positive opinions concerning the link between ingestion of xylitol and the maintenance of tooth remineralisation, as well its association with reduction in post-prandial glycaemic response (*EFSA Journal* 2011; 9(4), 2076 [25 pp.]). These positive opinions are related to the Article 13(1) claims as these are defined under Regulation (EC) 1924/2006.

REFERENCES

1. K.K. Mäkinen and E. Söderling. A quantitative study of mannitol, sorbitol, xylitol and xylose in wild berries and commercial fruits. *Journal of Food Science* 1980, 45, 367–374.
2. J. Washuett, P. Riederer and E. Bancher. A qualitative and quantitative study of sugar–alcohols in several foods. *Journal of Food Science* 1973, 38, 1262–1263.
3. O. Touster. The metabolism of polyols. In: H.L. Sipple and K.W. McNutt (eds). *Sugars in Nutrition*, Academic Press, New York; 1974, pp. 229–239.
4. E. Fischer and R. Stahel. Zur kenntnis der xylose. *Berichte der Deutschen Chemischen Fessellschaft* 1891, 24, 528–539.
5. O. Touster. Essential pentosuria and the glucuronate–xyluluose pathway. *Federation Proceedings* 1960, 19, 977–983.
6. G.M. Jaffe. Xylitol – a speciality sweetener. *Sugar y Azucar* 1978, 19, 36–42.
7. C. Aminoff, E. Vanninen and T.E. Doty. The occurrence, manufacture and properties of xylitol. In: J.N. Counsell (ed.). *Xylitol*, Applied Science Publishers, London; 1978.
8. J. Shaw. Medical progress – causes and control of dental caries. *New England Journal of Medicine* 1987, 317, 996–1004.
9. E. Newburn. Sugar and dental caries: a review of human studies. *Science* 1982, 217, 418–423.
10. W.J. Loesche. The rationale for caries prevention through the use of sugar substitutes. *International Dental Journal* 1985, 35, 1–8.
11. A. Scheinin, K.K. Mäkinen and K. Ylitalo. Turku sugar studies V. Final report on the effect of sucrose, fructose and xylitol on the caries incidence in man. *Acta Odontologica Scandinavica* 1975, 33(Suppl. 70), 67–104.
12. A. Scheinin, K.K. Mäkinen, E. Tammisalo and M. Rekola. Turku sugar studies XVIII. Incidence of dental caries in relation to a 1-year consumption of xylitol chewing gum. *Acta Odontologica Scandinavica* 1975, 33(Suppl. 70), 307–316.
13. A. Scheinin, J. Banoczy and J. Szoke. Collaborative WHO xylitol field studies in Hungary. I. Three year caries activity in institutionalised children. *Acta Odontologica Scandinavica* 1985, 43, 327–347.
14. P. Isokangas. Xylitol chewing gum in caries prevention. A longitudinal study on Finnish school children. *Proceedings of the Finnish Dental Society* 1987, 83(Suppl. 1), 1–117.
15. P. Isokangas, P. Alanen, J. Tiekso and K.K. Mäkinen. Xylitol chewing gum in caries prevention: a field study in children. *Journal of the American Dental Association* 1988, 117, 315–320.
16. D. Kandelman, A. Bar and A. Hefti. Collaborative WHO xylitol field study in French Polynesia. I. Baseline prevalence and 32-month caries increment. *Caries Research* 1988, 22, 55–62.
17. P. Isokangas, J. Tiekso, P. Alanen and K.K. Mäkinen. Long term effect of xylitol chewing gum on dental caries. *Community Dentistry and Oral Epidemiology* 1989, 17, 200–203.
18. D. Kandelman and G. Gagnon. A 24-month clinical study of the incidence and progression of dental caries in relation to consumption of chewing gum containing xylitol in school preventive programs. *Journal of Dental Research* 1990, 69, 1771–1775.
19. P. Isogangas, K.K. Mäkinen, J. Tiekso and P. Alanen. Long-term effect of xylitol chewing gum in the prevention of dental caries: a follow-up 5 years after termination of a prevention program. *Caries Research* 1993, 27, 495–498.
20. K.K. Mäkinen, C.A. Bennett and P.P. Hujoel. Xylitol chewing gums and caries rates: a 40-month cohort study. *Journal of Dental Research* 1995, 74, 1904–1913.
21. P. Alanen, P. Isokangas and K. Gutmann. Xylitol candies in caries prevention: results of a field study in Estonian children. *Community Dentistry and Oral Epidemiology* 2000, 28, 218–224.
22. J.N. Counsell (ed.). *Xylitol*, Applied Science Publishers, London; 1978.
23. L. Hyvoenen and P. Koivistoinen. Food technology evaluation of xylitol. *Advances in Food Research* 1982, 28, 373–403.
24. K.K. Mäkinen. Biochemical principles of the use of xylitol in medicine and nutrition with special consideration of dental aspects. *Experientia* 1978, 30, 1–160.
25. A. Bär. Safety assessment of polyol sweeteners – some aspects of toxicology. *Food Chemistry* 1985, 16, 231–241.
26. A. Bär. Toxicological aspects of sugar alcohols – studies with xylitol. In: *Low Digestibility Carbohydrates, Proceedings of a Workshop*, TNO-CIVO Institutes Zeist, The Netherlands; 27–28 November 1986, pp. 42–50.

27. H.R. Moskowitz. The sweetness and pleasantness of sugars. *American Journal of Psychology* 1971, 84, 387–405.
28. J. Gutschmidt and G. Ordynsky. Bestimmung des Süssungsgrades von Xylit. *Deutsche Lebensmittel Rundschav* 1961, 57, 321–324.
29. M.G. Lindley, G.G. Birch and R. Khan. Sweetness of sucrose and xylitol, structural considerations. *Journal of the Science of Food and Agriculture* 1976, 27, 140–144.
30. S. Yamaguchi, T. Yoshikawa, S. Ikeda and T. Ninomiya. Studies on the taste of some sweet substances. *Agricultural and Biological Chemistry* 1970, 34, 181–197.
31. C.K. Lee. Structural functions of taste in the sugar series: taste properties, of sugar alcohols and related compounds. *Food Chemistry* 1977, 2, 95–105.
32. L. Hyvoenen, R. Kurkela, P. Koivistoinen and P. Merimaa. Effects of temperature and concentration on the relative sweetness of fructose, glucose and xylitol. *Lebensmittel Wissenschaft und Technologie* 1977, 10, 316–320.
33. S.L. Munton and G.G. Birch. Accession of sweet stimuli to receptors. I. Absolute dominance of one molecular species in binary mixtures. *Journal of Theoretical Biology* 1985, 112, 539–551.
34. G.E. Demetrakopoulos and H. Amos. Xylose and xylitol. *World Review of Nutrition and Dietetics* 1978, 32, 96–122.
35. K.H. Baessler. Absorption, metabolism and tolerance of polyol sugar substitutes. *Pharmacology and Therapeutics in Dentistry* 1978, 3, 85–93.
36. J.F. Williams, K.K. Arora and J.P. Longenecker. The pentose pathway: a random harvest. *International Journal of Biochemistry* 1987, 19, 749–817.
37. T. Wood. Physiological functions of the pentose phosphate pathway. *Cell Biochemistry and Function* 1986, 4, 241–247.
38. N.Z. Baquer, J.S. Hothersall and P. McLean. Function and regulation of the pentose phosphate pathway in brain. *Current Topics in Cellular Regulation* 1988, 29, 265–289.
39. G. Grimble. Fibre, fermentation, flora, and flatus. *Gut* 1989, 30, 6–13.
40. A.A. Salyers and J.A.Z. Leedle. Carbohydrate metabolism in the human colon. In: D.J. Hentges (ed.). *Human Intestinal Flora in Health and Disease*, Academic Press, New York; 1983, pp. 129–146.
41. J.H. Cummings. Short chain fatty acids in the human colon. *Gut* 1981, 22, 763–779.
42. H. Ruppin, *et al.* Absorption of short chain fatty acids by the colon. *Gastroenterologica* 1980, 78, 1500–1507.
43. J.H. Cummings, *et al.* Short chain fatty acids in human large intestine, portal, hepatic and venous blood. *Gut* 1987, 28, 1221–1227.
44. C.L. Skutches, *et al.* Plasma acetate turnover and oxidation. *Journal of Clinical Investigation* 1979, 64, 708–713.
45. J.H. Cummings, H.N. Englyst and H.S. Wiggins. The role of carbohydrates in lower gut function. *Nutrition Reviews* 1986, 44, 50–54.
46. N.I. McNeil. The contribution of the large intestine to energy supplies in man. *American Journal of Clinical Nutrition* 1984, 39, 338–342.
47. S.E. Fleming and D.S. Aree. Volatile fatty acids: their production, absorption, utilisation and roles in human health. *Clinics in Gastroenterology* 1986, 15, 787–814.
48. S.S. Natah, K.H. Hussien, J.A. Tuominen and V.A. Koivisto. Metabolic response to lactitol and xylitol in healthy men. *American Journal of Clinical Nutrition* 1997, 65, 947–950.
49. N.U. Nguyen, *et al.* Carbohydrate metabolism and urinary excretion of calcium and oxalate after ingestion of polyol sweeteners. *Journal of Clinical Endocrinology and Metabolism* 1993, 77, 388–392.
50. Z.-H. Tong, W.-Z. Gu and Z. Gen. Effect on plasma glucose and insulin after xylitol loading in 30 normal adults. *Zhonghau Nei Ke Za Zhi* 1987, 26(7), 420–422.
51. W. Hassinger, *et al.* The effects of equal amounts of xylitol, sucrose and starch on insulin requirements and blood glucose levels in insulin-independent diabetics. *Diabetologia* 1981, 2, 37–40.
52. C.H. Mellinghoff. Ueber die Verwendbarkeit des Xylit als Ersatzzucker bei Diabetikern. *Klinische Wochenschrift* 1961, 39, 447.
53. S. Yamagata, *et al.* Clinical application of xylitol in diabetics. In: B.L. Horecker, K. Lang and Y. Takagi (eds). *Pentoses and Pentitols*, Springer-Verlag, Berlin; 1969, pp. 316–325.
54. R.S. Sandler, W.F. Stewart, J.N. Liberman, J.A. Ricci and N.L. Zorich. Abdominal pain, bloating, and diarrhoea in the United States: prevalence and impact. *Digestive Diseases and Sciences* 2000, 45(6), 1166–1171.

55. U.C. Dubach, E. Feiner and I. Forgo. Orate Verträglichkeit von Xylit bei stoffwechselgesunden Probanden. *Schweizerische medizinische Wochenschrift* 1969, 99, 190–194.
56. H. Förster. Tolerance in the human, adults and children. In: J.N. Counsell (ed.). *Xylitol*, Applied Science Publishers, London; 1978, pp. 43–66.
57. H.K. Akerblom, T. Koivukangas, R. Puukaa and M. Mononen. The tolerance of increasing amounts of dietary xylitol in children. *International Journal for Vitamin and Nutrition Research* 1982, 22, 53–66.
58. K.H. Bässler, W. Prellwitz, V. Unbehaun and K. Lang. Xylitstoffwechsel beim Menschen. *Zur Frage von Xylit als Zucker-Ersatz beim Diabetiker. Klinische Wohenschrift* 1962, 40, 791–793.
59. F. Amador and A. Eisenstein. The effects of oral xylitol administration in human subjects. Unpublished data. Cited by: M. Brin and O.N. Miller. The safety of oral xylitol. In: H.L. Sipple and K.W. McNutt (eds). *Sugars in Nutrition*, Academic Press, New York; 1974, pp. 591–606.
60. D.P. Mertz, V. Kaiser, M. Kloepfer-Zaar and H. Beisbarth. Serumkonzentrationen versehiedener Lipide und von Harnsäure während 2 wächiger Verabreichung von Xylit. *Klinische Wochenschrift* 1972, 50, 1107–1111.
61. H. Förster, S. Boecker and A. Walther. Verwendung von Xylit als Zucker-Austauschstoff bei diabetischen Kindern. *Fortschritte der Medizin* 1977, 95, 99–102.
62. H. Förster, R. Quadbeck and U. Gottstein. Metabolic tolerance to high doses of oral xylitol in human volunteers not previously adapted to xylitol. *International Journal for Vitamin and Nutrition Research* 1982, 22, 67–88.
63. S.J. Culbert, Y.M. Wang and H.A. Fritsche Jr. Oral xylitol in American adults. *Nutrition Research* 1986, 6, 913–922.
64. H. Schiweck and S.C. Ziesenitz. Physiological properties of polyols in comparison with easily metabolisable saccharides. In: T.H Grenby (ed.). *Advances in Sweeteners*, Blackie Academic and Professional, Glasgow; 1996, pp. 56-83.
65. Salford Symposium Consensus. Consensus statements from participants of the International Symposium on Low Digestible Carbohydrates. *British Journal of Nutrition* 2001, 85(Suppl. 1), S5.
66. Federation of American Societies for Experimental Biology. The evaluation of the energy of certain sugar alcohols used as food ingredients. 1994. Unpublished data.
67. Anonymous. De energetische waarde van suikeralcoholen. *Voeding* 1987, 48, 357–368.
68. E. Karimzadegan, A.J. Clifford and F.W. Hill. A rat bioassay for measuring the comparative availability of carbohydrates and its application to legume foods, pure carbohydrates and polyols. *Journal of Nutrition* 1979, 109, 2247–2259.
69. R. Mueller-Hess, *et al*. Effects of oral xylitol administration on carbohydrate and lipid metabolism in normal subjects. *Infusiontherapie* 1975, 2, 247–252.
70. F. Gehring. Cariogenic bacteria. In: J.N. Counsell (ed.). *Xylitol*, Applied Science Publishers, London; 1978, pp. 111–132.
71. A. Arens. *Oral Health – Diet and Other Factors*, Elsevier Science B.V., Amsterdam; 1999.
72. A.J. Rugg-Gunn and J.H. Nunn. *Nutrition, Diet and Oral Health*, Oxford University Press, Oxford; 1999.
73. A. Bär. Significance and promotion of sugar substitution for the prevention of dental caries. *Lebensmittel Wissenschaft und Technologie* 1989, 22, 46–53.
74. T. Imfeld. Efficacy of sweeteners and sugar substitutes in caries prevention. *Caries Research* 1993, 27(Suppl. 1), 50–55.
75. A. Bär. Caries Prevention with Xylitol. A review of the scientific evidence. *World Review of Nutrition and Dietetics* 1988, 55, 183–209.
76. H.J. Guelzow. Ueber den anaeroben Umsatz von Palatinit durch Mikro-organismen der menschlichen Mundhöhle. *Deutsche Zahnärztliche Zeitshcrift* 1982, 37, 669–672.
77. T.N. Imfeld. Identification of low caries risk dietary components. *Monographs in Oral Science* 1983, 11, 1–198.
78. M. Rekola. Acid production from xylitol products *in vivo* and *in vitro*. *Proceedings of the Finnish Dental Society* 1988, 84, 39–44.
79. A. Maguire, A.J. Rugg-Gunn and G. Wright. Adaptation of dental plaque to metabolise maltitol compared with other sweeteners. *Journal of Dentistry* 2000, 28, 51–59.
80. K.K. Mäkinen, E. Söderling, M. Hämäläinen and P. Antonen. Effect of long-term use of xylitol on dental plaque. *Proceedings of the Finnish Dental Society* 1985, 81, 28–35.
81. F. Gehring. Mikrobiologische Untersuchungen im Rahmen der 'Turku Sugar Studies'. *Deutsche Zahnärztliche Zeitshcrift* 1977, 32, 84–88.

82. J.S. Van der Hoeven. Cariogenicity of lactitol in program-fed rats (short communication). *Caries Research* 1986, 20, 441–443.
83. A. Scheinin and K.K. Mäkinen. Turku Sugar Studies. *Acta Odontologica Scandinavica* 1975, 33(Suppl. 70), 1–349.
84. L.G. Petersson, D. Birkhed and A. Gleerup. Caries-preventive effect of dentifrices containing various types and concentrations of fluorides and sugar alcohols. *Caries Research* 1991, 25(1), 74–79.
85. J.L. Sintes, *et al.* Enhanced anticaries efficacy of a 0.243% sodium fluoride/10% xylitol/silica dentifrice: 3-year clinical results. *American Journal of Dentistry* 1995, 8(5), 231–235.
86. J.L. Sintes, *et al.* Anticaries efficacy of a sodium monofluorophosphate dentifrice containing xylitol in a dicalcium phosphate dihydrate base. A 30-month caries clinical study in Costa Rica. *American Journal of Dentistry* 2002, 15(4), 215–219.
87. A. Cobanera, A. Morasso and E. White. Xylitol–sodium fluoride: effect on plaque. *Journal of Dental Research* 1987, 66, 814.
88. M. Svanberg and D. Birkhed. Effect of dentifrices containing either xylitol and glycerol or sorbitol on mutans streptococci in saliva. *Caries Research* 1991, 25(6), 449–453.
89. S. Twetman and L.G. Petersson. Influence of xylitol in dentifrice on salivary microflora of preschool children at caries risk. *Swedish Dental Journal* 1995, 19(3), 103–108.
90. L. Jannesson, S. Renvert and D. Birkhed. Effect of xylitol in an enzyme-containing dentifrice without sodium lauryl sulfate on mutans streptococci *in vivo*. *Acta Odontologica Scandinavica* 1997, 55(4), 212–216.
91. L. Jannesson, *et al.* Effect of a triclosan-containing toothpaste supplemented with 10% xylitol on mutans streptococci in saliva and dental plaque. A 6-month clinical study. *Caries Research* 2002, 36(1), 36–39.
92. K.K. Mäkinen, E. Söderling and I. Laikko. Zuckeralkohole (Polyole) als 'aktive' Zahnpastenbestandteile. *Oralprophylaxe* 1987, 9, 115–120.
93. K.K. Mäkinen, *et al.* Conclusion and review of the 'Michigan Xylitol Programme' (1986–1995) for the prevention of dental caries. *International Dental Journal* 1996, 46, 22–34
94. P.P. Hujoel, *et al.* The optimum time to initiate habitual xylitol gum-chewing for obtaining long-term caries prevention. *Journal of Dental Research* 1999, 78(3), 797–803.
95. L. Trahan. Xylitol: a review of its actions on mutans streptococci and dental plaque – its clinical significance. *International Dental Journal* 1995, 45(1 Suppl. 1), 77–92.
96. M. Cronin, J. Gordon, R. Reardon and F. Balbo. Three clinical trials comparing xylitol- and sorbitol-containing chewing gums for their effect on supragingival plaque accumulation. *Journal of Clinical Dentistry* 1994, 5, 106–109.
97. K.K. Mäkinen and A. Scheinin. Turku sugar studies VII – principle biochemical findings on whole saliva. *Acta Odontologica Scandinavica* 1975, 33(Suppl. 70), 129–171.
98. C. Mouton, A. Scheinin and K.K. Mäkinen. Effect of xylitol chewing gum on plaque quantity and quality. *Acta Odontologica Scandinavica* 1975, 33, 251–257.
99. C. Mouton, A. Scheinin and K.K. Mäkinen. Effect on plaque of a xylitol-containing chewing gum. *Acta Odontologica Scandinavica* 1975, 33, 27–31.
100. U. Harjola and H. Liesmaa. Effects of polyol and sucrose candies on plaque, gingivitis and lactobacillus index scores. *Acta Odontologica Scandinavica* 1978, 36, 237–242.
101. E. Söderling, K.K. Mäkinen and C.Y. Chen. Effect of sorbitol, xylitol and xylitol/sorbitol chewing gums on dental plaque. *Caries Research* 1989, 23, 378–384.
102. J.M. Gordon, M. Cronin and R.C. Reardon. Ability of a xylitol chewing gum to reduce plaque accumulation. *Journal of Dental Research* 1990, 69, 136.
103. E. Söderling, P. Isokangas and J. Tenovuo. Long-term xylitol consumption and mutans streptococci in plaque and saliva. *Caries Research* 1991, 25, 153–157.
104. L.M. Steinberg, F. Odusola and I.D. Mandel. Remineralizing potential, antiplaque and antigingivitis effects of xylitol and sorbitol sweetened chewing gum. *Clinical Preventive Dentistry* 1992, 14(5), 31–34.
105. K.P. Isotupa, S. Gunn and C.Y. Chen. Effect of polyol gums on dental plaque in orthodontic patients. *American Journal of Orthodontics and Dentofacial Orthopedics* 1995, 107(5), 497–504.
106. K.K. Mäkinen, K.P. Isotupa and T. Kivitompolo. Comparison of erythritol and xylitol saliva stimulants in the control of dental plaque and mutans streptococci. *Caries Research* 2001, 35, 129–135.

107. D. Simons, S. Brailsford, E.A. Kidd and D. Beighton. The effect of chlorhexidine acetate/xylitol chewing gum on the plaque and gingival indices of elderly occupants in residential homes. *Journal of Clinical Periodontology* 2001, 28(11), 1010–1015.
108. J.M. Tanzer. Xylitol chewing gum and dental caries. *International Dental Journal* 1995, 45(1 Suppl. 1), 77–92.
109. A. Scheinin and K.K. Mäkinen. Effect of sugars and sugar mixtures on dental plaque. *Acta Odontologica Scandinavica* 1972, 30, 235–257.
110. E. Söderling, L. Trahan, T. Tammiala-Salonen and L. Hakkinen. Effects of xylitol, xylitol–sorbitol, and placebo chewing gums on the plaque of habitual xylitol consumers. *European Journal of Oral Sciences* 1997, 105, 170–177.
111. T.H. Grenby, A.H. Bashaarat and K.F. Gey. A clinical trial to compare the effects of xylitol and sucrose chewing-gums on dental plaque growth. *British Dental Journal* 1982, 152, 339–343.
112. J. Bagg, T.W. MacFarlane and I.R. Poxton. Dental caries. In: J. Bagg, *et al*. (eds). *Essentials of Microbiology for Dental Students*, Oxford University Press, New York; 1999, pp. 247–258.
113. J.D. De Stoppelaar, J. Van Houte and O. Backer Dirks. The relationship between extracellular polysaccharide producing streptococci and smooth surface caries in 13 year old children. *Caries Research* 1969, 3, 190–199.
114. E. Söderling, L. Trahan, T. Tammiala-Salonen and L. Hakkinen. Effects of xylitol, xylitol–sorbitol, and placebo chewing gums on the plaque of habitual xylitol consumers. *European Journal of Oral Sciences* 1997, 105, 170–177.
115. L. Trahan, E. Söderling and M.F. Drean. Effect of xylitol consumption on the plaque-saliva distribution of mutans streptococci and the occurrence and long-term survival of xylitol-resistant strains. *Journal of Dental Research* 1992, 71, 1785–1791; Erratum *Journal of Dental Research* 1993, 72(1), 87–88.
116. F. Gehring, K.K. Mäkinen, M. Larmas and A. Scheinin. Turku sugar studies X: occurrence of polysaccharide-forming streptococci and ability of the mixed plaque microbiota to ferment various carbohydrates. *Acta Odontologica Scandinavica* 1975, 34(Suppl. 70), 223–227.
117. W.J. Loesche. The effect of sugar alcohols on plaque and saliva level of *Streptococcus mutans*. *Swedish Dental Journal* 1984, 8, 125–135.
118. J. Banoczy, M. Orsos, K. Pieihakkine and A. Scheinin. Collaborative WHO field studies in Hungary. IV. Saliva levels of *Streptococcus mutans*. *Acta Odontologica Scandinavica* 1985, 43, 367–370.
119. K.K. Mäkinen, E. Söderling and P. Isokangas. Oral biochemical status and depression of *Streptococcus mutans* in children during a 24- to 36-month use of xylitol chewing gum. *Caries Research* 1989, 23, 261–267.
120. K. Wennerholm, J. Arends and D. Birkhed. Effect of xylitol and sorbitol in chewing gums on mutans streptococci, plaque pH and mineral loss of enamel. *Caries Research* 1994, 28, 48–54.
121. D. Simons, E.A. Kidd, D. Beighton and B. Jones. The effect of chlorhexidine/xylitol chewing gum on cariogenic salivary microflora: a clinical trial in elderly patients. *Caries Research* 1997, 31, 91–96.
122. N.C. Goncalves, A. Valsecki Jr., S.L. Salvador and G.C. Bergamo. Effect of sodium fluoride mouth rinses containing xylitol and sorbitol on the number of *Streptococcus mutans* from human saliva. *Revista Panamericana de Salud Pública* 2001, 9(1), 30–34.
123. J.T. Autio. Effect of xylitol chewing gum on salivary *Streptococcus mutans* in preschool children. *Journal Dentistry for Children* 2002, 69(1), 81–86.
124. A. Maguire and A.J. Rugg-Gunn. Xylitol and caries prevention – is it a magic bullet? *British Dental Journal* 2003, 194(8), 429–436.
125. S. Assev, G. Vegarud and G. Rolla. Growth inhibition of *Streptococcus mutans* strain OMZ176 by xylitol. *Acta Pathologica et Microbiologica Scandinavica* 1980, 88, 61–63.
126. H. Tuompio, J.H. Meurman, K. Lounnatmaa and J. Linkola. Effect of xylitol and other carbon sources on the cell wall of *Streptococcus mutans*. *Scandinavian Journal of Dental Research* 1983, 91, 17–25.
127. L. Trahan, M. Bareil, L. Gauthier and C. Vadeboncoeur. Transport and phosphorylation of xylitol by a fructose phosphotransferase system in *Streptococcus mutans*. *Caries Research* 1985, 19, 53–63.
128. A.A. Scheie, O. Fejerskov, S. Assev and G. Rolla. Ultrastructural changes in *Streptococcus sobrinus* induced by xylitol, NaF and $ZnCl_2$. *Caries Research* 1989, 23, 320–327.
129. A. Pihlanto-Leppala, E. Söderling and K.K. Mäkinen. Expulsion mechanism of xylitol 5-phosphate in *Streptococcus mutans*. *Scandinavian Journal of Dental Research* 1990, 98, 112–119.
130. E. Söderling and A. Pihlanto-Leppala. Uptake and expulsion of 14C-xylitol by xylitol-cultured *Streptococcus mutans* ATCC 25175 *in vitro*. *Scandinavian Journal of Dental Research* 1989, 97, 511–519.

131. A.H. Rogers, K.A. Pilowsky, P.S. Zilm and N.J. Gully. Effects of pulsing with xylitol on mixed continuous cultures of oral streptococci. *Australian Dental Journal* 1991, 36, 231–235.
132. L. Trahan, S. Neron and M. Bareil. Intracellular xylitol-phosphate hydrolysis and efflux of xylitol in *Streptococcus sobrinus*. *Oral Microbiology and Immunology* 1991, 6, 41–50.
133. K.K. Mäkinen, P.L. Mäkinen and H.R. Pape Jr. Stabilisation of rampant caries: polyol gums and arrest of dentine caries in two long-term cohort studies in young subjects. *International Dental Journal* 1995, 45, 93–107.
134. E. Söderling, P. Isokangas, K. Pienihakkinen and J. Tenovuo. Influence of maternal xylitol consumption on acquisition of mutans streptococci by infants. *Journal of Dental Research* 2000, 79, 882–887.
135. P.W. Caufield, N.K. Childers, D.N. Allen and J.B. Hansen. Distinct bacteriocin groups correlate with different groups of *Streptococcus mutans* plasmids. *Infection and Immunity* 1985, 48, 51–56.
136. R.J. Berkowitz and H.V. Jordan. Similarity of bacteriocins of *Streptococcus mutans* from mother to infant. *Archives of Oral Biology* 1975, 20, 725–730.
137. P.W. Caufield, K. Ratanapridakul, D.N. Allen and G.R. Cutter. Plasmid-containing strains of *Streptococcus mutans* cluster within family and racial cohorts: implications for natural transmission. *Infection and Immunity* 1988, 56, 3216–3220.
138. B. Kohler, I. Andreen and B. Jonsson. The effect of caries-preventative measures in mothers on dental caries and the oral presence of the bacteria *Streptococcus mutans* and lactobacilli. *Archives of Oral Biology* 1984, 29, 879–883.
139. B. Kohler, I. Andreen and B. Jonsson. The earlier the colonization by mutans streptococci, the higher the caries prevalence at 4 years of age. *Oral Microbiology and Immunology* 1988, 3, 14–17.
140. B. Kohler and I. Andreen. Influence of caries-preventive measures in mothers on cariogenic bacteria and caries experience in their children. *Archives of Oral Biology* 1994, 39, 907–911.
141. P. Isokangas, E. S9Aderling, K. Pienihakkinen and P. Alanen. Occurrence of dental decay in children after maternal consumption of xylitol chewing gum, a follow-up from 0 to 5 years of age. *Journal of Dental Research* 2000, 79, 1885–1889.
142. M.L. Laitala. Dental health in primary teeth after prevention of mother–child transmission of mutans streptococci. A historical cohort study on restorative visits and maternal prevention costs. Dissertation at the University of Turku; 24 September 2010.
143. K. Foster-Powell, S.H.A. Holt and J.C. Brand-Miller. International table of glycemic index and glycaemic load values: 2002. *American Journal of Clinical Nutrition* 2002, 76, 5–56.
144. M. Gracey. Intestinal microflora and bacterial growth in early life. *Journal of Pediatric Gastroenterology and Nutrition* 1982, 1, 13–22.
145. I.R. Rowland. Nutrition and gut microflora metabolism. In: I.R. Rowland (ed.). *Nutrition, Toxicity and Cancer*, CRC Press, Boston, MA; 1991, pp. 113–136.
146. G. Livesey. Health potential of polyols as sugar replacers, with emphasis on low glycaemic properties. *Nutrition Research Reviews* 2003, 16, 163–191.
147. H. Younes, C. Demigne and C. Remesy. Acidic fermentation in the caecum increases absorption of calcium and magnesium in the large intestine of the rat. *British Journal of Nutrition* 1996, 75, 301–314.
148. P. Amman, R. Rizzoli and H. Fleisch. Influence of the disaccharide lactitol on intestinal absorption and body retention of calcium in rats. *Journal of Nutrition* 1988, 118, 793–795.
149. S. Salminen, *et al.* Gut microflora interactions with xylitol in the mouse, rat and man. *Food and Chemical Toxicology* 1985, 23(11), 985–990.
150. S. Salminen and E. Salminen. Lactulose, lactic acid bacteria, intestinal microecology and mucosal protection. *Scandinavian Journal of Gastroenterology* 1997, 32(Suppl. 222), 45–48.
151. B.J. Rolls and R.A. Barnett. *Volumetrics*, Harper Collins, New York; 2000.
152. B.J. Rolls, *et al.* Volume of food consumed affects satiety in men. *American Journal of Clinical Nutrition* 1998, 67, 1170–1177.
153. S.B. Roberts. High-glycemic index foods, hunger, and obesity: is there a connection? *Nutrition Reviews* 2000, 58, 163–169.
154. R.B. Shafer, A. Levine, J.M. Marlette and J.E. Morley. Effects of xylitol on gastric emptying and food intake. *American Journal of Clinical Nutrition* 1987, 45, 744–747.
155. J.E. Blundell, P.J. Rogers and T.C. Lambert. Investigation into the effects of xylitol on appetite control. Unpublished Research Report; 1991.
156. N.A. King, S.A.S. Craig, T. Pepper and J.E. Blundell. Evaluation of the independent and combined effects of xylitol and polydextrose consumed as a snack on hunger and food intake over 10 days. *British Journal of Nutrition* 2005, 93(6), 911–915.

157. M. Uhari, T. Kontiokari, M. Koskela and M. Niemela. Xylitol chewing gum in the prevention of acute otitis media: double blind randomised trial. *British Medical Journal* 1996, 313, 1180–1184.
158. M. Uhari, T. Kontiokari and M. Niemela. A novel use of xylitol sugar in preventing acute otitis media. *Pediatrics* 1998, 102, 879–884.
159. T. Kontiokari, M. Uhari and M. Koskela. Effect of xylitol on growth of nasopharyngeal bacteria *in vitro*. *Antimicrobial Agents and Chemotherapy* 1995, 39, 1820–1823.
160. M. Uhari, T. Tapiainen and T. Kontiokari. Xylitol in preventing acute otitis media. *Vaccine* 2001, 19, 144–147.
161. T. Kontiokari, M. Uhari and M. Koskela. Antiadhesive effects of xylitol on otopathogenic bacteria. *Journal of Antimicrobial Chemotherapy* 1998, 41, 563–565.
162. L. Vernacchio, R.M. Vezina and A.A. Mitchel. Tolerability of oral xylitol solution in young children: implications for otitis media prophylaxis. *International Journal Of Pediatric Otorhinolaryngology* 2007, 71(1), 89–94.
163. A. Emodi. Polyols: chemistry and applications. In: D. Lineback and G. Inglett (eds). *Food Carbohydrates*, AVI Publication, Westport, CT; 1982.
164. F. Ronda, M. Gomez, C.A. Blanco and P.A. Caballero. Effects of polyols and nondigestible oligosaccharides on the quality of sugar free sponge cakes. *Food Chemistry* 2004, 90(4), 549–555.
165. WHO/FAO. Summary of toxicological data of certain food additives. Twenty-first report of the Joint FAO/WHO Expert Committee on Food Additives, Geneva, WHO Technical Report Series, No. 617; 1977, pp. 124–147.
166. WHO/FAO. Summary of toxicological data of certain food additives and contaminants. Twenty-second report of the Joint FAO/WHO Expert Committee on Food Additives, Geneva, WHO Technical Report Series, No. 631; 1978, pp. 28–34.
167. WHO/FAO. Evaluation of certain food additives and contaminants, Twenty-seventh report of the Joint FAO/WHO Expert Committee on Food Additives, Geneva, WHO Technical Report Series, No. 696; 1983, pp. 23–24.
168. Report of the Scientific Committee for Food concerning Sweeteners. Opinion expressed 14 September 1984. Reports of the Scientific Committee for Food (Sixteenth Series); 1985.
169. Life Sciences Research Office. Health aspects of sugar alcohols and lactose. Report prepared for the Bureau of Foods, Food and Drug Administration, Washington DC under Contract No. FDA 223–2020 by the Life Sciences Research Office, Federation of American Societies for Experimental Biology, Bethesda, MD; 1986, p. 85.

Summary Table for Part Three A brief summary of the characteristics of some reduced-calorie bulk sweeteners.

The information given in this table is taken from published sources and from chapters in this book. The brief summaries provide the reader with a guide to the important reduced-calorie bulk sweetener characteristics. It is only by using reduced-calorie sweeteners in real applications will the reader appreciate the full potential of individual or blended sweeteners.

	Erythritol (polyol-monosaccharide)	Isomalt (polyol-disaccharide)	Lactitol (polyol-disaccharide)	Maltitol (powder) (polyol-disaccharide)	Mannitol (polyol-monosaccharide)	Sorbitol (polyol-monosaccharide)	Tagatose (monosaccharide)	Xylitol (polyol-monosaccharide)	Sucrose (disaccharide)
Molecular weight	122	344	344	344	182	182	180	152	342
Relative Sweetness[a] Sucrose = 1	0.7	0.4	0.4	0.9	0.5	0.6	0.9[b]	0.95	1
Cooling effect kcal/g (negative enthalpy) very strong >−35 Strong <−35>−25 Moderate <−25>−5 Weak <−5	Very strong	Weak	Moderate	Weak	Strong	Strong	Moderate	Very strong	Weak
Melting range (°C)	119–123	145–150	95–101	144–152	165–169	93–99	133–137	92–95	160–186

Aqueous solubility %w/w @ 25°C	37	25	57	60	20	70	~56	64	67
Hygroscopicity[c] Low >90% Moderate >80<90% High <80%	Low	Low	Low	Low	Low	High	Moderate	High	Low
The physical properties of viscosity, freezing point depression and osmotic pressure are the function of molecular weight. Generally, the higher the molecular weight, the higher the viscosity in solution, the lower the freezing point depression, the lower the osmotic pressure at equivalent concentrations and temperatures.									
Storage and processing stability	Very good	Very good	Very good	Very good	Very good	Very good	Conversion to various compounds atacidic and alkaline pHs. Decomposes more readily than sucrose at high temperatures	Very good	Hydrolyses at acidic and alkaline pH's. Caramelises at elevated temperatures

(Continued)

Summary Table for Part Three (Continued)

	Erythritol	Isomalt	Lactitol	Maltitol	Mannitol	Sorbitol	Tagatose	Xylitol	Sucrose
	Possible product labelling associated with the use of reduced-calorie bulk sweeteners include the following (not exhaustive): Diet, Light, Low carb, Reduced calorie, Low calorie, No-added sugar, Sugar free, Tooth-friendly, Reduced glycaemic response (diabetic suitability) and Prebiotic. It is advised that local expert opinion is obtained regarding the pertinent food regulations and labelling implications if these 'claims' are to be used on food products.								
Cariogenicity	Non-cariogenic	Non-cariogenic	Non-cariogenic	Non-cariogenic	Non-cariogenic	Non-cariogenic	Non-cariogenic	Non-cariogenic and cariostatic	Cariogenic
Energy value (USA) kcal/g	0.2	2	2	2.1	1.6	2.6	1.5	2.4	4
Energy value (EU) kcal/g		2.4	2.4	2.4	2.4	2.4		2.4	4
Energy value (Japan) kcal/g	0	2	2	2	2	3		3	4
Relative Glycaemic Response (RGR) versus glucose (100) Low <55 Moderate 55–70	Low	Low	Low	Low	Low	Low	Low	Low	Moderate

Regulatory status	Approved in Japan, Australia, New Zealand, Singapore, Korea, Russia, Israel, South Africa, Paraguay and Mexico	EU^d US^e	EU^d US^e	EU^d US^e	EU^d US^e	EU^d US^e	Self-affirmed GRAS (Generally Recognised As Safe) in Taiwan, the United States. Approved for use in Korea, Australia and New Zealand. Approved in Brazil and South Africa as a food ingredient. Approved as a Novel Food in the EU.	EU^d US^e

(Continued)

Summary Table for Part Three *(Continued)*

	Erythritol	Isomalt	Lactitol	Maltitol	Mannitol	Sorbitol	Tagatose	Xylitol	Sucrose
E Number		E953	E966	E965	E421	E420		E967	
ADI	Not specified (JECFA)	Not specified (JECFA)	Not specified (JECFA)	Not specified (JECFA)	Not specified (JECFA)	Not specified (JECFA)	Not specified (JECFA)	Not specified (JECFA)	

ADI, acceptable daily intake; GRAS, generally recognised as safe.

[a] Sweetness intensity depends on concentration, pH, temperature and the presence of other ingredients and is therefore application specific.
[b] Significant synergy and bitterness and aftertaste masking with intense sweeteners and flavours at use levels of 0.2–1%.
[c] Hygroscopicity – approximate relative humidity at which the product starts to take up moisture at 20BCC.
[d] EU – Approved as a Food Additive and its use is controlled by the Sweeteners in Food Regulations.
[e] US – Considered as food additive or GRAS.

REFERENCES

Soft Drinks World Supplement (Autumn 2004). Sweeteners.
T.H. Grenby. *Advances in Sweeteners*. Blackie Academic and Professional, Glasgow; 1996.
L. O'Brien Nabors. *Alternative Sweeteners*, 3rd edn revised and expanded. Marcel Dekker, New York; 2001.
S. Marie and J.R. Piggott. *Handbook of Sweeteners*. Blackie and Son, Glasgow; 1991.

Part Four
Other Sweeteners

17 New Developments in Sweeteners

Guy Servant and Gwen Rosenberg

Senomyx Inc., San Diego, CA, USA

17.1 SWEET TASTE MODULATORS

Consumers, medical experts and legislators are calling for more food and beverage offerings with significantly lower caloric content[1] and manufacturers are responding by reformulating an increasing number of products, replacing sugar or high-fructose corn syrup with several different types of zero-calorie sweeteners. However, many consumers believe that zero-calorie sweeteners have off-tastes or aftertastes and simply do not taste like sugar[2] and this is probably one of the biggest challenges that most diet brands are facing when trying to improve their respective market share. A new approach to sweetener alternatives has been pioneered by Senomyx Inc. located in San Diego, California. Senomyx is using novel technologies based on the science of taste receptors to discover and develop unique sweet taste modulators that can be used to significantly reduce the content of specific sweeteners in foods and beverages while maintaining the desired sweet taste.

17.2 SWEET MODULATOR TARGETS

The initial goals for Senomyx's sweet taste modulators are to reduce targeted sweeteners by at least 25% in foods and beverages while retaining the taste of the fully sweetened product. Unlike sweeteners, these modulators have no taste on their own at the recommended usage levels. Instead, they are intended to amplify or 'enhance' the taste of specific carbohydrate sweeteners such as sucrose or fructose, as well as non-nutritive sweeteners like sucralose (commonly referred to by the brand name Splenda™). Additional commercial objectives include a potency that allows use at low concentration, providing cost-effective flavour enhancement.

17.3 INDUSTRY NEED FOR REDUCED-CALORIE OFFERINGS

The overconsumption of rich foods and the lack of physical activity can cause conditions such as obesity, diabetes and cardiovascular diseases. More than a third of the American population is overweight,[3] costing tax payers more than $147 billion in annualised medical costs[4] and causing close to 10% of all deaths.[5] Organisations such as the CEO-led Healthy Weight Commitment Foundation, whose members include food and beverage manufacturers, retailers and restaurant owners, have pledged to use strategies such as product reformulation and innovation to help consumers maintain a healthy weight.[6]

Manufacturers have replaced some or all of the sugar or high-fructose corn syrup in products with a variety of zero-calorie sweeteners or sweetener blends. However, even after several decades of research, no single zero-calorie sweetener with a taste identical to sugar has been identified.[7] Many alternative sweeteners have to be used at high concentrations and these can display unwanted off-tastes (bitter, metallic, liquorice, cooling), inadequate temporal properties (slow onset and/or lingering of sweet taste), or a limited sweetness intensity at the use levels.[7–9] Senomyx has demonstrated that a low concentration of the appropriate modulator can enhance the sweetness of a product and enable a significant reduction in sucrose while preserving the natural sugar taste with no off-tastes. These highly desirable taste characteristics can meet the consumer demand for lower calorie products with an unchanged sweetness profile.

17.4 SWEET TASTE RECEPTORS

The most promising approach to identifying enhancers for a broad range of sweeteners and applications is to understand the molecular structure and function of the protein that serves as the human sweet taste receptor. Taste receptors, which are found on the surface of taste cells, interact with food ingredients and allow individuals to experience different taste sensations. Elucidation of the taste receptor construct allows rational design of new modulators that enhance the activity of the receptor when specific sweeteners are present.

Analysis of human lingual tissue has identified proteins expressed specifically in taste cells of the tongue and the soft palate.[10–14] The cells responsible for perception of sweet taste have specialised proteins that function as receptors for sweet tastants. Humans have only one type of sweet taste receptor. These 'G protein-coupled receptors' (GPCRs) are characterised by two subunits, called T1R2 and T1R3, each of which has a 'Venus flytrap' (VFT) configuration that opens and closes. When the VFT is in the open state, a sweetener can bind to the exposed pocket. The binding triggers the closing of the VFT and activates the receptor by stabilising the closed conformation. This sends a signal to the brain that is interpreted as a sweet taste sensation[15] (see Figure 17.1, left panel). Sucrose, for example, binds to a particular site in the VFT pocket of the T1R2 subunit of the receptor. When the sucrose level is low, few receptors are activated, resulting in a weak sweet taste. High levels of sucrose are needed to activate enough receptors to produce a strong sweet taste. Sweeteners other than sucrose may interact very differently with the sweet taste receptor. In fact, up to four distinct sweeteners binding domains have been identified on the sweet receptor molecule.[16–18] This explains the capacity of several dramatically diverse molecules, with equally divergent apparent affinities and potencies, to elicit a sweet taste.

17.4.1 Sweet taste modulator mechanism of action

GPCRs are now known to correspond to valid targets for allosteric modulation.[19,20] The savoury (umami) taste receptor, which also has a T1R3 subunit and shares other similarities

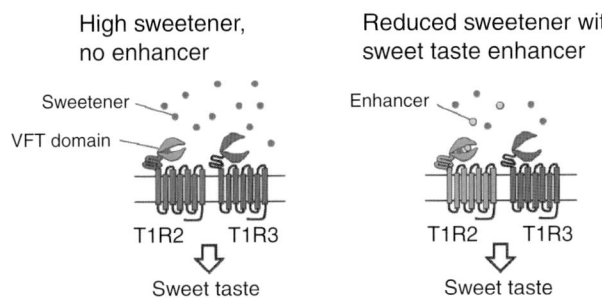

Fig. 17.1 Enhancers function as *positive allosteric modulators* of the receptor, act within the Venus flytrap (VFT) domain and effectively reduce the amount of sweetener needed to activate the receptor. Ideally, an enhancer will not activate the receptor in absence of a sweetener.

with the sweet taste receptor, is significantly enhanced by 5′-ribonucleotides.[21,22] Positive allosteric modulators (PAMs) have also been identified for the calcium sensing receptor and several other non-taste receptors.[23,24] PAMs can either act as pure enhancers (molecules enhancing the activity of an endogenous agonist but that lack intrinsic agonist activity) or as modulators with both enhancement and agonist activity.[25,26] In 2007, a team at Senomyx identified the first PAMs for the sweet taste receptor.[27] These molecules clearly enhance not only the receptor activity in the assay but also sweetness perception and they are significantly more effective than other sweetness enhancers at reducing caloric content in products without compromising taste.

Research at Senomyx indicated that these PAMs act within the VFT of the sweet taste receptor in a region near the sweetener binding site. The presence of a PAM can increase the receptor's affinity for a specific sweetener, thus lowering the amount of sweetener needed to trigger a sweet taste sensation. Therefore, sweet taste can be achieved using a Senomyx enhancer and a reduced level of the sweetener[15] (see Figure 17.1, right panel). For example, one of these sweet receptor PAMs, SE-2, significantly enhances the activity and potency of the zero-calorie sweetener sucralose in an *in vitro* assay designed to identify interactions between human sweet taste receptors and tastants.[27] In addition, in taste tests, SE-2 permits a four- to sixfold reduction is sucralose content while preserving the sweetness intensity.[27] Notably, SE-2 is not an agonist in the assay and it does not elicit any sweetness when tasted on its own. Further, evaluation revealed that SE-2 can even be used to reduce the off-tastes associated with high concentrations of sucralose in consumer products.[28]

17.4.2 Identification and evaluation of sweet taste modulators

Traditional discovery of new flavours is often achieved via taste-test evaluations of combinations of known chemical derivatives, oils and extracts of existing flavour compounds. In contrast, Senomyx has developed very sensitive and efficient *in vitro* assays that provide a new platform for identifying modulators that can provide sweet taste enhancement. These proprietary receptor-based 'high-throughput screening' assays are robotic systems that allow screening of thousands of samples per day, and hundreds of thousands of samples each year.

The receptor-based assay has been validated by characterising and quantifying the effects of more than 50 different known sweeteners.[29] This assay proved to be highly predictive, with

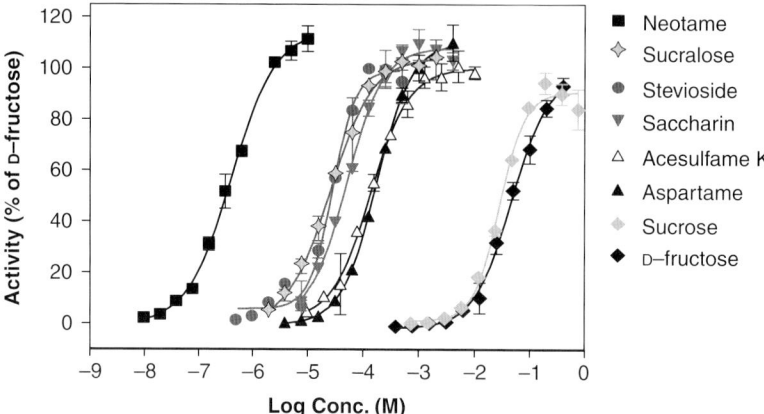

Fig. 17.2 Taste test results correlate with results in the *in vitro* assay.

higher potency sweeteners displaying higher apparent affinities than lower potency sweeteners[29] (see Figure 17.2). To find sweet receptor modulators, the assay was used to screen Senomyx's library of several hundred thousand samples in the presence of sub-optimal levels of carbohydrate sweeteners such as sucrose and fructose. Studies were also conducted with a number of commercially relevant non-caloric sweeteners including aspartame, rebaudioside A, and sucralose. SE-1 was identified while screening the sample library on sucralose. As expected for a PAM, SE-1 was not an agonist on its own but boosted the effect of a sub-optimal concentration of sucralose. As also expected for a receptor-mediated event, SE-1 effect on sucralose activity was dose dependent, very reproducible and could be observed only in the presence of sucralose. Even with concentrations as high as 75 μM, SE-1 produced little or no agonist activity by itself.[27] In addition, the effects of SE-1 were specific to the human sweet taste receptor; for example, it did not enhance the effect of L-glutamate (MSG) on the umami taste receptor. In addition, further evaluation revealed that SE-1 was surprisingly selective for sucralose as it could not enhance the activity of sub-optimal concentrations of other sweeteners including alitame, aspartame, saccharin, dulcin, stevioside, cyclamate, mogroside, perillartine, glyccyrhizic acid, NHDC and the sweet protein thaumatin.

During evaluation with human panellists, 100 μM SE-1 enhanced the sweetness intensity of a 100 ppm (251 μM) sucralose solution to that of a 300 ppm (753 μM) sucralose solution. Therefore, SE-1 permitted a reduction of the sucralose concentration by up to threefold while maintaining the sweetness intensity. SE-1 did not taste sweet by itself, confirming the *in vitro* assay results that indicated it behaved as a pure enhancer. SE-1 did exhibit a bitter off-taste when tasted on its own.[27]

17.4.3 Optimisation of sweet taste modulators

An important aspect of Senomyx's development of sweet taste modulators is the ability to optimise PAMs to create enhancers with the required physical and functional properties. Optimisation is an iterative process that involves chemical and structural modifications that are guided by screening assays and taste tests.

New Developments in Sweeteners **389**

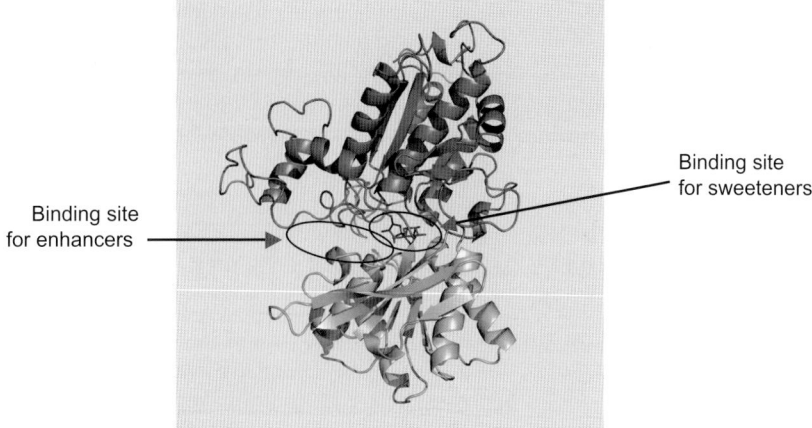

Fig. 17.3 Molecular modelling is used to better understand the binding sites for different sweeteners and to assist efforts to optimise sweetener-specific enhancers.

The remarkable sweetener selectivity demonstrated by the enhancers is related to their mechanism of action within the VFT.[15] Molecular modelling is used to better understand the binding sites for different sweeteners and to assist the efforts to optimise sweetener-specific enhancers (see Figure 17.3). Optimisation of SE-1, which included replacement of a thiol group by a ketone, produced SE-2, another potent enhancer of sucralose as described earlier (see Section 17.4.1). SE-2 was noticeably more effective than SE-1 and capable of boosting the sweetness intensities of sucralose solutions up to sixfold. Lower concentrations of sucralose were also evaluated with a fixed concentration of enhancer. Under these conditions, SE-2 permitted a $\geq 80\%$ reduction in the sucralose concentration while maintaining the desired level of sweetness intensity. Notably, when evaluated on its own, SE-2 did not taste sweet and had no bitterness. The absence of a bitter off-taste in SE-2 could be partially responsible for the improved apparent enhancement properties compared with SE-1 in taste tests. Since sucrose and sucralose are highly related molecules, it was reasoned that sucrose enhancers could potentially be optimised from the potent sucralose enhancers. Several analogues of SE-1 were synthesised, and led to the identification of SE-3 as a proof-of-concept sucrose enhancer *in vitro*.[27] SE-3 also showed promising effects in taste tests with human panellists.[27]

SE-1, SE-2 and SE-3 correspond to an entirely new type of efficient modulator of the human sweet taste receptor. Identification of these enhancers provided proof that PAMs for the sweet taste receptor may offer an alternative or complementary approach to lower the caloric content of food and beverages while maintaining the desired taste. Not only do the PAMs enhance sweetness at levels that are unprecedented but they also do not elicit the typical off-tastes or temporal issues encountered with some of the zero-calorie sweeteners.[27] In food products, enhancers such as SE-2 or its derivatives could reduce the amount of sucralose required to achieve a specific sweetness level, thereby potentially improving the palatability of sucralose-sweetened products.[8,9,30] More significantly, Senomyx's initial studies showed that further optimisation of SE-3 could lead to enhancers with increased efficacy and potency, permitting a valuable reduction in the caloric content of sucrose-sweetened consumer products. SE-4, for example, displays an advantageous specificity, allowing a reduction of up to 50% in the amount of sucrose while maintaining the sweetness of 100% sugar.[27,28] Importantly, at expected use levels SE-4 does not exhibit any bitterness or metallic off-taste nor does it impart any undesirable temporal effects. Product prototypes containing these enhancers are indistinguishable from the fully sweetened equivalent.[28,31]

17.5 COMMERCIALLY VIABLE SWEET TASTE MODULATORS

Senomyx's first commercially viable sweet taste modulator is S2383, an enhancer of sucralose. The worldwide market for sucralose, a high-intensity sweetener used in a wide variety of beverages and foods, is approximately $330 million, representing usage in categories that include confectionery, baked goods, desserts and dairy products, as well as over-the-counter (OTC) healthcare products and dietary supplements.[32]

S2383 was identified in 2007, and the magnitude of sweetness enhancement achieved with S2383 was significantly higher than that observed with previous sucralose PAMs. Taste tests demonstrated that the use of S2383 allowed a reduction of up to 75% of the sucralose in product prototypes without diminishing the sweet taste. S2383 exemplified Senomyx's ability to optimise a sweet taste modulator to provide a high degree of enhancement for a particular natural or artificial sweetener.

In August 2007, Senomyx initiated development phase activities for S2383, which included safety studies to support regulatory filings. This marked the first time that Senomyx commenced activities intended to lead to regulatory filings for a sweetness enhancer.[33]

One year later, in July 2008, Senomyx initiated development activities for S6973, a new PAM that displayed significant enhancement of sucrose. As with S2383, Senomyx's sweet-taste receptor screening assay was utilised in the discovery and optimisation of S6973. Taste tests demonstrated that S6973 did not have a sweet taste of its own, yet it increased the sweetness of product prototypes by up to twofold while maintaining the sweet taste of natural sugar[34] (see Figure 17.4). Initial sensory evaluation showed that S6973 was effective in enhancing the sweet taste of yogurt, cereal and cookie prototypes, as well as powdered and other beverages.

17.6 REGULATORY APPROVAL OF SWEET TASTE MODULATORS

In November 2008, Senomyx announced that the S2383 sucralose enhancer was determined to be generally recognised as safe (GRAS) under the provisions of the Federal Food, Drug and Cosmetic Act, administered by the US Food and Drug Administration (FDA). S2383 is the first sucralose PAM to be granted GRAS status. The GRAS designation allows S2383 to be incorporated into products in the United States and in numerous other countries. These products include a wide variety of beverages and foods such as confectionery, baked goods, desserts and dairy products, as well as OTC healthcare products and dietary supplements.[35]

The following year, in October 2009, Senomyx received GRAS determination for S6973, allowing its use in a large number of product categories. These include baked goods, cereals, gum, condiments and relishes, confectionery and frostings, frozen dairy offerings, fruit

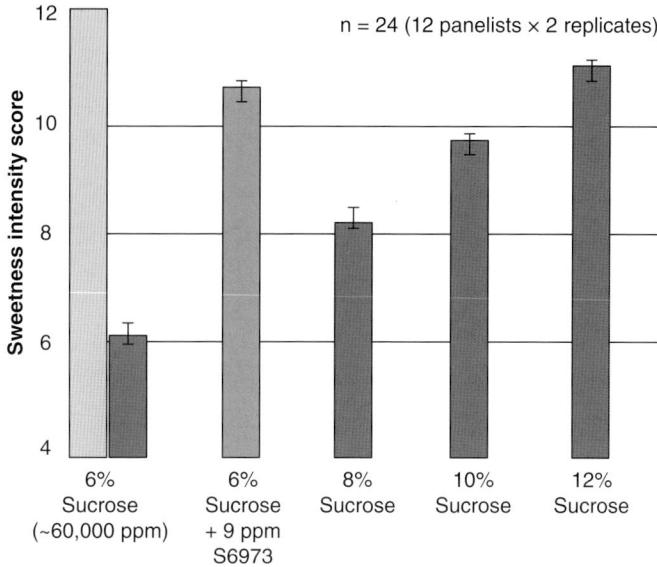

Fig. 17.4 In sensory tests, 9 ppm S6973 in 6% sucrose is as sweet as ∼12% sucrose. A 9 ppm S6973 sample without added sucrose is not sweet on its own.

ices, gelatines and puddings, hard and soft candy, jams and jellies, milk products such as flavoured milk, and sauces. Various individual product types are included in each of these broad categories.[36] The GRAS designation for S6973 was extended during the first quarter of 2010 to instant coffee and tea, and imitation dairy products including non-dairy creamers and whiteners.[37]

17.7 COMMERCIALISATION OF SWEET TASTE MODULATORS

Firmenich SA, a global leader in providing ingredients and flavour systems to major consumer companies, has exclusive worldwide rights to market S2383, Senomyx's extremely effective enhancer of sucralose, as either a stand-alone ingredient or as part of a flavour system for use in all food and beverage product categories. Firmenich is now processing orders for commercial quantities of S2383 in North America and their clients are evaluating the use of S2383 in beverages, cereal, dairy products, baked goods and confectionery products. By enabling the reduction of sucralose, S2383 may allow manufacturers to decrease their cost of goods and potentially improve the taste characteristics of certain products. Feedback from potential customers regarding the taste profile and other characteristics of S2383 has been very favourable.[38]

Firmenich also has exclusive worldwide rights to commercialise Senomyx's S6973 sucrose enhancer for virtually all food and specified beverage categories. S6973 represents a significant commercial opportunity; worldwide sales of sucrose are currently estimated to be approximately $100 billion per year.[32] By enabling a meaningful reduction in sugar content without changing the sweet taste, S6973 could allow manufacturers to offer appealing products with lower calories and improved nutritional profiles. Limiting sugar consumption is a priority for the many consumers who have dietary concerns, particularly the growing number of adults and children affected by obesity or diabetes.

In addition, S6973 is in an excellent position to benefit from other current market dynamics. These include the high price of sugar as well as the positive impact on sustainability and costs that can be achieved as a result of reduced water usage and lower transportation expenses.

Firmenich is continuing to introduce S6973 to clients for use in a wide spectrum of product categories, and evaluations by manufacturers have been very positive.[39]

17.8 FUTURE SWEET TASTE MODULATORS AND NEW SWEETENERS

PAMs for the sweet taste receptor are the first efficient sweetness enhancers ever described, and they correspond to a significant advancement in our effort to control caloric intake. Additional novel PAMs, from natural and synthetic sources and targeting other carbohydrate sweeteners are now being discovered and carefully evaluated.

Senomyx has continued to conduct high-throughput screening of additional sample libraries and optimisation of promising new PAMs for sucrose. In August 2011, S9632, a new sucrose enhancer, was advanced to the development phase and regulatory-focussed activities commenced. Taste tests have demonstrated that S9632 enables a reduction of up to 50% of the sugar in product prototypes without compromising taste. S9632 also possesses advantageous physical properties that appear to be beneficial for foods and a broader range of beverages than S6973, and its expected utilisation should be very complementary to S6973.[39]

Senomyx also has an ongoing research to discover and develop new flavour ingredients that amplify the sweet taste of fructose, a key component of high-fructose corn syrup. PAMs have been identified that demonstrated enhancement activity in Senomyx's proprietary fructose screening assays. This activity was confirmed in taste tests utilising high-fructose corn syrup, which is commonly used as a sweetener in many beverages. Optimisation of these enhancers, further screening, and additional taste tests are ongoing.[39]

In addition, Senomyx is using its sweet taste receptor assays to discover and develop novel no- or low-calorie natural high-intensity sweeteners. Activities include expansion of the company's natural products library with numerous additional samples, most recently from a variety of plant species. The natural screening and purification processes were validated by demonstrating that these methodologies can be used to identify and isolate a known natural sweetener from a complex plant extract. Multiple natural sample extracts have shown activity in the *in vitro* assay. These samples, which typically contain mixtures of three to ten or more components, are being purified in preparation for evaluation in taste tests.[40]

17.9 MODULATORS FOR OTHER TASTE MODALITIES

In addition to characterising the human sweet taste receptor, Senomyx has used its technologies to identify the structure and function of receptors associated with umami (savoury), bitter, and cooling taste sensations. As of 30 June 2011, Senomyx is the owner or exclusive licensee of 251 issued patents and several hundred pending patent applications related to proprietary taste receptor technologies in the United States, Europe and elsewhere. Technologies covered in the patents include taste receptor sequences and functions, screening assays, new flavour ingredients and product applications.[39]

17.10 SAVOURY FLAVOUR INGREDIENTS

Senomyx has developed a high-throughput screening assay based on the human savoury taste receptor. This was used to identify modulators that enhance the taste of naturally occurring glutamate and enable the reduction or elimination of added MSG and a related food additive, inosine monophosphate, or IMP. Two modulators, S336 and S807, are particularly effective in amplifying the savoury taste of glutamate.

Unlike the sweet receptor PAMs S2383 and S6973 that have no inherent sweetness, at optimal usage levels, both S336 and S807 impart a savoury taste sensation. In addition to use as enhancers, S336 and S807 can be combined with MSG or other ingredients to create new savoury blends and unique new flavours.

S336, S807, and two related savoury flavours were determined to be GRAS in March 2005. During 2007, the Chinese Ministry of Health granted official regulatory approval in China for S336 and S807. Also in 2007, both received a positive review by the Joint FAO/WHO Expert Committee on Food Additives, or JECFA. The JECFA determination facilitates the acceptance or approval of flavours for use in food in many countries. In 2010, the European Food Safety Authority, or EFSA, provided a 'favourable opinion' for S336 and S807, which means that no further evaluation is needed. Final regulatory approval and commercialisation in the European Union is contingent upon the ingredients being included in the EFSA List of Flavouring Substances, which has not yet been published.[40]

17.11 BITTER BLOCKERS

Although humans have only one type of sweet receptor and one type of bitter receptor, each person has more than 20 different T2R receptors that are associated with bitter tastes. Senomyx has screened its corporate library against a subset of the T2Rs and identified blockers for 18 different bitter receptors. The primary goals of these activities are to reduce or block bitter taste and to improve the overall taste characteristics of foods, beverages, and ingredients.[34]

Senomyx received GRAS regulatory status for S6821, which has demonstrated activity against bitter tasting foods and beverages that include soy and whey proteins, menthol, caffeine, cocoa, and Rebaudioside A (Stevia). S7958, a related bitter blocker with similar functionality, has alternative desirable physical properties that may be useful in these or other product applications. S6821 is currently under evaluation by Senomyx collaborators for potential future commercialisation.[39]

17.12 COOLING FLAVOURS

The receptor responsible for cooling taste sensations in humans, TRPM8, is unlike the GPCR receptors associated with sweet, savoury, and bitter tastes.[41] Senomyx is seeking novel cooling flavours that have advantages over currently available agents such as menthol and WS-3, which have some taste and usage limitations.

Senomyx developed a screening assay using TRPM8 that was used to identify new cooling flavours that mimic the time-intensity profile of agents that have rapid onset and short-acting cooling effects, as well as agents with slow onset and long-acting cooling effects. Several of the lead cooling flavour candidates have demonstrated approximately 10-fold greater potency than a commonly used agent in taste tests. These new cooling flavours are currently being evaluated in a variety of product prototypes.[42]

17.13 SALT TASTE MODULATORS

The protein that functions as the human salt taste receptor has proven elusive and is not yet known. Senomyx has assembled a proprietary database of proteins found in taste buds and is exploring the role of a number of these proteins that may be involved in salt taste perception. The goal of these activities is to identify the salt taste receptor and develop screening assays to facilitate the discovery of flavour ingredients that allow a significant reduction of sodium in foods and beverages yet maintain the salty taste desirable to consumers.[39]

17.14 CONCLUSIONS

Developments in the last 10 years, including the identification, isolation and preparation of taste receptors proteins, the ability to screen hundreds of thousands of compounds and then optimise identified molecules of interest, have transformed research in the field of understanding taste and identifying new compounds that will have a specific taste or modify the taste of other compounds. This technology will facilitate the identification of new sweeteners, sweetener enhancers and other taste modifying compounds and enable the formulation of great tasting products with improved nutritional profile in future years. The ability to provide satisfying sweet taste with lower calories is particularly important given new data that obesity now accounts for almost 21% of US healthcare costs, or $190.2 billion per year.[43]

REFERENCES

1. White House Task Force on Childhood Obesity Report to the President 2011, Recommendation 4.10, http://www.letsmove.gov/white-house-task-force-childhood-obesity-report-president.
2. G. Servant, *et al.* The sweet taste of true synergy: positive allosteric modulation of the human sweet taste receptor. *Trends in Pharmacological Sciences* 2011, 904, 1–6.
3. C.L. Ogden, *et al.* Prevalence and trends in overweight among US children and adolescents 1999–2000. *JAMA* 2002, 288, 1728–1732.
4. E.A. Finkelstein, *et al.* Annual medical spending attributable to obesity: payer-and service-specific estimates. *Health Aff (Millwood)* 2009, 28, w822–w831.
5. G. Danaei, *et al.* The preventable causes of death in the United States: comparative risk assessment of dietary, lifestyle, and metabolic risk factors. *PLoS Medicine* 2009, 6(4), e1000058. Doi: 10.1371/journal.pmed.1000058, http://www.plosmedicine.org/home.action.
6. Healthy Weight Commitment Foundation, Overview – http://www.healthyweightcommit.org/about/overview/.
7. G.E. DuBois. Validity of early indirect models of taste active sites and advances in new taste technologies enabled by improved models. *Flavour and Fragrance Journal* 2011, 26, 239–253. Doi: 10.1002/ffj.2042, http://onlinelibrary.wiley.com/journal/10.1002/(ISSN)1099-1026.

8. S.S. Schiffman and C.A. Gatlin. Sweeteners: state of knowledge review. *Neuroscience and Biobehavioral Reviews* 1993, 17, 313–345.
9. S.S. Schiffman, *et al.* Bitterness of sweeteners as a function of concentration. *Brain Research Bulletin* 1995, 36, 505–513.
10. M.A. Hoon, *et al.* Putative mammalian taste receptors: a class of taste-specific GPCRs with distinct topographic selectivity. *Cell* 1999, 96, 541–551.
11. M. Max, *et al.* Tas1r3, encoding a new candidate taste receptor, is allelic to the sweet responsiveness locus Sac. *Nature Genetics* 2001, 28, 58–63.
12. J.P. Montmayeur, *et al.* A candidate taste receptor gene near a sweet taste locus. *Nature Neuroscience* 2001, 4, 492–498.
13. E. Sainz, *et al.* Identification of a novel member of the T1R family of putative taste receptors. *Journal of Neurochemistry* 2001, 77, 896–903.
14. M. Kitagawa, *et al.* Molecular genetic identification of a candidate receptor gene for sweet taste. *Biochemical and Biophysical Research Communications* 2001, 283, 236–242.
15. F. Zhang, *et al.* Molecular mechanism of the sweet taste enhancers. *Proceedings of the National Academy of Sciences* 2010, 107(10), 4752–4757. www.pnas.org/cgi/doi/10.1073/pnas.0911660107.
16. M. Behrens, *et al.* Sweet and umami taste: natural products, their chemosensory targets, and beyond. *Angewandte Chemie: International Edition* 2011, 50, 2220–2242.
17. J.P. Slack. Functional method to identify tastants. US Patent Application 2009/0176266A1.
18. F. Zhang, *et al.* Molecular mechanism for the umami taste synergism. *Proceedings of the National Academy of Sciences* 2008, 105, 20930–20934.
19. P.J. Conn, *et al.* Allosteric modulators of GPCRs: a novel approach for the treatment of CNS disorders. *Nature Reviews: Drug Discovery* 2009, 8, 41–54.
20. K.J. Gregory, *et al.* Allosteric modulation of metabotropic glutamate receptors: structural insights and therapeutic potential. *Neuropharmacology* 2011, 60, 66–81.
21. X. Li, *et al.* Human receptors for sweet and umami taste. *Proceedings of the National Academy of Sciences of the United States of America* 2002, 99, 4692–4696.
22. G. Nelson, *et al.* An amino-acid taste receptor. *Nature* 2002, 416, 199–202.
23. H. Bräuner-Osborne, *et al.* Structure, pharmacology and therapeutic prospects of family C G-protein coupled receptors. *Current Drug Delivery* 2007, 8, 169–184.
24. P.J. Conn, *et al.* Allosteric modulators of GPCRs: a novel approach for the treatment of CNS disorders. *Nature Reviews: Drug Discovery* 2009, 8, 41–54.
25. T.W. Schwartz and B. Holst. Ago-allosteric modulation and other types of allostery in dimeric 7TM receptors. *Journal of Receptor and Signal Transduction Research* 2006, 26, 107–128.
26. D.W. Engers, *et al.* Synthesis, SAR and unanticipated pharmacological profiles of analogues of the mGluR5 ago-potentiator ADX-47273. *ChemMedChem* 2009, 4, 505–511.
27. G. Servant, *et al.* Positive allosteric modulators of the human sweet taste receptor enhance sweet taste. *Proceedings of the National Academy of Sciences* 2010, 107, 4746–4751.
28. R. Shigemura *et al.* Sweetener compositions and methods of making them. World Patent 2009/100333A2.
29. X. Li and G. Servant. Sweetness and Sweeteners. In: D.K. Weerasinghe and G.E. Dubois (eds). *Biology, Chemistry and Psychophysics*, Oxford University Press, Washington, DC; 2008, pp. 368–385.
30. C.T. Simons, *et al.* Sweetness and Sweeteners Biology. In: D.K. Weerasinghe and G.E. Dubois (eds). *Chemistry and Psychophysics*, Oxford University Press, Washington, DC; 2008, pp. 335–354.
31. R. Shigemura, *et al.* Composition comprising sweetness enhancers and methods of making them. World Patent 2010/014813A2.
32. Senomyx 10-K annual report filed with the US Securities and Exchange Commission on 3 March 2011.
33. Senomyx press release issued 9 August 2007, entitled, 'Senomyx Announces Second Quarter 2007 Financial Results'.
34. M. Zoller. Using Sweet and Umami Taste Receptors to Discover Taste Enhancers Keystone GPCR Meeting; 2008.
35. Senomyx press release issued 5 November 2008, entitled, 'Senomyx Receives Generally Recognized as Safe Determination for S2383 Sucralose Enhancer'.
36. Senomyx press release issued 23 October 2009, entitled, 'Senomyx Receives Generally Recognized as Safe (GRAS) Determination for S6973 Sucrose Enhancer'.
37. Senomyx press release issued 16 March 2010, entitled, 'Senomyx Announces Recent Progress in Sweet Enhancer and Bitter Blocker Programs'.

38. Senomyx press release issued 3 March 2011, entitled, 'Senomyx Announces Corporate Update and Fourth Quarter 2010 Financial Results'.
39. Senomyx press release issued 5 August 2011, entitled, 'Senomyx Announces Corporate Update and Second Quarter 2011 Financial Results'.
40. Senomyx 10-K annual report filed with the US Securities and Exchange Commission on 3 March 2011.
41. J. Chandrashekar et al. The receptors and cells for mammalian taste. *Nature* 2006, 444, 288–294.
42. Senomyx press release issued 29 April 2011, entitled, 'Senomyx Announces Corporate Update and First Quarter 2011 Financial Results'.
43. J. Cawley and C. Meyerhoefer. The medical care costs of obesity: an instrumental variables approach. *Journal of Health Economics* 2012, 31(1), 219–230.

18 Isomaltulose

Anke Sentko and Ingrid Willibald-Ettle

BENEO GmbH, Mannheim, Germany

18.1 INTRODUCTION

Palatinose™ is the trade name of isomaltulose, a functional carbohydrate composed of glucose and fructose and which is found naturally in small amounts in honey (<1%)[1,2] and in sugar cane extract.[3,4]

Isomaltulose is a disaccharide and although it has a similar composition to sugar it has different physiological properties. Isomaltulose is an isomer of sucrose and its chemical name is 6-*O*-α-D-glucopyranosyl-D-fructofuranose. In isomaltulose, the glucose and fructose moieties are linked by an α(1-6) glycosidic bond instead of the α(1-2) glycosidic bond found in sucrose. This different linkage results in completely different technological and physiological properties. The (1-6) linkage in isomaltulose is very stable and cannot be broken easily by the enzymes that occur naturally in the body. This difference results in different physiological properties such as tooth-friendliness and a low glycaemic blood glucose response. Isomaltulose is hydrolysed and absorbed slowly and completely in the small intestine. This unique metabolism provides energy in a balanced and sustained way without gastrointestinal distress.

Commercially, isomaltulose is derived from sucrose by enzymatic rearrangement (isomerisation) as shown in Figure 18.1. The isomerisation is catalysed by an immobilised enzyme preparation of non-viable cells of *Protaminobacter rubrum* followed by crystallisation. The enzyme and its production-organism were discovered by Südzucker in the 1950s.[5,6] The criteria for characterisation and purity and the corresponding analytical methods for isomaltulose are laid down in the Food Chemicals Codex.[7]

Isomaltulose can be used to partially or completely replace sugars or other readily available carbohydrates like maltodextrins in foodstuffs and by doing so the physiological profile of the product is changed.

18.2 ORGANOLEPTICAL PROPERTIES

The sweetness profile of isomaltulose is similar to sucrose without any aftertaste. Its sweetening power in a 10% solution is 0.48 related to a sucrose solution. In combination with intense sweeteners like Sucralose® or Stevia, a pleasant sweetness can be achieved.

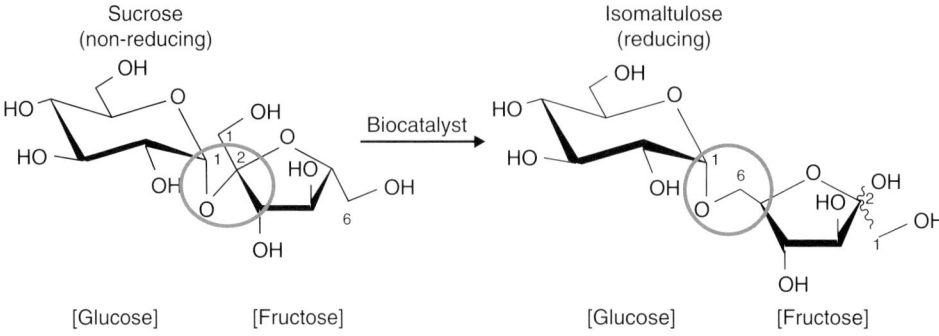

Fig. 18.1 Enzymatic rearrangement of sucrose to isomaltulose (structural formulae).
Source: Adapted from Schiweck.[5]

Isomaltulose is reported to mask the off-flavours of some intense sweeteners. Compared with sugar, it exhibits a slight cooling effect.

A range of potential sweetener combinations is shown in Table 18.1.

Table 18.1 Sweet functionality.

	Exemplary sweetening concepts	
Palatinose™ 99.97%	Sucralose 0.03%	10% Palatinose™
Palatinose™ 99.95%	Stevia[a] 0.05%	Solution and HIS to achieve an iso-sweetness of sucrose (10% solution)
Palatinose™ 99.986%	Stevia[a] 0.014%	6% Palatinose™ Solution and stevia to achieve an iso-sweetness of HFCS (6% solution)
Palatinose™ 99.930%	Stevia[a] 0.046%	4% Palatinose™ Acesulfam-K/Aspartam solution and HIS to achieve 0.024% (1:1) an iso-sweetness of sucrose (8% solution)
Palatinose™ 99.915%	Stevia[a] 0.059%	3% Palatinose™ Acesulfam-K/Aspartam solution and HIS to achieve 0.026% (1:1) an iso-sweetness of sucrose (8% solution)

Source: BENEO-Palatinit GmbH internal data.
[a]Total Rebaudioside A content: 95%.

18.3 PHYSICAL AND CHEMICAL PROPERTIES

18.3.1 Physical properties

18.3.1.1 Polymorphism/pseudopolymorphism

Like sucrose, isomaltulose is produced as a white crystalline material in different particle size distributions. Isomaltulose crystallises as a monohydrate;[8] no anhydrous crystalline phase has been found.[9] The water of crystallisation is so firmly bound that during milling or other shearing operations no release of this water is observed.[10]

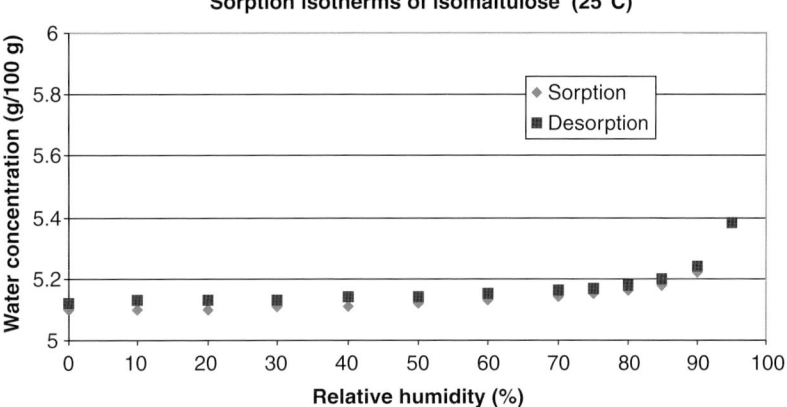

Fig. 18.2 Hygroscopicity of isomaltulose.
Source: Südzucker internal data.

18.3.1.2 Hygroscopicity

Isomaltulose has low hygroscopicity as shown by the sorption isotherms in Figure 18.2. No significant water pick-up takes place at a storage temperature at 25°C and a relative humidity (RH) of approximately 85%. Therefore, isomaltulose is ideal for moisture-sensitive applications such as chocolate and in blends with other hygroscopic ingredients it functions as an anti-caking agent.

18.3.1.3 Melting point

Like sorbitol hydrate, isomaltulose hydrate melts in its water of crystallisation. Its melting temperature (123–124°C) is lower than that of sucrose (186°C). The properties of isomaltulose crystals and amorphous isomaltulose are summarised in Table 18.2.[8–12]

Table 18.2 Physico-chemical properties of isomaltulose.

	Isomaltulose	Sucrose
Molecular mass (g/mol)	360,318 (hydrate)	342,303
Crystal group	Rhombic $P2_1P2_1P2_1$ [1]	Monoclinic [3]
specific rotation	103 … 104°	186 ± 4° [3]
melting temperature (°C)	123 … 124	46.41 [3]
heat of fusion (kJ/mol)	61 [4]	65 [3]
heat of solution (kJ/mol)	23.46 crystal [2]	6.22 crystal [2]
	−17.25 amorphous [2]	−13,79 amorphous [2]
glass transition temperature (°C)	62 [2]	65 [3]

[1] Dreissig.[8]
[2] Gehrich.[9]
[3] Bubník.[11]
[4] Südzucker. internal data.
Source: Reproduced with permission from Sentko and Bernard.[10]

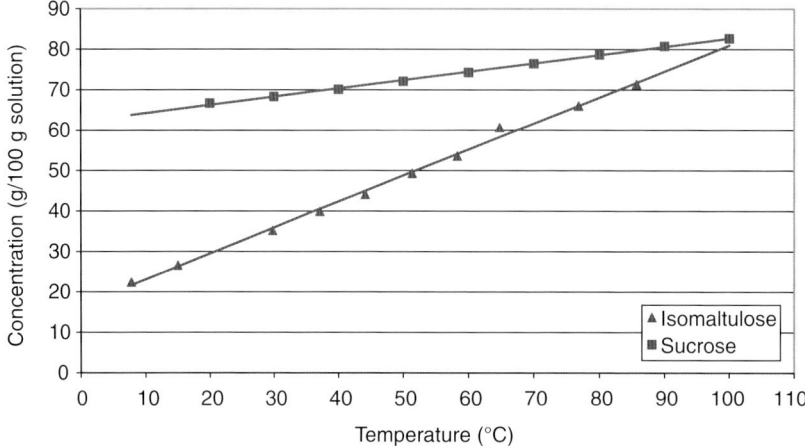

Fig. 18.3 Solubility of isomaltulose.
Source: Adapted from Schiweck.[5]

18.3.1.4 *Solubility*

The solubility of isomaltulose, as shown in Figure 18.3, although lower than that of sucrose, is adequate for most applications. The viscosity and density of isomaltulose solutions are very similar to those of sucrose.[12] Similarly, the vapour pressure (water activity) and the freezing point depression of aqueous isomaltulose solutions are similar to those of sucrose solutions.[10]

18.3.2 Chemical properties

18.3.2.1 *Acidic hydrolysis*

Isomaltulose is very acid stable compared with sucrose (see Figure 18.4). Even at pH 1 with HCl, a 10% isomaltulose solution is stable for more than 30 minutes when held at 95°C,

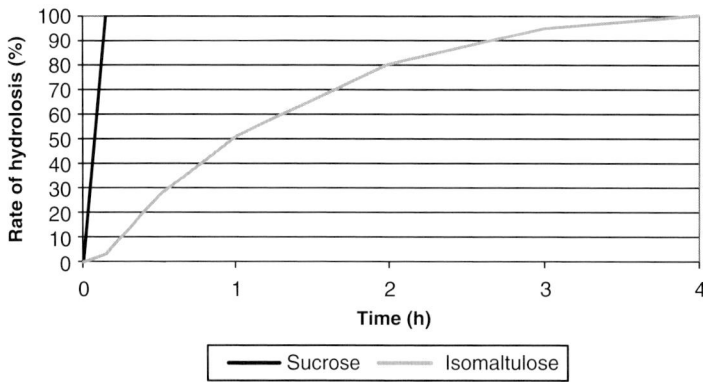

Fig. 18.4 Acid stability of isomaltulose at elevated temperature (pH 1 with HCl, 10% solution).
Source: Südzucker internal data.

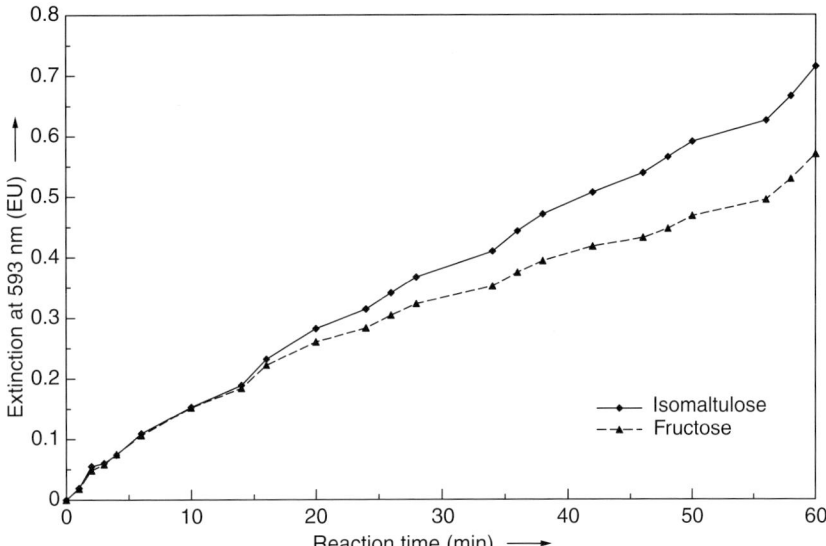

Fig. 18.5 Anti-oxidative potential of isomaltulose and fructose in a Fe(II)/Fe(III) system. *Source*: Südzucker internal data.

whereas a 10% sucrose solution is almost completely hydrolysed under these conditions. Similarly, it has been shown that an 8% (w/w) isomaltulose solution at pH 3 (typical value for beverages) is stable for at least nine months at room temperature (see Figure 18.8). One benefit of using isomaltulose in this application is that the osmolarity of the beverage is kept constant.

18.3.2.2 Heat stability

Owing to the lower melting temperature and the higher stability of the glycosidic linkage, isomaltulose shows less caramelisation during heat treatment than sucrose. Isomaltulose as a reducing sugar is subjected to Maillard reactions.

18.3.2.3 Anti-oxidative effect

With regard to anti-oxidative behaviour, isomaltulose is one of the most active saccharides. Figure 18.5 shows the difference between the reductive potential of isomaltulose versus fructose with respect to ferric salts at 60°C. Although this property is by far less developed in sugars than in classical anti-oxidative agents, a significant difference can be observed compared with other saccharides. The concentration of saccharides in food products is high compared to classical anti-oxidative compounds, and thus, the anti-oxidative capacity could be sufficient to increase the shelf life of a finished product containing isomaltulose.[10]

18.4 MICROBIOLOGICAL PROPERTIES

The microbial stability of products prepared with isomaltulose is very high. Two recent studies showed that isomaltulose could not be fermented by lactic acid bacteria or by the most common yeast used in the brewery industry (see Figure 18.6).[13,14]

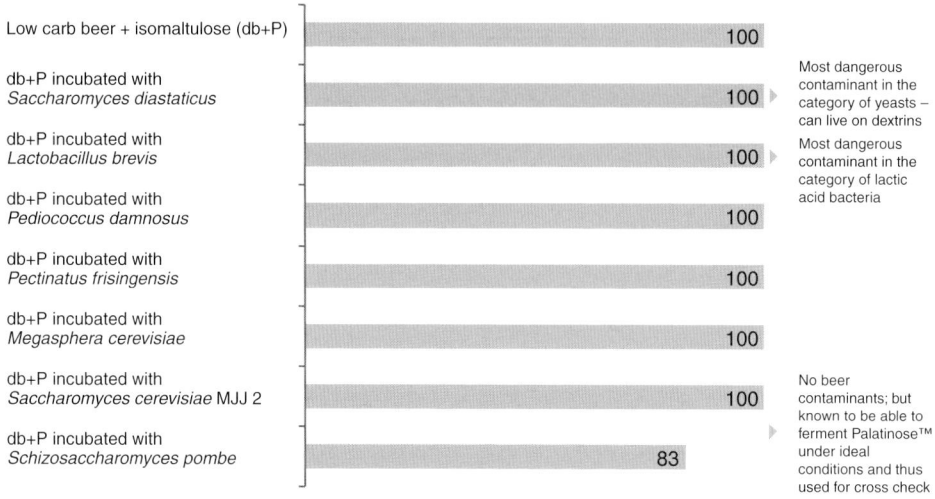

Fig. 18.6 Isomaltulose concentration (%) in beer after fermentation.
Source: Adapted from Pahl *et al.*[13]

18.5 PHYSIOLOGICAL PROPERTIES

Isomaltulose is a carbohydrate and thus a macronutrient. More than 55% of our energy intake should come from carbohydrates according to nutrition recommendations in many countries. From a food chemical point of view – and this is what is most often taken into account in food labelling requirements – carbohydrates are differentiated into mono- and disaccharides (sugars), oligosaccharides and polysaccharides. For food labelling purposes, the total amount of carbohydrates might be broken down into 'sugars' and 'starches'. From a physiological point of view – and this is what the consumer is interested in – the group of carbohydrates is more complex. Physiological differentiation starts with the way a product is digested and degree of digestibility, and availability to the body. There are carbohydrates that are fully digestible (i.e. fully hydrolysed to their monomers and absorbed in the small intestine; examples are sucrose, glucose, maltodextrin and cooked starches), those that are low-digestible (e.g. polyols) and there are non-digestible carbohydrates (i.e. they escape digestion and absorption in the small intestine, they reach the large intestine where those carbohydrates are largely used as substrates for the gut microflora (e.g. many types of dietary fibres including resistant starch). Others pass through the whole digestive system and are excreted from the body without breakdown (e.g. dietary fibres like cellulose).[15]

Fully digestible carbohydrates are assumed to provide energy very quickly to the body, reflected by a rapid increase in the blood glucose response curve after oral intake as shown in the response after intake of sucrose, glucose or wheat bread. Starches give a lower blood glucose profile compared with these products. Therefore, they are less available in that a certain amount reaches the large intestine and escapes absorption in the small intestine.

The unique feature of the disaccharide isomaltulose is related to its way of entering the metabolism and providing energy. Although it is completely hydrolysed and absorbed in the small intestine, this occurs gradually, four to five times slower than sugar.

Similar to disaccharides such as sucrose, maltose and isomaltose, isomaltulose is hydrolysed by the disaccharidase sucrose/isomaltase into its constituents, glucose and fructose.[16,17] The slower hydrolysis and absorption (as glucose and fructose) of isomaltulose results in a slower rise and lower maximum increase in blood glucose and corresponding insulin response compared with sucrose.[18,19]

Although the breakdown and absorption of isomaltulose is slower than sucrose, it is nevertheless completely metabolised in the small intestine and no significant amounts reach the large intestine. This was demonstrated in a study with healthy subjects using an end ileostomy technique. Fifty grams of isomaltulose was incorporated in a meal (breakfast) or in another trial taken as a drink, in both cases after overnight fasting. When taken as a drink, an apparent digestibility of 95.5% was determined and the apparent absorption was estimated to be 93.6%. Intake as a meal resulted in an apparent digestibility of 98.8% and apparent absorption of 96.1%. This demonstrates virtually complete digestion and absorption of isomaltulose in the small intestine, independent of the mode of intake.[19] While isomaltulose is slowly released, it is nevertheless fully available, resulting in an energy conversion factor for food labelling purposes of 4 kcal/g, which is the general value for available carbohydrates.

18.5.1 Dental health

Isomaltulose is virtually unused by the oral flora as a substrate for fermentation. No significant amounts of acids are produced, and therefore, the demineralisation process initiated when the oral environment falls below pH 5.7 (which is known to be caused by fermentation of carbohydrates) does not occur after the consumption of isomaltulose. Plaque pH telemetry, an *in vivo* method developed at the University of Zürich, enables the observation of plaque pH changes on the tooth surface during and after the ingestion of a carbohydrate-based food. This has been established as the basic method for the substantiation of dental claims in several countries including the United States and the European Union (EU).[20,21] The pH telemetry curve of isomaltulose in Figure 18.7 shows that the ingestion of isomaltulose, unlike that of sucrose, is not followed by a decrease in plaque pH below the value of 5.7.[22] Therefore, isomaltulose is tooth-friendly and can be used in the production of tooth-friendly foods if appropriate consideration is given to the food matrix (e.g. acidity, sources of fermentable carbohydrates).

Isomaltulose is included in the list of non-cariogenic carbohydrate sweeteners in the context of the US Food and Drug Administration (FDA) health claim approval on carbohydrate sweeteners and dental caries.[21,23]

In the EU, European Food Safety Authority (EFSA) evaluated isomaltulose in the context of their claims evaluation process related to claims on dental health. They concluded that consumption of food containing isomaltulose instead of other sugar(s) may help maintain tooth mineralisation by decreasing tooth demineralisation.[20] In Japan, isomaltulose has FOSHU (Food for Specified Health Uses) status owing to its dental properties.[24,25]

18.5.2 Effect on blood glucose and insulin

The slow release of isomaltulose owing to its slow hydrolysis and absorption is reflected in its blood glucose response and insulin response curves. Compared with sucrose, the blood glucose response of isomaltulose is lower, more balanced and prolonged (i.e. extended over a longer period of time) and still delivering energy while the blood glucose response curve after sugar intake would already be below the baseline. These basic characteristics were

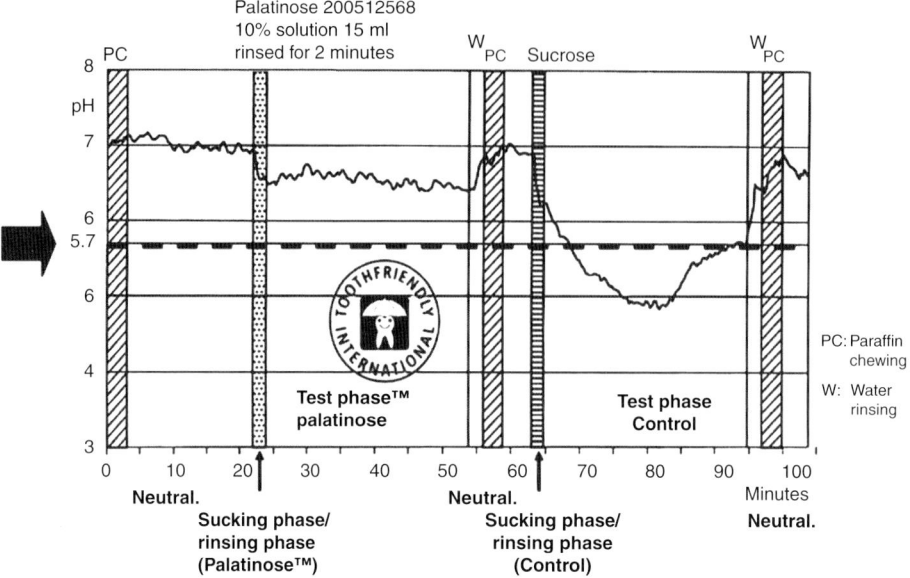

Fig. 18.7 pH Telemetry curve of isomaltulose demonstrating tooth-friendliness.
Source: Adapted from Van Loveren.[22]

demonstrated in a number of blood glucose response studies including, for instance, the glycaemic index (GI) study at Sydney University[26] and the study by Holub *et al.*[19]

The GI and insulinaemic index (II) of isomaltulose were determined at Sydney University based on an intake of 50 g isomaltulose. Using the same amount of glucose as reference, a GI of 32 and an II of 30 were found.[27]

Therefore, isomaltulose is a disaccharide carbohydrate with a very low GI and can be used in the production of carbohydrate-based foods with a low or reduced blood glucose response.[28] These physiological characteristics of isomaltulose were evaluated by EFSA in the EU claims approval process and confirmed in a positive opinion in which EFSA concluded that the consumption of foods and drinks containing isomaltulose instead of highly available sugars induces a lower blood glucose rise after meals compared to sugar-containing foods or drinks. EFSA proposed a 30% reduction of traditional sugars in this context.[20] On a long-term perspective, a diet based primarily on slowly available carbohydrates, leading to lower blood glucose and insulin levels throughout the day, may be beneficial to health.[28] Isomaltulose is also thought to be supportive in metabolic control. Findings from a 4-week intervention study by Holub *et al.*[19] show that a daily intake of 50 g isomaltulose with various foods as part of a normal diet is well tolerated and may even contribute to improved carbohydrate metabolism parameters over a longer period of time as a consequence of its lower and prolonged glycaemic response. In this study, adults with impaired fat metabolism (hyperlipidaemia) consumed 50 g isomaltulose or sucrose daily with various foods as part of a controlled diet. The regular isomaltulose consumption was well tolerated with no adverse effects on blood lipids (cholesterol, LDL, HDL, TAG, apo) or cardiovascular risk markers (oxidised LDL, NEFA). Additionally, carbohydrate metabolism parameters, namely fasting blood glucose levels and insulin resistance, were significantly lower after 4 weeks of isomaltulose intake compared with the parameters at the beginning. No such significant differences were seen with sucrose.

18.5.3 Effect on fat oxidation

As a result of its more gradual blood glucose supply at a low glycaemic level and the corresponding low insulin profile, the ingestion of isomaltulose is associated with less suppression of fat oxidation in comparison with readily available carbohydrates. Higher levels of fat oxidation with isomaltulose in comparison to conventional high glycaemic carbohydrates have been observed in physically active people as well as in overweight persons under mostly sedentary conditions.

Findings with physically active people showed that the intake of isomaltulose-containing sports drinks before and during an intense endurance-type activity on a bicycle ergometer led to a higher contribution of fat utilisation to total energy expenditure than the consumption of respective drinks with the high glycaemic carbohydrate maltodextrin.[29] With isomaltulose, the post-prandial increase in blood glucose and insulin levels was lower and concentrations of free fatty acids were higher than with maltodextrin. In parallel, the respiratory quotient (RQ), determined from gas exchange in breath using indirect calorimetry, was significantly lower throughout the entire test period, indicating a higher contribution of fat oxidation to the total energy expenditure. Also, Achten et al.[30] reported a higher contribution of fat oxidation during physical activity with isomaltulose in comparison with sucrose during a 150-minute exercise of moderate intensity by trained men. An increased contribution of fat oxidation is particularly relevant in endurance-type physical activity because it can potentially spare carbohydrates sources in glycogen stores and therefore improve endurance performance.

West et al.[31] investigated the effect of isomaltulose (75 g) in combination with a 75% reduced insulin dose in physically active type-1 diabetes mellitus subjects. More balanced blood glucose profile and increased fat oxidation rates were found. In further studies, the same research group at Swansea University demonstrated that the intake of isomaltulose 30 minutes before running improved glycaemia spared carbohydrate combustion during running and promoted fat oxidation without increasing ketogenesis during or for 3 hours after exercise.[32]

In addition, overweight people may profit from a higher contribution of fat oxidation in energy metabolism. In a study at the University of Freiburg with overweight people, with insulin resistance, the intake of meals containing isomaltulose in comparison with conventional readily available carbohydrates (i.e. a sucrose-glucose syrup combination) resulted in lower rises in blood glucose and lower daily insulin levels as well as in an increase in fat oxidation of up to 28%.[33]

The effect of isomaltulose on fat oxidation in the overweight under mostly sedentary conditions has been investigated in the study by van Can et al.,[34] which was conducted with ten healthy, overweight people in a randomised, single-blind cross-over design. In this study, the ingestion of an isomaltulose-containing drink (75 g in 400 ml) with a standardised mixed meal at lunch time caused lower glucose and insulin responses in comparison with sucrose and subsequently less inhibition of post-prandial fat oxidation. Fat oxidation rates were higher by about 14%.

18.5.4 Gastrointestinal tolerance

The gastrointestinal tolerance of isomaltulose is similar to sucrose.[18,19] As isomaltulose is almost completely digested and absorbed in the small intestine, no gastrointestinal distress occurs, as this sign of physiological overload is related to nutrients that are non- or low-digestible. Even when taken in high amounts during physical activity, no gastrointestinal discomfort was reported. Other nutritive sweeteners or carbohydrates with a low glycaemic response have this characteristic property because they are low or non-digestible. Non-digestible carbohydrates reach the large intestine where they are fermented and being osmotically active would be expected to cause gastrointestinal distress at some point. It needs to be emphasised that isomaltulose, being fully digested, causes no gastrointestinal distress.

18.6 TOXICOLOGICAL EVALUATIONS

The chemical nature of isomaltulose, the fact that it is broken down and absorbed as glucose and fructose (like sucrose but in a slower manner) together with its known physiology clearly demonstrate that isomaltulose is safe for consumption. Biochemical and toxicological studies were carried out[18] and the conclusion from these was that the use of isomaltulose causes no health concerns. This is also reflected in several assessments of the safety of isomaltulose in the legislative processes for approval from various major authorities, resulting in the confirmation of the food status of isomaltulose and its use in food in general.

18.7 APPLICATIONS

Owing to the outstanding properties of isomaltulose and its unique features – tooth-friendly, low glycaemic and low insulinaemic – it is ideal for use in a broad range of food applications where it can replace sugar on a weight per weight basis. The primary applications are beverages and tooth-friendly confectionery. Different types of isomaltulose are available under the brand name Palatinose™ produced by Beneo-Palatinit GmbH. Isomaltulose has been applied most recently in ready-to-drink (RDT) drinks like sports drinks, instant beverages, table-top application, sports and clinical nutrition, tablets, chocolates, chewing gum and coated confectionery. Existing processing equipment can be used for all applications without requiring major changes. Only formula and process parameter modifications are recommended to optimise specific processes and products.[35] Table 18.3 summarises the applications of isomaltulose (Palatinose™).

18.7.1 Beverage applications

18.7.1.1 RTD beverages

Owing to the chemical stability of isomaltulose, it is most suitable in RTD beverages such as sports drinks, soft drinks, near water or tooth-friendly beverages.[36] Isomaltulose exhibits no hydrolysis in low pH beverages for at least one year. Figure 18.8 demonstrates the stability of an isomaltulose RTD beverage compared with a sucrose based formulation.

Beverages, like sports drinks with a pH value of ~ 3, currently on the market contain sucrose which hydrolyses into glucose and fructose under these acidic conditions. This leads to an increase in the number of osmoactive ingredients in the drink and the isotonic balance is destroyed. In contrast, isomaltulose remains stable under these conditions and can act as an ingredient in isotonic, hypotonic and hypertonic beverages, as it helps to maintain the osmolality of the product.[37] This applies equally to water, fruit or dairy based products.[10]

The use of isomaltulose in soft drinks results in a sucrose-like natural sweet perception without any aftertaste. As its sweetening power is about 50% of sucrose, the sweetness may be adjusted by using a combination with other sugars like fructose or intense sweeteners.

Table 18.3 The range of isomaltulose applications.

Palatinose™ PST-N	Crystalline: 90% < 0.71 mm	• Functional beverages • Sports nutrition • Dairy products • Beer and beer specialties • Meal replacement • Clinical and special nutrition • Chocolate, cereals and bars • Confections
Palatinose™ PST-PF	Powder: 90% < 0.1 mm	• Powder drinks and blends • Coated products • Confections • Granulates and agglomerates
Palatinose™ PST-PA	Powder: 90% < 0.05 mm	
Palatinose™ PAP-N	Crystalline: 90% < 0.71 mm (additional purity)	Tooth-friendly products • Chocolate • Drinks • Coating • Confections
Palatinose™ PAP-PF	Powder: 90% < 0.1 mm (additional purity)	Tooth-friendly products • Powder drinks and blends • Coated products • Confections • Granulates and agglomerates
Palatinose™ PAP-PA	Powder: 90% < 0.05 mm (additional purity)	
Palatinose™ DC	Agglomerate	• Tablets

Source: BENEO-Palatinit GmbH internal data.

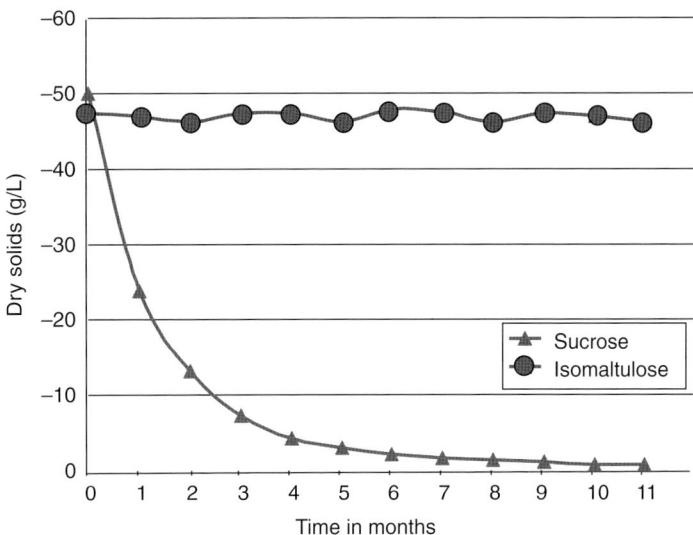

Fig. 18.8 Isomaltulose and sucrose in sports drinks after 11-months storage at 25°C and pH 2.5–2.7. *Source*: Südzucker internal data.

Isomaltulose-based soft drinks are characterised by the same natural sweetness profile as those with sugar and have a fully rounded mouthfeel.[10]

18.7.1.2 Dairy drinks

Like lactose in milk and many other sugars, isomaltulose will take part in Maillard reactions. This might result in colouration at certain high temperatures and long heating periods. However, if isomaltulose is added to, for example, milk or yogurt, this issue can be managed. For sterilisation of, for example, special or clinical nutrition products, UHT can effectively be used to avoid colouration. Generally, shorter heating periods at higher temperatures are favoured when isomaltulose is used.[38]

Because of its higher bacterial and chemical stability compared with sucrose, isomaltulose can be used as a sweetener in dairy products containing active Lactobacillus cultures with acidophilus and bifidus bacteria. These bacteria cannot break down isomaltulose and the sweetness level remains unchanged on storage.[10]

18.7.1.3 Speciality beer

Isomaltulose displays a high resistance to fermentation by most yeasts and bacteria. This is useful in the production of beer such as alcohol-reduced beers, alcohol-free beer and beer mixes, for example shandy and malt-based beverages with a non-fermentable functional bulk sweetener to increase final extract (see Figure 18.9). This results in increased palate fullness, body and an optimised true-to-type sensorial profile.[13] In speciality beers isomaltulose cannot

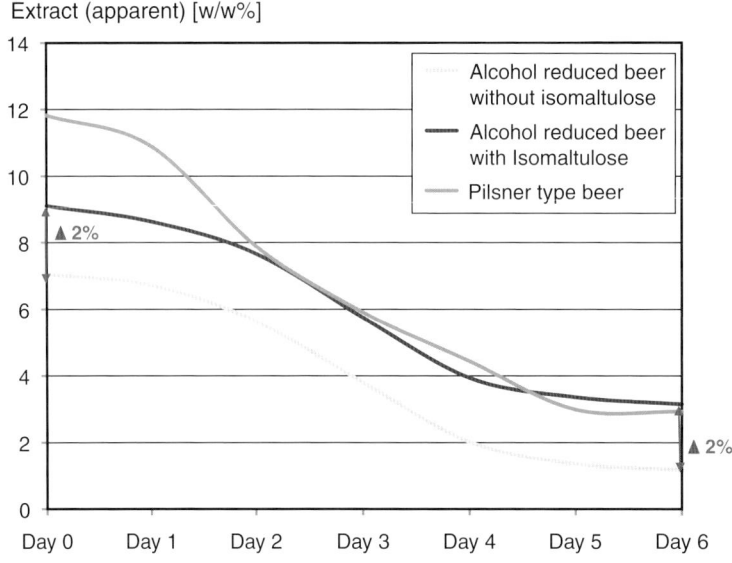

Fig. 18.9 Use of isomaltulose in beer to provide body.
Source: Adapted from Pahl *et al.*[13]

be fermented by a wide range of beer contaminants including *Lactobacillus brevis* and *Saccharomyces diastaticus*. If isomaltulose is the predominant carbohydrate in the beer, an increase in microbiological stability can be achieved. Parameters like foam stability and stability against turbidity remain unchanged. In the case of malt-based beverages, the optimum nutritive benefits of isomaltulose are predominant.[14]

Additionally, isomaltulose can be used to enhance the stability of food products that are sensitive to oxygen. This complies with the findings in different solid and liquid matrixes, for example beer, where isomaltulose significantly reduces the formation of aging products like E-2-Nonenal.[13]

18.7.1.4 Instant drinks

In these applications, isomaltulose – being a very low hygroscopic and free-flowing powder (Palatinose™ PAP-PF) or agglomerate (Palatinose™ DC) – hardly forms lumps and significantly reduces the water absorption in blends with fructose, for example. This is reflected in the moisture sorption curves of powder blends stored at 25°C/65% RH (see Figure 18.10).[10]

The glass transition temperature of isomaltulose is, at 62°C, considerably higher than fructose (10°C) or glucose (35°C). This is important during spray drying or in mixing operations where carbohydrates such as fructose, glucose may develop amorphous regions. The amorphous region leads to increased hygroscopicity affecting the product stability, especially if the glass transition temperature is low compared with room temperature. Furthermore, isomaltulose develops a very low portion of amorphous material during these treatments and the agglomerated isomaltulose combines very low hygroscopicity with excellent flowability. The water activity of isomaltulose and sucrose are about the same.[38]

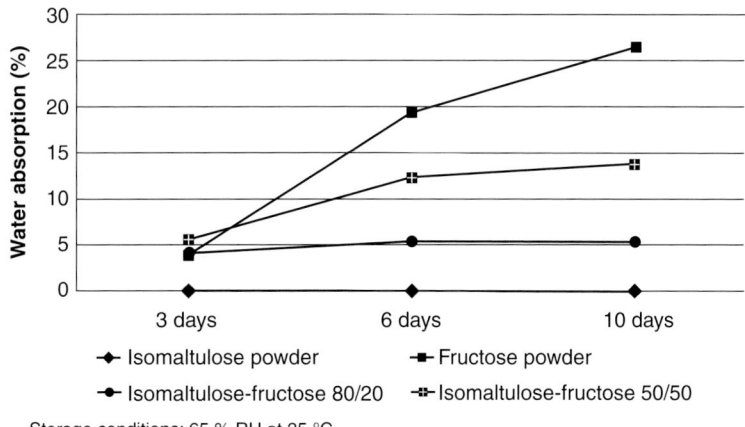

Fig. 18.10 Water uptake of isomaltulose compared with fructose and isomaltulose/fructose blends at 25°C/65% RH.
Source: Südzucker internal data.

18.7.2 Confectionery applications

18.7.2.1 Chocolate

Isomaltulose is very suitable for the manufacture of tooth-friendly chocolate and being fully digested it does not have the laxative effect of polyols. This is a big advantage for chocolate products, especially for children. Owing to its stable bound water of crystallisation, isomaltulose can be used in the traditional conching process and the production process is essentially the same as for conventional chocolate.[38] Other fermentable carbohydrate sources need to be considered and excluded from the recipe. For example, in the production of tooth-friendly milk chocolate, milk powder should be replaced by casein, as even in lactose-reduced milk powder the residual content of fermentable carbohydrates is still too high to produce tooth-friendly chocolate. Isomaltulose provides a full-bodied mouthfeel and a crispy snap that are comparable with conventional chocolate.[10]

18.7.2.2 Chewing gum

Another most interesting application of isomaltulose, especially for children, is in centre filled tooth-friendly chewing gum giving a filling that has no laxative effect. Chewing gum sticks, slabs, pellets, gum balls and coatings can be manufactured with isomaltulose[12] as it combines low hygroscopicity with the appropriate solubility to give a cost-effective coating process. Isomaltulose crystals remain stable in the chewing gum mass, leading to a soft and smooth texture.

When isomaltulose is used in gum, the stickiness of the gum mass during the production process is low. As the water uptake and the drying out of the gum during storage is also low, chewing gum with isomaltulose shows good stability with a long shelf life.[12] The slow dissolution rate of isomaltulose supports long-lasting flavour release and mild sweetness.

18.7.2.3 Coated products

Isomaltulose is an ideal ingredient for coatings, as it can be used at moderate process temperatures in hard-coating applications. The coating procedure itself depends on the type of coating equipment and on the type of centre.[12] Coated products with isomaltulose have an excellent sugar-like crunch and very good colour stability.

The low hygroscopicity of isomaltulose results in excellent shelf life of the finished products that can be packed without individual wrapping in fliptop boxes.

Isomaltulose coatings are also stable against abrasion forces experienced during the coating process. This reduces damage and results in well-defined corners and edges on the coated products.

Owing to its pronounced recrystallisation property, the coating of isomaltulose is a comparatively short process similar to sugar or even shorter.

18.7.2.4 Compressed tablets/lozenges

Isomaltulose is an interesting ingredient in direct compressible (DC) applications for the production of compressed tablets and lozenges. It is available in DC quality grades possessing both excellent flowability and compressibility. Because of its low hygroscopicity, isomaltulose tablets have good storage stability, helping to protect moisture-sensitive ingredients. Isomaltulose tablets exhibit organoleptic properties similar to sugar. Its physiological properties, delivering glucose as a source of energy over a longer period of time, make it suitable for energy products like 'dextro-energy' tablets.

18.7.2.5 Gummies

In gummies, isomaltulose can be applied in combination with glucose syrup or in tooth-friendly formulations with resistant maltodextrins (e.g. Nutriose®) at levels of about 10% w/w. Owing to their low hygroscopicity, the gummies stay dry and do not tend to stick together.[12]

18.7.2.6 Chewy candies

Non-cariogenic chewy candies with a good flavour profile can be produced with isomaltulose at levels between 30% and 35%. For the production of tooth-friendly chewy candies, a combination of isomaltulose with polydextrose can be used resulting in a fudge-like texture.[12]

18.7.3 Other applications

18.7.3.1 Extruded products

In extruded products, isomaltulose can be used in both frostings and glazes, and in both cases, isomaltulose increases the bowl life and the crispiness of the extruded flakes or balls (see Figure 18.11).[12] In general, the low glycaemic response of such products can be optimised by the replacement of sugar by isomaltulose.

18.7.3.2 Ice cream

Isomaltulose can replace sucrose in an ice cream recipe in a 1:1 ratio where it gives a similar texture to the sugar-based product. Isomaltulose depresses the freezing point to the same degree as sugar leading to similar melting behaviour as sugar-based ice cream. The advantage is that a low glycaemic product with high stability can be obtained.

18.7.3.3 Water-based ice cream (sorbet)

In this application, a tooth-friendly sherbet can be developed by replacing the sugar with isomaltulose. In combination with tagatose, this results in a sorbet with a good texture and good melting properties.[12]

18.7.3.4 Baked goods

Isomaltulose is used in baked goods for its nutritional benefit to create low glycaemic products in combination with sugar or fructose. It tends to reduce browning in the finished product compared with fructose, whilst maintaining volume and texture. Although isomaltulose is less sweet than fructose, this can be compensated by the addition of intense sweeteners or by combination with fructose.[10]

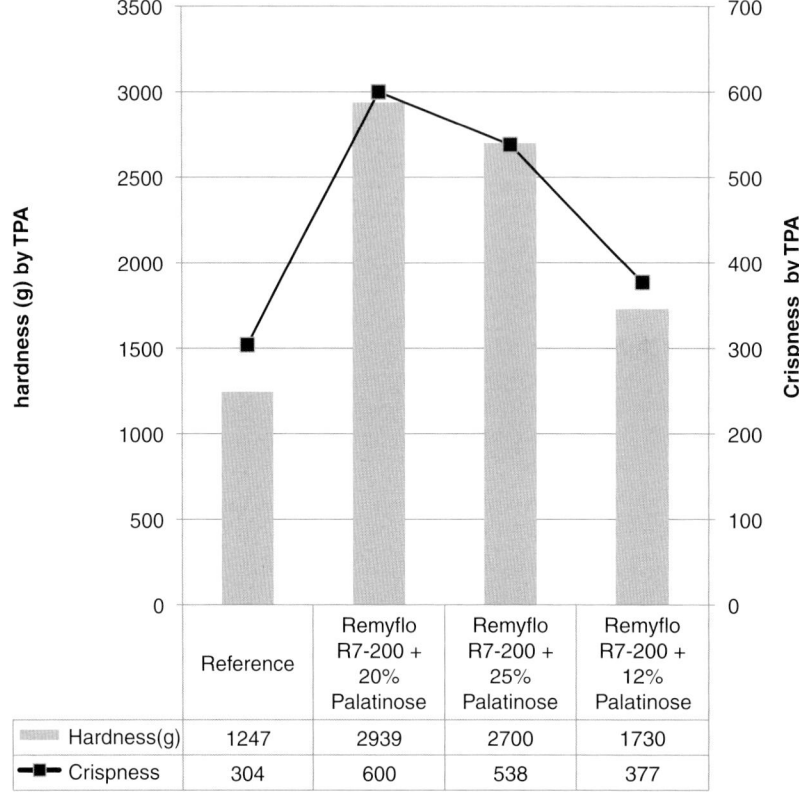

Fig. 18.11 Effect of isomaltulose on the bowl life and crispiness of extruded cereals.
Source: Südzucker internal data.

18.7.3.5 Fondant, icings, glazes

Isomaltulose is most suitable for application in icings owing to its tendency to crystallise quickly, although in combination with liquid components such as glucose syrup, the crystallisation speed can be controlled. Fondant for fillings or icings and glazings for decoration purposes as well as the protection of bars and baked goods avoiding from drying out are known applications. Icings and glazing dry very quickly owing to the rapid crystallisation of isomaltulose. This is of benefit as the baked goods, for example donuts, do not stick to the packaging material and have an extended shelf life.[10, 12]

18.7.3.6 Clinical nutrition

In clinical nutrition, isomaltulose is very beneficial owing to its low glycaemic and insulinaemic properties in the creation of calorie dense, diabetes-specific enteral products featuring a unique carbohydrate blend for enhanced glycaemic control.[39] The overall digestion of isomaltulose is essentially completed in the small intestine and no significant amounts of isomaltulose reach the large intestine.[40] Thus, isomaltulose provides the same amount of calories as all digestible carbohydrates and is equally well tolerated.

18.8 REGULATORY STATUS

Isomaltulose is a food/food ingredient like sugar or flour and is permitted in many countries. In Japan, isomaltulose has been available since 1985 and it is approved as a FOSHU ingredient owing to its dental health properties.[24,25] In the United States, a GRAS file was submitted by Südzucker/BENEO-Palatinit and this has been accepted by the FDA.[41] In 2008, the US FDA added isomaltulose to the list of nutritive sweeteners that do not promote tooth decay and a related health claim was approved in 2008.[21] Some countries have approval processes for food and food ingredients that have not previously been consumed in that country (i.e. Novel Food approvals). Dossiers were submitted in the EU and in its member states[42] and in Australia/New Zealand,[43] and isomaltulose was approved without any limitations. In Europe, the key physiological characteristics of isomaltulose, that is its reduced glycaemic response and its positive contribution to dental health have been positively evaluated in an EFSA opinion within the context of the EU health claims approval process.[20] Many countries have confirmed the food status of isomaltulose and the possible use in food for the benefit of the consumer based on the physiological properties of the product.

18.9 CONCLUSIONS

Isomaltulose (sold under the trade name Palatinose™) is a functional carbohydrate composed of glucose and fructose. It is derived from sugar by enzymatic rearrangement that leads to a stable (1-6) bond between the glucose and the fructose. The unique physical–chemical and the physiological properties, tooth-friendliness and low glycaemic response result from this strong linkage. As isomaltulose is hydrolysed and absorbed slowly and completely in the small intestine, energy is provided in a balanced and sustained way without gastrointestinal distress. With these key physiological properties, isomaltulose can contribute to healthy nutrition and isomaltulose opens up new opportunities in foods owing to its specific technological advantages. Its high stability in acidic environments and the unchanged osmolarity on storage, making stable isotonic beverages possible and its low hygroscopicity in powder applications are typical application examples.

Key application areas are beverages, instant products, sports and clinical nutrition and tooth-friendly confectionery.

REFERENCES

1. I.R. Siddiqua and B. Furgala. 1967. Isolation and characterization of oligosaccharides from honey. *Journal of Apicultural Research* 1967, 6(3), 139–145.
2. J.A. Gómez Bárez, R.J. Garcia-Villanova, S. Elvira Garcia, T. Rivas Palá, A.M. Gonzálas Paramás and J. Sánchez Sánchez. Geographical discrimination of honeys through the employment of sugar patterns and common chemical quality parameters. *European Food Research and Technology* 2000, 210, 427–444.
3. G. Egglestoneand and M. Grisham. Oligosaccharides in cane and their formation on cane deterioration 2003, 849(16), 211–232. ACS Symposium Series.
4. I. Takazoe. New Trends in Sweeteners in Japan. *International Dental Journal* 1985, 35, 58–65.
5. H. Schiweck. Palatinit® – Herstellung, technologische Eigenschaften und Analytik palatinithaltiger Lebensmittel. *Alimenta* 1980, 19. 5–16.
6. R. Weidenhagen and S. Lorenz. Palatinose (6-O-alpha-D-glucopyranosyl-D-fructofuranose), ein neues bakterielles Umwandlungsprodukt der Saccharose. *Zeitschrift für die Zückerindustrie. Fachorgan für Technik, Rübenbau und Wirtschaft* 1957, 7, 533–534.

7. Food Chemical Codex, 7th edn, Monograph on Isomaltulose; 2010.
8. W. Dreissig and P. Luger. Die Strukturbestimmung der Isomaltulose. *Acta Crystallographica* 1973, B29, 514–521.
9. K. Gehrich. *Phasenverhalten einiger Zucker und Zuckeraustauschstoffe*. PhD thesis, TU Braunschweig; 2002.
10. A. Sentko and J. Bernard. Isomaltulose. In: Lyn Nabors (ed.). *Alternative Sweeteners*. CRC Press, New York; 2012, pp. 423–438.
11. Z. Bubnik, P. Kadlec, D. Urban and M. Bruhns. *Sugar Technologists Manual*, 8th edn, Bartens Verlag, Berlin; 1995.
12. Südzuckre AG Mannheim/Ochsenfurt, CRDS, 2010, 2011, *Internal Data*.
13. R. Pahl, F.J. Methner, J. Schneider, J. Kowalczyk, S. Hausmanns and A. Radowski. Study on the applicability of isomaltulose (Palatinose™ in beer and beer specialities, and its remarkable results. *Brewing Science* 2008, 62, 49–55.
14. R. Pahl, T. Dörr, A. Radowski and S. Hausmanns. Isomaltulose: a new, non fermentable sugar in beer and beer specialities. *Beverage Technology* 2010, 2, 1–4.
15. J. Gray. Carbohydrates: nutritional and health aspects. *ILSI Europe Concise Monograph Series*; 2003.
16. A. Dahlqvist, S. Auricchio, G. Semenzaand and A. Prader. Human intestinal disaccharidases and hereditary disaccharide intolerance. The hydrolysis of sucrose, isomaltose, Palatinose (isomaltulose) and a 1,6-alpha-oligosaccharide (iso-malto-oligosaccharide) preparation. *Journal of Clinical Investigation* 1963, 42, 556.
17. P.S.J. Cheetham. The human sucrase-isomaltase complex: physiological, biochemical, nutritional and medical aspects; Chapter 5. In: C.K. Lee and M.G. Lindley (eds). *Developments in Food Carbohydrates*. Applied Science Publishers, London; 1982, pp. 107–140.
18. B.A. Lina, D. Jonker and G. Kozianowski. Isomaltulose (Palatinose): a review of biological and toxicological studies. Food and Chemical Toxicology *2002*, 40(10), 1375–1381.
19. I. Holub, A. Gostner, S. Theis, L. Nosek, T. Kudlich, R. Melcher and W. Scheppach. Novel findings on the metabolic effects of the low glycaemic carbohydrate isomaltulose (Palatinose). *British Journal of Nutrition* 2010, 103(12), 1730–1737.
20. EFSA; Scientific Opinion on the substantiation of health claims related to the sugar replacers xylitol, sorbitol, mannitol, maltitol, lactitol, isomalt, erythritol, D-tagatose, isomaltulose, sucralose and polydextrose and maintainance of tooth mineralisation by decreasing tooth demineralization (ID 463, 464, 563, 618, 647, 1182, 1591, 2907, 2921, 4300), and reduction of post-prandial glycaemic responses (ID 617, 619, 669, 1590, 1762, 2903, 2908, 2920) pursuant to Article 13(1) of Regulation (EC) No. 1924/2006. *EFSA Journal* 2011, 9(4), 2076.
21. US FDA; 21 CFR 101.80; http://www.accessdata.fda.gov/scripts/cdrh/cfdocs/cfcfr/CFRSearch.cfm?fr=101.80; 2 June 2011.
22. C. Van Loveren. Oral and dental health – revention of dental caries, erosion, gingivitis and periodontitis. ILSI Europe Concise Monograph Series, ILSI Europe; 2009.
23. Food and Drug Administration (United States of America), 21CFR§101.80 Food Labeling: Health Claims: Dietary non-cariogenic carbohydrate sweeteners and dental caries. Federal Register, Vol. 73, No. 102, pp. 30299–30301, 27 May 2008.
24. Japanese Ministry of Health, Labour and Welfare (MHLW): http://www.mhlw.go.jp/english/topics/foodsafety/fhc/02.html, 2 June 2011.
25. T. Shimizu. Health Claims on functional foods: the Japanese regulations and an international comparison. *Nutrition research Reviews* 2003, 16, 241–252.
26. Sydney University's Glycaemic Research Service (SURiRS). Glycaemic Index Report – Isomaltulose; 2002.
27. University of Sydney; http://www.glycemicindex.com/, 2 June 2011.
28. G. Livesey, R. Taylor, T. Hulshof and J. Howlett. Glycemic response and health – a systematic review and meta-analysis: relations between dietary glycemic properties and health outcomes. *American Journal of Clinical Nutrition* 2008, 87(Suppl), 258S–68S.
29. D. Koenig, W. Luther, V. Polland and A. Berg. Carbohydrates in sports nutrition – impact of the glycemic index. *AgroFood* 2007, 18(5), 9–10.
30. J. Achten, R.L. Jentjens, F. Brouns and A.E. Jeukendrup. Exogenous oxidation of isomaltulose is lower than that of sucrose during exercise in men. *Journal of Nutrition* 2007, 137(5), 1143–1148.

31. D.J. West, J.K. Stephens, S.C. Bain, L.P. Kilduff, S. Luzio, R. Still and R.M. Bracken. A combined insulin reduction and carbohydrate feeding strategy 30 min before running best preserves blood glucose concentration after exercise through improved fuel oxidation in type 1 diabetes mellitus. Journal of Sports Sciences 2011, 29(3), 279–289.
32. D.J. West, J.W. Stephens, S.C. Bain and R.M. Bracken. A combined insulin reduction & carbohydrate feeding strategy employed 30 minutes before exercise increases lipid oxidation but not ketogenesis in T1DM individuals. *Poster & Abstracts, Diabetes UK Annual Professional Conference, 30 March – 1 April 2011, London.*
33. G. Kozianowski. Physiological functionalities of the novel low glycemic carbohydrate isomaltulose (Palatinose™). *Annals of Nutrition and Metabolism* 2007, 51(Suppl 1), 157.
34. J.G. Van Can, T.H. Ijzerman, L.J. van Loon, F. Brouns and E. Blaak. Reduced glycaemic and insulinaemic responses following isomaltulose ingestion: implications for postprandial substrate use. *British Journal of Nutrition* 2009, 102(10), 1408–1413.
35. BENEO-Palatinit GmbH, Technical Sheets; 2011.
36. M. Cegelski, T. Dörr, A. Radowski and J. Schneider. Kohlenhydrate in Sportgetränken, Untersuchungen von im Markt befindlichen Spotgetränken. *Brauwelt* 2009, 7, 176–180.
37. M. Cegelski, T. Dörr, A. Radowski and J. Schneider. Kohlenhydrate in Sportgetränken, sensorische Eigenschaften und Osmolalitätsstabilität. *Brauwelt* 2009, 6, 134–139.
38. BENEO-Palatinit GmbH, Information Brochure, Figures, Facts, Benefits; 2010.
39. Abbott Laboratories Inc, *Glucerna 1.2 Cal Monograph*; 2008.
40. H.L. Lazar, et al. Tight glycemic control in diabetic coronary artery bypass graft patients improves perioperative outcomes and decreases recurrent ischemic events. *Circulation* 2004, 109, 1497–1502.
41. FDA GRAS Notice inventory, GRN 0184, submitted by Südzucker/BENEO-Palatinit; http://www.accessdata.fda.gov/scripts/fcn/fcnNavigation.cfm?filter=&sortColumn=%263%5C%2C3K%24Y%3D%0A&rpt=grasListing&displayAll=false&page=4; 3 June 2011.
42. European Commission Decision 2005/581/EC of 25 July 2005 authorising the placing on the market of isomaltulose as a novel food or novel food ingredient; 2005. http://eur-lex.europa.eu/LexUriServ/LexUriServ.do?uri=CELEX:32005D0581:EN:NOT.
43. Australia New Zealand Food Standards Code, Amendment No. 92, Gazette of the Commonwealth of Australia No. FSC 34; 2007. http://www.foodstandards.gov.au/foodstandards/changingthecode/gazettenotices/amendment922august203626.cfm.

19 Trehalose

Takanobu Higashiyama and Alan B. Richards

Hayashibara International Inc., Broomfield, CO, USA

19.1 INTRODUCTION

Trehalose is a disaccharide that occurs widely in nature and has been known to play an important role as a 'cell protectant' in many plant and animal species for over a half of a century.[1–3] Hundreds, if not thousands, of organisms have specifically evolved biosynthetic pathways to produce trehalose prior to life cycles involving exposure to heat, dehydration or freezing. The presence of relatively high concentrations of trehalose in an organism protects biological structures from degradation. Owing to this functionality, trehalose has been termed by one group of workers as 'nature's sugar of choice for preservation'.[4]

Trehalose has probably been consumed as part of the diet since humans first appeared on earth, because it is found in many different species that are believed to have been part of the primitive diet.[2,3,5] A modern diet contains several foods that contain trehalose, including mushrooms, yeast and yeast products, honey and shell fish.[2,3,5]

Trehalose consists of two glucose molecules bound by an $\alpha,\alpha(1\text{-}1)$ glycosidic linkage.[1] This linkage combines the reducing ends of both glucose molecules giving a symmetrical structure with low reactivity (Figure 19.1). Considering the functional biological properties of trehalose in a wide variety of organisms, trehalose has been identified as a sugar that might have unique properties compared with other sugars commonly used in the food industry. The main production method up to the 1990s was the extraction from yeast, which was not highly efficient. The process limited the production capacity and resulted in a price that was too expensive to be used in the cost-sensitive environment of the food industry.

In 1994, the Japanese biotechnology company, Hayashibara, identified enzymes from soil micro-organisms that specifically produced trehalose using starch as the substrate. This manufacturing method reduced the price of trehalose from approximately $300 per kilogram to about one hundredth of this and consequently lead to interest from the food industry.[5] Trehalose has been marketed in Japan as a food ingredient since 1995. Research in foods has shown trehalose to have a variety of beneficial properties, such as extension of shelf life, maintenance of texture and shape, and modification of taste.

In Japan, many companies are currently using trehalose in a wide range of commercial products under the brand name TREHA™. One reason for this is that since trehalose is found naturally in many foods in the human diet, it is considered to be natural and safe.

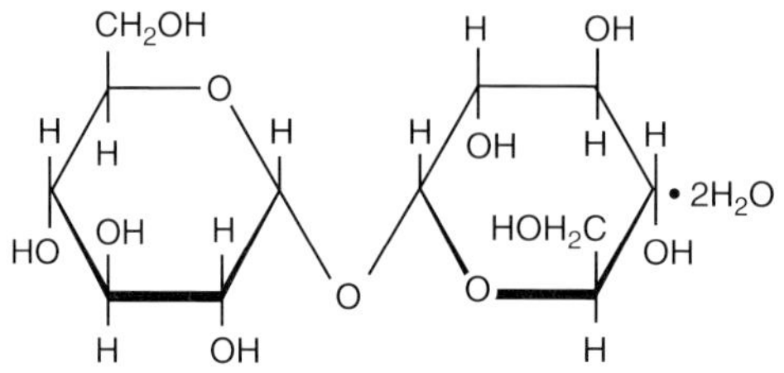

Fig. 19.1 Schematic structure of α,α-1,1 trehalose dihydrate.

The safety has been confirmed by a number of published studies. Trehalose is approved for use in foods in most countries and in addition to its growing use in food products, it is also sold into the cosmetic and pharmaceutical industries.

19.2 TREHALOSE IN NATURE

Trehalose occurs widely in nature and its biologic functions appear to be diverse. For example, trehalose is found in a number of insects where it is the energy source for flight.[6,7] In some micro-organisms, it is used as a structural component or a metabolic intermediate.[2,3] The most wildly known effect of trehalose is its ability to stabilise and protect biologically active molecules in many organisms, especially during cycles of either freezing, heat or dehydration.[1,2] One of the best-known examples of this protective function is the resurrection plant (*Selaginella lepidophylla*) that is found in the desert of the Southwestern United States and Northern Mexico. When the plant senses a lack of water, it synthesises trehalose to a concentration of about 12.5% dry weight.[8] The plant can be completely dried for several years, but upon addition of water it returns to normal activity (Figure 19.2). Brine shrimp, baker's and brewer's yeast and many other organisms also accumulate relatively high concentrations

Fig. 19.2 *Selaginella lepidophylla* or more commonly known as the 'resurrection plant' can endure years of almost total dehydration and then revive with the addition of water. One of the most important biologic activities that prevents the irreversible damage of important biologic structures is the synthesis of trehalose as the plant is in the process of dehydration.
Source: Courtesy of Hayashibara International Inc.

of trehalose that, as previously described, allows survival after exposure to heat, cold or dehydration.[9–11]

19.3 PRODUCTION

After trehalose was serendipitously shown to be present in yeast, a number of methods were developed to produce trehalose in commercially feasible amounts from this source.[5,12–15] Unfortunately, none of the methods allowed production of trehalose in an economically viable manner. In the early 1990s, Hayashibara Company, Ltd started to screen a wide range of micro-organisms for enzymatic systems that could synthesise trehalose from inexpensive raw materials. While a number of enzyme systems were isolated, the most promising involved two enzymes from a non-pathogenic, soil bacteria from the *Arthrobacter* genus. The two novel enzymes were named maltooligosyl-trehalose synthase (MTSase) and maltooligosyl-trehalose trehalohydrolase (MTHase). The combination of both enzymes enables the production of trehalose directly from starch.[16–18] As a first step, starch that can be obtained from a number of sources is treated with commonly used starch-hydrolysing enzymes (Figure 19.3). These enzymes debranch the amylopectin to produce amylose and shorten the chain length of the amylose. The MTSase recognises the reducing end of the terminal D-glucose units of amylose molecules and converts the natural $\alpha(1\text{-}4)$ linkage to an $\alpha,\alpha(1\text{-}1)$ linkage by intramolecular transglycosylation. This creates an amylose molecule with a trehalose unit as the terminal residue. The second enzyme, MTHase, then hydrolyses the $\alpha(1\text{-}4)$ glycosidic bond between the second and third D-glucose molecules, releasing the trehalose molecule into solution. This combination of multiple enzymatic reactions can repeatedly act on $\alpha(1\text{-}4)$-glucans to produce trehalose with yields of more than 80%.

Further processing of the crude trehalose solution is carried out using decolourisation, filtration and ion exchange followed by a final crystallisation step. The crystals are washed and granulated, making a product similar to table sugar (Figure 19.4), and it can also be finely

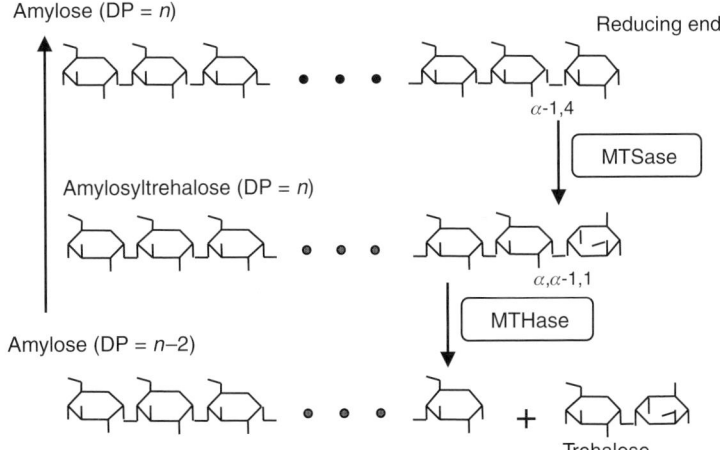

Fig. 19.3 Schematic of the trehalose production process from starch using two unique enzymes, maltooligosyl-trehalose synthase (MTSase) and maltooligo-trehalose trehalohydrolase (MTHase), as well as other enzymes common to the food industry.

Fig. 19.4 Rhombohedral crystals of trehalose produced by the Hayashibara enzymatic method.
Source: Courtesy of Hayashibara International Inc.

ground to make a microcrystalline grade. These processes and the relatively inexpensive raw materials result in a highly efficient manufacturing method, giving a product that has $\geq 98\%$ purity which can be used in price sensitive food applications, as well as in the cosmetic and pharmaceutical industries.[19,20] In Japan, trehalose is used in many traditional Japanese food products (Figure 19.5). At present, Hayashibara is the sole manufacturer of trehalose with a plant capacity in excess of 40,000 metric tons per year.

19.4 METABOLISM, SAFETY AND TOLERANCE

Since trehalose is a relatively new food ingredient this section will be slightly more detailed than might be warranted for more established sweetening agents. A detailed description of the safety and tolerance of trehalose can be found in the review and safety article by Richards *et al.*[3]

Fig. 19.5 Several traditional Japanese sweets made using trehalose to improve shelf life and texture.
Source: Courtesy of Hayashibara International Inc.

Trehalose is enzymatically hydrolysed in the small intestine, by the trehalose-specific enzyme, trehalase, into two D-glucose molecules.[21] The enzyme is located on the microvilli of the enterocytes. The glucose units released from the enzymatic process are subsequently absorbed and metabolised by the same physiological mechanisms as for maltose, sucrose and lactose.[21,22] There are no gender or age associated differences in trehalase activity.[23] There have been only a limited number of reports in the literature regarding intolerance to trehalose (and/or low trehalase activity) and it appears that it is less common (<1%) than lactose intolerance.[3,5,22] Intolerance to trehalose results in self-limiting loose stools, similar to what is seen in lactose intolerance or overconsumption of sugar alcohols. One research group reported that trehalase and lactase activity in the small intestine are similar, and that intolerance occurs when the activity of either one is <6 IU/g protein, whereas another group used trehalase activity of <5 IU/g as the limit for malabsorption.[24,25] In one case of trehalose intolerance, a number of relatives were also shown to be intolerant.[26] Additionally, low trehalase values (<8 IU/g protein) were reported in approximately 14% of 97 individuals in Greenland.[25] However, reduced lactase concentrations were observed in approximately 60% of this same population. Therefore, it appears that this condition is genetic in nature. No other specific ethnic group has been identified that has a high number of individual members intolerant to trehalose.

While no specific studies have been reported on consumption of high concentrations of trehalose in a human diet, several studies in Western populations have been published in which 50 g of trehalose has been consumed after fasting.[3,5] These data have shown little if any untoward effect in normal subjects. In addition to the human safety reports, a comprehensive series of toxicity studies were performed and reviewed by the Food and Drug Administration (FDA), Joint FAO/WHO Expert Committee on Food Additives (JECFA), European Novel Food Authority and other regulatory agencies.[3]

19.5 REGULATORY STATUS

The first regulatory approval of trehalose for food was in the United Kingdom in 1991 where its use at levels up to 5% as a cryopreservative were permitted. Trehalose was approved in Japan in 1995 as a food additive. It was affirmed in the United States as generally recognised as safe (GRAS) in 2000 and later that year the US FDA gave a letter of no objection to a GRAS Notice (GRN 000045). JECFA reviewed and approved trehalose in 2000, with an ADI (acceptable daily intake) as not specified. Regulatory approval as a novel food or food ingredient was granted in Europe in 2001. Trehalose is now approved for use by most regulatory agencies around the world.

19.6 PROPERTIES

Examination of the important role of trehalose in nature has resulted in the expectation that it might have unique and significant benefits in food systems. Table 19.1 lists several of the physical properties of trehalose that make it unique compared with other saccharides. As previously described, chemically, trehalose consists of two glucose units linked by an $\alpha,\alpha(1\text{-}1)$ glycosidic bond (Figure 19.1). It is interesting that trehalose is one of only a few symmetrical molecules found in nature and is the only common disaccharide.

Table 19.1 Properties of trehalose.

Properties	Conditions			References
Melting temperature		97 (dihydrate)	(°C)	20
		210.5 (anhydride)		
Heat of fusion		57.8 (dihydrate)	(kJ/mol)	20
		53.4 (anhydride)		
Specific rotation	5% (w/v), 20°C	+199	(°)	20
Glass transition temperature		115	(°C)	5
Solubility	20°C	68.9	(g/100 g)	20
	50°C	140.1		
	90°C	602.9		
Viscosity (25°C)	10% (w/w)	1.1	(mPa·s)	5
	20%	1.78		
	30%	2.95		
Sweetness	As sucrose (10%)	45	%	5
Osmotic pressure	10% (w/w)	298	(mOsm)	5
	20%	690		
	30%	1,229		
pH stability	100°C, 24 h, 4%(w/w)	3.5–10		20
Heat stability	120°C, pH 6.0, 10% (w/w)	>99%		20
Maillard reaction	With glycine	No reaction		20
	With peptone	No reaction		20
Hygroscopicity		Non-hygroscopic under RH 90%		20

Trehalose has three main physical properties resulting from this unique structure. The first is related to its structure. Trehalose is a very stable molecule, compared with other sugars and this results from two features in its structure.[3,5] First, the $\alpha,\alpha(1\text{-}1)$ glycosidic bond is inherently very stable and second, when the two glucopyranose rings are linked in this manner, there is no exposed glucosyl residue as in many other saccharides.[1] Therefore, trehalose is a non-reducing sugar. These two features result in a molecule that is stable in the presence of acid, alkaline and/or heat. Owing to this stability, trehalose does not caramelise like many sugars and additionally does not participate in Maillard reactions with amino acids, which are dependent on a reducing moiety (Table 19.1). The glycosidic bond is cleaved specifically by trehalase; therefore, it will not be hydrolysed by other carbohydrate-associated hydrolytic enzymes.

The second physical property that makes trehalose unique is its very hydrophilic molecular nature. Interestingly, during production and concentration by either the aqueous alcohol or the enzymatic routes, the trehalose crystal forms as a dihydrate.[1,2] The chemical formula and molecular weight are $C_{12}H_{22}O_{11}\cdot 2H_2O$ and 378.33 daltons, respectively. The dihydrate crystal of trehalose has low hygroscopicity and does not accumulate water from the atmosphere even at a relative humidity of approximately 90% (Table 19.1). Since trehalose has only one intermolecular hydrogen bond, there are more sites available to form hydrogen bonds with water or other molecules.[27] Two mechanisms have been proposed by which trehalose can protect biological systems in a dry environment. The first is by hydrogen bonding to the biological substance in place of water, thus effectively protecting it from denaturation from heat and dehydration.[5] This is called the water replacement theory. The second possible mechanism is by trapping a layer of water between the biological system and a layer of trehalose (water entrapment theory). This physical property may be potentially important in a number of food systems.[5]

The third important physical feature of trehalose is its high glass transition temperature, which is the highest of all disaccharides (Table 19.1). This allows trehalose to maintain a glassy (amorphous) structure without recrystallisation under a wide range of environmental conditions. The glass state enables trehalose to form a 'cast' around biomolecules, thereby protecting the molecules because of its high viscosity and low molecular mobility.[28] Owing to this glass-forming property, trehalose can be used to stabilise proteins and lipids against the denaturation caused by desiccation and freezing.[29–31] This is the third hypothesised mechanism, whereby trehalose can stabilise biological systems.

19.7 APPLICATION IN FOOD

19.7.1 Technical properties

The establishment of the large-scale production of trehalose has stimulated the development of new and varied applications in food systems. In addition, as described previously, trehalose is non-reducing, acid and heat resistant and has properties that can stabilise various ingredients to prevent deterioration of food quality. The close interaction with water molecules and the ability of trehalose to form multiple hydrogen-bonds has been shown to modify the water activity in food products, extending shelf life in various products being stored under heated, room temperature, refrigerator or frozen conditions. According to recent study reports, trehalose can stabilise carbohydrates and proteins. These technical features enable trehalose to add value to food products.

19.7.2 Stabilisation of carbohydrates

In Japan, rice-based, traditional confections are very common and these are made and sold locally throughout Japan. The products have a relatively long shelf life and do not require either refrigeration or added preservatives owing to their high-sugar content that reduces the water activity and avoids microbial spoilage. The current trend in Japan is a preference toward lower sweetness and manufactures are therefore trying to replace sugar with less sweet substitutes. Trehalose has a lower and milder sweetness than sucrose while maintaining low water activity (Table 19.1).[5]

During storage, starch-based products commonly become harder owing to starch retrogradation, which is described as the recrystallisation process of gelatinised starch.[32] Saccharide molecules can interact with the glucose chains in the starch matrix, slowing starch retrogradation and trehalose has been shown to inhibit retrogradation more effectively than other saccharides. Trehalose has been incorporated into many starch-based products in Japan. As with reduced water activity in rice starch products described earlier, trehalose is also used to inhibit starch retrogradation in products using other starches. In addition to Japanese confections, trehalose is commonly used in starch-based products such as western confections, noodles and bread.[5] The effect of trehalose on the texture of cooked rice has also been investigated.[33] The taste preference of freshly cooked rice increased with the addition of 2–4% trehalose per weight of raw rice, when compared with a control rice or when glucose or sucrose were added. The texture of cooked rice during storage at 20°C or at freezer temperatures was also maintained by the addition of 3% trehalose on a dry weight basis (dwb). Additionally, trehalose was shown to improve the quality of bread and delay staling during storage.[34] Trehalose was added at 3%, 7% and 11% of the flour (dwb) in a basic dough recipe. The results demonstrated an increase in specific volume, decreased hardness of bread crumb, reduced rate of bread firming, and improvement in bread sensory quality compared with the standard recipe.[34]

19.7.3 Stabilisation of proteins

In general, saccharides have the ability to stabilise proteins during cycles of both freeze-thaw and dehydration-rehydration.[35] Among the saccharides evaluated, trehalose was the most consistent in being able to maintain delicate protein structures and stabilise sulphur-containing amino acids. This results in the inhibition of the development of undesirable odours and flavours. The ability of trehalose to stabilise frozen egg white was examined. Trehalose was added to egg whites at 5%, mixed, and held frozen for 5 days. Upon thawing, there was little denaturation of the egg white compared with the control or other sweeteners, as measured by turbidity.[5] In a study to understand the ability of trehalose to stabilise proteins in dried preparations, pulverised carrots were mixed with 10% (w/w) of two sugar alcohols and four sugars.[36] The samples were dried for 64 hours at 40°C and then stored for 7 days. The samples were then analysed for superoxide dismutase (SOD)-like activity. The amount of SOD-like activity for the control, sucrose, glucose, maltose and trehalose were approximately 11%, 28%, 30%, 32% and 65%, respectively, of the activity before treatment.[36] The effects of two methods of pre-treatment with trehalose or sucrose were also examined. Carrots and potatoes were sliced (1 mm), steam-blanched and soaked in a 20% solution of the appropriate disaccharide.[37] For comparison, the carrots and potatoes slices were first coated with the sugars and then steam-blanched. Both pre-treatment methods using trehalose resulted in dried samples that had less shrinkage, better colour properties and better cell reconstruction properties than those of the non-pre-treatment control or sucrose samples.[37]

The solubility of freeze-dried water-soluble myofibril protein is low. When trehalose (0.5%) is added or when heat (70°C) is applied before freeze-drying, the solubility is only slightly increased.[38] However, when both heat and trehalose are applied, they synergistically achieve solubilisation of more than 80% of the freeze-dried water-soluble myofibril proteins. In this instance, it appears that the heating does not cause the denaturisation and subsequent insolubility, but rather the trehalose prevents denaturation of the proteins during the freeze-drying process.[38] The use of probiotic micro-organisms has become an increasingly important feature in the food industry. Freeze-drying is the preferred method of preserving the desired organisms but can result in low viability. The addition of trehalose (4%) or sucrose (4%), and/or skimmed milk powder significantly increased the survival of *Lactobacillus salivarius*.[39] Samples were freeze-dried, stored at −85°C and evaluated for viability every 7 days for a total of 49 days. At the end of the storage period, the combination of sucrose and skimmed milk gave a survival rate of approximately 42%, whereas the addition of trehalose to skimmed milk, sucrose or skimmed milk with sucrose resulted in viabilities of approximately 72%, 80% and 83%, respectively.[39]

The cryoprotective effect of trehalose on surimi (minced fish muscle) during frozen storage has also been investigated.[40] Surimi may lose functional properties or texture owing to denaturation and/or aggregation of the protein and cryoprotectants are required to help maintain the quality during frozen storage. Trehalose, at an 8% (w/w) concentration, was compared with an 8% commercial blend of sucrose/sorbitol, and the same concentration of sodium lactate. Using tilapia surimi, standard assays demonstrated that trehalose was more effective in preventing protein denaturation than the other ingredients. The use of trehalose also provided the greatest gel-forming ability as evidenced by the increased breaking and deformation forces of the surimi following storage at $-18°C$ for 24 weeks.[40] Additionally, the low sweetness of trehalose makes it an attractive alternative to standard cryopreservatives for surimi.

Plant cells can also be frozen and thawed with greater viability and potential for subsequent growth using trehalose.[41] In the first experiment, a carrot and two tobacco cell lines were frozen in a 40% trehalose solution. However, no viable cells were recovered. The authors then used a pre-treatment of 5% or 10% trehalose for various periods of time (1–6 days), depending on the cell type, with a cryopreservation fluid of 40% trehalose. After thawing, the cells were able to grow on appropriate medium. In addition, a common cryoprotectant (DMSO (5%)) was used in combination with 10% or 20% trehalose, but while post-thaw viability occurred, the cells did not subsequently grow.[41]

Increased protein degradation and decreased viability of cells during freezing or freeze-drying is partially the result of the formation of ice crystals, and subsequent disruption of protein and cell morphology. One reason for the protective effect of trehalose on proteins and cells, along with vitrification and water replacement/entrapment, might be explained by the fact that trehalose at high concentrations is approximately twice as effective as sucrose in suppressing the growth rate of ice crystals.[42] It also significantly suppresses morphological instability of ice crystals that can result in the degradation of protein or cell structure. This property has been suggested to be related to the increased affinity of water for trehalose compared with sucrose.[42]

19.7.4 Stabilisation of flavours and aromas

Trehalose can be used as a bulking agent to replace sucrose on a one to one basis. However, it is less sweet than sucrose having a sweetness equivalent to 45% of sucrose at a 10% concentration, and if trehalose is used to replace sucrose, it does not add sweetness that can change or interfere with the original flavour and/or taste of the food.[5] An additional benefit is that trehalose has remarkable chemical stability (see Section 19.6). The Maillard reaction can potentially result in production of unwanted tastes or odours in food processing; however, trehalose is not involved in caramelisation or Maillard reactions. In addition, the high glass transition temperature improves the physical stability of ingredients during dehydration processes such as spray drying or freeze drying. The retention of aroma components in strawberry and apricot purees by the addition of trehalose and sucrose during freeze drying and foam-mat drying was investigated.[43,44] The sugar concentrations were 8% and 4% (wet basis), respectively. The aroma components that included esters, carbonyl compounds, terpenoids, several alcohols and acids were identified by manual headspace solid-phase microextraction followed by gas chromatography analysis. The addition of trehalose to dehydrated strawberry and apricot purees resulted in the lowest loss of total aroma as well as individual fruit volatiles when compared with sucrose. In the freeze-dried strawberry purees trehalose retained just under 90% of the original aroma, whereas sucrose retained about 40%.[43] The foam-mat products made with trehalose and sucrose had much less residual total aroma than the freeze-dried preparations, 32% and 13%, respectively; however, the ratio of retention of trehalose to sucrose between the two drying methods were similar. The total aroma numbers for apricot purees was consistent with the data from the strawberries, except that sucrose retained a higher level of aroma in the freeze dried apricot samples (64%) than the strawberry, and the retained aromas in foam-mat samples with trehalose were lower with the apricot purees.[44] In both studies, it was concluded that after rehydration, trehalose-containing samples retained both total and individual aroma components, better than sucrose or the control samples, without the presence of off flavours.[43,44]

In another study 30% trehalose or sucrose were added to strawberry puree and the retained organic volatiles measured.[45] The samples containing trehalose had significantly greater amounts of most, but not all retained volatiles. It was also noted in a separate study that the addition of trehalose to freeze-dried strawberry purees maintained a good sweetness to sourness profile, and retained a fresher strawberry aroma and flavour, than samples containing sucrose.[45] In a further study, the sensory quality of strawberry cream filling made by evaporation and freeze-drying was evaluated.[46] Trehalose was used as partial replacement for sucrose at 3%, 5% and 10%, and the samples stored for 6 months at room temperature. The partial addition of trehalose had a dose-dependent beneficial effect on colour and anthocyanin retention. While trehalose had a positive effect, in general, on the fruity ester flavour of the samples, it was not dose dependent and was conditional on the addition of strawberry aroma to the samples.[46] The positive effects noted with trehalose were greater in the freeze-dried preparations than in the samples made using evaporation.

19.8 PHYSIOLOGICAL PROPERTIES

In addition to the technical features described previously, several useful physiological properties of trehalose have been reported. Trehalose is hydrolysed into two molecules of glucose by a specific enzyme, trehalase which is found on the surface of the brush border in the small intestine.[47,48] The metabolism of ingested trehalose resembles that of maltose, and both are absorbed as glucose and provide about 4 kcal/g.[5] However, the profile of metabolism appears to be different from other saccharides and this might provide some beneficial effect that other saccharides do not possess. Several reports have indicated that consumption of trehalose results in a lower glycaemic and insulinemic response than other mono- and disaccharides.

In one study, eight trained male cyclists consumed 75 g of trehalose, glucose or galactose after an overnight fast on three separate days.[49] After 45 minutes, the subjects started an exercise routine consisting of 20 minutes of submaximal steady-state exercise and then a subsequent maximal power period for approximately 40 minutes. The results demonstrated that the ingestion of trehalose led to a lower glucose (GI index 67) and insulin response prior to exercise, and consumption of trehalose also reduced the rebound hypoglycaemia when compared with glucose. No differences between treatments were noted in time-trial performance of the subjects.[49] In another exercise study, nine trained subjects consumed 1.1 g of trehalose or maltose per minute, while a control group drank an equivalent volume of water on one of three separate occasions. The subjects consumed the various liquids while cycling for 150 minutes at 55% of their maximal power output.[50] The overall results revealed that trehalose ingestion is accompanied by a much reduced and delayed glucose peak, and the trehalose was oxidised at a rate approximately 27% lower than maltose. Total fat oxidation was significantly reduced ($p < 0.05$) compared with the water trial, but tended to be greater ($p < 0.06$) than when maltose was consumed.[50]

Examination of the use of trehalose in overweight populations has also been examined. Ten healthy but overweight subjects fasted overnight and consumed drinks in the morning containing either 75 g of glucose or trehalose in 400 mL of water on two separate occasions.[51] Blood and other samples were obtained immediately before and for 3 hours after consumption. On the same day, the subjects consumed a lunch in which 25% of the total energy of the meal was provided by the glucose or trehalose beverage. Experimental samples were again collected before and for 3 hours after the meal. The results showed that ingestion of trehalose, compared with glucose, significantly reduced ($p < 0.05$) the peak blood glucose and insulin concentrations in both the morning and afternoon periods. Total blood glucose (area under the curve, AUC) in the morning tended to be less ($p = 0.08$) after trehalose ingestion than glucose, whereas the value after the meal was significantly less ($p < 0.05$). The AUC for insulin blood concentrations at both the morning and after meal samplings were significantly less ($p < 0.01$) for the trehalose than the glucose group. Triacylglycerol values were lower ($p < 0.05$) in the trehalose group in the morning, but significantly greater ($p < 0.05$) than the glucose group in the afternoon.[51] This study demonstrated a reduction in the glycaemic and insulinemic response to trehalose compared with glucose. Another study using 21 healthy but overweight males was performed to assess the glycaemic and insulinemic response to ingestion of 75 g of trehalose, trehalose/fructose (50/25 g) or glucose.[52] Each subject consumed each of the three solutions on three separate occasions. Blood samples were collected at 0, 30, 60, 90 and 120 minutes after ingestion. The reason for the addition of fructose was to make the product sweeter, making the taste more like sucrose and glucose, but with a lower glycaemic and insulinemic profile. Peak blood glucose values and AUC for trehalose and trehalose/fructose ingestion were significantly less ($p < 0.02$) than after glucose consumption. The reductions in the AUC were 19% and 34% after trehalose and trehalose/fructose, respectively. The same comparisons for insulin were also significant ($p < 0.03$), with the reductions in AUC for trehalose and trehalose/fructose being 29% and 34%, respectively.[52] In addition to the ability of trehalose to be used in foods and sports drinks to provide a reduced glycaemic and insulinemic response, it can also be used to provide a milder more palatable sweetness because the sweetness of trehalose is approximately 40%, 45% or 65% that of high-fructose corn syrup (HFCS), sucrose and glucose, respectively.[3,5,49,52]

Owing to the differences in the insulin response to trehalose, compared with other saccharides, a study was carried out to evaluate the effect of trehalose when provided in conjunction with a high-fat diet (HFD) in a mouse model.[53] The HFD was previously reported to induce insulin resistance.[54] In the trehalose study, mice were fed the HFD and provided drinking water containing 2.5% trehalose, glucose, maltose, fructose or HFCS, *ad libitum*. Two control groups consisted of the HFD with water, and a standard diet with water. After 8 weeks of feeding, the effects of trehalose on the hypertrophy of adipocytes, insulin resistance, and various other biological variables were assessed. Mesenteric adipocyte size in the trehalose/HFD group was significantly less ($p < 0.05$) than the glucose-, maltose-, HFCS-, and fructose-HFD groups. Trehalose intake significantly ($p < 0.05$) suppressed the elevation of fasting insulin secretion compared with maltose and glucose. Furthermore, it reduced insulin levels compared with other groups using an oral glucose tolerance test. Mechanistic analyses suggest that trehalose exhibits its suppressive effects by decreasing insulin secretion and down-regulating mRNA expression of factors that can disrupt insulin function. These data suggest that trehalose may provide a beneficial approach for people with metabolic syndrome.[53]

The cariogenicity of trehalose versus sucrose was studied in *in vitro*, animal and human experiments and demonstrated a favourable effect of trehalose on dental health.[55] Unlike sucrose, trehalose was not converted to water-insoluble glucan (WIG) by the enzyme found in cariogenic bacteria. When trehalose was mixed with sucrose, it reduced the amount of WIG produced in a dose-dependent manner. The fermentation time, as measured by pH change, of trehalose by *Streptococcus mutans* and *Streptococcus sobrinus* was greater than twice that of sucrose, and the production of lactic acid by the same bacteria were about 24 and 59% less than sucrose. Specific pathogen free rats were infected with *S. mutans* and fed a diet containing trehalose, sucrose or a mixture of both for 60 days. Animals consuming trehalose had a significantly lower ($p < 0.01$) caries score than those given sucrose. Interestingly animals receiving the mixture of trehalose and sucrose, while not statistically significant, had a 21.3% reduction in the caries score. Twelve human subjects rinsed their mouths with 10 mL of a 10% solution of trehalose, sucrose and sorbitol on separate days. Plaque samples were gathered at 0, 3, 7, 11, 20 and 30 minutes after rinsing. The trehalose plaque samples never reached the critical pH for subsequent caries, whereas sucrose showed evidence of standard cariogenicity. The sample with sorbitol did not show any pH change. These results suggest that trehalose has low cariogenic potential and may reduce the cariogenicity of other sugars.[55]

Two preliminary animal studies reported that trehalose may have the potential to improve bone metabolism and inhibit osteoporosis.[56,57] The effects of trehalose on bone resorption were studied using a mouse model of osteoporosis caused by estrogen deficiency. Trehalose was orally administered to the mice five times a week for 4 weeks, and the differences in body and bone weights, uterine weight, calcium and phosphorous content, bone morphology, and factors associated with bone resorption were compared with control groups. While body weights were the same for each group, there was a significant inhibition of bone weight loss ($p < 0.05$) in the group treated with 100 mg/kg trehalose compared with the non-treatment group.[56] The concentrations of calcium and phosphorus were also greater ($p < 0.05$) in the trehalose treated group; however, the calcium to phosphorous ratio was not different, suggesting that trehalose maintains the normal chemical balance. Furthermore, the internal morphologic structure (trabeculae) of the bones in the trehalose treatment group were only marginally affected when compared with those of the control. The number of bone marrow osteoclasts, which are the cells that resorb bone, were significantly reduced ($p < 0.05$) from that observed in the control group. Trehalose is unlikely to have exerted these effects via an estrogen-like function, as its administration had no effect on uterine weight.[56] Rather, it is likely that trehalose caused an inhibition of the secretion of a soluble factor from the bone marrow that stimulated osteoclast formation.[57] Diets high in sucrose have been shown to change calcium balance in humans, and reduce the weight and strength of bones in rats.[58] In an animal study, high doses of trehalose were shown not to have the same negative effect on bone length or weight as did high-sucrose consumption. Examination of various factors suggest that this effect might result from high-dose trehalose helping to maintain an effective estradiol level in the body, which results in more normal bone development.[58] The mechanism of these affects are currently unknown.

19.9 CONCLUSIONS

Since trehalose has multi-functional effects not possessed by other saccharides, it is being employed for a range of unique applications in the food industry. It has also been demonstrated that trehalose has physiological effects that may have benefit for human health. The uses of trehalose are not limited to the food industry, but have been employed in a wide variety of cosmetic products, including moisturisers, and in commercial pharmaceutical products to help stabilise proteinaceous medical therapies.[59] Although initially regarded as a rare sugar because of its cost, large-scale production has made trehalose as accessible as other common saccharides. The novel properties of trehalose will probably result in far wider application in foods in the future.

REFERENCES

1. G.G. Birch. Trehalose. In: M.L. Wolfrom and R.S. Tyson (eds). *Advances in Carbohydrate Chemistry*, Vol. 18, Academic Press, New York; 1963, pp. 201–225.
2. A.D. Elbein. The metabolism of α,α-trehalose. *Advances in Carbohydrate Chemistry and Biochemistry* 1974, 30, 227–256.
3. A.B. Richards, S. Krakowka, L.B. Dexter, H. Schmid, A.P. Wolterbeek, D.H. Waalkens-Berendsen, S. Arai and M. Kurimoto. Trehalose: a review of properties, history of use and human tolerance, and results of multiple safety studies. *Food and Chemical Toxicology* 2002, 40(7), 871–898.
4. J.H. Crowe, J.F. Carpenter and L.M. Crowe. The role of vitrification in anhydrobiosis. *Annual Review of Physiology* 1998, 60, 73–103.
5. A.B. Richards and L.B. Dexter. Trehalose. In: L.O. Nabors (ed.). *Alternative Sweeteners*, Marcel Dekker, New York; 2001, pp. 423–461.
6. G.R. Wyatt and G.F. Kale. The chemistry of insect hemolymph. II. Trehalose and other carbohydrates. *Journal of General Physiology* 1957, 40(6), 833–847.
7. A. Becker, P. Schloder, J.E. Steele and G. Wegener. The regulation of trehalose metabolism in insects. *Experientia* 1996, 52(5), 433–439.
8. R.P. Adams, E. Kendall and K.K. Kartha. Comparison of free sugars in growing and desiccated plants of *Selaginella lepidophylla*. *Biochemical Systematics and Ecology* 1990, 18, 107–110.
9. J.S. Clegg. The origin of trehalose and its significance during the formation of encysted dormant embryos of *Artemia salina*. *Comparative Biochemistry and Physiology* 1965, 14, 135–144.
10. Y.M. Newman, S.G. Ring and C. Colaco. The role of trehalose and other carbohydrates in biopreservation. *Biotechnology and Genetic Engineering Reviews* 1993, 11, 263–294.
11. P. Van Dijck, D. Colavizza, P. Smet and J.M. Thevelein. Differential importance of trehalose in stress resistance in fermenting and nonfermenting Saccharomyces cerevisiae cells. *Applied and Environmental Microbiology* 1995. 61(1), 109–115.
12. E.M. Koch and F.C. Koch. The presence of trehalose in yeast. *Science* 1925, 61(1587), 570–572.
13. A. Steiner and C.F. Cori. The preparation and determination of trehalose in yeast. *Science* 1935, 82(2131), 422–423.
14. L.C. Stewart, N.K. Richtmyer and C.S. Hudson. The preparation of trehalose from Yeast. *Journal of the American Chemical Society* 1950, 72(5), 2059–2061.
15. T. Suzuki, K. Tanaka and S. Kinoshita. The extracellular accumulation of trehalose and glucose by bacteria grown on n-alkanes. *Agricultural and Biological Chemistry* 1969, 33(2), 190–195.
16. K. Maruta, T. Nakada, M. Kubota, H. Chaen, T. Sugimoto, M. Kurimoto and Y. Tsujisaka. Formation of trehalose from maltooligosaccharides by a novel enzymatic system. *Bioscience, Biotechnology and Biochemistry* 1995, 59(10), 1829–1834.
17. T. Nakada, K. Maruta, H. Mitsuzumi, M. Kubota, H. Chaen, T. Sugimoto, M. Kurimoto and Y. Tsujisaka. Purification and characterization of a novel enzyme, maltooligosyl trehalose trehalohydrolase, from *Arthrobacter* sp. Q36. *Bioscience, Biotechnology and Biochemistry* 1995, 59(12), 2215–2218.
18. T. Nakada, K. Maruta, K. Tsusaki, M. Kubota, H. Chaen, T. Sugimoto, M. Kurimoto and Y. Tsujisaka. Purification and properties of a novel enzyme, maltooligosyl trehalose synthase, from *Arthrobacter* sp. Q36. *Bioscience, Biotechnology and Biochemistry* 1995, 59(12), 2210–2214.
19. T. Sugimoto. Production of trehalose by enzymatic starch saccharification and its applications. *Shokuhin Kogyo (Food Industry)* 1995, 38(10), 34–39 (Japanese).
20. M. Kubota. Trehalose production. In: M. Ohnishi (ed.). *Glycoenzymes*., Japan Scientific Societies Press, Tokyo; 2000, pp. 217–225.

21. A. Dahlqvist. Specificity of the human intestinal disaccharidases and implications for hereditary disaccharide intolerance. *Journal of Clinical Investigations* 1962, 41, 463–470.
22. A. Dahlqvist. Enzyme deficiency and malabsorption of carbohydrates. In: H. Sipple (ed.). *Sugars in Nutrition*, Academic Press, New York; 1974, 187–214.
23. D. Welsh, J.R. Poley, M. Bhatia and D.E. Stevenson. Intestinal disaccharidase activities in relation to age, race and mucosal damage. *Gastroenterology* 1978, 75, 847–855.
24. R. Bergoz, M.C. Vallotton and E. Loizeau. Trehalase Deficiency. Prevalence and relation to single-cell protein food. *Annals of Nutrition and Metabolism* 1982, 26(5), 291–295.
25. E. Gudmand-Høyer, H.J. Fenger, H. Skovbjerg, P. Kern-Hansen and P.R. Madsen. Trehalase deficiency in Greenland. *Scandinavian Journal of Gastroenterology* 1988, 23, 775–778.
26. J. Madzarovova-Nohejlova. Trehalose deficiency in a family. *Gastroenterology* 1973, 65(1), 130–133.
27. G.M. Brown, D.C. Rohrer, B. Berking, C.A. Beevers, R.O. Gould and R. Simpson. The crystal structure of α,α-trehalose dihydrate from three independent X-ray determinations. *Acta Crystallography* 1972, B28, 3145–3158.
28. D. Kilburn, S. Townrow, V. Meunier, R. Richardson, A. Alam and J. Ubbink. Organization and mobility of water in amorphous and crystalline trehalose. *Nature Materials* 2006, 5, 632–635.
29. L.M. Crowe, J.H. Crowe, A. Rudolph, C. Womersley and L. Appel. Preservation of freeze-dried liposomes by trehalose. *Archives of Biochemistry and Biophysics* 1985, 242(1), 240–247.
30. J.F. Carpenter and J.H. Crowe. An infrared spectroscopic study of the interactions of carbohydrates with dried proteins. *Biochemistry* 1989, 28(9), 3916–3922.
31. L. Kreilgaard, S. Frokjaer, J.M. Flink, T.W. Randolph and J.F. Carpenter. Effects of additives on the stability of *Humicola lanuginosa* lipase during freeze-drying and storage in the dried solid. *Journal of Pharmaceutical Science* 1999, 88(3), 281–290.
32. S. Ikeda, T. Yabuzoe, T. Takaya and K. Nishinari. Effects of sugars on gelatinisation and retrogradation of corn starch. In: T.L. Barsby, A.M. Donald and P.J. Frazier (eds). *Starch: Advances in Structure and Function*, Royal Society of Chemistry, London; 2001, pp. 67–76.
33. T. Hirata. Effects of trehalose on texture of cooked rice. *Bulletin of Hiroshima Prefectural Technology Research Institute Food Technology Research Center* 2009, 25, 1–4 (Japanese, English abstract).
34. J-C. Zhou, Y-F. Peng and N. Xu. Effect of Trehalose on Fresh Bread and Bread Staling. *Cereal Foods World* 2007, 52(6), 313–316.
35. J.H. Crowe, J.F. Carpenter, L.M. Crowe and T.J. Anchordoguy. Are freezing and dehydration similar stress vectors? A comparison of modes of interaction of stabilizing solutes with biomolecules. *Cryobiology* 1990, 27, 219–231.
36. H. Aga, T. Shibuya, H. Chaen, S. Fukuda and M. Kurimoto. Stabilization by trehalose of superoxide dismutase-like activity of various vegetables. *Nippon Shokuhin Kagaku Kogaku Kaishi* 1998, 45(3), 210–215 (Japanese, English abstract).
37. T. Aktas, S. Fujii, Y. Kawano and S. Yamamoto. Effect of pretreatments of sliced vegetables with trehalose on drying characteristics and quality of dried products. *Food and Bioproducts Processing* 2007, 85(C3), 178–183.
38. Y. Ito, S. Toki, T. Omori, H. Ide, R. Tatsumi, J. Wakamatsu, T. Nishimura and A. Hattori. Physiochemical properties of water-soluble myofibrillar proteins prepared from chicken muscle. *Animal Science Journal* 2004, 75, 59–65.
39. G. Zayed and Y.H. Roos. Influence of trehalose and moisture content on survival of Lactobacillus salivarius subjected to freeze-drying and storage. *Process Biochemistry* 2004, 39, 1081–1086.
40. A. Zhou, S. Benjakul, K. Pan, J. Gong and X. Liu. Cryoprotective effects of trehalose and sodium lactate on tilapia (*Sarotherodon nilotica*) surimi during frozen storage. *Food Chemistry* 2006, 96(1), 96–103.
41. I.S. Bhandal, R.M. Hauptmann and J.M. Widholm. Trehalose as cryoprotectant for the freeze preservation of carrot and tobacco cells. *Plant Physiology* 1985, 78(2), 430–432.
42. T. Sei, T. Gonda and Y. Arima. Growth rate and morphology of ice crystals growing in a solution of trehalose and water. *Journal of Crystal Growth* 2002, 240, 218–229.
43. D. Komes, T. Lovric, K.K. Ganic and L. Gracin. Study of trehalose addition on aroma retention in dehydrated strawberry puree. *Food Technology and Biotechnology* 2003, 41(2), 111–119.
44. D. Komes, T. Lovric, K.K. Ganic, J.G. Kljusuric and M. Banovic. Trehalose improves flavour retention in dehydrated apricot puree. *International Journal of Food Science and Technology* 2005, 40, 425–435.
45. M.V. Galmarini, C.M. van Baren, M.C. Zamora and J. Chirife. Trehalose as a drying aid of fruit products: influence on physical properties, sensory characteristics and volatile retention. In: R. Filip

(ed.). *Multidisciplinary Approaches on Food Science and Nutrition for the XXI Century*, Transworld Research Network, India; 2011, pp. 113–130.
46. M. Kopjar, V. Pilizota, J. Hribar, M. Simcic, E. Zlatic, N.N. Tiban. Influence of trehalose addition and storage conditions on the quality of strawberry cream filling. *Journal of Food Engineering* 2008, 87, 341–351.
47. A. Dahlqvist and A. Brun. A method for the histochemical demonstration of disaccharidase activities: application to invertase and trehalase in some animal tissues. *Journal of Histochemistry and Cytochemistry* 1962, 10(3), 294–302.
48. A. Dahlqvist and C. Nordstrom. The distribution of disaccharidase activities in the villi and crypts of the small-intestinal mucosa. *Biochimica et Biophysica Acta* 1966, 113(3), 624–626.
49. R.L.P.G. Jentjens and A.E. Jeukendrup. Effects of pre-exercise ingestion of trehalose, galactose and glucose on subsequent metabolism and cycling performance. *European Journal of Applied Physiology* 2003, 88(4–5), 459–465.
50. M.C. Venables, F. Brouns and A.E. Jeukendrup. Oxidation of maltose and trehalose during prolonged moderate-intensity exercise. *Medicine and Science in Sports and Exercise* 2008, 40(9), 1653–1659.
51. J.G.P. van Can, T.H. Ijzerman, L.J.C. van Loon, F. Brouns and E.E. Blaak. Reduced glycaemic and insulinaemic responses following trehalose ingestion: implications for postprandial substrate use. *British Journal of Nutrition* 2009, 102(10), 1395–1399.
52. K.C. Maki, M. Kanter, T.M. Rains, S.P. Hess and J. Geohas. Acute effects of low insulinemic sweeteners on postprandial insulin and glucose concentrations in obese men. *International Journal of Food Sciences and Nutrition* 2009, 60(Suppl 3), 48–55.
53. C. Arai, N. Arai, A. Mizote, K. Kohno, K. Iwaki, T. Hanaya, S. Arai, S. Ushio and S. Fukuda. Trehalose prevents adipocyte hypertrophy and mitigates insulin resistance. *Nutrition Research* 2010, 30(12), 840–848.
54. S. Koya-Miyata, N. Arai, A. Mizote, Y. Taniguchi, S. Ushio, K. Iwaki and S. Fukuda. Propolis prevents diet-induced hyperlipidemia and mitigates weight gain in diet-induced obesity in mice. *Biological and Pharmaceutical Bulletin* 2009, 32(12), 2022–2028.
55. T. Neta, K. Takada and M. Hirasawa. Low-cariogenicity of trehalose as a substrate. *Journal of Dentistry* 2000, 28(8), 571–576.
56. Y. Nishizaki, C. Yoshizane, Y. Toshimori, N. Arai, S. Akamatsu, T. Hanaya, S. Arai, M. Ikeda and M. Kurimoto. Disaccharide-trehalose inhibits bone resorption in ovariectomized mice. *Nutrition Research* 2000, 20(5), 653–664.
57. C. Yoshizane, N. Arai, C. Arai, M. Yamamoto, Y. Nishizaki, T. Hanaya, S. Arai, M. Ikeda and M. Kurimoto. Trehalose suppresses osteoclast differentiation in ovariectomized mice: correlation with decreased in vitro interleukin-6 production by bone marrow cells. *Nutrition Research* 2000, 20(10), 1485–1491.
58. S. Takayama, S. Inoue, M. Tanaka-Kataoka, K. Iwaki, M. Ikeda, M. Kurimoto, T. Ohta and S. Fukuda. Trehalose suppresses weight loss of bone induced by high-dose sucrose diet: possible involvement of estradiol. *ITE Letters* 2007, 8(10), 69–75.
59. S. Ohtake and Y.J. Wang. Trehalose: current use and future applications. *Journal of Pharmaceutical Sciences* 2011, 100(6), 2020–2053.

Part Five
Bulking Agents – Multi-Functional Ingredients

20 Bulking Agents – Multi-Functional Ingredients

Michael Auerbach[1] and Anne-Karine Dedman[2]

[1] Active Nutrition, DuPont Nutrition & Health, New York, USA
[2] Active Nutrition, DuPont Nutrition & Health, Paris, France

20.1 INTRODUCTION

Sugar is a multi-functional ingredient that comes in a variety of forms and can therefore be used in many different food products. It has a range of properties that, either individually or in combination with other ingredients, makes it an important ingredient in modern food production. Table 20.1 indicates some of the many functional properties of sugar in food. It is expected that a sugar replacement ingredient should have the same or similar attributes to ensure optimal organoleptic, microbiological and functional performance.

Sweetening solutions have been highlighted in Parts Two to Four of this book, and in liquid products such as soft drinks or sweeteners for coffee or tea, sugar can be replaced directly by a high-intensity sweetener with water providing the bulk phase where appropriate. When formulating solid foods without sugar, the recipe requires careful manipulation to ensure that all the performance criteria are met for the specific food product and this normally means the inclusion of a bulking agent. It is very rare that one sweetener or bulking agent will replace sugar directly in an application because of the variety of functions required. The best solution is often a mixed ingredient approach.[1]

Speciality carbohydrates have been developed commercially that allow great flexibility when replacing sugar in formulations and these complement the use of high-intensity sweeteners, polyols and other sweeteners, including sugar itself.

The majority of these speciality carbohydrate ingredients are non-digestible oligo- or polysaccharides, these products are lower in calories than sugar, have soluble dietary fibre and, often, prebiotic properties. Large polysaccharides (molecular weight (MW) > 5000) tend to be unsuitable as direct sugar replacers because of their low solubility and high viscosity in use and are more commonly classified and used as insoluble dietary fibre ingredients. Materials falling into this category include cellulose, hemicelluloses, pectins, gums, mucilages and lignins, and these materials are often very complex mixtures of carbohydrate polymers. As a result of their complexity, they are chemically somewhat ill defined. They are, however, common in our diets and as isolates have been used mainly as insoluble dietary fibre sources. Although bulking agents by definition, these products are limited in practical terms regarding their sugar-replacing properties because of their large size and often complicated chemical structure, which affects functionality.

Table 20.1 Functional properties of sugar in food.

	Sweetness	Flavour/aroma	Volume	Texture	Shelf life	Fermentation	Freezing point depression	Colour	Moisture retention
Beverages	•	•						•	
Preserves	•	•			•			•	
Jam/marmalade	•	•	•	•	•				
Sauces/dressings	•	•		•	•				
Confectionery	•	•	•	•	•			•	
Dairy products	•		•	•			•		
Bakery products	•	•	•	•		•		•	•
Pharmaceuticals/non-food	•					•			•

This chapter will concentrate on the oligo- and polysaccharide materials that are most applicable as sugar alternatives and that have sugar-like qualities in food applications. The following products will be covered in detail: gluco-polysaccharides (including polydextrose, resistant starches and maltodextrins) and fructo-oligosaccharides (FOS) (including inulin).

It is important to recognise, however, that as well as being important bulking agents supporting high-intensity sweeteners, polyols or sugar in formulations, these particular polymeric materials have important physiological benefits when used alone in formulations. These physiological benefits are related, on the whole, to their soluble dietary fibre and potential prebiotic properties, and new clinical studies are providing evidence of their positive health benefits. They are all, by definition, calorie reduced and in some applications they can even replace or partially replace fat. Their unique chemistry permits other technological benefits in formulations and they can therefore be considered truly multi-functional.

20.2 GLUCO-POLYSACCHARIDES

20.2.1 Polydextrose

20.2.1.1 Description

Polydextrose is a low-molecular-weight randomly bonded polysaccharide of glucose with a calorific value of 1 kcal/g.[2-6] It is prepared by the bulk melt polycondensation of glucose and sorbitol with small amounts of food grade acid *in vacuo*. All possible glycosidic linkages with the anomeric carbon of glucose are present: α and β (1-2), (1-3), (1-4) and (1-6), although (1-6) linkages predominate (see Figure 20.1). It has an average degree of polymerisation (DP) of 12 and an average molecular weight of 2000. The US Food and Drug Administration (FDA) studied polydextrose extensively for safety prior to its 1982 approval as a food additive under 21 Code of Federal Regulations (CFR) 172.841.[7] It has attained significant use as a low-calorie bulking agent to replace sugar in reduced-calorie foods.[8] The low-calorie content of

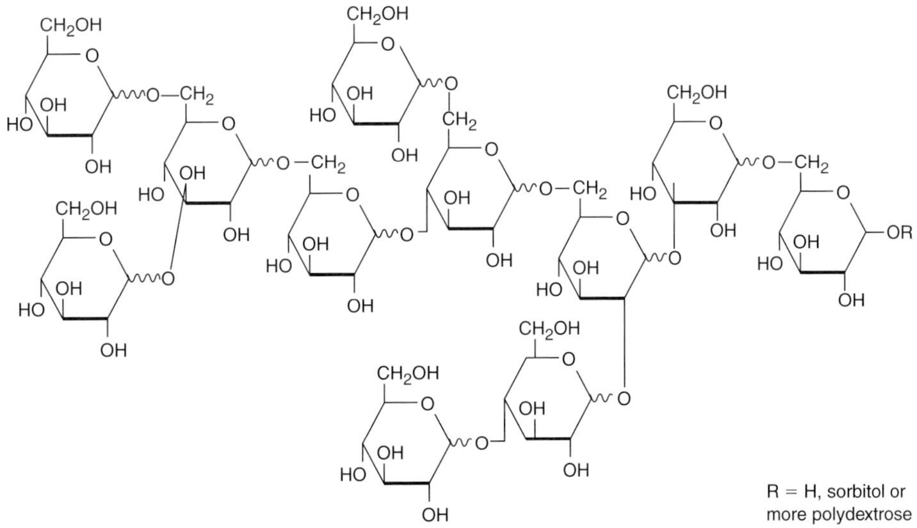

Fig. 20.1 Chemical structure of polydextrose.

Table 20.2 Attributes of polydextrose and Litesse.

	Polydextrose	**Litesse**	**Litesse Two**	**Litesse *Ultra***
Taste	Tartaric acid	Bland neutral	Clean mildly sweet	Very clean mildly sweet
Colour	Cream	Cream	Cream	White
pH range (10% w/w aqueous solution)	2.5–3.5	3.0–4.5	3.5–5.0	4.5–6.5
Maillard reaction	Yes	Yes	Yes	No

polydextrose is a result of its indigestibility in the small intestine and incomplete fermentation in the large intestine. This property has led to acceptance of polydextrose as a dietary fibre in many countries.[9] Polydextrose has been developed through technological advances to meet a wide range of application needs. Pfizer initially developed polydextrose followed by an improved family of polydextrose products under the brand name, Litesse®. These products are now part of the Dupont™ Danisco® range of ingredients and a competitive polydextrose range of products is also commercialised by Tate and Lyle under the name Sta-Lite®.

20.2.1.2 Organoleptic qualities

Although Litesse possesses many of the functional properties of other carbohydrates such as sugar, glucose syrups and maltodextrin, essentially it is not sweet. However, Litesse can be used to balance and reduce the sweetness level of products and is also suitable for savoury applications or in products that require bulk, with a less sweet profile.

When used in combination with sugars and polyols, Litesse has an enhancing effect on the perceived sweetness of the product and very often, a good sweetness level can be achieved without the need for intense sweeteners.

20.2.1.3 Physical and chemical properties

Polydextrose and the Litesse family of products range from a bland, neutral powder through to a colourless, mildly sweet liquid. Table 20.2 summarises the different attributes of the Litesse grades and Table 20.3 summarises the properties common to all grades of Litesse and polydextrose.

Litesse and polydextrose are available in the following forms: powdered, granulated (for quicker dissolution and reduced dusting) and in solution (70% w/w).

Table 20.3 Properties common to all grades of polydextrose and Litesse.

Formula	Molecular weight range	Structure	Melting point range	Optical character
$(C_6H_{12}O_6)_x$	182–5000	Amorphous	90–110°C	Inactive
Solubility in ethanol % w/w	**Solubility in water % w/w**	**Viscosity (70% w/w solution) @ 25°C**	**Heat of solution**	**Relative sweetness (sucrose = 1)**
Insoluble	>80%	1800 cps	8 cal/g	approx. 0.05

Table 20.4 Relative stability of Litesse versus fructo-oligosaccharides (FOS) (inulin average degree of polymerisation (DP) = 10) as measured by percentage increase in free monomers at pH 2.6, 100°C over 5 hours.

Free monomers (%)	Time in hours		
	0	1	5
Litesse Two (randomly branched polymer)	0.02	0.02	2.28
Litesse *Ultra* (randomly branched polymer)	0.08	0.3	2.38
FOS (regular, linear polymer)	0.2	45.42	100

Solubility

Litesse polydextrose has a higher water solubility than most carbohydrates and polyols, allowing solutions of greater than 80% w/w concentration at 25°C. This influences the perceived mouthfeel and texture of the product in foods. Solubility can also influence the flavour release from certain food systems such as hard candy. Litesse has a very low solubility in ethanol and is not soluble in glycerine or propylene glycol.

Stability

Litesse is also very stable in solution. Polydextrose and the Litesse family are complex glucose polymers containing all types of glycosidic bonds. The polymers are composed of randomly cross-linked glucose, predominately in the $\beta(1,6)$ form and these bonds are more than two to four times more resistant to hydrolysis than $\alpha(1,2)$, $\alpha(1,3)$ or $\alpha(1,4)$ bonds. Table 20.4 indicates how bonding type and molecular shape can affect the acidic hydrolysis rate of carbohydrate polymers. Litesse is relatively stable at low pH and high temperature compared with linear polymers such as FOS.

Model systems containing polydextrose have indicated very good stability against hydrolysis over a broad range of pH and temperature making it ideal for use in many beverage applications, even those of lower pH (see Table 20.5).

Melting properties and glass transition

Litesse polydextrose powder is an amorphous glass with an anhydrous glass transition temperature (T_g) of 110°C. This is significantly greater than the T_g of most other sugars and is partly a function of molecular weight. Heating above the T_g leads to a flowable melt that,

Table 20.5 Process stability of Litesse.

Temperature	pH	Increase in free glucose (%)
Process stability		
70°C (Pasteurisation, 10 min)	3 and 7	No significant hydrolysis
142°C (UHT 6–10 s)	3 and 7	No significant hydrolysis
Storage stability ~3 months		
40°C	3	Maximum 2%
20°C	3	Maximum 0.8%
25°C	3	No significant hydrolysis
220°C	3	No significant hydrolysis

No significant hydrolysis would be expected at any storage temperature when pH is >4.0.

after cooling, produces a clear glass with a brittle texture. The high T_g of Litesse can be useful to help raise the composite glass transition temperature (T'_g) of a food.

Litesse is a mixture of branched molecules of varying molecular weight, so it will not crystallise, and polydextrose can be used to stabilise foods by preventing sugar and polyol crystallisation (e.g. in hard candies).

Litesse can protect the structure of frozen and thawed materials and this can be useful in applications such as frozen dairy desserts, frozen dough and surimi to reduce deleterious changes in texture, structure and chemical composition on storage. Polydextrose can act by interrupting sugar recrystallisation and starch retrogradation, providing structure and raising the composite T'_g of the food. The T_g values of lactose ($-28°C$), sucrose ($-32°C$), fructose ($-42°C$), glucose ($-43°C$) and sorbitol ($-43.5°C$) are all lower than polydextrose ($110°C$). This means that replacement of these sugars with polydextrose raises the composite T'_g of food. Freezer storage stability improves when the difference between T'_g and storage temperature (typically, $-18°C$ for a domestic freezer) is minimised or if T'_g exceeds storage temperature.

Heat of solution
The heat of solution of Litesse polydextrose is 8 cal/g, as measured by solution calorimetry. Blends of Litesse with polyols can produce a similar heat of solution to sucrose and this is particularly advantageous in products such as sugar-free chocolate or fruit-flavoured hard candies where 'warmer' flavour notes are desirable.

Moisture management
Moisture management is an important property in the development of new and existing foods as moisture influences texture, flavour, shelf life, consumer acceptability and food safety.

At low concentrations (below 60% w/w), sucrose lowers water activity (A_w) slightly more than Litesse, while at higher concentrations, Litesse is more effective at reducing A_w. This is because sucrose crystallises at high concentrations and these crystals do not interact with the water to lower A_w.

Litesse helps to retain moisture, texture and shelf life in a range of product applications, including confectionery, baked goods and reformed meat products.

Viscosity
Litesse solutions behave as Newtonian fluids. Polydextrose solutions have a higher viscosity than sucrose or sorbitol at equivalent temperatures and concentration and this characteristic enables polydextrose to provide the desirable mouthfeel and textural qualities so important when replacing sugar and fat.

Freezing point depression
The freezing point depression (FDP) of a frozen dessert depends on the soluble constituents of the product and will therefore vary with product composition. This function is important in achieving creamy, palatable frozen desserts. FPD is a colligative property (smaller molecular weight carbohydrates have a greater depressing effect). Modification of textural qualities, such as softness and hardness, can be achieved by using a combination of higher molecular weight polydextrose with lower molecular weight sugars or polyols.

20.2.1.4 Physiological properties

Energy value

The energy value of polydextrose is generally recognised as being 1 kcal/g, determined from many animal and human studies. The calorie value of radio-labelled polydextrose and various cellulose polymers in 46 male Wistar rats was studied.[5] Total polydextrose utilisation (total for digestion and fermentation) was 27.4% of the ingested dose. The energy value of polydextrose was thus calculated as 0.274×17 kJ/g (energy value of carbohydrates); that is, 4.7 kJ/g ($= 1.1$ kcal/g).

Similarly, four metabolic studies examined the fate of ^{14}C on ingestion or intravenous administration of [^{14}C] polydextrose in rats.[3] The studies concluded that polydextrose given orally was not absorbed and that the majority of the unabsorbed polydextrose (approximately 60%) was eliminated in the faeces. Part of the ingested polydextrose (approximately 30%) was fermented in the lower part of the intestine by the intestinal flora, thereby producing short-chain fatty acids (SCFA) and carbon dioxide (CO_2).

Other authors have studied and compared the calorie values of several bulking agents, including polydextrose, in rats. Sixty-five Sprague–Dawley rats were fed a diet comprising 65% standard feed and 35% bulking agent (starch, silica, wheat fibres, polydextrose, isomalt or FOS).[6] The results demonstrated that consumption of polydextrose, isomalt and FOS exerted a laxative effect. The weight gain seen in rats consuming polydextrose was between that observed following consumption of starch (highest value) and that observed following consumption of silica (lowest value) ($p < 0.05$). The quantities of fat, protein, ash and water were highest in the carcasses of rats fed with starch and lowest in the carcasses of rats fed with silica. The quantities of other test substances, including polydextrose, were between these values. The caloric value of the carbohydrates tested was calculated from the energy value of the carcasses, which in turn was calculated from body composition. According to this method, the calorie value of polydextrose was 0.77 kcal/g, while that of wheat fibre was 0.88 kcal/g, that of isomalt was 1.64 kcal/g and that of FOS was 1.48 kcal/g.

Figdor and Bianchine[4] studied the gastrointestinal (GI) fate of polydextrose (10 g) in four men and determined its calorie value. Radioactivity was measured in exhaled CO_2, urine and stools; 15.97% of the ingested radioactivity was recovered in the form of $^{14}CO_2$, 1.41% in the urine and 50.07% in the faeces. The caloric value of 1 kcal/g was calculated using the energy conversion factor for acetate to CO_2 (0.6) and for expired $^{14}CO_2$ resulting from metabolism of [^{14}C] polydextrose (26.6%).

Achour et al.[2] conducted a clinical study in seven non-obese healthy male subjects to determine the energy value of polydextrose (30 g per day) and study the GI effects during a prolonged period of consumption (22 days). A small fraction of ingested radioactivity was recovered in the urine ($4 \pm 1\%$), in flatus ($<1\%$), in short-chain acids in the faeces ($<1\%$) and in bacteria (3–4%). The remaining radioactivity in the faeces ($33 \pm 3\%$ for PD1 and $32 \pm 4\%$ for PD2) was considered as representing intact polydextrose. The calorie value of polydextrose was 4.01 kJ/g (0.95 kcal/g) where respiratory $^{14}CO_2$ was used for the determination and 6.1 kJ/g (1.45 kcal/g) where the calculation was based on the quantities of volatile fatty acids (VFAs) produced.

Note: PD1 is 7-day acute feeding period and PD2 is 21-day chronic feeding period.

According to Auerbach et al.[10], all the studies used to determine the energy value of polydextrose are concordant and show this value to be 1 kcal/g. They also demonstrate the indigestibility of polydextrose in the small intestine as well as its ability to undergo fermentation in the colon.

Toleration

Partially digestible carbohydrates and related compounds exert overall beneficial effects on GI function when ingested at low to moderate doses.[11] These effects include improved bowel function (e.g. prevention of constipation), enhanced environment for the growth of such beneficial bacteria as *Lactobacillus* and *Bifidobacterium* species and thereby decreased formation of harmful bacterial metabolites, and regeneration of colonic mucosa.[12] Fermentation by colonic flora yields the SCFA acetic, propionic and butyric and small amounts of the gases CO_2, hydrogen (H_2) and methane (CH_4). The host rapidly and almost completely absorbs these products. The SCFA are utilised via established metabolic pathways and are energetically fully available. The gaseous products are largely expired in the breath. This is a normal process for metabolism of non-absorbed oligosaccharides (e.g. raffinose), resistant starch, non-digested plant gums and soluble fibres, secreted glycoproteins and mucopolysaccharides.[13]

However, excessive consumption of fermentable carbohydrates can lead to GI distress such as bloating, abdominal cramps (colic), flatus/gas, soft stools, borborygmi and, in extreme cases, diarrhoea in sensitive individuals.[14] Ingredients causing similar effects include, but are not limited to, polyols (e.g. lactitol, sorbitol and maltitol), lactulose, FOS, and many ingredients known collectively as dietary fibre (e.g. cellulose, cereal bran, xanthan and guar gum). These GI manifestations are transient and cease promptly upon cessation or reduction of intake of the responsible food, although subjects often develop improved tolerance over time. Therefore, these high-dose effects are not symptoms of toxicity but rather represent a normal consequence of overloading the GI tract.

Nine clinical studies in adults and children were conducted with polydextrose to evaluate the extent of such symptoms. The Joint FAO/WHO Expert Committee on Food Additives (JECFA) in 1987 and the EU Scientific Committee for Food (SCF) in 1990 allocated polydextrose an acceptable daily intake (ADI) 'not specified' after reviewing these studies. This means that neither agency found it necessary to stipulate an upper level of safe intake because excessive consumption is a matter of tolerance rather than safety. Therefore, polydextrose is permitted for use in any food at any level without restriction other than good manufacturing practice (GMP) in most markets. The United States allows polydextrose use as a food additive in most foods and requires that:

> The label and labelling of food, a single serving of which would be expected to exceed 15 g of the additive, shall bear the statement: 'Sensitive individuals may experience a laxative effect from excessive consumption of this product'.

JECFA concluded that the threshold for laxation was at an intake of about 90 g polydextrose per person per day or 50 g as a single dose.[15] In comparison, a daily dose of 20 g was found acceptable for sugar alcohols by the European Commission (EC) SCF in 1985. The SCF rapporteur pointed out, however, that such estimates are only provided as a guide and should not be used to establish maximum levels of use.[16]

Metabolism

The scientific data show that polydextrose is partly fermented in the colon in the same way as other dietary fibres. In general, animal data[3,5] and human data[2,4,17,18] suggest that the quantity of polydextrose fermented is of the order of 30–50%, which is similar to that for certain other types of dietary fibre (e.g. cellulose). This low degree of fermentation, in spite of the high level of solubilisation of polydextrose, appears to be due to the highly branched polymerised structure. This results in a slower fermentation rate by the intestinal flora, resulting in beneficial effects in the distal colon. Colonic fermentation of polydextrose produced gases (H_2, CO_2 and CH_4) and organic acids, mainly lactic acid and SCFA (acetic acid, propionic acid and butyric acid), as for other types of dietary fibre. One study[19] measured SCFA produced by fermentation of 17 carbohydrates by slurries of mixed faecal bacteria. Polydextrose was found to have a molar ratio of acetate, propionate and butyrate of 61:25:14. The ratio for inulin was 72:19:8, for oligofructose 78:14:8, for starch 58:17:25, for pectin 75:16:10 and for arabinogalactan 68:24:8. A high ratio of butyrate is desirable, as it is thought to be particularly beneficial to colonocyte health.[20–22] A high ratio of propionate is desirable, as it may inhibit cholesterol synthesis.

Production of SCFA leads to a decrease in intestinal pH, and studies show that the presence of polydextrose results in a faecal or intestinal pH fall.[17,23–25] This change can improve the composition of gut microflora by promoting growth of specific beneficial bacterial strains such as *Bifidus* and *Lactobacillus* and/or diminishing detrimental bacteria such as *Clostridium perfringens*. This is often termed a prebiotic effect. Studies on polydextrose show such effects.[17,19,24] Also, one study showed that polydextrose reduced the levels of certain putrefactive/carcinogenic substances (indole and *p*-cresol) produced by bacterial fermentation. Inulin and oligofructose also show marked increases in *Bifudus* and *Lactobacillus*.[19]

Polydextrose has been shown to increase faecal volume, weight and moisture, soften stools and decrease transit time.[18,24,26,27]

The effect of polydextrose on carbohydrate and lipid metabolism is similar to that of other low-viscosity indigestible bulking agents. Litesse has a low glycaemic index[5,7–9] compared with glucose.[28] Also, it has been shown that addition of polydextrose to a glucose solution will slightly reduce subsequent post-prandial glycaemia.[24]

Studies on polydextrose and lipid metabolism have been performed in animals,[18,29–31] in healthy human volunteers[24,32–34] and in hypercholesterolemic subjects.[31] The authors reported varying reductions in fasting plasma triglyceride levels and reduced blood cholesterol.

Finally, an animal study showed that polydextrose ingestion increased calcium absorption and bone mineralisation.[35]

20.2.1.5 Applications

Litesse is compatible with all other food ingredients and is used in many food categories. As a multi-functional speciality carbohydrate, it offers many physiological and functional benefits.

Litesse allows the development of food products with a wide variety of nutritional improvements such as fibre fortification, calorie reduction, reduced glycaemic load as well as sugar and fat reduction. The functional properties of Litesse deliver products with a taste and texture profile that is similar to the standard product.

Confectionery
The combination of the high water solubility and the high solution viscosity of Litesse facilitates the manufacture of sugar-free and reduced-sugar candy of excellent eating quality. Litesse can also be used to make reduced-calorie and reduced-sugar hard and chewy candies, caramels, and pectin and gelatine jellies. Litesse is amorphous and does not crystallise at low temperatures or high concentrations, so it can be used to control the crystallisation of polyols and sugars, and therefore the structure and texture of the final product. This is analogous

to conventional sugar confectionery production where glucose syrups are used to prevent or control sucrose crystallisation although selection of the appropriate Litesse form offers greater flexibility in terms of colour and taste depending on the extent of Maillard reaction desired. Its non-cariogenic properties can be useful in tooth-friendly confectionery. Litesse® II and Litesse® Ultra™ are the grades of choice for confectionery applications.

Combinations of Litesse Two and Litesse *Ultra* can also be used to adjust the flavour profile of confectionery products as Litesse gives a cooked sugar flavour to sugar-free and reduced sugar products.

Litesse *Ultra* is a particularly useful ingredient in the production of sugar-free and reduced-calorie hard candy and offers both product and processing improvements when used at low levels of addition to polyol mixes. While acknowledging that some polyols work very well in this application, there are still some technical issues associated with their use in hard-candy systems. From production trials using Litesse *Ultra* in combination with polyols, the following general conclusions have been drawn:

- Increasing proportions of Litesse *Ultra* increase the molten mass viscosity, improving the handling characteristics of sugar-free hard candy. Cooling times are reduced and the products may be processed using conventional stamping and depositing technology.
- Low-level additions of Litesse *Ultra* reduce the graining of isomalt candies and the cold flow of other polyols such as maltitol (formulation dependent).
- The stability of Litesse *Ultra* combinations may be equivalent or better than sugar/glucose candy.
- Clear, transparent products with excellent flavour release are possible.
- Greater than 50% calorie reductions are possible.

Chocolate

Factors impacting the taste and mouthfeel of chocolate include the sweetness, melting point, heat of solution and solubility of the sugar substitute. Sweetness can easily be manipulated by the appropriate addition of intense sweeteners. Products with low solubility, however, will give rise to a gritty mouthfeel and those with a significant negative heat of solution will give rise to a cooling effect, which is not desirable in chocolate. Litesse may be used in combination with polyols to overcome some of these taste considerations by providing a warm, creamy texture in the chocolate matrix without contributing a mouth-cooling effect or scratchy aftertaste. Litesse also helps to balance the flavour profile of sugar-free chocolate through the formation of small amounts of caramel during processing. Its low residual acidity ensures that the delicate cocoa and sweet flavours are brought forward and maintained. Conventional conching temperatures are possible with Litesse, so there are no technical compromises. As with other food applications, Litesse in chocolate can be used to manufacture products that range from sugar free to reduced sugar, and when used in combination with fructose or xylitol the sweetness synergy with Litesse enables reduced sugar chocolate products to be made without intense sweeteners.

Frozen dairy desserts

Litesse is used in take-home, frozen novelties and soft-serve ice creams to replace all or part of the sugars, ensuring frozen dairy desserts with excellent mouthfeel, smoothness and taste.

Textural qualities such as hardness, softness and meltdown can be modified to varying degrees when using Litesse in combination with sugars or polyols.

Litesse has a higher solution viscosity than simple sugars and polyols at equivalent concentration and temperature, and improves the creaminess of frozen dairy desserts, making it possible to replace part of the fat content.

Litesse combines well with other ingredients used in frozen dairy dessert such as emulsifiers, stabilisers, flavours, sugars, polyols and intense sweeteners.

Cultured dairy products
Litesse in yoghurt improves the creaminess, mouthfeel, taste and flavour of yoghurt white base, when used a low levels (3% w/w).

Fruit preparations with Litesse lead to improved creaminess and mouthfeel in fruit yoghurts.

The ability of Litesse to increase the mouthfeel and body of cultured dairy products is also seen in drinking yoghurts and cream cheese, for example, and is particularly noticeable in fat-free or low-fat applications.

Beverages and dairy drinks
Litesse is used in dairy drinks; neutral or flavoured, neutral or low pH, pasteurised or ultra-high temperature (UHT), and in many other clear, beverage formats.

It will improve the mouthfeel, giving the taste experience of a product of a much higher-fat content and this is particularly noticeable in low-fat dairy drink applications. Litesse is also added to these formats as a source of dietary fibre as it is very soluble, forming clear solutions and is very stable over shelf life.

Fruit spreads and fruit fillings
Litesse is used in fruit spreads and fruit fillings to replace all or part of sugars. Its high water solubility and high viscosity is used to reduce the sugar content and the calorie value.

Litesse helps prevent migration of moisture from fruit fillings into dough and pastry increasing the shelf life in a combination product.

Pasta and noodles
Fibre enhancement of noodle and pasta products is possible with Litesse as well as some process improvement benefits to the mechanical properties of the dough. The addition of Litesse to the dough improves firmness that can aid forming of noodle or spaghetti strands or pasta shapes. The texture of the cooked product is not significantly changed by the addition of Litesse and 95% of the added Litesse remains in the pasta or noodles after cooking.

20.2.1.6 Safety

Numerous safety studies were performed on polydextrose prior to its filing for approval as a food additive in 1979 by Pfizer in the United States and many other countries. These are listed in the following text and summarised in detail by Burdock and Flamm.[7] The reports refer to two types of polydextrose products: Type A (PD-A) is the weakly acidic native condensation polymer, and Type N (PD-N) is Type A neutralised with potassium carbonate.

Acute studies
Seven acute toxicity studies were conducted – two in mice, two in rats and three in dogs – with six oral and one intravenous. As no deaths occurred, the LD_{50} was greater than the highest dose tested (19–47 g/kg body weight).

Table 20.6 Polydextrose mammalian toxicology studies: effect levels.

Species	Route	Sample	Doses[a]	Study	Lowest effect level (mg/kg)	Highest no-effect level (mg/kg)
Monkey	Gavage	PD-N	1, 2, 10	3-month gavage study		10,000
Dog	Diet	PD-A	50%	3-month feeding study	10,000[b]	
Dog	Diet	PD-N/A	50%	6-month feeding study	10,000[b]	
Dog	Diet	PD-A	16.7%, 33%	13-month feeding study	3,340	
Dog	Diet	PD-N	10%, 20%	24-month toxicity study	4,000	2,000
Dog	Diet	PD-N	50%	24-month toxicity study	10,000[b]	
Mouse	Diet	PD-A	5%, 10%	18-month carcinogenicity study		15,000[c]
Rat	Diet	PD-A	1, 2, 10	3-month dietary study		10,000
Rat	Diet	PD-A	5%, 10%	24-month carcinogenicity study	5,000	2,500
Rat	Diet	PD-A	1, 2, 4 g/day	Segment I study		10,000[c]
Rat	Gavage	PD-A	1, 2, 4 g/day	Segment II study		10,000[c]
Rat	Gavage	PD-A	1, 2, 4 g/day	Segment III study		10,000[c]
Rat	Diet	PD-A	5%, 10%	Three-generation study		10,000[c]
Rabbit	Gavage	PD-A	3, 6, 12 g/day	Segment II study		3,000[c]

[a]Dose in mg/kg body weight per day if not given as % of diet or g per day.
[b]Lowest effect level in this single dose study.
[c]Highest level tested.

Genotoxicity studies
No mutagenicity, cytogenicity or dominant lethality was seen in any of the following studies:

- Ames spot test and quantitative plate assay with five strains of *Salmonella typhimurium*.
- Host-mediated assay with *S. typhimurium* in CD-1 albino mice.
- *In vitro* and *in vivo* cytogenicity studies with human lymphocytes.
- Dominant lethal assay in mice.

Subchronic and chronic studies
Fourteen mammalian subchronic and chronic studies were carried out with polydextrose. These are listed in Table 20.6.

These studies indicated a no-effect level (NOEL) of 2500–10,000 mg/kg body weight per day. The only significant effect observed was the appearance of laxation at high doses in the diet, with the beagle dog being the most susceptible.

After review of the study reports, both the JECFA (1987) and the EC/SCF gave polydextrose a 'not specified' ADI.[15]

Clinical toleration studies
Nine clinical studies were conducted with polydextrose to evaluate its GI toleration. Seven other clinical studies on polydextrose were reported in the literature where different properties

were evaluated and GI toleration was also noted. These are summarised in Table 20.7 and discussed in greater detail in Flood et al.[39]

The first seven of these studies were reviewed by JECFA at its 31st meeting in 1986. They concluded in their 1987 report (WHO Technical Report Series 759 (13)) that:

> [S]tudies in man have demonstrated that polydextroses, when administered at very high doses, exert a laxative effect, with a mean laxative threshold of 90 g per day or 50 g as a single dose.

This view was echoed by the EC/SCF in 1990.

These studies showed that polydextrose is better tolerated than most other poorly digestible carbohydrates, like the polyols, primarily because it is a much higher molecular weight polymer, so has a lower osmotic effect. The practical no-effect dose for laxation is 50 g per day according to Flood et al.[39]

20.2.1.7 Cariogenicity

Polydextrose[40] and hydrogenated polydextrose[41] (Litesse *Ultra*) have been shown to be non-cariogenic by the indwelling oral plaque pH telemetry method.

20.2.1.8 Regulatory status

Polydextrose is an approved food additive in the United States for a broad range of uses in 21 CFR 172.841 and in the European Union as a miscellaneous food additive for use at *quantum satis*. It is approved for use in foods in 75 countries, 74 of whom permit use of the 1 kcal/g energy value for labelling. Polydextrose has been evaluated by the JECFA which allocated an ADI 'not specified'.

Due to its minimal digestibility and low-calorie content, polydextrose also functions as a dietary fibre. Labelling of polydextrose as dietary fibre is currently approved in 31 countries, including Argentina, China, Indonesia, Mexico, Poland, Australia (FSANZ 2004), Czech Republic, Japan, New Zealand, Singapore, Belgium, Finland, Korea, Norway, Spain, Brazil, France, Malaysia, Philippines, the United Kingdom, Vietnam and the United States.

20.2.1.9 Method of analysis

The Association of Official Analytical Chemists (AOAC) method (985.29) for determination of total dietary fibre (TDF) in food is an enzyme-gravimetric method that is accepted in most countries. An ethanol precipitation step is designed to precipitate non-digested polysaccharides and leave the simple sugars and other small molecules in solution. However, the supernatant contains resistant oligosaccharides, and other resistant polysaccharides, preventing them from being measured in the TDF content of foods. Studies indicate that polysaccharides such as polydextrose, inulin, pectin, arabinan and arabinogalactan are not quantitatively precipitated by 80% ethanol.[42–45] Therefore, several new AOAC methods have been developed to quantify these components. They include methods for polydextrose (2000.11), fructans/inulin (997.08, 999.03), galacto-oligosaccharide (GOS) (2001.02) and resistant maltodextrin (2001.03). These methods can be used in conjunction with the enzyme-gravimetric methods for fibre measurement and are in concurrence with recent definitions developed by AACC International, National Academy of The Sciences/Institute of Medicine (IOM), Australia New Zealand Food Authority (ANZFA) and Codex Alimentarius.

Table 20.7 Polydextrose clinical toleration studies.

Author	Date	Site	Top dose	Subjects	Duration
Alter[23]	1974	Pfizer (CT)	150 g/day	20 male adults	3 weeks
Knirsh[23]	1974	Pfizer (CT)	79 g/day	57 male adults	10 days
McMahon[23]	1974	Tulane University (LA)	50 g/day	10 type-2 diabetes	Single dose
Raphan-a[23]	1975	Pfizer (CT)	130 g/day	21 adults (11 male/10 female)	10 days[a]
Raphan-b[23]	1975	Pfizer (CT)	60 g/day	51 adults (31 male/20 female)	12 weeks
Bunde[23]	1975	Hill Top Residents (OH)	55 g/day	58 children 2–16 years old	6 weeks
Scrimshaw/Young[23]	1977	MIT (MA)	50 g/day	16 adults (11 male/5 female)	8 weeks[a]
Beer[36]	1989	University of Texas	58 g/day	24 male adults	Single dose[a]
Curtis[37]	1990	Harris Labs (NE)	40 g/day	200 female adults	Single dose[a]
Achour et al.[2]	1994	Paris	30 g/day	7 adult males	21 days[a]
Zhong et al.[24]	1999	PR China	12 g/day	120 adults (66 male/54 female)	28 days[a]
Tomlin/Read[26]	1988	Sheffield UK	30 g/day	12 adult males	10 days[a]
Nakagawa et al.[27]	1990	Hino Japan	10 g/day	22 female adolescents	5 days[a]
Saku et al.[38]	1991	Fukuoka Japan	15 g/day	51 adults (25 male/26 female)	60 days[a]
Endo et al.[17]	1991	Saitama Japan	15 g/day	8 adults (6 male/2 female)	14 days[a]
Liu/Tsai[33]	1994	Taipei Taiwan	10 g/day	10 university students	18 days[a]

[a]No diarrhoea reported.

20.3 RESISTANT STARCHES AND RESISTANT MALTODEXTRINS

Resistant starch was discovered during the process of developing a procedure for measuring dietary fibre as non-starch polysaccharide (NSP), that is the non-α-glucan polysaccharides of plant material.[46] Heat processing of certain starchy foods caused a fraction of the starch to become resistant to digestive enzymes. On the basis of the probable rate of digestion in the small intestine, starches can be classified into three categories: (1) rapidly digested starch, (2) slowly digested starch and (3) resistant starch (Table 20.8).[47]

Resistant starch was further classified into three types (RS I, RS II and RS III) on the basis of intrinsic factors of the starch that render it indigestible.[48,49]

Similar types of indigestible components can be found in chemically modified starches (starch hydrolysates such as dextrins, maltodextrin and glucose/corn syrup). These are termed resistant maltodextrins and form a fourth classification of resistant starches, RS IV.

Different resistant starches can be manufactured by different physical processes. By changing the processing conditions, the starch structure can be remodelled to have different chemical properties and hence functionality in foods, depending on the polymer size, types of bond and degree of branching.

The degree of resistance to digestion and fermentation rate of resistant starches is a function of structural differences. The manufacturing process and source of starch is also of crucial importance. Many studies in the literature discuss resistant starch collectively and it has generally been observed that the commercial promotion of resistant starches relies on the use of general data, rather than differentiating between the type of starch or process used.

RS II, RS III and RS IV are generally used as sources of dietary fibre in food formulations. RS II and RS III are naturally present in food products but can be manufactured and isolated in purer forms, although these products are generally not very soluble, and have limited application as sugar alternatives. Their main application is as fibre-enrichment ingredients and limited sugar reduction functionality in low-moisture systems such as baked goods and cereal products.

The chemical modifications used to produce RS-IV-resistant maltodextrins, however, produce ingredients that are relatively soluble with more 'sugar-like' properties. Fibersol-2® and Nutriose® FB are commercial examples of this type of product.

Table 20.8 Definitions and examples of resistant starch and resistant maltodextrins.

	Definition	Examples
RS I	Starch physically inaccessible	Partially milled grains, seeds and legumes
RS II	Raw starch granules or high amylose starch unmodified	Native potato starch, banana starch
RS III	Non-granular retrograded starch or crystalline starch (mainly retrograded amylose)	Cooked and cooled potato, bread, ready to eat breakfast cereals
RS IV	Chemically modified cross-linked starch	Fibersol-2 NutrioseFB

RS, resistant starch.

20.3.1 Fibersol-2

20.3.1.1 Description

Starch hydrolysates containing indigestible components are termed *resistant maltodextrins*. One such material that is produced by a combination of hydrolysis and transglucosidation is Fibersol-2.

Fibersol-2 is a commercial example of a resistant maltodextrin (RS IV) and is produced by a combination of heat and enzymatic treatment of cornstarch.[50] It is produced commercially by Matsutani Chemical Industry Co. Ltd, in a two-part process.

In the first reaction, cornstarch is heated with a small amount of hydrochloric acid under low-moisture conditions. During this reaction, the cornstarch is hydrolysed by transglucosidation. In the second reaction, the solution is hydrolysed by amylase. The material is then refined to separate out impurities, and spray dried.

The average molecular weight of Fibersol-2 is 2000 and it is composed not only of $\alpha(1\text{-}4)$ and $\alpha(1\text{-}6)$ glucosidic bonds, but also $(1\text{-}2)$ and $(1\text{-}3)$ linkages as well as some levoglucosan. The highly branched nature of Fibersol-2 means that it is only partially hydrolysed by human digestive enzymes. Fibersol-2 is a white powder and contains about 90% dietary fibre.

The viscosity of Fibersol-2 is lower than that of a conventional maltodextrin. In solution it is very clear and stable and does not become cloudy or show signs of retrogradation when stored for long periods of time. It has also very good anti-acid properties and can be cooked and sterilised at high temperature in food applications owing to its stability in heat processes.

Owing to its relatively low average molecular weight, low viscosity, and low calorie value (1–1.5 kcal/g), Fibersol-2 can be used as a bulking agent in combination with high-intensity sweeteners. It forms clear aqueous solutions with a clean taste and a relative sweetness of 0.1 compared with sucrose.[51]

20.3.1.2 Physiological properties

Studies have clearly shown that significant amounts of dietary starch may escape digestion in the small intestine and pass into the colon. There the starch is fermented by anaerobic bacteria and the resulting SCFA are absorbed. Resistant starch that reaches the large intestine may share some of the characteristics and health benefits attributed to dietary fibre such as amelioration of diabetes, cardiovascular disease and colon cancer. Replacement of digestible starch with resistant starch in the diet of humans results in significant reduction in postprandial glycaemia and insulinaemia, and in the subjective sensation of satiety.

The amount of starch reaching the large intestine is important in determining the extent of fermentation and effect on colonic function. The source of starch also appears to affect absorption.

Fibersol-2 is only partially digested and absorbed in the upper GI tract. When it reaches the large intestine, it is partly fermented by bacteria producing SCFA. According to a rat study conducted by Tsuji and Gordon,[52] the caloric value of Fibersol-2 is 1.2 kcal/g (after taking into account an estimate for the energy utilisation of SCFA and adding the energy value from 10% of Fibersol-2 digested in the small intestine). Matsutani suggests a calorific value of 1.5 kcal/g for Fibersol-2. It is estimated that 90% of Fibersol-2 reaches the lower intestine and that approximately half of that is fermented by intestinal bacteria.[53] Fibersol-2 has a modified structure that leads to different fermentation patterns to other polyglucoses.

According to Kishimoto et al.,[54] *in vitro* fermentation produces an SCFA ratio of 44:18:38 (acetate:proprionate:butyrate).

Fibersol-2 increases the proportion of *Bifidobacteria* in the microbiota but does not seem to have an effect on *Bacteroides*.[55]

Human and rat studies have shown that the ingestion of RS IV increases the faecal frequency and weight but no significant results were reported regarding transit time.[52–55]

Post-prandial glucose levels
Following loading tests in rats using various saccharides, co-administration of Fibersol-2 led to the lowering of blood glucose levels (to 70% peak value) found after sucrose or maltose administration. It was also discovered that the glucose attenuation effect was specific to disaccharides and starch. Wakabayashi[56] and Wakabayashi et al.[57,58] proposed that Fibersol-2 is able to inhibit the absorption of disaccharides by reversibly blocking the co-operative disaccharide-related transport system.

Fat metabolism
It has been shown that Fibersol-2 lowered total cholesterol levels by lowering low-density lipoprotein (LDL) levels, without changing high-density lipoprotein (HDL) levels. Furthermore, ingestion of Fibersol-2 lowered serum triglyceride levels in both healthy (control) adults and type-2 diabetic patients.[51]

20.3.1.3 Applications

As a dietary fibre source, Fibersol-2 can be used in processed food products to make a dietary fibre fortification claim such as 'good source of fibre' or 'rich in fibre'. This type of claim has been employed in Japan and many other countries and used for processed food products, including fruit beverages, vegetable beverages, soups, cereals and bread.

In Japan, there is an approval programme called 'Foods for Specified Health Use' (FOSHU), whose products are certified through application to the Ministry of Health and Welfare in a similar manner to medical drug applications. Supporting data of safety, physiochemical properties and test methods must be provided. FOSHU products, which must contain the approved ingredients and must provide medically and nutritionally effective physiological functions with an effective intake, can be marketed with health claims such as bowel regularity, blood glucose control effect, lowering cholesterol effect and blood-pressure-lowering effect. Presently, FOSHU products containing Fibersol-2 as the effective ingredient, in the form of beverages, powdered beverages, cookies and sausages have been approved and marketed (*note*: polydextrose is also listed under FOSHU for 'maintains improved bowel function').

Conventional maltodextrins have been used in sport drinks to maintain blood glucose levels during activity. However, maltodextrins and other digestible saccharides promote additional secretion of insulin and interrupt the metabolism of fat, an important energy source in the body. The addition of Fibersol-2 to sports drinks containing maltodextrins may help to maintain stamina during exercise by maintaining blood glucose levels, without the additional secretion of insulin, while encouraging the effective metabolism of fat.[51]

Owing to its low average molecular weight, low viscosity and low calorie value, Fibersol-2 can be used as a bulking agent in conjunction with high-potency sweeteners such as aspartame, sucralose or acesulfame K, or as an ingredient for formulating low-calorie foods.[51]

20.3.2 Nutriose FB

20.3.2.1 Description

Nutriose FB, a registered trademark (and commercial product) of Roquette Freres, is a further example of a chemically modified maltodextrin. Nutriose FB is made by chemical dextrinisation of wheat starch, followed by refining, purification and drying. This results in an agglomerated soluble dextrin with high dietary fibre and low sugar content. The product contains a significant level of indigestible (1-2) and (1-3) glucose linkages, that contribute to its high dietary fibre content.

Animal studies with Nutriose FB show that about 15% is hydrolysed in the small intestine and 75% is fermented in the colon with the production of SCFA and a commensurate reduction in colonic pH. Owing to its slow, progressive digestion through the gut, Nutriose FB is well tolerated at up to 45 g per day without discomfort. It has a low glycaemic response, of 25% that of glucose, and has a caloric value of 2 kcal/g.

Nutriose FB contains 55% fibre as determined by AOAC method 985.29, which has been recognised by the FDA. However, it is 85% soluble fibre by AOAC method 2001.03 (enzyme high-performance liquid chromatography (enzyme-HPLC) method)). Nutriose FB was designed to contain very low amounts of mono- and disaccharides (<0.5%), which makes it useful as a bulking agent in the formulation of sugar-free, no-sugar-added and reduced-sugar products. Its reduced-calorie content and ability to partially replace fat in many applications facilitate the formulation of reduced-calorie foods. Owing to its low fermentability, Nutriose FB is non-cariogenic and can be formulated into foods that do not promote tooth decay.

The average DP of Nutriose FB is 16, which results in physical behaviour similar to that of a 10–15 dextrose equivalent (DE) maltodextrin in many applications.

Like other bulking agents of this type, Nutriose FB has a low sweetness.

The high stability and solubility (>70% w/w in aqueous solution) of Nutriose FB makes it a versatile ingredient in food products for fortifying soluble dietary fibre. It is thermally stable at the pH of most food systems, even during UHT pasteurisation. It will not break down during extrusion or through freeze–thaw cycling. It is also stable at low pH, enabling its use in high-acid fruit-based fillings and beverages, where FOS would break down to simple sugars. Nutriose FB is not consumed by yeast and so is compatible in most baking applications to give the full fibre content in the finished product.[59]

20.3.2.2 Safety-resistant maltodextrins

Several toxicology and tolerance studies have been reported[60] and these are summarised in the following text.

Acute studies
Wakabayashi et al.[61] studied the acute toxicity of one Matsutani RMD – Fibersol-2 – in male mice given a single dose of 5, 10 and 20 g/kg body weight. After a 7-day observation period, there were no deaths; hence, the LD_{50} was greater than 20 g/kg. It was also noted that all of the mice in the 20 g/kg group and most in the 10 g/kg group exhibited diarrhoea 2–3 hours after administration.

Genotoxicity studies
An Ames mutagenicity study was performed with Fibersol-2 using four strains of *S. typhimurium* with and without metabolic activation with S-9 mix.[61] No mutagenicity was observed.

Subchronic and chronic studies
A subchronic dietary study in rats was performed with Fibersol-2 at 5%, 10% and 20% of the diet for 5 weeks.[62] No remarkable observations were noted, except for a significant drop in serum cholesterol levels in all test groups versus the controls.
   No chronic studies were reported.

Human clinical studies
Several human toleration studies have been conducted with Fibersol-2 (see Table 20.9). No significant adverse effects were reported in any of these studies and no incidences of diarrhoea. Several of these studies demonstrated reduced post-prandial peak glucose levels versus controls.

20.3.2.3 Regulatory status

In the United States, 'resistant maltodextrins' are considered maltodextrins and are thus generally recognised as safe (GRAS) substances under 21 CFR 184.1444. The product literature for Fibersol-2 indicates that this product is an indigestible dextrin with soluble dietary fibre properties.[74] The United States allows manufacturers to label foods with terms such as 'resistant' or 'indigestible', but they must clearly label the legally approved name of

Table 20.9 Human toleration studies for Fibersol-2.

Investigator	Year	Subjects	Dose
Nomura et al.[63]	1992	5 NIDDM	60 g/day for 12 weeks
Matsuoka et al.[64]	1992	2 male	30 g/day for 4 weeks
		2 male	30 g/day for 8 weeks
		8 male	30 g/day for 4 weeks then 15 g/day for 4 weeks
Fujiwara and Matsuoka[65]	1993	3 male/2 female NIDDM	30 g/day for 16 weeks
Tokunaga and Matsuoka[66]	1999	32 male/8 female	5.12 g single dose
Tokunaga and Matsuoka[66]	1999	10	15.4 g/day for 1 month
Sinohara et al.[67]	1999	26 male/13 female	5 g single dose
Uno et al.[68]	1999	17 male/17 female	5 g single dose
Wakabayashi et al.[69]	1999	5 male	10 g single dose
Mizushima et al.[70]	1999	25 male	9.8 g single dose
Mizushima et al.[70]	1999	3	29.4 g/day for 1 month
Mizushima et al.[71]	2000	Not indicated	29.4 g/day for 12 weeks
Kishimoto et al.[72]	2000	27 male	5 g single dose
Kishimoto et al.[72]	2000	10	15 g/day for 3 months
Kajimoto et al.[73]	2001	8 male/8 female	18.3 g/day for 13 weeks

NIDDM, non-insulin dependant diabetes mellitus.

the corresponding starches. Matsutani also affirms that Fibersol-2 is a FOSHU ingredient in Japan used for the following:

- intestinal regularity (1992);
- moderating post-prandial blood glucose levels (1994);
- lowering serum cholesterol levels (1998);
- lowering triglyceride levels (1998).

Recommended intake amount: 3–10 g per serving.

Currently, in the European Union, chemically modified starches are regulated as modified starches under European parliament and Council Directive 95/2/EC.

20.3.2.4 Method of analysis AOAC 2001.03 for RS IV

The standard Prosky method includes a precipitation of the fibre in ethanol. However, like polydextrose, Fibersol-2 is not precipitated at this stage and would be lost in the filtrate. So, the 2001.03 method adds extra steps to analyse the filtrate content by HPLC to determine low-molecular-weight soluble dietary fibre.

20.4 FRUCTO-OLIGOSACCHARIDES

20.4.1 Inulin and low-molecular-weight FOS

20.4.1.1 Description

Inulin and low-molecular-weight FOS are interrelated, but it is important to understand the difference. Generic FOS are composed of linear chains of fructose units linked by β(2-1) bonds and often terminated by a glucose unit. Inulin contains chains of 3–60 fructose units but low-molecular-weight FOS have between 2 and 7 fructose units and are generally obtained from the partial enzymatic hydrolysis of inulin (see Figure 20.2). The polymeric distribution of inulin varies depending on factors such as the source and growing condition of the raw materials. Since they are both polymers of fructose, they are also considered fructans.

Inulin is found in more than 36,000 plants, including leeks, onions, artichokes, garlic and wheat. Inulin is also present in large amounts in chicory roots from which it is extracted by hot water. On a commercial scale, the FOS (when used in this text, FOS refers to low-molecular-weight FOS (2–7 fructose units)) may be produced either from sucrose through the transfructosylating action of fungal fructofuranosidase or from chicory inulin by partial hydrolysis with endoglycosidases.

The different commercial forms of inulin and FOS differ with regard to their granulation, sugar content and average DP. Standard inulin products have an average DP of around 10. The higher DP forms of inulin (DP > 23) are less soluble but have the unique ability to form gels under certain conditions. This property varies with the concentration of inulin (which typically needs to be >25%) and the DP of the inulin polymer chain (the longer the chain, the less inulin is needed to form a gel). This property allows inulin to replace fat in foods such as yoghurts, cheese spreads and frozen desserts.

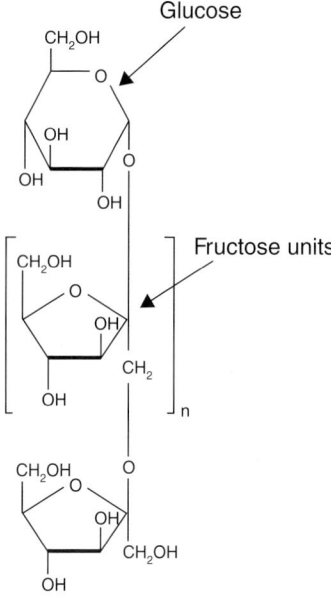

Fig. 20.2 The chemical structure of inulin ($n = 3–60$) and fructo-oligosaccharides ($n = 2–7$).

Inulin and FOS are available in many commercial forms, including powder, granulated, instant, gel and low sugar.

20.4.1.2 Organoleptic quality: FOS[75]

The sweetness intensity of FOS depends on the composition of the product. Where a product contains higher levels of the natural sugars present such as glucose, fructose and sucrose, these will have higher sweetness. Commercial products vary from 30% to 65% of the sweetness of sucrose. The sweetness profile is comparable to that of sucrose and FOS can be used to replace sugar and glucose syrups in many applications. Some aromas such as fruit flavours are often more pronounced when used in combination with FOS than they are when mixed with sugar.

FOS has a synergistic effect when combined with intense sweeteners. The resulting mixture has a sweetening profile that more closely resembles that of sugar than intense sweeteners alone. Combinations of FOS and intense sweeteners are used particularly in fruit preparations.

20.4.1.3 Physical and chemical properties: FOS[75]

Acid stability
In very acidic conditions and with significant exposure times and temperatures, FOS may be hydrolysed to fructose. This makes the use of FOS in very acid products with long shelf life, such as soft drinks, very difficult. In most other applications, hydrolysis can be limited as in industrial fruit preparations and fresh fruit juices.

Solubility
FOS has a higher solubility than sucrose. FOS will not crystallise, precipitate or leave a dry or sandy feeling in the mouth. The moderate reducing power of the different FOS products can give rise to slight browning reactions during baking.

20.4.1.4 Organoleptic quality: inulin[76]

Inulin is only slightly sweet, without any aftertaste or off-flavour.

20.4.1.5 Physical and chemical properties of inulin[76]

Acid stability
At pH values of less than 4 and with significant exposure times and high temperatures, inulin and FOS hydrolyse into fructose and glucose. In very acid conditions and in products with a long shelf life, such as soft drinks, the use of inulin and FOS is difficult. In most other cases, a solution can be found, either by modifying the process conditions or by using inulin gels that have greater resistance to acid hydrolysis.

Solubility and thermal stability
Inulin has limited solubility in water (approximately 10% at room temperature) but is highly stable even when heated to high temperatures used for sterilisation.

Inulin, like FOS, is susceptible to hydrolysis, but the degree of hydrolysis depends on time, temperature, product composition and pH. This property can affect the suitability of inulin as a sugar replacer in some applications. Despite this, inulin works very well in many food applications.

20.4.1.6 Physiological properties of inulin and FOS

The majority of inulin and FOS are not digested in the stomach or small intestine. The indigestibility of fructans is due to the fact that, in the human small intestine, there is no enzyme that is able to hydrolyse the β (1-2) glycosidic linkages.

Ellegard *et al.*[77] have used an ileostomy model and the studies have shown that 88% of the ingested doses of FOS or inulin were recovered in the ileostomy effluent, supporting the conclusion that chicory FOS are practically indigestible in the small intestine of man.

Dietary fibre
Inulin and FOS are resistant to digestion in the upper part of the intestinal tract and are subsequently fermented in the colon. They have a bulking effect due to the increase in microbial biomass that results from their fermentation, but they also have other important functional attributes.

Energy value
Various methodologies have been used to estimate the caloric value of inulin and FOS in human subjects. According to Roberfroid[78], because only a part of the energy of these dietary carbohydrates is salvaged, their available energy content is only 40–50% that of digestible carbohydrate, giving an energy value of 1–1.5 kcal/g.

Fermentation of FOS
It has been demonstrated in human studies that inulin and FOS are fermented by bacteria colonising the large bowel. This has been supported by a large number of *in vitro* and *in vivo* studies.

Molis *et al.*[79] used an intubation technique with six human volunteers who were fed 20.1 g FOS per day in three doses over an 11-day period. Most of the ingested FOS (89%) was not absorbed in the small intestine and none was excreted in the stools, indicating that the 89% reaching the colon was completely fermented by the gut microflora. A small fraction of FOS was recovered in urine (0.12%).

Increases in breath H_2 measurements over time indicated a high fermentation rate by colonic bacteria. Twenty-four human volunteers in a 5-week study using a glucose control were fed 5 g and 15 g FOS per day. Breath H_2 measurements showed a significant increase of 24-hour integrated excretion of breath H_2 for the highest dose of FOS (15 g per day). For the lowest dose (5 g per day), an increase was observed but this did not reach a significant difference versus the control. No changes were observed in SCFA in the faeces, molar proportions of SCFA, faecal pH and faecal weight, and no FOS was recovered in the faeces. This led the authors to conclude that FOS were fully metabolised in the large intestine with the level of fermentation appearing to be dose dependent.[80]

Further, 12 healthy human volunteers were given a control diet or diet supplemented with FOS, inulin or GOS at an intake level of 15 g per day. Levels of breath H_2 were measured for 4 hours after ingestion. The overall H_2 production in 4 hours was significantly higher on the FOS and inulin diet than on the control diet.[81]

Fermentation of inulin and FOS
In vitro fermentation studies comparing human faecal inoculums of inulin with rye, wheat and oat brans showed that inulin was fermented at a much faster rate than the brans. Ninety-nine percent of inulin was fermented during the first 4 hours whereas only 56%, 41% and 64% of the original rye, wheat and oat brans, respectively, were fermented after 24 hours.[82] Gas production was also measured: the initial rate of total gases and H_2 production was much quicker for inulin than for the bran products and the amount of total gases and H_2 produced by fermentation of inulin was much higher than for the bran products.

In human fermentation profile studies, it has been shown that the molar ratios of SCFA for inulin was 60:25:15. FOS produced higher ratios of acetate and lower butyrate and proprionate than inulin (70:15:15).[19,81,83–85]

No difference in faecal pH was observed in most of the studies, which showed that the SCFA produced by the fermentation of inulin and FOS are mainly utilised by the body.[81,82,84,86,87–89]

In general, studies show that FOS is completely fermented and that inulin is fermented very quickly on entering the large intestine.

Toleration
The digestive tolerance of FOS and inulin are well reported in human studies, and Briet *et al.*[90] have shown that digestive tolerance did not improve after 12 or 15 days of regular administration of FOS. Doses under 20 g per day resulted in only minor complaints, doses over 40 g per day resulted in abdominal rumbling and bloating and doses over 50 g per day resulted in abdominal cramps and diarrhoea.

In a 12-week, double-blind study, subjects suffering from irritable bowel syndrome (IBS) were given 20 g FOS powder per day ($n = 52$) or placebo glucose ($n = 46$). After a 4–6-week treatment, IBS symptoms improved more in the placebo group than the FOS group. After completion of the study, there were no significant differences between the two groups.[91]

In a study on 12 healthy humans,[81] the consumption of 15 g of FOS and GOS per day did not show any great increase in GI complaints. Most subjects reported more cases of flatulence, but as some flatulence complaints were also recorded on the control diet it was concluded that FOS and GOS were well tolerated.

Where ileostomy patients were fed a control diet supplemented with 17 g of inulin, 17 g of oligofructose (FOS) or 7 g of sucrose, for periods of 3 days, no subjects noticed any difference between the three diets indicating that doses of 17 g of inulin and oligofructose (FOS) were well tolerated.[77]

In other studies[86,90,92] it has been shown that at doses of 10–15 g per day, some secondary effects start to appear but the complaints were in general significant only for doses over 20 g per day.

Most results including studies that consider the transit time and quality of stools on ingestion of FOS or inulin are slightly contradictory: most studies did not find any effect on stools quality for FOS apart from Menne *et al.*,[87] who found an increase in frequency, quantity and softness.

Mineral absorption

In rats, the ingestion of inulin was shown to significantly increase the absorption of important minerals such as calcium and magnesium.[93] This could have important consequences for the prevention of osteoporosis. It has been shown that supplementing the diet daily with 15 g of FOS or 40 g of inulin caused a significant increase in the apparent absorption of calcium.[79] This was also confirmed in a later study.[94]

Bifidogenic effect

Several studies in human volunteers, animals and *in vitro* systems have shown that inulin specifically stimulates *Bifidobacteria*, although most studies used FOS. Relatively good evidence exists that FOS and inulin increase the bifidobacterial counts. However, effects on other types of bacteria were less evident (including the effect on *Bacteroides*). The underlying mechanism is most probably based on the chemical structure of inulin, more specifically the β(2-1) bonding that is not hydrolysed by any mammalian digestive enzyme but that can be hydrolysed by *Bifidobacterium*. This effect is called the 'bifidogenic effect' or more generally, 'prebiotic' function. Three to fifteen grams of inulin per day for a few weeks are enough to cause a significant increase in the population of intestinal *Bifidobacteria*.[95]

Furthermore, inulin decreases the incidence of pre-cancerous colonic lesions (aberrant crypt foci) induced in experimental animals and could therefore significantly contribute to the prevention of colon cancer.[96] A synergistic effect was observed when FOS and inulin were combined.[97]

Glucose responses

Twelve healthy volunteers received either 20 g of FOS per day or sucrose for 4 weeks in a double-blind crossover design. FOS did not modify fasting plasma glucose and insulin concentrations.[89] Mean basal hepatic glucose production was lower after FOS than after sucrose consumption. However, neither insulin suppression of hepatic glucose production nor insulin stimulation of glucose uptake measured by hyperinsulinaemic clamp was significantly different between the two periods. Yamashita *et al.*[98] showed a reduction in fasting glycaemia in diabetic patients after 14 days of FOS intake.

Inulin and FOS produce very low glycaemic responses compared with glucose.

In a double-blind trial of middle-aged men and women with elevated cholesterol and triglyceride levels, supplementation with inulin (10 g per day for 8 weeks) significantly reduced insulin concentrations (suggesting an improvement in blood glucose control) and significantly lowered triglyceride levels were observed.[99]

However, other studies with type-2 diabetics and healthy people showed no effect on blood sugar levels, insulin secretion or blood lipids.[81,89,98,100] Owing to these conflicting results, more research is needed to determine the effect of FOS and inulin on diabetes and lipid levels.

20.4.1.7 Applications of FOS[75]

Dairy products

Using FOS to replace sugars in ice creams, sorbets and dairy desserts is simple and often FOS syrups offer the best solution. The resulting products have excellent mouthfeel and taste, and hardness and melting characteristics are comparable with reference products. In combination with inulin, the fat content can also be reduced. In sorbets the fruit taste is enhanced and the 'scoopability' is improved. FOS combined with intense sweeteners often masks any residual aftertaste. Depending on the recipe, this can help formulate reduced sugar, no sugar added and calorie reduced products, and formulations suitable for diabetics.

FOS liquids or powder can easily be added to almost all dairy drinks. This allows the upgrading of such products with an additional nutritional claim in the area of prebiotics, synbiotics and dietary fibre effects.

Biscuits and bakery products

FOS syrups and powder are elegant replacements for carbohydrate in baked foods. This leads to reduced sugar products, products fit for diabetics and also fibre-enriched products. The resulting products may also have improved crunchiness. In breakfast cereals, FOS can be incorporated in the extruded material or used in coatings. To correct for the lower sweetness of FOS, intense sweeteners or combinations with fructose can be used. The baking process may lead to a slightly browner appearance after baking.

Fruit preparations for low-fat yoghurts

Low-fat yoghurts are a rapidly growing market segment. In these products, fruit preparations with a new combination of FOS and intense sweeteners have made a significant breakthrough. The acid hydrolysis during the production process of the fruit preparations can be minimised and the acid hydrolysis during the shelf life of the chilled final products is negligible. Fruit preparations containing up to 20% FOS are easily produced.

Cereal bars

In cereal bars, FOS allows the reduction of the amount of carbohydrates and calories and an increase in the dietary fibre content. FOS acts as a binder and has a positive influence on the structure, taste, and shelf-life freshness of the bars. Many examples on the market show the success of FOS in these applications.

20.4.1.8 Applications of inulin[76]

Chocolate

Inulin can be used for sugar substitution in chocolates. No major processing changes are required. The resulting products can be sugar free, reduced sugar or suitable for diabetics.

Combinations with polyols or fructose give excellent results. Dark chocolate, milk chocolate, chocolate spreads and chocolate/hazelnut spreads can be produced containing inulin.

Baked goods and breakfast cereals
There are almost unlimited possibilities for adding inulin to baked goods and breakfast cereals. Recipes are available for cakes, cookies, crackers, bread fillings and breakfast cereals where fibre enrichment is important. The resulting products may have improved structure and crispiness.

Fat replacer
Inulin has the potential to form gel-like structures when sheared in water under specific conditions and these structures are food imitations of fat, with the same mouthfeel and texture. This is the basis for the use of inulin as a fat replacer.

Inulin can be used as a fat substitute in many applications, including yellow fat spreads, meat products, frozen dairy desserts, dairy and vegetable creams and reduced fat cheeses.

The addition of 2–5% inulin to a beverage (above pH 4 and with standard heat treatments) provides the body and mouthfeel in skim and fat-free milk products normally associated with full-fat formulations. Inulin can also minimise bitter aftertastes, and reduce sweetener and flavour usage by 15–20% because of its enhanced flavour delivery.

20.4.1.9 Safety

Safety studies for FOS are summarised in the following text.

Acute studies[101]
Four acute studies were conducted – two in mice and two in rats – by single dose oral gavage at 3, 6 and 9 g/kg body weight. No deaths or any other signs of abnormality were seen, so the LD_{50} would be greater than 9 g/kg body weight.

Genotoxicity studies[102]
The following mutagenicity/genotoxicity studies were conducted:

- microbial reverse mutation assay with five strains of *S. typhimurium* and one strain of *Escherichia coli*;
- mouse lymphoma mammalian cell mutation assay;
- unscheduled DNA synthesis study in human epithelioid cells.

The investigator reported that FOS with a mean DP of 3.5 showed no evidence of genotoxic potential.

Subchronic and chronic studies
Tokunaga *et al.*[103] and Clevenger *et al.*[102] reported no significant adverse effects in subchronic and chronic toxicity and carcinogenicity studies in rats at doses up to 2170 mg/kg body weight per day. These studies were all conducted with 3.5 DP FOS, and are listed in Table 20.10.

A dose-dependent increase in pituitary adenomas was seen in the 24-month rat study. No other significant issues were seen in this or the other studies with FOS in the diet at up to 20%.[104–106]

Table 20.10 Fructo-oligosaccharides (FOS) mammalian toxicology studies.

Species	Route	Doses	Study
Rat	Gavage	1.5, 3.0, 4.5 g/kg	6-week feeding study (28)
Rat	Diet	5%, 10%	6-week feeding study (28)
Rat	Diet	0.8%, 2.0%, 5.0%	24-month carcinogenicity study (4)
Rat	Diet	20%	Developmental toxicity study (8)
Rat	Diet	5%, 10%, 20%	Reproductive toxicity study (24)
Rat	Gavage	3, 6 g/kg	Diarrhoea test (16)

Toleration studies

As inulin and FOS are considered dietary fibre, overconsumption can lead to adverse GI symptoms and diarrhoea. It was noted in the aforementioned rat feeding studies that the animals showed soft stools and diarrhoea at concentrations above 5% of the diet. Half of the rats in the aforementioned 3 and 6 g/kg dose groups had diarrhoea after 6 hours of administration, but not as extensively as control groups fed the same amount of sorbitol or maltitol.

Carabin and Flamm[107] reviewed 12 clinical toleration studies on FOS (see Table 20.11). They concluded that daily consumption of up to 20 g inulin is well tolerated.

20.4.1.10 Regulatory status

Inulin is a GRAS substance in the United States under proposed rule 21 CFR 170.36 as GRN 118 (5/5/03). FOS is also a GRAS substance in the United States under the same rule as GRN 44 (11/22/00). These materials have various approvals in other countries as a food, food ingredients or novel food.

As they are incompletely digested, inulin and FOS are considered dietary fibre in many countries, including Australia,[116] New Zealand, France and the United States.

Table 20.11 Inulin/fructo-oligosaccharides (FOS) clinical toleration studies.

Investigator	Date	Sample	Top dose	Subjects	Diarrhoea
Alles et al.[80]	1996	Oligofructose	15 g/day	24 adult males	
Briet et al.[90]	1995	FOS	50 g/day	14 (8/6 male/female)	Yes
Cadranel and Coussement[108]	1995	Oligofructose	9 g/day	43 children 10–13 years old	
Davidson et al.[109]	1998	Inulin	18 g/day	21 adults	
Garleb et al.[110]	1996	FOS	31 g/day	27 adult males	Yes
Gibson et al.[95]	1995	Oligofructose	15 g/day	8 (7/1 male/female)	
Kleeson et al.[111]	1997	Inulin	40 g/day	10 constipated females	
Molis et al.[79]	1996	FOS	20 g/day	6 (3/3 male/female)	
Pedersen et al.[112]	1997	Inulin	14 g/day	64 adult females	
Rumessen et al.[113]	1998	Inulin	30 g/day	10 (5/5 male/female)	Yes
Stone-Dorshow et al.[114]	1987	FOS	15 g/day	15 adult males	
Van den Heuvel et al.[115]	1999	Oligofructose	15 g/day	12 males 14–16 years old	

FOS, fructo-oligosaccharides.

20.4.1.11 *Method of analysis*

A specific AOAC method, the 'fructan method' AOAC 997.08 allows for the determination of FOS and inulin in foods.

REFERENCES

1. S. Alonso and C. Setser. Functional replacements for sugars in foods. *Trends in Food Science & Technology* 1994, 5, 139–146.
2. L. Achour, B. Flourie, F. Briet, P. Pellier, P. Marteau and J.C. Rambaud. Gastrointestinal effects and energy value of polydextrose in healthy non-obese men. *American Journal of Clinical Nutrition* 1994, 59(6), 1362–1368.
3. S.K. Figdor and H.H. Rennhard. Caloric utilization and disposition of [14C] polydextrose in the rat. *Journal of Agricultural and Food Chemistry* 1981, 29(6), 1181–1189.
4. S.K. Figdor and J.R. Bianchine. Caloric utilization and disposition of [14C] polydextrose in man. *Journal of Agriculture and Food Chemistry* 1983, 31(2), 389–393.
5. N.C. Juhr and J. Franke. A method for estimating the available energy of incompletely digested carbohydrates in rats. *Journal of Nutrition* 1992, 122(7), 1425–1433.
6. G.S. Ranhotra, J.A. Gelroth and B.K. Glaser. Usable energy value of selected bulking agents. *Journal of Food Science* 1993, 58(5), 1176–1178.
7. G.A. Burdock and W.G. Flamm. A review of the studies of the safety of polydextrose in food. *Food and Chemical Toxicology* 1999, 37(2–3), 233–264.
8. H. Mitchell, M.H. Auerbach and F.K. Moppett. Polydextrose. In: L.O. Nabors (ed.). *Alternative Sweeteners*, 3rd edn, Marcel Dekker, New York; 2001, pp. 499–518.
9. S.A.S. Craig, J.F. Holden, J.P. Troup, M.H. Auerbach and H.I. Frier. Polydextrose as soluble fiber and complex carbohydrate. In: S.S. Cho, L. Prosky and M. Dreher (eds). *Complex Carbohydrates in Foods*, Marcel Dekker, New York; 1999, pp. 229–247.
10. M.H. Auerbach, S.A.S. Craig, J.F. Howlett and K.C. Hayes. Caloric availability of polydextrose. *Nutrition Reviews* 2007, 65(12), 544–549.
11. J.H. Cummings. Fermentation in the human large intestine: evidence and implications for health. *Lancet* 1983, 1(8335), 1206–1209.
12. S.A. Kripke, A.D. Fox, J.M. Berman, R.G. Settle and J.L. Rombeau. Stimulation of intestinal mucosal growth with intracolonic infusion of short-chain fatty acids. *Journal of Parenteral Enteral Nutrition* 1989, 13(2), 109–116.
13. A.A. Saylers. Breakdown of polysaccharides by human intestinal bacteria. *Journal of Environmental Pathology and Toxicology* 1985, 5, 211–231.
14. R.S. Sandler, W.F. Stewart, J.N. Liberman, J.A. Ricci and N.L. Zorich. Abdominal pain, bloating, and diarrhea in the United States: prevalence and impact. *Digestive Diseases and Sciences* 2000, 45(6), 1166–1171.
15. JECFA. Evaluation of certain food additives and contaminants. In: *Thirty-First Report of the Joint FAO/WHO Expert Committee on Food Additives*, World Health Organization Technical Report Series 759, Geneva; 1987.
16. J. van Esch. Regulatory aspects of low-digestible carbohydrates. In: D.C. Leegwater, V.J. Feron and R.J.J. Hermus (eds). *Low Digestibility Carbohydrates*, Pudoc, Wageningen; 1987, pp. 128–133.
17. K. Endo, *et al.* Effects of high cholesterol diet and polydextrose supplementation on the microflora, bacterial enzyme activity, putrefactive products, volatile fatty acid (VFA) profile, weight, and pH of the feces in healthy volunteers. *Bifidobacteria Microflora* 1991, 10, 53–64.
18. T. Oku, Y. Fujii and H. Okamatsu. Polydextrose as dietary fiber: hydrolysis by digestive enzyme and its effect on gastrointestinal transit time in rats. *Journal of Clinical Biochemistry and Nutrition* 1991, 11, 31–40.
19. X. Wang and G.R. Gibson. Effects of the *in vitro* fermentation of oligofructose and inulin by bacteria growing in the human large intestine. *Journal of Applied Bacteriology* 1993, 75(4), 373–380.
20. N.I. McNeil, J.H. Cummings and W.P. James. Short chain fatty acid absorption by the human large intestine. *Gut* 1978, 19(9), 819–822.

21. J.H. Cummings. Production and metabolism of short-chain fatty acids in humans. In: *Report of 10th Ross Conference on Medical Research*, Ross Laboratories, Sanibel Island, FL, 1991, p. 11.
22. W.M. Scheppach. Short-chain fatty acids are a trophic factor for the human colonic mucosa in vitro. In: *Report of 10th Ross Conference on Medical Research*, Ross Laboratories, Sanibel Island, FL, 1991, p. 90.
23. Polydextrose Food Additive Petition, Pfizer, Report No. 9A3441; 1978.
24. Z. Jie, et al. Studies on the effects of polydextrose intake on physiologic functions in Chinese people. *American Journal of Clinical Nutrition* 2000, 72(6), 1503–1509.
25. M. Yoshioka, Y. Shimomura and M. Suzuki. Dietary polydextrose affects the large intestine in rats. *Journal of Nutrition* 1994, 124(4), 539–547.
26. J. Tomlin and N.W. Read. A comparative study of the effects on colon function caused by feeding ispaghula husk and polydextrose. *Alimentary Pharmacology and Therapeutics* 1988, 2(6), 513–519.
27. Y. Nakagawa, H. Okamatsu and Y. Fujii. Effects of polydextrose feeding on the frequency and feeling of defecation in healthy female volunteers. *Journal of Japan Society of Nutrition and Food Science* 1990, 43, 95–101.
28. K. Foster-Powell, S.H. Holt and J.C. Brand-Miller. International table of glycemic index and glycemic load values: 2002. *American Journal Clinical Nutrition* 2002, 76(1), 5–56.
29. M. Choe, J.D. Kim and J.S. Ju. Effects of polydextrose and hydrolysed guar gum on lipid metabolism of normal rats with different levels of dietary fat. *Korean Journal of Nutrition* 1992, 25, 211–220.
30. S. Ogata, K. Fujimoto, R. Iwakiri, C. Matsunaga, Y. Ogawa, T. Koyama and T. Sakai. Effect of polydextrose on absorption of triglyceride and cholesterol in mesenteric lymph-fistula rats. *Proceedings of the Society for Experimental Biology and Medicine* 1997, 215(1), 53–58.
31. A. Pronczuk and K.C. Hayes. Hypocholesterolemic effect of dietary polydextrose in gerbils and humans. Nutrition Research 2006, 26(1), 27–31.
32. K. Saku, K. Yoshinaga, Y. Okura, H. Ying, R. Harada and K. Arakawa. Effects of polydextrose on serum lipids, lipoproteins, and apolipoproteins in healthy subjects. *Clinical Therapeutics* 1991, 1(3/2), 254–258.
33. S. Liu and C.E. Tsai. Effects of biotechnically synthetized oligosaccharides and polydextrose on serum lipids in the human. *Journal of the Chinese Nutritional Society* 1994, 20, 1–12.
34. U. Schwab, A. Louheranta, A. Torronen and M. Uusitupa. Impact of sugar beet pectin and polydextrose on fasting concentration of plasma glucose and serum total and lipoprotein lipids, and postprandial glycemia in middle aged subjects with abnormal glucose metabolism *European Journal of Clinical Nutrition* 2006, 60, 1073–1080.
35. H. Hara, T. Suzuki, T. Kasai, Y. Aoyama and A. Ohta. Ingestion of guar-gum hydrolysate partially restores calcium absorption in the large intestine lowered by suppression of gastric acid secretion in rats. *British Journal of Nutrition* 1999, 81, 315–321.
36. W.H. Beer. *Gastrointestinal Tolerance to Multiple Confectionery Products Containing Polydextrose*, University of Texas Health Science Center, San Antonio, TX; 1989.
37. G.L. Curtis. *Effect of 'Light' Confections On Gastrointestinal Tolerability Compared to a Conventional Confection*. Harris Laboratories, Lincoln, NE; 1991.
38. K. Saku, K. Yoshinaga, Y. Okura, H. Ying, R. Harada and K. Arakawa. Effects of polydextrose on serum lipids, lipoproteins and apolipoproteins in healthy subjects. *Clinical Therapeutics* 1991, 13, 254–258.
39. M.T. Flood, M.H. Auerbach and S.A.S. Craig. A review of the clinical toleration studies of polydextrose in food. *Food and Chemical Toxicology* 2004, 42, 1531–1542.
40. H.R. Muhleman. Polydextrose – ein kalorienarmer zuckerersatzstoff, zahnmedizinische prufungen. *Swiss Dentistry* 1980, 2(3), 29–32.
41. L. Stoesser, W. Tietze, R. Heinrich-Weltzien, C. Kruger, J.C. Griffiths and M.H. Auerbach. Polydextrose – ein 'zahn-freundlicher' kohlenhydrat-fullstoff. *Oralprophylaxe und Kinderzahnheilkunde* 2005, 27, 144–149.
42. E. Manas and F. Saura-Calixto. Ethanolic precipitation: a source of error in dietary fiber determination. *Food Chemistry* 1993, 47, 351–355.
43. R. Amado and H. Neukon. Minor constituents of wheat flour: the pentosans. In: R.D. Hill and L. Munck (eds). *New Approaches to Research on Cereal Carbohydrates*, Elsevier, Amsterdam, The Netherlands; 1985, pp. 217–230.
44. O. Larm, O. Theander and P. Aman. Structural studies on a water-soluble arabinan from rapeseed (Brassica rapus). *Acta Chemica Scandinavica Series B-Organic Chemistry and Biochemistry* 1975, B29, 1011–1014.

45. S.A.S. Craig, J.F. Holden, J.P. Troup, M.H. Auerbach and H.I. Frier. Polydextrose as soluble fiber: physiological and analytical aspects. *Cereal Foods World* 1998, 43(5), 370–376.
46. H.N. Englyst, S.M. Trowell, D.A.T. Southgate and J.H. Cummings. Dietary fibre and resistant starch. *American Journal of Clinical Nutrition* 1987, 46, 873–874.
47. H.N. Englyst, S.M. Kingman and J.H. Cummings. Classification and measurement of nutritionally important starch fractions. *European Journal of Clinical Nutrition* 1992, 46(Suppl. 2), S33–S50.
48. P. Snow and K. O'Dea. Factors affecting the rate of hydrolysis of starch in food. *American Journal of Clinical Nutrition* 1981, 34, 2721–2727.
49. P. Wursch, S.D. Vedovo and B. Koellreutter. Cell structure and starch nature as key determinants of the digestion rate of starch in legume. *American Journal of Clinical Nutrition* 1986, 43, 25–29.
50. US Patent, 5620873 and 5358729.
51. K. Ohkuma and S. Wakabyashi. Fibersol-2: a soluble non-digestible starch derived dietary fibre. In: B.V. Mc Cleary and L. Prosky (eds). *Advanced Dietary Fibre Technology* Blackwell Publishing Ltd., Oxford; 2001, pp. 509–523.
52. K. Tsuji and D.T. Gordon. Energy value of a mixed glycosidic linked dextrin determined in rats. *Journal of Agricultural and Food Chemistry* 1998, 46, 2253–2259.
53. S. Wakabyashi, M. Satouchi, Y. Nogami, K. Ohkuma and A. Matsouka. Effect of indigestible dextrin on cholesterol metabolism in rat. *Journal of Japanese society of Nutrition and Food Science* 1991, 44, 471–478.
54. Y. Kishimoto, S. Wakabayashi and H. Takeda. Hypocholesterolemic effect of dietary fibre: relation to intestinal fermentation and acid excretion. *Journal of Nutritional Science and Vitaminology* 1995, 41, 11–161.
55. M. Satouchi, S. Wakabayashi, K. Ohkuma, K. Fujiwara and A. Matsuoka. Effects of indigestible dextrin on bowel movements. *Japanese Journal of Nutrition* 1993, 51, 31–37.
56. S. Wakabayashi, Y. Ueda and A. Matsuoka. Effects of indigestible dextrin on sugar tolerance: II. *Journal of Japanese Diabetes Society* 1992, 35, 873–880.
57. S. Wakabayashi, Y. Ueda and A. Matsuoka. Effects of indigestible dextrin on blood glucose and insulin levels after various sugar loads in rats. *Journal of Japanese Society of Nutrition and Food Science* 1993, 46, 131–137.
58. S. Wakabayashi, Y. Kishimoto and A. Matsuoka. Effects of indigestible dextrin on glucose tolerance in rats. *Journal of Endocrinology* 1995, 144, 533–538.
59. Nutriose®FB. Roquettte America Inc, Technical Bulletin.
60. Food Safety Australia New Zealand (FSANZ). Final Assessment Report Application A491–Resistant Maltodextrin as Dietary Fiber, Canberra, Australia; 2004.
61. S. Wakabayashi, M. Satouchi, Y. Ueda and K. Ohkuma. Acute toxicity and mutagenicity studies of indigestible dextrin and its effect on bowel movement of the rat. *Journal of Food Hygienic Society of Japan* 1992, 33, 557–562.
62. Matsutani Chemical Industry (ed.) Review: Safety of Pine Fibre; 1990.
63. M. Nomura, Y. Nakajima and H. Abe. Effects of long-term administration of indigestible dextrin as soluble dietary fiber on lipid and glucose metabolism. *Journal of Japan Society of Nutrition and Food Science* 1992, 45, 21–25.
64. A. Matsuoka, M. Saito and S. Nagano. Continuous administration tests of indigestible dextrin I: study of the effects of the improvement of fat metabolism in healthy volunteers. *Journal of Japanese Clinical Nutrition* 1992, 80(2), 167–172.
65. K. Fujiwara and A. Matsuoka. Continuous administration tests of indigestible dextrin II: study on the effects of the improvement of fat metabolism in patients with non-insulin-dependant diabetes mellitus. *Journal of Japanese Clinical Nutrition* 1993, 83(3), 301–305.
66. K. Tokunaga and A. Matsuoka. Effects of a FOSHU (food for specified health use) containing indigestible dextrin as a functional component on glucose and fat metabolisms. *Journal of Japanese Diabetes Society* 1999, 42, 61–65.
67. H. Sinohara, H. Tsuji and A. Seto. Effects of indigestible dextrin-containing green tea on blood glucose level in healthy human subjects. *Journal of Nutritional Food* 1999, 2(1), 52–56.
68. K. Uno, K. Takagi, M. Akaza, N. Takagi, N. Yoshio and I. Maeda. Effect of indigestible dextrin containing tofu on blood glucose level in healthy human subjects. *Journal of Nutritional Food* 1999, 2(4): 25–31.
69. S. Wakabayashi, Y. Kishimoto, S. Nanbu and A. Matsuoka. Effect of Indigestible dextrin on postprandial rise in blood glucose levels in man. *Journal of Japanese Association of Dietary fiber research* 1999, 3, 13–19.

70. N. Mizushima, Y. Chiba, S. Katsuyama, Y. Daigo and C. Kobayashi. Effect of indigestible dextrin-containing soft drinks on blood glucose level in healthy human subjects. *Journal of Nutritional Food* 1999, 2(4), 17–23.
71. N. Mizushima, Y. Chiba, S. Katsuyama and C. Kobayashi. Effect of long-term ingestion of indigestible dextrin-containing soft drinks on safety and blood glucose levels. *Journal of Nutritional Food* 2000, 3(3), 75–82.
72. Ki Y. Shimoto, S. Wakabayashi and K. Yuba. Effects of instant miso-soup containing indigestible dextrin on moderating the rise of postprandial blood glucose levels, and safety of long term administration. *Journal of Nutritional Food* 2000, 3(2), 19–27.
73. O. Kajimoto, C. Yoshimura, F. Morimoto, M. Henmi, K. Ohki, T. Takahashi and H. Takeuchi. Safety of a long-term intake of a tea beverage containing indigestible dextrin. *Journal of Nutritional Food* 2001, 4(2), 19–26.
74. Matsutani Chemical Industry. Basic Properties of Fibersol-2; 1998.
75. www.orafti.com and Innovate with Raftilose. Orafti, Technical Bulletin.
76. www.orafti.com and Innovate with Raftiline. Orafti, Technical Bulletin.
77. L. Ellegard, H. Anderson and I. Bosaeus. Inulin and oligofructose do not influence the absorption of cholesterol, or the excretion of cholesterol, Ca, Mg, Zn, Fe, or bile acids but increases energy excretion in ileostomy subjects. *European Journal of Clinical Nutrition* 1997, 51(1), 1–5.
78. M.B. Roberfroid. Functional fibres: inulin and oligofructose. *International Food Ingredients* 2000, 3.
79. C. Molis, B. Flourie, F. Ouame, M.F. Gailing, S. Lartigue and A. Guibert. Digestion, excretion and energy value of fructo-oligosaccharides in healthy humans. *American Journal of Clinical Nutrition* 1996, 64, 324–328.
80. M.S. Alles, G.A.J. Hautvast, F.M. Nagengast, R. Hartemink, K.M.J. Van Laere and J.B.M.J. Jansen. Fate of fructo-oligosaccharides in the human intestine. *British Journal of Nutrition* 1996, 76, 211–221.
81. W. Van Dokkum, B. Wezendonk, T.S. Srikumar and E.G.H.M. van den Heuvel. Effect of nondigestible oligosaccharides on large-bowel functions, blood lipid concentrations, and glucose absorption in young healthy male subjects. *European Journal of Clinical Nutrition* 1999, 53, 1–7.
82. S. Karpinnen, K. Liukkonen, A.M. Aura, P. Forssell and K. Poutanen. In vitro fermentation of polysaccharides of rye, wheat and oat brans and inulin by human faecal bacteria. *Journal of Food Science of Food and Agriculture* 2000, 80(10), 1469–1476.
83. H. Younes, C. Coudray, J. Bellanger, C. Demigne, Y. Rayssiguier and C. Remesy. Effects of two fermentable carbohydrates (inulin and resistant starch) and their combination on calcium and magnesium balance in rats. *British Journal of Nutrition* 2001, 86(4), 479–485.
84. J.M. Campbell, G.C. Fahey and B.W. Wolf. Selected indigestible oligosaccharides affect large bowel mass, cecal and fecal short chain fatty acids, pH and microflora in rats. *Journal of Nutrition* 1997, 127, 130–136.
85. M. Kim and H.K. Shin. The water soluble extract of chicory influences serum and liver lipid concentrations, caecal short chain fatty acid concentrations and faecal lipid excretion in rats. *Journal of Nutrition* 1998, 128(10), 1731–1736.
86. Y. Bouhnik, K. Vahedi, L. Achour, A. Attar, J. Salfati, J. Pochart, P. Marteau, B. Flourie, F.R.J. Bornet and J.C. Rambaud. Short chain fructo-oligosaccharide administration dose-dependently increases faecal bifidobacteria in healthy humans. *Journal of Nutrition* 1999, 129(1), 113–116.
87. E. Menne, N. Guggenbuhl and M. Roberfroid. Fn-type chicory inulin hydrosylate has a prebiotic effect in humans. *Journal of Nutrition* 2002, 130(5), 1197–1199.
88. Y. Guidoz, F. Rochat, G. Perruisseau-Carrier, I. Rochat and E.J. Schiffrin. Effects of oligosaccharides on the faecal flora and non-specific immune system in elderly people. *Nutrition Research* 2002, 22, 13–25.
89. J. Luo, S.W. Rizkalla, C. Alamowitch, A. Boussairi, A. Blayo, J.L. Barry, A. Laffite, F. Guyon, F.R.J. Bornet and G. Slama. Chronic consumption of short chain fructooligosaccharides by healthy subjects decreased basal hepatic glucose production but had no effect on insulin-stimulated glucose metabolism. *American Journal of Clinical Nutrition* 1996, 63, 939–945.
90. F. Briet, L. Achour, B. Flourie, L. Beaugerie, P. Pellier and C. Franchisseur. Symptomatic response to varying levels of fructo-oligosaccharides consumed occasionally or regularly. *European Journal of Clinical Nutrition* 1995, 49, 501–507.
91. M. Olesen and E. Gudmand-Hoyer. Efficacy, safety and tolerability of fructooligosaccharides in the treatment of irritable bowel syndrome. *American Journal of Clinical Nutrition* 2000, 72(6), 1570–1575.

92. J. Luo, M. Van Yperselle, S.W. Rizkalla, F. Rossi, F.R.J. Bornet and G. Slama. Chronic consumption of short chain fructooligosaccharides does not affect basal hepatic glucose production or insulin resistance in Type 2 diabetics. *Journal of Nutrition* 2000, 130(6), 1572–1577.
93. N. Delzenne, J. Aertssens, H. Verplaetse, M. Roccaro and M. Roberfroid. Effects of fermentable fructo-oligosaccharides on mineral, nitrogen and energy digestive balance in the rate. *Life Sciences* 1995, 57(17), 1579–1587.
94. C. Coudray, J. Bellanger, C. Castiglia-Delavaud, C. Remesy, M. Vermeorel and Y. Rayssingnuier. The effects of soluble or partly soluble dietary fibres supplementation on absorption and balance of calcium, magnesium, iron, and zinc in healthy young men. *European Journal of Clinical Nutrition* 1997, 51, 375–380.
95. G.R. Gibson, E.R. Beatty, X. Wang and J. Cummings. Selective stimulation of bifidobacteria in the human colon by oligofructose and inulin. *Gastroenterology* 1995, 108, 975–982.
96. D.S. Reddy, R. Hamid and C.V. Rao. Effect of dietary oligosaccharide and inulin on colonic preneoplastic aberrant crypt foci inhibition. *Carcinogenesis* 1997, 18(7), 1371–1374.
97. I.R. Rowland, C.J. Rumney, J.T. Coutts and L.C. Lievense. Effect of *Bifidobacterium longum* and inulin on gut bacterial metabolism and carcinogen induced aberrant crypt foci in rats. *Carcinogenesis* 1998, 19(2), 281–285.
98. K. Yamashita, K. Kawai and M. Itakura. Effect of fructo-oligosaccharides on blood glucose and serum lipids in diabetic subjects. *Nutrition Research* 1984, 4, 961–964.
99. K.G. Jackson, G.R.J. Taylor, A.M. Clohessy and C.M. Williams. The effect of the daily intake of inulin on fasting lipid, insulin and glucose concentrations in middle-aged men and women. *British Journal of Nutrition* 1999, 82, 23–30.
100. M. Roberfroid. Dietary fibre, inulin and oligosofructose. A review comparing their physiological effects. *Critical Reviews in Food Science and Nutrition* 1993, 33, 103–148.
101. U. Takeda and T. Niizato. Acute and subacute safety tests. 1st Neosugar Research Conference, Tokyo, Japan, 20 May 1982.
102. M.A. Clevenger, D. Turnbull, H. Inone, M. Enomoto, J.A. Allen, L.M. Henderson and E. Jones. Toxicological evaluation of neosugar: genotoxicity, carcinogenicity and chronic toxicity. *Journal of the American College of Toxicology* 1988, 7, 643–662.
103. T. Tokunaga, T. Oku and N. Hosoya. Influence of chronic intake of new sweetener fructooligosaccharide (neosugar R) on growth and gastrointestinal function of the rat. *Journal of Nutritional Science and Vitaminology* 1986, 32, 111–121.
104. G.S. Ranhotra, J.A. Gelroth and B.K. Glaser. Usable energy value of selected bulking agents. *Journal of Food Science* 1993, 58, 1176–1178.
105. T. Oku, T. Tokunaga and N. Hosoya. Nondigestibility of a new sweetener, 'Neosugar', in the rat. *Journal Nutrition* 1984, 114, 1574–1581.
106. N. Hosoya, B. Dhorranintra and H. Hidaka. Utilization of (U-14C) fructooligosaccharides in man as energy resources. *Journal of Clinical Biochemistry and Nutrition* 1988, 5, 67–74.
107. I.G. Carabin and W.G. Flamm. Evaluation of safety of inulin and oligofructose as dietary fiber. *Regulatory Toxicology and Pharmacology* 1999, 30, 268–282.
108. S. Cadranel and P. Coussement. Tolerance study with oligofructose for school children. In: *Proceedings First Orafti Research Conference*; 1995, p. 217. Orafti, Tienen, Belgium.
109. M.H. Davidson, K.C. Maki, C. Synecki, S.A. Torri and K.B. Drennan. Effects of dietary inulin on serum lipids in men and women with hypercholesterolemia. *Nutrition Research* 1998, 18, 503–517.
110. K.A. Garleb, J.T. Snook, M.J. Marcons, B.W. Wolf and W.A. Johnson. Effect of fructooligosaccharide containing enteral formulas on subjective tolerance factors, serum chemistry profiles and faecal bifidobacteria in healthy adult male subjects. *Microbial Ecology in Health and Disease* 1996, 9, 279–285.
111. B. Kleessen, B. Sykura, H.J. Zunft and M. Blaut. Effects of inulin and lactose on fecal microflora, microbial activity and bowel habit in elderly constipated persons. *American Journal of Clinical Nutrition* 1997, 65, 1397–1402.
112. A. Pedersen, B. Sandstrom and H.M.M. van Amelsvoort. The effects of inulin on blood lipids and gastrointestinal symptoms in healthy females. *British Journal of Nutrition* 1997, 78, 215–222.
113. J.J. Rumessen, S. Bode, O. Hamberg and E. Gudmand-Hoyer. Fructans of Jerusalem artichokes: intestinal transport, absorption, fermentation, and influence on blood glucose, insulin, and C-peptide responses in healthy subjects. *American Journal of Clinical Nutrition* 1990, 52, 675–681.

114. T. Stone-Dorshow and M.D. Levitt. Gaseous response to ingestion of a poorly absorbed fructooligosaccharide sweetener. *American Journal of Clinical Nutrition* 46, 61–65.
115. E.G.H.M. van den Heuvel, T. Muys, W. van Dokkum and G. Schaafsma. Oligofructose stimulates calcium absorption in adolescents. *American Journal of Clinical Nutrition* 1999, 69, 544–548.
116. Australia New Zealand Food Authority (ANZFA). Final Assessment Report Application A277–Inulin and FOS as Dietary Fiber, Canberra, Australia; 2000.

Summary Table for Part Five A brief summary of the characteristics of some bulking agents.

The information given in this table is taken from published sources and from chapters in this book. The brief summaries are an attempt to provide the reader with a guide to the important bulking agent characteristics. It is only by using bulking agents in real applications will the reader appreciate the full potential of individual bulking agents and the blends with sweeteners.

	Polydextrose commercial example – Litesse®	Resistant maltodextrin commercial example – Fibersol-2®	Resistant maltodextrin commercial example – Nutriose® FB	Inulin commercial example – Raftiline®	FOS commercial example – Raftilose®
Type of polymer	Branched gluco-polysaccharide	Resistant maltodextrin – branched gluco-polysaccharide	Resistant maltodextrin – branched gluco-polysaccharide	Linear fructo-oligosaccharide	Linear fructo-oligosaccharide
Degree of polymerisation (DP)	12 (average)	12 (average)	16 (average)	2–60 (range) 10 (average) High performance >23 (average)	2–7 (range)
Available forms	Powder, granular and syrup	Powder and agglomerated solid	Agglomerated solid	Powder, granular, gel (instant/low sugar/high performance)	Powder and syrup
Aqueous solubility % w/w @ 25°C	>80	approx. 70	>70	12 (DP = 10-12) <5 (DP = 23)	75

Storage and processing	Good	Good	Good	Hydrolysis may occur at pH > 4. Stable in most food heat processes. High performance inulin (DP = 23) has better resistance to acid hydrolysis	Hydrolysis may occur at pH < 4. Stable in most food heat processes
Browning	No browning with Litesse Ultra. Some Maillard reaction with other types of Litesse	Some Maillard reaction	Some Maillard reaction	Some browning reactions at baking temperatures	Some browning reactions at baking temperatures
Viscosity	Higher than sucrose or sorbitol at equivalent concentration and temperature <10 cps @ 30°C, 30% w/w solution	Lower than conventional maltodextrin (DE-10) – higher than sucrose at equivalent concentration and temperature 15 cps @ 30°C, 30% w/w solution	Lower than conventional maltodextrin – higher than sucrose at equivalent concentration and temperature	Low dispersibility in solution. Gels can be formed producing fat-like cremes	Lower than maltodextrin – higher than sucrose at equivalent concentration and temperature

Possible product labelling associated with the use of bulking agents include the following (not exhaustive): Diet, Light, Low carb, Reduced calorie, Low calorie, No-added sugar, Sugar free, Tooth-friendly, Reduced glycaemic response (diabetic suitability), Prebiotic and Fibre. It is advised that local expert opinion is obtained regarding the pertinent food regulations and labelling implications if these 'claims' are to be used on food products.

(Continued)

Summary Table for Part Five *(Continued)*

Relative glycaemic response (RGR) Glucose = 100	<7	<15	25	<4	<2
Dietary fibre content and analytical method	Approx. 90% AOAC 2000.11 (polydextrose)	Minimum 90% AOAC 2001.03 (resistant maltodextrin)	55% AOAC 985.29 85% AOAC 2001.03 (resistant maltodextrin)	>90% AOAC 997.08 (fructan method)	>90% AOAC 997.08 (fructan method)
Bifidus stimulation 'prebiotic'	Yes	Yes	Yes	Yes	Yes
Caloric value	1 kcal/g	EU 1–1.5 kcal/g US 4 kcal/g Japan 0.5 kcal/g	2 kcal/g	1 kcal/g	1.5 kcal/g
Regulatory status, approval and labelling	EU: Miscellaneous Food additive: E 1200 US: Food additive/GRAS Japan: Food FOSHU – status approved	EU: Dextrin/resistant dextrin US: Self GRAS as maltodextrin/digestion resistant-maltodextrin/ resistant maltodextrin Japan: Indigestible dextrin FOSHU – status approved	EU: Dextrin/resistant dextrin US: Self GRAS as maltodextrin/digestion resistant maltodextrin/ resistant maltodextrin Japan: Indigestible dextrin	EU: Food ingredient US: GRAS notified (GRN: 118 (5/3/03)) Japan: Food	EU: Food ingredient US: GRAS notified (GRN: 44 (11/22/00)) Japan: Food

FOS, low-molecular-weight fructo-oligosaccharide; GRAS, Generally Recognized As Safe; FOSHU, food for specified health use; AOAC, Association of Official Analytical.

Index

Note: Page numbers with italicised *b*'s, *f*'s and *t*'s refer to boxes, figures and tables, respectively.

acesulfame K, 93–112. *See also* high-potency sweeteners
  acceptable daily intake, 111
  analytical methods, 111
  appearance, 98
  applications, 100–10
    bakery products, 104
    beverages, 100–10
    cereals, 104
    chewing gum, 104–7
    cosmetics, 109–10
    dairy products, 103–4, 105*t*
    delicatessen products, 107–8
    ice cream, 103–4, 106*t*
    jams/marmalades/preserves/canned fruit, 107
    pharmaceuticals, 109
    sweets, 104–7
    table-top sweeteners, 108–9
    tobacco products, 110
  approvals, 112
  blends with other sweeteners, 95–8, 101–3, 122
  characteristics of, 208–12*t*
  chemical structure, 94*f*
  compatibility with flavours, 98
  dental effects of, 43
  history, 93–4
  organoleptic properties, 94–8
  pharmacology, 110
  physical and chemical properties, 98–100
  physiological properties, 100
  production of, 93
  purity criteria, 112
  regulatory status, 112
  safety assessments of, 111
  as single sweetener, 94–5
  solubility, 98
  stability, 99–100
    in contact with food constituents, 100
    in solution, 99
    storage, 99
    temperature, 99
  synthesis of, 94*f*
  toxicology, 110–11
acidogenicity, 30
acute otitis media, 365–6
Advantame, 132–4. *See also* high-potency sweeteners
  chemical name, 133
  discovery of, 132
  regulatory status, 133–4
  safety of, 133–4
  sensory properties, 133
  solubility, 133–4
  stability, 133
  structure, 133*f*
  synthesis of, 133
Ajinomoto Company of Japan, 117–19, 132
ancient diets, 4–5
Apatococcus lobatus, 215
appetite, 80–81
aromas, stabilisation of, 425–6
aspartame, 117–27. *See also* high-potency sweeteners
  analysis of, 126
  applications, 125–6
    confectionery, 126
    manufacturing, 125
    soft drinks, 125
    yoghurts, 125–6
  blends with other sweeteners, 95–8, 101–3, 119–20
  characteristics of, 208–12*t*

Sweeteners and Sugar Alternatives in Food Technology, Second Edition.
Edited by Dr Kay O'Donnell and Dr Malcolm W. Kearsley.
© 2012 John Wiley & Sons, Ltd. Published 2012 by John Wiley & Sons, Ltd.

aspartame (*Continued*)
  chemical name, 117
  dental effects of, 44
  glycaemic response, 14
  hydrolysis products, 121*t*
  losses under different processing conditions, 122*t*
  metabolites, 124*f*
  physicochemical properties of, 120–23
    dry form, 121
    liquid form, 121–3
    solubility, 120
    stability, 121
  physiological properties, 123–5
    effect on blood glucose, 125
    oral health, 124
    phenylketonuria, 124
    weight loss and maintenance, 125
  regulatory status, 127
  relative sweetness of, 119
  safety of, 126–7
  salt, 120
  sensory properties, 119–20
  sucrose equivalence, 119
  synergy and flavour enhancement, 119–20
  synthesis of, 117–18
aspartame-acesulfame salt, 208–12*t*
aspartic acid, 123
Audrieth-Sveda process, 154*f*

bakery products
  acesulfame K in, 104
  erythritol in, 237–9
  fructo-oligosaccharides in, 459
  inulin in, 460
  isomalt in, 267
  isomaltulose in, 411
  lactitol in, 288–9
  maltitol in, 304–5
  maltitol syrups in, 328
  sorbitol in, 344
  sucralose in, 178
  xylitol in, 368–9
beer, 408–9
beverages
  acesulfame K in, 100–103
  erythritol in, 230–31
  isomaltulose in, 406–8
  polydextrose in, 445
Bifidobacterium, 67–8, 283–4, 442, 458
bifidogenic effect, 458
bile, 65

BioVittoria, 198
bitter blockers, 393
blood glucose, 125, 131, 403–4
brazzein, 200–201. *See also* natural high-potency sweeteners
breakfast cereals, 268–9, 460
breast-fed infants, 64–5
bulk sweeteners, reduced-calorie, 213, 378–82*t*
  characteristics of, 378–82*t*
  dental effects of, 32–43
  D-tagatose, 38
  erythritol, 32, 215–40
  isomalt, 32–3, 243–71
  lactitol, 34, 275–92
  maltitol, 34–6, 295–306
  maltitol syrups, 309–29
  mannitol, 37–8, 331–46
  sorbitol, 36–7, 331–46
  xylitol, 38–42, 347–70
bulking agents, 435–62
  in caloric reduction, 81*t*
  characteristics of, 468–70*t*
  dental effects of, 47–8
  Fibersol-2, 450–52
  fructo-oligosaccharides, 47–8, 454–62
  fructose polymers, 47–8
  gluco-polysaccharides, 437–48
  inulin, 454–62
  long-term glycaemic control with, 15–17
  Nutriose FB, 452–4
  polydextrose, 47, 437–48
  resistant maltodextrin, 449–54
  resistant starches, 449–54

Calorie Control Council, 87
calorie-free food, 85
calories
  control of, 77–8
  energy density, 84
  labeling, 85–7
  legislation, 85–7
  reduction in foods, 78–80
candies
  chewy, 411
  erythritol in, 235
  hard, 253–9, 324, 343
  isomalt in, 253–9
  isomaltulose in, 411
  maltitol syrups in, 324
  mannitol in, 343
  sorbitol in, 343
canned fruit, 107

caramels, 325–6
carbohydrate sweeteners, 46
carbohydrates, 5–6
  digestion of, 221–2
  stabilisation of, 423–4
cariogenicity, 30–32
  animal experiments, 31
  clinical trials, 31–2
  definition of, 30
  enamel slab experiments, 31
  estimation of, 30–32
  incubation experiments, 30–31
  plaque pH experiments, 31
cereal bars
  fructo-oligosaccharides in, 459
  isomalt in, 269–70
  speciality carbohydrates in, 80f
cereals, acesulfame K in, 104
Chana dahl, 5
chewing gum
  acesulfame K in, 104–7
  erythritol in, 231–4
  isomalt in, 263–4
  isomaltulose in, 410
  lactitol in, 289
  maltitol syrups in, 327
  mannitol in, 342
  sorbitol in, 342
  xylitol in, 38–41, 367
chewy candies, 411
chocolates
  erythritol in, 234–5
  inulin in, 460
  isomalt in, 259–61
  isomaltulose in, 410
  maltitol in, 303–4
  polydextrose in, 444
  sorbitol in, 344
  xylitol in, 368
coated products, 410
Codex Alimentarius, 87, 112
composite foods, 305
compressed tablets, 410
confectionery
  aspartame in, 125
  erythritol in, 235
  isomaltulose in, 410–11
  lactitol in, 289
  maltitol syrups in, 324–5, 327
  polydextrose in, 443–4
  sucralose in, 178
  xylitol in, 366–7

cooling flavours, 393–4
cosmetics, acesulfame K in, 109–10
COX-2 gene, 69
cultured dairy products, 445
cyclamate, 151–63. *See also* high-potency
    sweeteners
  absorption, 158
  applications, 158–9
  blends with other sweeteners, 95–7, 101–3
  characteristics of, 208–12t
  dental effects of, 44
  distribution, 158
  excretion, 158
  history, 154
  manufacture of, 154f
  metabolism, 158
  organoleptic properties, 154–6
    blends with other sweeteners, 155–6
    flavour profile, 154
    sweetness potency, 154
    sweetness synergy, 155–6
    temporal profile, 154–5
  physicochemical properties of, 156–7
  physiological properties, 158
  regulatory status, 160–63
  safety of, 159–60
  solubility, 157
  stability, 157

dairy products
  acesulfame K in, 103–4, 105t
  fructo-oligosaccharides in, 459
  isomaltulose in, 408
  maltitol in, 304
  maltitol syrups in, 327–8
  polydextrose in, 445
  sucralose in, 178
  xylitol in, 368
decayed missing and filled teeth (DMFT), 28
delicatessen products, 107–8
dental caries, 27–32
  aetiology of, 28–9
  cariogenicity, 30–32
  control/prevention of, 29–30
  measurement index, 28
  prevalence, 27–8
  prevention of, 359–61
dental plaque, effect on
  acesulfame K, 43
  cyclamate, 44
  D-tagatose, 38
  erythritol, 32

dental plaque, effect on (*Continued*)
  fructose polymers, 47–8
  inulins, 48
  isomalt, 32–3
  isomaltulose, 46
  lactitol, 34
  maltitol, 34–6
  maltodextrin, 48
  mannitol, 37–8
  Neotame, 44
  polydextrose (PDX), 47
  saccharin, 44–5
  sorbitol, 36–7
  sucralose, 45–6
  thaumatin, 46
  xylitol, 38–42, 361
Dermatophagoides pteronyssinus, 190
diabetics
  lactitol for, 285, 301–2
  polyols for, 318–19
  xylitol for, 355
Diet, Nutrition and Prevention of Chronic
    Diseases (report), 29
dietary fat, 5–6
dietary fibre, 64
digestive health, 63–71
  dietary fibre in, 64
  endogenous prebiotics in, 64
  gut health, 64
  microbiota, 53
  milk oligosaccharides in, 64–5
  prebiotics in, 64–9
  secreted substrates in gut, 65
  synbiotics in, 69–70
digestive tolerance. *See* gastrointestinal tolerance
disaccharide, glycaemic responses to, 12t
D-tagatose, 38. *See also* reduced-calorie bulk
    sweeteners
  caloric value, 79t
  cariogenicity of, 38
  characteristics of, 378–82t
  dental effects of, 38
  glycaemic response, 15, 17
  relative sweetness of, 79t

early diet, 4
edible ices, 103–4
enamel slab experiments, 31
endogenous prebiotics, 64
energy density, 84
energy intake, 82–3
energy-free food, 86–7

energy-reduced food, 86–7
erythritol, 215–40. *See also* reduced-calorie bulk
    sweeteners
  applications, 228–39
    bakery products, 237–9
    beverages, 230–31
    candies, 235
    chewing gum, 231–4
    chocolate, 234–5
    fondant, 236
    lozenges, 236–7
    table-top sweeteners, 228–9
  bioavailable calorie, 147
  caloric value, 79t
  cariogenicity of, 32
  characteristics of, 215–17, 378–82t
    anti-oxidant, 217
    high-digestive tolerance, 217
    natural, 216–17
    no glycaemic or insulinaemic response, 216
    non-caloric, 216
    non-cariogenic, 217
  dental effects of, 32
  glycaemic response, 9t, 14
  history, 215
  manufacture of, 217
  organoleptic properties, 218–19
    cooling effect, 219f
    sweetness intensity, 218
    sweetness profile, 218
    synergy with other sweeteners, 219
  physicochemical properties of, 219–20
    boiling point elevation, 220
    freezing point depression, 220
    heat of solution, 219f
    hygroscopicity, 220
    melting point, 220, 222f
    solubility, 220, 221f
    stability, 219–20
    viscosity, 220
    water activity, 220
  physiological properties, 221–8
    anti-oxidant properties, 227–8
    caloric value, 225
    dental health, 226–7
    digestion of carbohydrates, 221–2
    digestive tolerance, 225
    glycaemic response, 225–6
    insulinaemic response, 225–6
    metabolism, 222–5
  regulatory status, 239
  relative sweetness of, 79t, 218f

safety of, 239
specification tests, 240*t*
Escherichia coli WP2, 190
European Food Safety Authority (EFSA), 78, 127
extruded products, 411

fat oxidation, 405
fat replacement, 460
Fibersol-2, 450–52. *See also* bulking agents
   applications, 451–2
   characteristics of, 468–70*t*
   fat metabolism, 451
   manufacture of, 450
   molecular weight, 450
   physical properties of, 450–51
   post-prandial glucose levels, 451
   viscosity, 450
fibre, 64
flavour profile analysis (FPA), 141
flavours, stabilisation of, 425–6
fondant, 236, 412
Food Additive Regulation, 85
Food and Agriculture Organization (FAO), 77
food labelling, 85–7
free sugars, 29
frozen carbonated beverages (FCBs), erythritol in, 231
frozen desserts
   lactitol in, 289–90
   polydextrose in, 444–5
   xylitol in, 368
fructans, 15
fructo-oligosaccharides (FOS), 17, 67–8, 71, 454–2. *See also* bulking agents
   acid stability, 455
   applications, 459
      bakery products, 459
      cereal bars, 459
      dairy products, 459
      low-fat yoghurts, 459
   characteristics of, 468–70*t*
   chemical composition of, 456
   commercial forms, 454–5
   method of analysis, 462
   organoleptic properties, 455
   physicochemical properties of, 455–6
   physiological properties, 456–9
      bifidogenic effect, 458
      dietary fibre, 456
      energy value, 456
      fermentation, 457
      glucose responses, 458–9
      mineral absorption, 458
      tolerance, 457–8
   plant sources, 454
   regulatory status, 461
   safety of, 460–61
   solubility, 456
   structure, 455*f*
   toleration studies, 461
fructosamine, 15–17
fructose
   glycaemic response, 14
   heat of solution, 219*f*
   melting point, 222*f*
fructose polymers, dental effects of, 47–8
fruit fillings, 445
fruit spreads, 268, 445

G protein-coupled receptors (GPCRs), 138, 386
galacto-oligosaccharides (GOS), 65–6, 68, 71
gastrointestinal tolerance, 18–19
   erythritol, 217, 225
   fructo-oligosaccharides (FOS), 457–8
   and glycaemic response, 18–19
   inulin, 71, 457–8
   mannitol, 341
   Nutriose FB, 453*t*
   polydextrose, 442
   sorbitol, 341
   trehalose, 420–21
   xylitol, 356
glazes, 412
gluco-polysaccharides, 437–48
glucostatic theory, 84
glycaemia
   and dietary fat, 5–6
   lowering, 3, 15
glycaemic index (GI), 7–11
   assay, 7–11
   classifications, 13*t*
   high-GI diets, 85
   isomaltulose, 404
   low-GI diets, 3, 6–7
   maltitol, 301
   maltitol syrups, 318
   mannitol, 340–41
   polyglycitol, 318
   sorbitol, 340–41
   sugars and alternatives, 13–15
   xylitol, 364
glycaemic load (GL), 7–11
   classifications
   high-GL diets, 85

glycaemic load (GL) (*Continued*)
  low-GL diets, 3, 6–7
  reduction in, 18–19
  sugars and alternatives, 13–15
glycaemic response
  in 19th century, 5–6
  and adverse outcomes, 7
  in ancient times, 4–5
  in future nutrition, 6–7
  and gastrointestinal tolerance, 18–19
  measurement/expression of, 7–11
  and satiety, 84–5
  true value, 9*t*
glycated haemoglobin (HbA$_{1c}$), 15–17
glycated protein markers, 16
gummies, 411
gut health, 64

hard candies, 253–9, 324, 343
hard coatings, 367
hard panning, 304, 326
Health Promotion Act, 85
Healthy Weight Commitment Foundation, 386
hepatic encephalopathy, 285
high-glycaemic carbohydrates, 16–17
high-potency sweeteners, 93–203
  acesulfame K, 43, 93–112
  Advantame, 132–4
  aspartame, 44, 117–27
  cyclamate, 44, 151–63
  dental effects of, 43–7
  natural, 185–203
  Neotame, 44, 127–32
  saccharin, 44–5, 139–51
  sucralose, 45–6, 167–81
Hoechst AG, 94
Holland Sweetener Company, 117
human milk oligosaccharides (HMOs), 65
hunter-gatherers, diet of, 4
hydrogenated disaccharides, glycaemic responses to, 12*t*
hydrogenated monosaccharide, glycaemic responses to, 12*t*
hydrogenated polydispersed saccharides, glycaemic responses to, 12*t*
hydrogenated starch hydrolysates (HSH), 309
hygroscopicity
  erythritol, 220
  isomalt, 249, 250*f*
  isomaltulose, 399
  lactitol, 280
  maltitol, 298

maltitol syrups, 313–14
mannitol, 338
polyglycitol, 313–14
sorbitol, 338
xylitol, 354

ice cream, 104, 106*t*
  isomaltulose in, 411
  lactitol in, 289–90
icings, 412
inflammatory bowel disease (IBD), 68
instant drinks, 409
insulin, 403–4
insulinaemia, 3
  elevation of, 15
intestinal microbiota, 53
inulin, 65–6. *See also* bulking agents
  acid stability, 456
  applications, 459–60
    baked goods, 460
    breakfast cereals, 460
    chocolate, 459–60
    fat replacer, 460
  caloric value, 79*t*
  characteristics of, 468–70*t*
  chemical composition of, 456
  commercial forms, 454–5
  dental effects of, 48
  method of analysis, 462
  organoleptic properties, 456
  physicochemical properties of, 456
  physiological properties, 456–9
    bifidogenic effect, 458
    dietary fibre, 456
    energy value, 456
    fermentation, 457
    glucose responses, 458–9
    mineral absorption, 458
    tolerance, 71, 457–8
  plant sources, 454
  regulatory status, 461
  relative sweetness of, 79*t*
  solubility, 456
  stability, 456
  structure, 455*f*
  toleration studies, 461
isomalt, 243–71. *See also* reduced-calorie bulk sweeteners
  applications
    baked goods, 267
    breakfast cereals, 268–9
    cereal bars, 269–70

chewing gum, 263–4
chocolates, 259–61
compressed tablets, 266–7
fruit spreads, 268
hard candies, 253–9
low boilings, 261–3
muesli, 270
pan coating, 264–6
caloric value, 79t
cariogenicity of, 32–3
characteristics of, 378–82t
dental effects of, 32–3
glycaemic response, 9t, 15, 17
heat of solution, 219f
history, 243
manufacture of, 243–4
melting point, 222f
organoleptic properties, 244–5
   sweetness potency, 244
   sweetness profile, 245
   synergy with other sweeteners, 245
physicochemical properties of, 245–51
   acid and enzymatic hydrolysis, 246–7
   boiling point elevation, 248–9
   heat of solution, 247–8
   hygroscopicity, 249, 250f
   melting range, 248–9
   solubility, 247
   stability, 245
   viscosity, 247, 248f
   water activity, 250
physiological properties, 252–3
regulatory status, 271
relative sweetness of, 79t
safety of, 270–71
ISOMALT DC, 266–7
ISOMALT GS, 261–8
ISOMALT LM chocolates, 259–61
ISOMALT ST, 267
ISOMALT ST-PF, 261–3
isomalto-oligosaccharides (IMO), 67
isomaltulose, 397–413
   applications, 406–12
      baked goods, 411
      beverages, 406–8
      chewing gum, 410
      chewy candies, 411
      chocolate, 410
      clinical nutrition, 412
      coated products, 410
      compressed tablets, 410
      confectionery, 410–11
      dairy drinks, 408
      extruded products, 411
      fondant, 412
      glazes, 412
      gummies, 411
      ice cream, 411
      icings, 412
      instant drinks, 409
      lozenges, 410–11
      sorbet, 411
      speciality beer, 408–9
   chemical composition of, 397
   dental effects of, 46
   enzymatic rearrangement of sucrose to, 398f
   glycaemic response, 14–15
   manufacture of, 397
   microbiological properties, 401–2
   organoleptic properties, 397–8
   physicochemical properties of, 308–401
      acidic hydrolysis, 400–401
      anti-oxidative effect, 401
      heat stability, 401
      hygroscopicity, 399
      melting point, 399
      polymorphism/pseudopolymorphism, 398
      solubility, 400
   physiological properties, 402–6
      blood glucose effects, 403–4
      dental effects, 403
      fat oxidation effects, 405
      insulin effects, 403–4
   regulatory status, 413
   sweet functionality, 398t
   toxicological evaluation of, 405–6

jams, 107
Joint Expert Commission for Food Additives (JECFA), 181

ketchup, 328–9

lactitol, 275–92. *See also* reduced-calorie bulk sweeteners
   applications, 287–90
      baked goods, 288–9
      chewing gum, 289
      confectionery, 289
      frozen desserts, 289–90
      ice cream, 289–90
      preserves, 290
      tablets, 290
   caloric value, 79t

lactitol (*Continued*)
  cariogenicity of, 34
  chemical composition of, 275
  dental effects of, 34
  glycaemic response, 9*t*
  health benefits of, 282–7
    diabetes, 285
    prebiotic, 67, 282–5
    tooth-protective properties, 285–7
    treatment of hepatic encephalopathy, 285
  heat of solution, 219*f*
  history, 275
  manufacture of, 275
  melting point, 222*f*
  metabolism of, 281–2
  organoleptic properties, 275–6
  physicochemical properties of, 276–81
    boiling point elevation, 280
    heat of solution, 279
    hygroscopicity, 280
    solubility, 277–8
    stability, 276–7
    viscosity, 278–9
    water activity, 280–81
  physiological properties, 281–2
  regulatory status, 291
  relative sweetness of, 79*t*
  structure, 276*f*
Lactobacillus, 283–4, 442
lactulose, 66–7
Law of Mass Action, 141
laxatives, 345
light food, 87
lipbalm, 110
lipstick, 110
Litesse, 438, 443–5, 468–70*t*
lo han guo. *See* mogroside
low sugar food, 85
low-boiling technology, 261–3
low-calorie food, 85–6
low-energy food, 86–7
low-glycaemic carbohydrates, 16–18
low-sugar food, 86, 305
lozenges, 236–7, 410–11
Lycasin, 33, 35–6

maltitol, 295–306. *See also* reduced-calorie bulk sweeteners
  applications, 302–5
    bakery products, 304–5
    chocolate, 303–4
    composite foods, 305
    dairy products, 304
    sugar-free panning, 304
  caloric value, 79*t*
  cariogenicity of, 34–6
  characteristics of, 378–82*t*
  dental effects of, 34–6
  glycaemic response, 9*t*, 14–15
  heat of solution, 219*f*
  labelling claims, 305–6
  legal status, 306
  manufacture of, 296–7
  melting point, 222*f*
  physicochemical properties of, 297–9
    chemical reactivity, 298
    compressibility, 298
    cooling effect, 298
    heat of solution, 298
    humectancy, 298
    hygroscopicity, 298
    molecular weight, 298–9
    solubility, 299
    sweetness, 299
  physiological properties, 299–302
    calorific value, 299–300
    dental aspects, 300
    diabetic suitability, 300
    glycaemic index, 301
    laxative effects, 301–2
    sweetness, 322–3
  purity requirements, 295
  relative sweetness of, 79*t*
  structure, 297
maltitol syrups, 309–29. *See also* reduced-calorie bulk sweeteners
  applications, 323–9
    aerated confectionery, 324–5
    bakery products, 328
    caramels, 325–6
    chewing gum, 327
    dairy products, 327–8
    hard candies, 324
    ketchup, 328–9
    sugar-free or reduced sugar confectionery, 327
    sugar-free panning, 326
  cariogenicity of, 25–6
  glycaemic response, 14
  hydrogenation, 310–11
  legal status, 329
  Lycasin, 33, 35–6
  physicochemical properties of, 312–13
    chemical reactivity, 312–13

cooling effect, 313
heat of solution, 313
humectancy, 313
hygroscopicity, 313–14
molecular weight, 314–15
solubility, 315–16
viscosity, 316
physiological properties, 316–23
calorific value, 316–17
dental effects of, 33, 35–6, 317–18
diabetic suitability, 318–19
glycaemic index, 318
toleration, 319–21
production of, 310
safety of, 329
structures, 312*f*
maltodextrin
dental effects of, 48
glycaemic response, 14
resistant, 449–54
maltooligosyl-trehalose synthase (MTSase), 419–20
maltooligosyl-trehalose trehalohydrolase (MTHase), 419
mannitol, 331–46. *See also* reduced-calorie bulk sweeteners
applications, 341–5
chewing gum, 342
chocolate, 344
hard candies, 343
panning, 344
pharmaceutical tabletting, 344
tabletting, 343
caloric value, 79*t*
cariogenicity of, 37–8
characteristics of, 378–82*t*
dental effects of, 37–8
glycaemic response, 14
heat of solution, 219*f*
hydrogenation, 335
legal status, 345–6
manufacture of, 333–5
melting point, 222*f*
physicochemical properties of, 337–9
chemical reactivity, 337
compressibility, 337
cooling effect, 338
humectancy, 338
hygroscopicity, 338
molecular weight, 338–9
solubility, 339
viscosity, 339

physiological properties, 339–42
calorific value, 339
dental aspects, 340
diabetic suitability, 340
glycaemic response, 340–41
sweetness, 341–2
tolerance, 341
relative sweetness of, 79*t*
safety of, 336–7
storage, 335
structure, 335–6
marmalades, 107
Maumee process, 130–31
metabolic diseases, 3
microbiota, 53
microflora, 53
milk oligosaccharides, 64–5
mogroside, 197–9. *See also* natural high-potency sweeteners
applications, 199
chemical composition of, 197
physicochemical properties of, 199
physiological properties, 199
plant source, 197
regulatory status, 199
safety of, 199
sensory properties, 198*f*
structure, 198*f*
Momordica grosvenorii, 197
monatin, 201–3. *See also* natural high-potency sweeteners
Moniliella pollinis, 217
monosaccharides, glycaemic responses to, 12*t*
Monsanto Chemical Company, 139
mouthwash, 110, 344
mucins, 65
muesli, 270
Mycobacterium tuberculosis, 154

natural high-potency sweeteners, 185–203
brazzein, 200–201
mogroside, 197–9
monatin, 201–3
rebaudioside A, 191–7
steviol glycosides, 191–7
thaumatin, 187–91
Neohesperidine DC, dental effects of, 46
Neotame, 127–32. *See also* high-potency sweeteners
applications, 131–2
blood glucose effects, 131
characteristics of, 208–12*t*

Neotame (*Continued*)
  chemical name, 127
  chemical structure, 128*f*
  dental effects of, 44
  forms of, 132
  in oral health, 131
  physicochemical properties of, 130–31
  physiological properties, 131
  regulatory status, 132
  safety of, 132
  sensory properties, 128–9
    synergy and flavour enhancement, 129
    taste, 128–9
    temporal properties, 129
  solubility, 130
  stability, 130–31
  synthesis of, 128*f*
no added sugar, 306
non-digestible polysaccharides, glycaemic responses to, 12*t*
non-milk extrinsic sugars, 29
noodles, 445
Nutrasweet, 117
NutraSweet Company, 117, 127
Nutrinova Nutrition Specialties & Food Ingredients GmbH, 94
Nutriose FB, 452–4
  characteristics of, 468–70*t*
  chemical composition of, 452
  manufacture of, 452
  method of analysis, 453–4
  regulatory status, 453–4
  tolerance studies, 453*t*
  toxicity studies, 452–3
nutritional labelling, 85–7

obesity, and sugar intake, 78
otitis media, 365–6
overweight people, fat oxidation in, 405

Palatinose, 46, 243, 397
pan coating, 264–6
panning, 344
PARNUT Directive, 86
pasta, 445
pharmaceuticals
  acesulfame K in, 109
  sucralose in, 179
phenylalanine, 123
phenylketonuria, 124
plaque pH, 30–32, 226–7, 403

PMC Specialties Group, Inc., 139
polydextrose. *See also* bulking agents
polydextrose (PDX), 437–48
  applications, 443–5
    beverages, 445
    chocolate, 444
    confectionery, 443–4
    cultured dairy products, 445
    dairy drinks, 445
    frozen desserts, 444–5
    fruit spreads and fillings, 445
    pasta and noodles, 445
  caloric value, 79*t*
  cariogenicity of, 447
  characteristics of, 468–70*t*
  clinical toleration studies, 448*t*
  dental effects of, 47
  description, 437–8
  genotoxicity studies, 446
  glycaemic response, 17
  manufacture of, 437
  method of analysis, 447
  organoleptic properties, 438
  physicochemical properties of, 438–40
    freezing point depression, 440
    glass transition, 440
    heat of solution, 219*f*, 440
    melting properties, 222*f*, 439–40
    moisture management, 440
    solubility, 439
    stability, 439
    viscosity, 440
  physiological properties, 441–3
    energy value, 441
    metabolism, 442–3
    tolerance, 442
  as prebiotic, 67, 69
  regulatory status, 447
  relative sweetness of, 79*t*
  safety of, 445–7
  structure, 437*f*
  toxicity studies, 445
polyglycitol, 309–29. *See also* reduced-calorie bulk sweeteners
  glycaemic response, 14
  physicochemical properties of, 312–13
    chemical reactivity, 312–13
    cooling effect, 313
    heat of solution, 313
    humectancy, 313
    hygroscopicity, 313–14
    molecular weight, 314–15

solubility, 315–16
viscosity, 316
physiological properties, 316–23
  calorific value, 316–17
  dental aspects, 317–18
  diabetic suitability, 318–19
  glycaemic index, 318
  toleration, 319–21
production of, 310–11
structures, 312f
polymorphism, 398
polyols, 14–15, 17
  in caloric reduction, 81t
  laxation of, 319–21
  and polyols, 318–19
  sweetness of, 322–3
positive allosteric modulators (PAMs), 387, 392
prebiotics, 64–5
  vs. dietary fibre, 64
  endogenous, 64
  fructo-oligosaccharides, 65–6
  galacto-oligosaccharides, 66
  health benefits of, 67–8, 282–5
  inulin, 65–6
  isomalto-oligosaccharides, 67
  lactitol, 67
  lactitol as, 282–5
  lactulose, 66–7
  polydextrose, 67
  safety considerations, 70–71
  soy-oligosaccharides, 67
  xylo-oligosaccharides, 67
preserves, 107
  lactitol in, 290
probiotics, 69–70
proteins, stabilisation of, 424–5
Protococcus vulgaris, 215
pseudopolymorphism, 398

Raftiline, 468–70t
Raftilose, 468–70t
rebaudioside A, 191–7. *See also* natural high-potency sweeteners
  applications, 195–6
  chemical composition of, 191
  dose-response relationships of, 194f
  physiological properties, 195
  plant source, 191
  regulatory status, 197
  sensory properties, 193–5
  solubility, 195
  stability, 195

structure, 192f
sweetness potency, 194t
taste quality, 194t
reduced-calorie bulk sweeteners, 213, 378–82t
  characteristics of, 378–82t
  dental effects of, 32–43
  D-tagatose, 38
  erythritol, 32, 215–40
  isomalt, 32–3, 243–71
  lactitol, 34, 275–92
  maltitol, 34–6, 295–306
  maltitol syrups, 309–29
  mannitol, 37–8, 331–46
  polyglycitol, 309–29
  sorbitol, 36–7, 331–46
  xylitol, 38–42, 347–70
reduced-calorie food, 85
reduced-sugar confectionery, 327
reduced-sugar food, 85, 86, 306
remineralisation, 39–40
Remsen–Fahlberg process, 139f
resistant maltodextrin, 48, 449–54
resistant starches, 449–54
rudimentary diet, 4

S2383, 390
S336, 393
S6821, 393
S6973, 390
S7958, 393
S807, 393
saccharin, 139–51. *See also* high-potency sweeteners
  absorption, 146
  applications, 147–9
  blends with other sweeteners, 143–4
  characteristics of, 208–12t
  dental effects of, 44–5
  distribution, 146
  excretion, 146
  flavour profile, 141–2
  history, 130
  manufacture of, 130–31
  metabolism, 146
  organoleptic properties, 140–44
  physicochemical properties of, 144–6
  physiological properties, 146–7
  regulatory status, 151, 152–3t
  safety of, 149–51
  solubility, 144–5
  stability, 145–6
  sweetness potency, 140–41

saccharin (*Continued*)
  sweetness synergy, 143–4
  temporal profile, 142–3
Salmonella typhimurium, 190
salt taste modulators, 394
satiety, 80–81
  definition of, 80
  and energy intake, 82–3
  and glycaemic response, 84–5
sausages, 343–4
savoury flavour ingredients, 393
secretory immunoglobulin A (SIgA), 65
Selaginella lepidophylla, 418
Senomyx Inc., 385, 387, 390, 392–4
short-chain fatty acids (SCFAs), 71, 283–4
short-chain fructo-oligosaccharides (sc-FOS), 47–8
Siraitia grosvenorii, 197–9
soft drinks
  aspartame in, 125
  Neotame in, 132
  sucralose in, 176, 177f
soft panning, 304, 326
sorbet, 411
sorbitol, 331–46. *See also* reduced-calorie bulk sweeteners
  applications, 341–5
    baked goods, 344
    chewing gum, 342
    chocolate, 344
    cooked sausages, 343–4
    hard candies, 343
    laxatives, 344
    mouthwash, 344
    over-the-counter products, 344
    pharmaceutical tabletting, 344
    surimi, 343
    tabletting, 343
    toothpaste, 344
    vitamin C manufacture, 344
  caloric value, 79t
  cariogenicity of, 36–7
  characteristics of, 378–82t
  dental effects of, 36–7
  glycaemic response, 9t, 14–15
  heat of solution, 219f
  hydrogenation, 335
  legal status, 345–6
  manufacture of, 331–2
  melting point, 222f
  physicochemical properties of, 337–9
    chemical reactivity, 337
    compressibility, 337
    cooling effect, 338
    humectancy, 338
    hygroscopicity, 338
    molecular weight, 338–9
    solubility, 339
    viscosity, 339
  physiological properties, 339–42
    calorific value, 339
    dental aspects, 340
    diabetic suitability, 340
    glycaemic response, 340–41
    sweetness, 341–2
    tolerance, 341
  powder, 332–3
  relative sweetness of, 79t
  safety of, 336–7
  storage, 335
  structure, 335–6
  syrups, 333
soy-oligosaccharides (SOS), 67
speciality beer, 408–9
Splenda, 385
Stevia rebaudiana, 191–3
steviol glycosides, 191–7
stevioside, 191–7. *See also* natural high-potency sweeteners
  applications, 195–6
  chemical composition of, 191
  dose-response relationships of, 194f
  physiological properties, 195
  plant source, 191
  regulatory status, 197
  safety of, 196
  sensory properties, 193–5
  solubility, 195
  stability, 195
  structure, 192f
  sweetness potency, 194t
  taste quality, 194t
Streptococcus mutans, 28, 34, 45, 100, 175, 362–3
sucralose, 167–81. *See also* high-potency sweeteners
  analytical methods, 179
  applications, 175–9, 177t
    bakery products, 178
    beverages, 175–6, 177f
    confectionery, 178
    dairy products, 178
    pharmaceuticals, 179
  blends with other sweeteners, 95–7, 101–3
  characteristics of, 208–12t

dental effects of, 45–6
flavour profile, 170*f*
glycaemic response, 14
history, 167–8
organoleptic properties, 168–70
physicochemical properties of, 170–74
  refractive index, 171
  stability, 172–4
  viscosity, 171
physiological properties, 174–5
production of, 168
regulatory status, 181
relative sweetness of, 169*t*
safety of, 179–81
structure, 168*f*
sweet taste modulators, 385
sucrose
  characteristics of, 378–82*t*
  glycaemic response, 9*t*
  heat of solution, 219*f*
  melting point, 222*f*
sugar alternatives, glycaemic responses to, 12*t*, 13–15
sugar-free confectionery, 327
sugar-free food, 85–7, 305
sugar-free panning, 304, 326
sugars
  caloric contribution of, 77
  and dental caries, 29
  influence on health, 3
Sunett®, 94
surimi, 343
sweet taste modulators, 385
  commercialisation of, 391–2
  commercially available, 390
  development of, 392
  identification and evaluation of, 387–8
  mechanism of action, 386–7
  optimisation of, 388–90
  regulatory approval of, 390–91
  targets, 385
sweet taste receptors, 386
sweeteners
  in caloric reduction, 81*t*
  and energy intake, 82–3
  glycaemic responses to, 12*t*, 13–15
  high-potency, 43–7, 93–203
    acesulfame K, 43, 93–112
    aspartame, 44, 117–27
    cyclamate, 44, 151–63
    dental evidence for, 47
    natural, 185–203
    Neotame, 44, 127–32

saccharin, 44–5, 139–51
sucralose, 45–6, 167–81
influence on health, 3
long-term glycaemic control with, 15–17
reduced-calorie, 32–43, 213, 378–82*t*
  dental evidence for, 42–3
  D-tagatose, 38
  erythritol, 32, 215–40
  isomalt, 32–3, 243–71
  lactitol, 34, 275–92
  maltitol, 34–6, 295–306
  maltitol syrups, 309–29
  mannitol, 37–8, 331–46
  sorbitol, 36–7, 331–46
  xylitol, 38–42, 347–70
and satiety, 82–3
and weight management, 83–4
sweetness potency, 140–41
synbiotics, 69–70

T2Rs receptors, 148
table-top sweeteners, 108–9, 228–9
tablets, 290
tabletting, 343, 345
tagatose, 38. *See also* reduced-calorie bulk sweeteners
  caloric value, 79*t*
  cariogenicity of, 38
  characteristics of, 378–82*t*
  dental effects of, 38
  glycaemic response, 15, 17
  relative sweetness of, 79*t*
Thaumatococcus daniellii, 186*t*, 187
thaumatin, 187–91. *See also* natural high-potency sweeteners
  amino acids in, 188*t*
  applications, 189–90
  bitterness reduction by, 190*t*
  dental effects of, 46
  dose-response relationships of, 189*f*
  physicochemical properties of, 189
  physiological properties, 189
  plant sources, 187
  regulatory status, 191
  safety of, 190
  sensory properties, 188
tobacco products, acesulfame K in, 110
tolerance. *See* gastrointestinal tolerance
toothpaste, 109–10, 344
trans-galacto-oligosaccharides (TOS), 66
7-transmembrane domain (TMD) proteins, 138

trehalose, 417–29
  applications, 423–6
    stabilisation of carbohydrates, 423–4
    stabilisation of flavours and aromas, 425–6
    stabilisation of proteins, 424–5
  chemical composition of, 417
  glycaemic response, 14
  history, 417
  metabolism, 420–21
  natural sources of, 418–19
  physiological properties, 426–8
  production of, 419–20
  properties, 421–3
  regulatory status, 421
  safety of, 420–21
  structure, 418f
  technical properties, 423
  tolerance, 420–21
Trentepohlia jolithus, 215
TRPM8, 393–4
type-2 diabetes, 17

umami, 386

Venus flytrap domains (VFDs), 138, 386
viscosity
  erythritol, 220
  isomalt, 247, 248f
  lactitol, 278–9
  maltitol syrups, 316
  mannitol, 339
  polydextrose, 440
  polyglycitol, 316
  sorbitol, 339
  sucralose, 171
  xylitol, 352, 353f
vitamin C, 344

weight loss, 6, 125
weight management, 83–4
  aspartame in, 125
  calorie control in, 77–8
white cheese, 103
with no added sugar, 86
World Health Organization (WHO), 77

xylitol, 347–70. *See also* reduced-calorie bulk
    sweeteners
  additive effects, 40–41
  applications, 366–9
    baked goods, 36–9
    chewing gum, 367
    chocolate, 368

    confectionery, 366–7
    dairy products, 368
    frozen desserts, 368
    hard coatings, 367
    non-food, 369
  caloric value, 79t
  characteristics of, 378–82t
  delivery vehicles, 42
  dental effects of, 38–42, 357–63
    acidogenicity, 359
    caries prevention, 38–9, 359–61
    effects on mutans streptococci, 362–3
    plaque reduction, 361
    prevention of mother-to-child transmission
        of cariogenic bacteria, 363
    remineralisation, 363
  history, 347
  manufacture of, 347–8
  natural sources of, 348t
  organoleptic properties, 348–50
    sweetness, 348–9
    sweetness synergy, 349–50
  physicochemical properties of, 350–54
    boiling point elevation, 352, 353f
    heat of solution, 219f, 350
    hygroscopicity, 354
    melting point, 222f
    solubility, 351–2
    stability, 351
    viscosity, 352, 353f
    water activity, 354
  physiological properties, 354–66
    caloric value, 356–7
    diabetic suitability, 355
    glycaemic response, 9t
    low glycaemic response, 14, 364
    metabolism, 354–5
    prevention of acute otitis media, 365–6
    safety effects, 364–5
    tolerance, 356
    in reduction of mutans streptococci in plaque,
        40b
  regulatory status, 370
  relative sweetness of, 79t
  safety of, 369–70
  uses for, 42
xylo-oligosaccharides (XOS), 67

yoghurts
  acesulfame K in, 103–4
  aspartame in, 125
  fructo-oligosaccharides in, 459
  polydextrose in, 445